VOLUME ONE HUNDRED AND FORTY ONE

CURRENT TOPICS IN DEVELOPMENTAL BIOLOGY

Evolutionary Developmental Biology

CURRENT TOPICS IN DEVELOPMENTAL BIOLOGY

"A meeting-ground for critical review and discussion of developmental processes"
A.A. Moscona and Alberto Monroy (Volume 1, 1966)

VOLUME ONE HUNDRED AND FORTY ONE

Current Topics in DEVELOPMENTAL BIOLOGY

Evolutionary Developmental Biology

Edited by

SCOTT F. GILBERT
Department of Biology, Swarthmore College, Swarthmore, PA, United states;
Institute of Biotechnology, University of Helsinki, Helsinki, Finland

ELSEVIER

ACADEMIC PRESS
An imprint of Elsevier

Academic Press is an imprint of Elsevier
50 Hampshire Street, 5th Floor, Cambridge, MA 02139, United States
525 B Street, Suite 1650, San Diego, CA 92101, United States
The Boulevard, Langford Lane, Kidlington, Oxford OX5 1GB, United Kingdom
125 London Wall, London, EC2Y 5AS, United Kingdom

First edition 2021

ISBN: 978-0-12-814968-3
ISSN: 0070-2153

For information on all Academic Press publications
visit our website at https://www.elsevier.com/books-and-journals

Publisher: Zoe Kruze
Editorial Project Manager: Naiza Ermin Mendoza
Production Project Manager: Denny Mansingh
Cover Designer: Victoria Pearson

Typeset by SPi Global, India

Contents

Contributors

Ehab Abouheif
Department of Biology, McGill University, Montreal, QC, Canada

Arhat Abzhanov
Department of Life Sciences, Imperial College London, Berkshire; Natural History Museum, London, United Kingdom

Detlev Arendt
European Molecular Biology Laboratory, Heidelberg, Germany

Paola Y. Bertucci
European Molecular Biology Laboratory, Heidelberg, Germany

Thomas C.G. Bosch
Zoological Institute, Christian-Albrechts-University Kiel, Kiel, Germany

Marianne E. Bronner
Division of Biology and Biological Engineering, California Institute of Technology, Pasadena, CA, United States

Kimberly L. Cooper
Division of Biological Sciences, University of California San Diego, La Jolla, CA, United States

Alexander Fodor
Biology Department, University of Washington, Seattle; Friday Harbor Laboratories, Friday Harbor, WA, United States

Lisa Hanna
Department of Biology, McGill University, Montreal, QC, Canada

William L. Hatleberg
Department of Biological Sciences, Carnegie Mellon University, Pittsburgh, PA, United States

Caleb C. Hazelwood
Department of Biology, Duke University, Durham, NC, United States

Veronica F. Hinman
Department of Biological Sciences, Carnegie Mellon University, Pittsburgh, PA, United States

Linda Z. Holland
Marine Biology Research Division, Scripps Institution of Oceanography, University of California San Diego, La Jolla, CA, United States

Nicholas D. Holland
Marine Biology Research Division, Scripps Institution of Oceanography, University of California San Diego, La Jolla, CA, United States

Anna M. Kudla
Department of Biology, Duke University, Durham, NC, United States

Shigeru Kuratani
Laboratory for Evolutionary Morphology, RIKEN Center for Biosystems Dynamics Research (BDR); Evolutionary Morphology Laboratory, RIKEN Cluster for Pioneering Research (CPR), Kobe, Hyogo, Japan

Antonella Lauri
European Molecular Biology Laboratory, Heidelberg, Germany

Jiatai Liu
Biology Department, University of Washington, Seattle; Friday Harbor Laboratories, Friday Harbor, WA, United States

Christopher J. Lowe
Hopkins Marine Station, Department of Biology, Stanford University, Pacific Grove, CA, United States

Margaret McFall-Ngai
Pacific Biosciences Research Center, Kewalo Marine Laboratory, University of Hawai'i at Mānoa, Honolulu, HI, United States

Kenneth Z. McKenna
Division of Biological Sciences, University of California San Diego, La Jolla, CA, United States

Zachary S. Morris
Museum of Comparative Zoology and Department of Organismic and Evolutionary Biology, Harvard University, Cambridge, MA, United States

H. Frederik Nijhout
Department of Biology, Duke University, Durham, NC, United States

Jan Stundl
Division of Biology and Biological Engineering, California Institute of Technology, Pasadena, CA, United States

Billie J. Swalla
Biology Department, University of Washington, Seattle; Friday Harbor Laboratories, Friday Harbor, WA, United States

Lindsay Turner
Biology Department, University of Washington, Seattle; Friday Harbor Laboratories, Friday Harbor, WA, United States

Günter P. Wagner
Department of Ecology and Evolutionary Biology, Yale University, New Haven, CT, United States

Preface

Scott F. Gilbert[a,b,*]
[a]Department of Biology, Swarthmore College, Swarthmore, PA, United States
[b]Institute of Biotechnology, University of Helsinki, Helsinki, Finland
*Corresponding author: e-mail address: sgilber1@swarthmore.edu

I. Fulfilling the prophecy

This is the first volume of *Current Topics in Developmental Biology* devoted to evolution. It's taken a long time to get here, for development and evolution have had an extensive and convoluted history together (see Baedke & Gilbert, 2020). For the first generation of evolutionary biologists, embryology had been the model and motor of evolution. Darwin (1874) saw embryology as developmental anatomy, and, following von Baer, he argued that embryology provided "ever the safest guide in classification." Similarly, Müller (1869) believed that for delineating relationships between divergent organisms, "above all things, a thorough knowledge of development was indispensable." Conserved embryonic and larval forms were critical for intuiting relationships. All crustaceans had the nauplius larvae in common, and all chordates had embryos containing a notochord and pharyngeal pouches. Remarkably, when Kowalevsky (1866, 1867) discovered notochords and pharyngeal pouches in certain deuterostome invertebrates, he was able to unite what had been regarded as the two great branches of the animal kingdom—the vertebrates and invertebrates. To Haeckel (1866), evolution came about when the most advanced form of an embryo developed a bit further, adding a new terminal stage to the previous ones.

But at the turn of the last century, Haeckel's student, Wilhelm Roux, wanted to push the boundaries of embryology further and to make embryology an experimental science. In the founding document of developmental mechanics, Roux (1894) urged embryologists to focus on the physiological mechanisms of development, rather than the anatomical changes in embryos. Embryology was to cease being a handmaiden for evolutionary biology and to pursue its own questions—by what physiological mechanisms does an egg become an embryo and the embryo becomes an adult? How do cells differentiate and migrate?

Roux thought that the developmental anatomical approach to evolution had reached a dead-end. It could not distinguish homologous structures from convergent ones, making homology a matter of opinion, not science. Moreover, its notion of progressive evolution was mired in contradictions. He advised embryologists to go from the seashore and forest into the laboratory and to study *Entwicklungsmechanik* (developmental physiology). Moreover, once embryologists knew how organisms were created (ontogeny), that knowledge could be applied to uncover how evolution (phylogeny) occurred. Roux (1894) proposed adding developmental mechanics to natural selection to yield "insight into actually operant and hence immediate formative causes of the prehistoric life of organisms." Indeed, this experimental embryology would provide the mechanisms by which evolution occurred. It would give a causal analysis to evolution.

> *In accordance with the double course of development, viz., the* phyletic *and the* ontogenetic, *developmental mechanics must look for the causes, or* modi operandi, *of each of these two courses; hence, an ontogenetic and phylogenetic developmental mechanics are to be perfected...But in consequence of the intimate causal connections existing between ontogeny and phylogeny, many of the conclusions drawn from the study of ontogeny will throw light on phylogenetic processes.*

At the same time as embryology was leaving evolutionary biology, genetics was budding off from embryology. Embryologists such as William Bateson and Thomas Hunt Morgan (both of whom had studied hemichordate development!) focused their attention on how nuclear genes directed phenotype production. In so doing, they formulated chromosomal genetics. Population-based and mathematically inclined evolutionary biology teamed up with this new genetic mode of analysis to form the *Modern Synthesis*. Morgan and Bateson were some of the first to hypothesize that evolution involved alterations in genes, and they declared that the genetic approach to evolution, focusing on different alleles, had now superseded the embryological approach (Gilbert, 1998). The Modern Synthesis of evolution and genetics left embryology out of the framework of evolutionary biology and concentrated on the mathematics of natural selection (Amundson, 2005; Hamburger, 1980). "The study of mechanisms of evolution," said Morgan's student, Dobzhansky (1951), "falls within the province of population genetics."

But population genetics could only be done within a species, and it had to work with existing variations to explain the survival of the fittest. What it could not explain was the conundrum that so perplexed Darwin—how did the variations arise in the first place? How did one get the *arrival* of the fittest? The phyletic part of developmental mechanics had to wait until the techniques of molecular biology enabled gene products to be localized within embryos and gene sequences to be compared between species. By looking at changes in gene expression, we could see how alterations in the placement, timing, and amount of gene expression (as well as changes in the gene being expressed) could change between species to create new morphologies (Arthur, 2004). And these could be correlated with changes in the relationships between transcription factors and their enhancers as well as the relationships between paracrine factors and their receptors. The phyletic component of Roux's developmental mechanics, consisting of discovering the heritable changes in ontogenetic development, is evolutionary developmental biology "evo-devo." Developmental biology is finally fulfilling Roux's prophecy, returning to evolutionary biology to provide the mechanics of evolution. We have returned, and it's about time. This volume of the *Current Topics in Developmental Biology* is testimony to that return.

II. A book of changes

One of the best statements of the questions evolutionary developmental biology asks (and what the present volume tries to explain) comes from the author Salman Rushdie (1988):

> *How does newness come into the world? How is it born? Of what fusions, translations, conjoinings is it made? How does it survive, extreme and dangerous as it is? What compromises, what deals, what betrayals of its secret nature must it make to stave off the wrecking crew, the exterminating angel, the guillotine?*

This is a remarkable time to review the field and its answers. The first journals dedicated to evo-devo—*Evolution and Development* and *Journal of Experimental Zoology B: Molecular and Developmental Evolution*—were first published in 2000. The chapters in this volume demonstrate how much we have learned about evolution and development as evo-devo "came of age."

Of course, the morphogenesis of a *CTDB* volume is a very problematic affair. It never turns out how the editor intends. In addition to the usual problems of finding contributors to write authoritative chapters, this volume had to contend with a pandemic of SARS-CoV-2 as well as extensive wildfires and floods that wreaked havoc with the lives of authors. Several proposed topics and chapters that did not make it into the book; and due to changes in authors and coauthors, the content of several chapters has changed. I wholeheartedly and most sincerely thank the scientists who gave their increasingly precious time to write their respective chapters.

They performed a heroic job. And I fully understand why others had pressures on them such that they could not contribute. So I apologize to the readers for the lack of plants and the paucity of protostomes in this volume. Similarly, we will not be emphasizing the deep homologies of axis development between arthropods and vertebrates. These are areas of excellent ongoing research, but for reasons beyond the editor's control, won't be in this volume. These lacunae were not intended. Such are the combinatorial, stochastic, and environmental changes that make evolution and its books unpredictable.

This volume is intended to highlight some of the places where evo-devo meets the unknown. I often tell my students that what we learn in biology classes is actually the "recent history of biology." If it is published in articles or textbooks, it's history. So this volume is not intended to celebrate the rich findings of evo-devo over the past decades, its explanations of evolution through changes in limb development, heart development, or axis formation. It is not going to look at the stories we can now tell about how snakes elongated their anterior–posterior axes, expanded their rib cages, and lost their limbs. Nor will we discuss the equally Biblical narrative about how humans lost their facial hair and traded hair follicles for sweat glands. Rather, in keeping with the "Current Topics" idea, we focus on some ongoing problems in the field, the skin of that Pascalian balloon (Pascal, 1655/1973), where the greater the volume of knowledge, the larger its contact with the unknown.

This volume is a remarkable testimony to the important questions being asked and answered by this field. There are two general, albeit overlapping, groupings of chapters in this volume: The developmental origin of vertebrates and the roles of environmental agents in development.

The first group begins with an essay that goes directly to the theoretical core of evolutionary developmental biology: homology and novelty. Homology has been considered the most central and most problematic concept for all biology, and whether at the morphological or molecular level, determination of homology defines the parameters of our interpretations about evolution and our identification of novelty (Hall, 1994; Wake, 1994). In this volume, Kenneth Z. McKenna, Günter P. Wagner, and Kimberly L. Cooper have pooled their resources to propose a new unit of homology, the Character Identity Module, a set of interactions between Gene Regulatory Networks (GRNs), kernels, and *Hox* genes, that produce a multicellular unit capable of some degree of self-organization. This is a paper that demands careful reading, as it presents an important model in which to place our data and our ideas about innovation, novelty, and continuity, and it provides specific, and sometimes counterintuitive examples of these.

What follows from this essay are five chapters concerning the diversification of deuterostomes and the origin of chordates. This has been a fundamental research program for evo-devo, and one of its founding problems. While Darwin (1874) rejoiced that in Kowalevsky's research, "we have at last gained a clue to the source whence the Vertebrata were derived," the antievolutionist Agassiz (1874) complained, "One could hardly open a scientific journal or any popular essay without meeting some allusion to the Ascidians as our ancestors." I'm thrilled to say that this volume is a collecting net full of incredible deuterostomes and their distinctive features.

Echinoderms—sea urchins, sand dollars, and sea stars—have been used as model organisms since the days of Roux and Driesch. More recently, they have provided remarkable insights into the logic of developmental systems—feedback loops, double-negative gates, and other features of gene regulatory networks. In an approach similar to the first paper, William L. Hatleberg and Veronica F. Hinman explicate the GRN concept and expand it by placing it into temporal–spatial contexts. They show how the levels of complexity build up from interacting modules at all levels to allow for the evolutionary changes seen in echinoderms. The GRN is a critical node, forming a context for gene interactions, and is itself being contextually determined by agents from below and above. Hatleberg and Hinman provide examples of how this new understanding of hierarchy could be useful in future evolutionary studies (such as understanding circuit co-option events or determining the homology or convergence of GRNs between species).

Christopher J. Lowe then takes us into the evolution of the other ambulacra clade, the hemichordates. With this foray into enteropneusts, he shows that the hemichordates pose fascinating challenges to the straightforward attributions of homology from gene expression data. While hemichordates share very close patterning similarities to vertebrates in ectodermal patterning, they have morphologically very divergent body plans that make reconstructing ancestral morphology from patterning similarities far more complicated than one might expect.

Next, two chapters detail the evolution of the invertebrate chordates. Linda Z. Holland and Nicholas D. Holland take us into the cephalochordate lineage to discuss how the interactions of genomics with detailed microanatomy provide critical insights into how the chordates evolved their levels of brain and head organization. Cephalochordates (amphioxus or lancelets) split from the nonchordate lineages over 500 mya, but have been evolving exceptionally slowly. Developmental gene expression patterns have been seen as phenotypic characters, and together with microanatomy and comparative genomics, they have proved useful to infer homologies between amphioxus and vertebrates. This has provided evidence that the modern amphioxus is an excellent proxy or model for the long-extinct ancestral chordate. This ancestor most likely had a brain lacking a telencephalon and segmental muscles extending the full length of the body, although some argue that segmented head muscles were not present in this ancestor but are an amphioxus invention. Next, Alexander Fodor, Jiatai Liu, Lindsay Turner, and Billie Swalla take us down the other chordate path, detailing the evolution and development of the Tunicata, those urochordates with their notochords and hearts, *sans* vertebrae. Tunicates arrive at their highly modified adult body plan by a radical metamorphosis of the chordate tadpole. The small genomes and ability to genetically manipulate ascidian eggs and embryos have contributed greatly to our understanding of the development and evolution of genetic networks, and the homologies of the heart fields and notochordal GRNs, combined with the sequencing of tunicate genomes, make these strange animals a fascinating portal into the origins of the vertebrates.

From here, Jan Stundl, Paola Y. Bertucci, Detlev Arendt, and Marianne E. Bronner provide insights into the origin of two of the most remarkable innovations in vertebrate development, the neural crest and sensory placodes. These tissues form where the epidermal ectoderm meet the neural ectoderm, and they give rise to the peripheral nervous system, pigment cells, bone, and cartilage. The authors also show that there are many indications of cells at the neural plate border of invertebrate embryos that share properties with vertebrate neural crest cells and cranial placodes. The authors provide new evidence from GRNs and anatomical studies suggesting that the lateral neural border actually existed in bilaterian ancestors prior to the origin of vertebrates and that it became a developmental hot-spot for the acquisition of new cell types and evolutionary change.

These cells form the "new head" of vertebrates (Gans & Northcutt, 1983), and two papers look at how these structures form. First, Shigeru Kuratani looks at the evolution of the jaw, showing the profound differences and similarities between the cyclostome group of jawless fish and the gnathostomes from which we arise. And those placodes and cephalic crest cells mentioned in the previous paper become extremely important in the origin of the jaw. The differences between the cyclostomes and jawed vertebrates can be traced to the distribution patterns of nasal and adenohypophyseal placodes. These displacements result in new distribution patterns of cephalic neural crest cells. This finding helps establish homologies of chondrocranial elements between the lamprey and hagfish, a long-standing question of the comparative and evolutionary morphology of vertebrates.

The development of the remainder of the face is discussed by Zachary S. Morris and Arhat Abzhanov. Specifically, their discussion is about the face as a module of the amniote head. Although species with very different skull shapes share a similar face shape at the earliest stages of development, they change into shapes recognizably different. Using data from bats, birds, and crocodylians, Morris and Abzhanov generate a model for the generation of amniote facial disparities based on differential proliferation of the cranial neural crest cells in the anterior growth zone, coupled with different rates of their skeletal differentiation.

It is quite remarkable to see how the detailed anatomy, gene expression patterns, and intercellular interactions of extant organisms can be used to delineate patterns of evolution. It also shows the important interactions taking place between developmental biology and paleontology. Moreover, whereas population genetics can say what *might* have happened; the combination of developmental biology and paleontology, following Roux's directions, now gives us a handle on the mechanics of what actually did happen. As Faulkner (1951) wrote, "The past isn't dead. It isn't even past."

The second group of papers goes beyond the path that Roux had pioneered. When Roux got developmental biologists out of the forest and seashore and into the laboratories, this constricted embryology. It enabled us to study only those animals whose development could occur in laboratories. Developmental biology came to study these model organisms that did not need seasonal environmental cues to develop nor specific environmental factors to metamorphose (Bolker, 1995). And this made us believe that all organisms were like our model systems. Now, however, with new techniques, we can return to the forest and seashore and see how the environment is critical for normal development. Thus, this second group of papers concerns an area of evolutionary developmental biology where the environment has agency not only in selecting, but also in producing, the phenotype (Gilbert, 2001; Hall, 1999; Hertwig, 1894). Following Van Valen's (1976) dictum that "evolution is the control of development by ecology," ecological evolutionary developmental biology (eco-evo-devo) looks at how environmental agents can not only select phenotypes, but also can help create them. If environmental factors can generate new phenotypes (as has been shown by dietary, temperature, and predator-induced polyphenisms), then hereditable changes in how these agents are perceived should cause selectable variants for natural selection.

Lisa Hanna and Ehab Abouheif bridge this and the earlier section with an "eco-evo-devo perspective" of wing polyphenisms and eusociality in ants. Here, they put gene regulatory networks into an evolutionary, developmental, and ecological context, focusing on the ability of the environment to regulate components of the insect wing gene regulatory network to produce winged and wingless worker females. Paleontological evidence suggests that wing polyphenism and eusociality evolved simultaneously in ants, and the evidence from developmental biology shows that by preventing some workers from participating in mating flights, wing polyphenism may have contributed to the origin or establishment of their eusocial condition. Finally, wing polyphenism in ants may have evolved from an ancestral developmental potential present in the Hymenoptera for the generation of sexual wing dimorphism.

H. Frederik Nijhout, Anna M. Kudla and Caleb C. Hazelwood then focus on the molecular mechanisms by which environmentally induced phenotypes could be selected and transmitted. Here, environmentally induced phenotypes can become part of the standard genetic repertoire of the organism and become insensitive to environmental cues (genetic assimilation); or they can be selected to become plastic and environmentally sensitive at particular thresholds (genetic accommodation). The authors demonstrate that developmental systems are nonlinear, and they discuss the difficulties involved in understanding genotype–phenotype relationships in such systems. Moreover, they show that these nonlinear properties produce systems in which the same phenotype can be produced by many different genotypes, and that, in the cases of developmental plasticity, a single genotype to produce different phenotypes. This has important consequences in evolution, and the insulin signaling network, which is involved in growth and size regulation, is used as an example wherein a single genotype produces two alternative optimal phenotypes in different environments.

The last two papers, by Margaret McFall-Ngai and Thomas C.G. Bosch, take us into a new realm of eco-evo-devo, that of the holobiont or metaorganism, the organism composed of many species. Development, it seems, is a team sport. We not only use developmental signals produced by the embryonic cells derived from the zygote; but also we use developmental signals coming from the microbes in our environment (Hadfield & Bosch, 2020; McFall-Ngai, 2002). These microbial-derived signals are normal and expected in development, and embryogenesis does not occur properly without them. As Haraway (2008) says, "To be one is always to become with many." Codevelopment is the norm.

The first of these papers looks at two of the most powerful model systems in this field—the cnidarian *Hydra* and the squid *Euprymna*. The development of neither is complete without its symbionts, and the relative simplicity of these interactions can be studied more readily than those of vertebrates, whose baroque collections of microorganisms make such interactive development more difficult to elucidate. Although we have long been aware that all animal body plans arose in oceans rich with microbes, until about 15 years ago, we lacked the technical capacity to identify these microbes and determine their influence on embryonic and postembryonic development. The second paper is a fitting closing to this volume: a reflection on the entire concept of developmental biology and what it studies. Given the centrality of symbiosis and that all animals develop in a microbial environment, we need to reconsider how we think about multicellularity, organogenesis, gut formation, and life-history strategies. We need to think about the origins of the immune system and neural patterning not just from the zygotic view, but also from the view of the holobiont. The nervous system and immune system may not be merely host systems, but systems that are those of the holobiont/metaorganism.

There is a term in the arts, *ostranenie*, that may characterize how one is made to feel after reading these chapters. *Ostranenie* is the artistic technique of defamiliarization. It is the representation of familiar things in an unfamiliar way, so that audiences see that item and its world in a new manner. Viewing organisms as processes, as biomes as well as individuals, should change our perspectives on organisms, development, and evolution.

III. A personal reflection

It's been a wonderful experience to edit this volume. It's been 25 years since Rudy Raff, John Opitz, and I published a paper suggesting ways that development and evolution could be restructured together. In that paper, we mentioned the need to synthesize a framework involving "embryology, macroevolution, and homology." This is precisely the strength of the present volume. Indeed, the Character Identity Module postulated by Wagner and colleagues in this the first chapter may provide a molecular analog to the morphogenetic fields discussed a quarter of a century ago as a way of uniting the worlds of molecular and morphological homology (Gilbert, Opitz, & Raff, 1996; see also Levin, 2012). That paper had its detractors. Ruse (2006) even dismissed it as "hogwash." He and others claimed that we had placed all of the creativity for evolution in development and none in natural selection. My ideas have evolved since then, and my current thinking (Gilbert, 2019) is that "development displays the creativity of the artist. Natural selection displays the creativity of the curator." Perhaps, one can view the Modern Synthesis in its own terms, as a local fitness maximum.

That 1996 paper also confirmed the view of historians that the Modern Synthesis was a contraction as well as an integration (Amundson, 2005; Hamburger, 1980; Provine, 1989; Smocovitis, 1996). While the Modern Synthesis integrated evolutionary biology with genetics, it left out ecology, development, and paleontology as contributors to evolutionary biology. In this volume, we see how these three perspectives have joined together in a *Salon des Refusés* to provide startling insights into the causes of life's biodiversity.

I foresee amazing things for the future of evo-devo. I had once mused, "that sooner or later, the term 'evo–devo' will be abandoned, because at that time it will have become synonymous with 'evolutionary biology'" (Gilbert, 2009, p. 332). Time will tell; for the current explanatory modes of evo-devo and even eco-evo-devo will no doubt be superseded by others. We may have many more molts to go before reaching maturity.

Acknowledgments

All who read this volume should be grateful to the contributors who wrote their chapters during the coronavirus pandemic year of Zoom teaching, lockdowns, student crises, political turmoil, and demonstrations against racism. The patience, forbearance, and tolerance of the Elsevier publishers, especially Paul Wassarman and Shellie Bryant, have been essential for this book to come into existence. I wish to thank Brian Hall for his excellent discussions in formulating this preface and for his enduring influence on the field. I also wish to thank Joana Formosinho for bringing ostranenie to my attention. Funds for this work have been provided by the Swarthmore College Faculty Research fund.

References

Agassiz, L. (1874). Evolution and the permanence of type. *The Atlantic Magazine*, (January, 1874), 92–101. (p. 96).

Amundson, R. (2005). *The changing role of the embryo in evolutionary thought: Structure and synthesis*. Cambridge: Cambridge University Press.

Arthur, W. (2004). *Biased embryos and evolution*. Cambridge: Cambridge University Press.

Baedke, J., & Gilbert, S. F. (2020). *Evolution and development. Stanford encyclopedia of philosophy*. https://plato.stanford.edu/entries/evolution-development/.

Bolker, J. A. (1995). Model systems in developmental biology. *BioEssays*, *17*, 451–455.

Darwin, C. (1874). *The descent of man and selection in relation to sex 2 vols*, (2nd ed.). London: John Murray.
Dobzhansky, T. (1951). *Genetics and the origin of species* (3rd ed.). New York: Columbia University Press.
Faulkner, W. (1951). *Requiem for a nun*. NY: Random House.
Gans, C., & Northcutt, R. G. (1983). Neural crest and the origin of vertebrates: A new head. *Science, 220*, 268–273.
Gilbert, S. F. (1998). Bearing crosses: The historiography of genetics and embryology. *American Journal of Medical Genetics, 76*, 168–182.
Gilbert, S. F. (2001). Ecological developmental biology: Developmental biology meets the real world. *Developmental Biology, 233*, 1–12.
Gilbert, S. F. (2009). Bio. [autobiographical sketch]. *Evolution and Development, 11*, 331–332.
Gilbert, S. F. (2019). Evolutionary transitions revisited: Holobiont evo-devo. *Journal of Experimental Zoology Molecular and Developmental Biology, 332*, 307–314.
Gilbert, S. F., Opitz, J., & Raff, R. A. (1996). Resynthesizing evolutionary and developmental biology. *Developmental Biology, 173*, 357–372.
Hadfield, M., & Bosch, T. (Eds.). (2020). *Cellular dialogues in the holobiont*. Boca Raton, FL: CRC Press.
Haeckel, E. (1866). *Generelle morphologie der organismen [General morphology of the organisms]*. Berlin: G. Reimer.
Hall, B. K. (1994). *Homology: The hierarchical basis of comparative biology*. Boca Raton: Academic Press.
Hall, B. K. (1999). *Evolutionary developmental biology* (2nd ed.). Dordrecht: Kluwer.
Hamburger, V. (1980). Embryology and the modern synthesis in evolutionary theory. In E. Mayr, & W. Provine (Eds.), *The evolutionary synthesis: Perspectives on the unification of biology* (pp. 97–112). NY: Cambridge University Press.
Haraway, D. J. (2008). *When species meet* (p. 4). Minneapolis: U. Minnesota Press.
Hertwig, O. (1894). *Zeit- und Streitfragen der Biologie I. Präformation oder Epigenese? Grundzüge einer Entwicklungstheorie der Organismen. Gustav Fischer Jena. Translated by P. C. Mitchell as The biological problem of today: Preformation or epigenesis?*. New York: Macmillan.
Kowalevsky, A. (1866). Entwicklungsgeschichte der einfachen Ascidien. In *Mem. L'acad.*
Kowalevsky, A. (1867). Entwickelungsgeschichte des *Amphioxus lanceolatus*. In *Mémoires de l'Académie impériale des sciences de St. Petersbourg. 5e série* (Ser. VII) 11, 1–17 + pl.I–III).
Levin, M. (2012). Morphogenetic fields in embryogenesis, regeneration, and cancer: Non-local control of complex patterning. *Biosystems, 109*, 243–261.
McFall-Ngai, M. (2002). Unseen forces: The influence of bacteria on development. *Developmental Biology, 242*, 1–14.
Müller, F. (1869). *Facts and arguments for Darwin* (trans. W. S. Dallas). London: John Murray.
Pascal, B. (1655/1973). *The physical treatises of pascal*. New York: Octagon.
Provine, W. B. (1989). Progress in evolution and meaning in life. In M. H. Nitecki (Ed.), *Evolutionary progress* (pp. 49–74). Chicago: University of Chicago Press.
Roux, W. (1894). Einleitung. Roux's Arch. Entwicklungsmech. Org. Translated (W. M. Wheeler) as The problems, methods and scope of developmental mechanics. In *Biological lectures of the marine biology laboratory, woods hole* (pp. 149–190). Boston: Ginn.
Ruse, M. (2006). Forty years a philosopher of biology: Why EvoDevo makes me still excited about my subject. *Biological Theory, 1*, 35–37.
Rushdie, S. (1988). *The satanic verses*. London: Viking Penguin.
Smocovitis, V. B. (1996). *Unifying biology: The evolutionary synthesis and evolutionary biology*. Princeton: Princeton University Press.
Van Valen, L. M. (1976). Energy and evolution. *Theory of Evolution, 1*, 179–229.
Wake, D. B. (1994). Comparative terminology. [review of] homology: The hierarchical basis of comparative biology (B. K. Hall, ed.). *Science, 265*, 268–269.

CHAPTER ONE

A developmental perspective of homology and evolutionary novelty

Kenneth Z. McKenna[a], Günter P. Wagner[b,*], and Kimberly L. Cooper[a]
[a]Division of Biological Sciences, University of California San Diego, La Jolla, CA, United States
[b]Department of Ecology and Evolutionary Biology, Yale University, New Haven, CT, United States
*Corresponding author: e-mail address: gunter.wagner@yale.edu

Contents

Abstract

The development and evolution of multicellular body plans is complex. Many distinct organs and body parts must be reproduced at each generation, and those that are traceable over long time scales are considered homologous. Among the most pressing and least understood phenomena in evolutionary biology is the mode by which new homologs, or "novelties" are introduced to the body plan and whether the developmental changes associated with such evolution deserve special treatment. In this

Current Topics in Developmental Biology, Volume 141
ISSN 0070-2153
https://doi.org/10.1016/bs.ctdb.2020.12.001

chapter, we address the concepts of homology and evolutionary novelty through the lens of development. We present a series of case studies, within insects and vertebrates, from which we propose a developmental model of multicellular organ identity. With this model in hand, we make predictions regarding the developmental evolution of body plans and highlight the need for more integrative analysis of developing systems.

1. Introduction to the concepts of homology and evolutionary novelty

All living things have the ability to reproduce largely faithful copies of themselves with each character reproduced in a predictable location and in a predictable form from generation to generation. This is remarkable when you consider the diversity of outwardly observable traits across species. Unlike a copy machine, which creates a copy from an existing template, reproduction of characters occurs by recreating them from scratch in each generation and occurs with a shuffling of the parental genetic information and new mutations that together produce genetic variation, which alters characters over long time scales.

Characters that are reproducible over such long time scales, despite alterations, are considered homologous (Roth, 1984; Van Valen, 1982; Wagner, 1989). Homologies are evolutionarily traceable markers of lineages in so far as we can infer the presence of said character in the most recent common ancestor. However, each character evolved deeper in the phylogenetic tree as an addition to the existing body plan, and therefore is also novel with respect to its lineage's history. Conceived in this way, evolutionary novelty refers to the generation of a new character and is only identifiable by the absence of homologies in common ancestors (Müller & Wagner, 1991).

While the definitions of homology and novelty in evolutionary biology have been accepted for many years, their interpretation has also caused some confusion in the field. For example, the term novelty has recently been used more liberally to refer to highly derived characters that are not additions to the basic body plan (Hallgrímsson et al., 2012). Can a trait be novel and simultaneously homologous to a character in a common ancestor? This discordance may undermine our pursuit of understanding how differences in development drive the evolution of a new character compared to the elaboration of an existing one, therefore inhibiting our ability to produce a theoretical framework for the evolution of development.

In this chapter, we propose a developmental model of multicellular organ and/or structural identity. We then use this model to make predictions regarding the developmental evolution of the body plan as it relates to highly-derived characters and to novel characters. To make these points, we utilize case studies from insects and vertebrates, oftentimes alternating between the two. We then propose the concept of Character Identity Modules in relation to gene network co-option and evolution of novelty. We recognize that in an effort to be clearly understood, we have generalized concepts for which there may be exceptions. We end with a series of questions arising from this conceptual framework that will hopefully foster new research avenues.

2. Part 1: The concept of likeness in biology

2.1 What is morphological homology?

We can easily grasp the concept that all life on Earth can be traced to a single origin. However, the number of branch points that produced the diversity of life today and the total number of extinctions along the way are almost unfathomable as the concept of infinity. This creates a problem for the student of evolution. Evolutionary biologists are interested in understanding how the diversity of extant forms evolved, a task that requires some knowledge of what came before. In a perfect world, the fossil record would have preserved every branching point for every extant species, tethering descendants in a seamless transition of morphological transformations. Unfortunately, we have been left with a notoriously fragmentary fossil record that has made it difficult to follow the evolution of organismal characters.

To skirt the problem of an incomplete fossil record, naturalists often identify and group species based on inferred common ancestry and then assess morphological congruence of characters. The similarity of a trait due to shared descent from a common ancestor with the trait is the simplest definition of homology (Owen, 1848; Roth, 1984; Van Valen, 1982; Wagner, 1989).

However, homology is not as simple as the outward appearance of similarity. For instance, if given a skeleton of two closely related vertebrate species, how would you identify a homologous structure, such as the tibia? First, you may recognize the entire arrangement of the skeleton and its proportionality. Anteriorly there is a set of bones that make up the forelimb and posteriorly, a set of bones that make up the hindlimb. Aware that the tibia belongs to the hindlimb, you may move your inquiry to that set of bones. Furthermore, you know that the tibia lies between the femur and the bones that comprise the foot. Upon identifying this bony element, you can compare the differences in its size and shape between the two closely related species, all the while appreciating that this is the same bone.

We learn a few things through this exercise: first, we recognize that the body of an organism has a certain set of rules by which the elements are arranged. In other words, organisms have distinct body plans (see, e.g., Woodger, 1945). Second, based on that body plan, we can appreciate that each element within the plan has regional identity manifest in part by common occurrence with other elements (Roth, 1988). This is implicit in the task of distinguishing forelimb from hindlimb and tibia from femur. Finally, by comparing the skeleton of the two related species, we can appreciate that a homologous character can differ in its morphology. Thus, the form of a homologous character can exist in different states (Wagner, 2007).

2.2 Differentiating character identity from character state

In most cases, identification of homology relies first upon morphological analysis to formulate primary homology hypotheses based on the comparison of the position and structure of characters. Primary homology hypotheses are then tested for phylogenetic coherence (DiFrisco, 2019; Jardine, 1969; Owen, 1848; Remane, 1956; Roth, 1988). For reasons that will be clear later in this essay, we think that position is of more fundamental importance than structural similarity for multicellular organs and body parts. These aspects of homology represent two distinct properties of a character: (1) character identity (Wagner, 2007, 2014): the entity appears in the given location relatively consistently in the majority of organisms that share a common ancestor; and (2) character state (Wagner, 2007, 2014): the entity can be geometrically and/or functionally different from the same thing in other organisms that share common descent. The distinction between character identity and character state is therefore similar by analogy to the identity of a genetic locus versus the allelic variation at that locus (Wagner, 2014).

The differences in character identity and character state have underlying developmental significance. Character identity reflects the relative reliability of the position and general form, whereas character state encompasses the quantitative differences. The former represents the conservation of patterning mechanism during early embryonic development (Roth, 1988), while the latter can be explained by divergence of processes underlying growth and patterning within the character. Along these lines of thought, the study of morphological evolution can be broken down into two categories:

(1) Phylogenetic questions of homology, which seek answers to questions regarding the evolution of character identity and regional organization. These studies are often accompanied by broader questions regarding the evolution of body plans (Müller & Wagner, 1991). (2) The second category encompasses studies in which the evolution of the size, shape, and function of a character are the primary focus, and therefore aims to elucidate how character states evolve.

The tetrapod forelimb skeleton is a paradigmatic example of homology in vertebrates. Although the shape and relative proportion of forelimb segments and the number of digits vary, all tetrapod limbs are built on an inherited architecture (Owen, 1848). Originating from the pectoral girdle, the forelimb begins proximally with the humerus, or stylopod, followed by the radius and ulna of the zeugopod and terminates distally with the autopod made up of small wrist bones that connect to a distal collection of metacarpal and phalangeal elements (Owen, 1848). If skeletons were made of clay, it would be easy to imagine transforming the limb of one species into another by tugging the finger bones, for example, to make the bat wing. Thus, homology of the tetrapod forelimb is recognized because the character exists in the same relative location affixed to the pectoral girdle, and the anatomical relationships of the bony elements within the limb are conserved.

A slightly different example of homology also appears in iterations within an organism when a similar structure is present in different regions of the body, referred to as serial homology (Roth, 2000; Van Valen, 1993). The development of repetitive structures can occur because the ancestral condition of a structure was segmented or because a novel mutation resulted in derived duplication of an ancestrally non-segmented structure. The vertebrate axial skeleton is an example of the former. Within the vertebral column there are obvious structural differences between the cervical, rib-bearing thoracic, and non-rib-bearing lumbar vertebrae (Owen, 1848; Slijper, 1946). Upon inspection of vertebrate phylogeny, one can see how these repetitive elements transformed from a non-differentiated state to the derived state in amniotes, which display differentiated vertebral identities (Jones et al., 2018).

Although the numbers of vertebrae at the cervical, thoracic, lumbar, sacral, and caudal regions differ widely between species, families, and orders, their developmental specification is well conserved. All vertebrae arise from a segmented somitic mesoderm that was present in the common ancestor of vertebrates and the basal chordate, *Amphioxus* (Dequéant & Pourquié, 2008; Gomez et al., 2008; Keynes & Stern, 1988). The serial development and organization of the vertebral skeleton reveals that variation in the number and identity of vertebrae is an elaboration of an ancestrally repetitive trait (Dequéant & Pourquié, 2008; Gomez et al., 2008). Together, the morphological and developmental continuity of these structures has helped to begin to unravel their evolution as serial homologs.

These examples demonstrate that it is often very possible to identify morphological similarity by descent within and between related organisms. But can the highly derived elaboration of a trait obscure what is homology and what is not? In the sections that follow, we analyze challenges to the simplest definitions of homology.

2.3 Introducing new homologs: The concept of evolutionary novelty

In the cases above, the similarity of characters is outwardly recognizable, both within the organism and between organisms that share common descent. However, recognizing homology is not always this simple and is quite difficult in many cases. For example, a fragmented fossil record may hinder our ability to trace a character from modest beginnings to a highly elaborated trait (Darwin, 1859). Another confounding factor is the evolution of body plans themselves (Valentine, Erwin, & Jablonski, 1996). Take for instance the transformation from single-celled eukaryotes, such as an ancestor perhaps similar to cellular slime molds, to vertebrates with a multitude of bony and fleshy elements. Such a transformation required the evolution of new characters with specific forms and functions. To account for the evolution of characters and body plans, evolutionary biologists have formulated the concept of evolutionary novelty (Mayr, 1960; Moczek, 2008; Müller & Wagner, 1991; Peterson & Müller, 2013).

Novelty can be difficult to identify even on smaller evolutionary timescales, because it relies upon our *inability* to draw morphological congruence to a character in the last common ancestor. A common issue mentioned throughout the literature is determining at what threshold a character can be considered novel. This problem can be best-understood with an analogy to water. At very low temperatures water is a solid, but as temperature rises above a certain threshold the solid is qualitatively transformed into liquid, and that liquid will transform to gas if the temperature passes yet another critical threshold. Quantitative changes in the activity of individual water molecules that are identical in each phase lead to qualitatively different states of matter—solid, liquid, and gas. Some characters we observe may be as subtly different from the ancestor as a melting ice cube, while others appear as striking as the difference between ice and water vapor.

It is certainly possible that quantitative changes in the development of a character can surpass a threshold that yields something qualitatively different. Take for instance the insect wing. The fossil record of insects and arthropods at large has not produced any intermediates between extinct wingless and winged arthropods (Kukalova-Peck, 1978). Similarly, no living species clearly bridges the gap between extant wingless arthropods and winged insects (Kukalova-Peck, 1978; Wigglesworth, 1973). This evidence would suggest that the wing is a novel addition. But is this perception of novelty accurate? Or is the wing an exaggeration of a pre-existing character that is present in extant arthropods, and we simply fail to draw the connection with incomplete information?

In Crustacea, the closest living arthropods to insects, the structure that most closely resembles a wing is an outgrowth of the dorsal portion of the leg, called the epipod (Fig. 1) (Averof & Cohen, 1997; Niwa et al., 2010). To call this structure a wing is a vivid stretch of the imagination, but it does share spatial continuity with the insect wing, developing as a dorsal outgrowth from the proximal-most leg segments (Clark-Hachtel & Tomoyasu, 2020; Kukalova-Peck, 1978). Developmental studies in *Drosophila* have

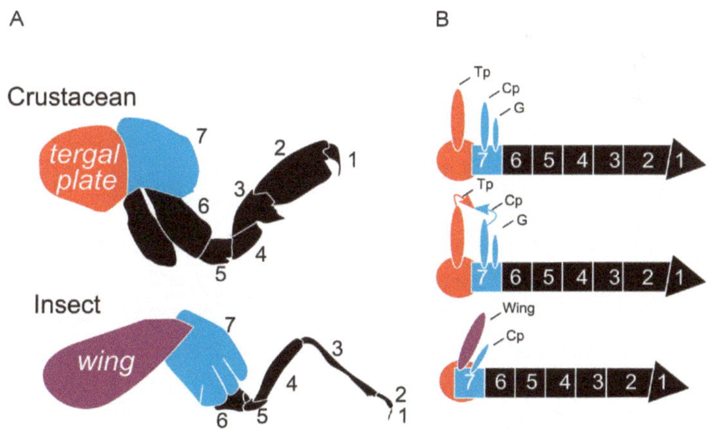

Fig. 1 Evolution of the insect wing. (A) Morphological analysis of the arthropod leg shows a conserved proximo-distal pattern of segments. The structure that shares regional and structural similarity to the insect wing is the crustacean tergal plate. (B) Developmental analyses suggest the wing may have evolved via the union of the epipods of the tergal plate (Tp) and the coxal plate (Cp). Panel (A) modeled after (Bruce & Patel, 2018) and panel (B) modeled after (Clark-Hachtel & Tomoyasu, 2020).

shown that the wing precursor cells originate in close proximity to the leg precursors and subsequently migrate dorsally (Cohen, Simcox, & Cohen, 1993; Williams, Bell, Carroll, 1991). This suggests that modern wings may share some developmental relationship with the legs. More recent studies have found that wings share a great deal of transcriptional regulation with multiple epipods: that of the tergum and that of the coxal plate of the leg (Bruce & Patel, 2018; Clark-Hachtel & Tomoyasu, 2020; Niwa et al., 2010). Both epipods express the wing selector gene *vestigial* (Averof & Cohen, 1997; Linz & Tomoyasu, 2018; Williams et al., 1991). The new hypothesis that has formed is that the wing evolved from the merger of these two epipods (Bruce & Patel, 2018; Clark-Hachtel & Tomoyasu, 2020). Interestingly, the tergal epipod and coxal epipod are present in apterygote (wingless) insects, and developmental analysis shows the presence of these same wing selector genes in each (Niwa et al., 2010). In light of these findings, the insect wing may in fact be homologous to the multiple crustacean epipods.

The insect wing highlights an important problem in morphological evolution: What in some cases can look like an addition to the body plan can in fact be an elaboration of a pre-existing part. This is a critical distinction, because dramatic phenotypic change represents evolution of character state whereas novelty represents the evolution of a new character identity. However, both the extreme elaboration of a shared character and a de novo addition have been considered evolutionary novelties, raising confusion as to where to draw the line in considering which characters are truly "novel." We briefly break down two prominent definitions of novelty below.

2.4 Camp 1: "Novelties are key innovations"

Evolution can occur rapidly when major adaptive changes in the body plan occur. A "key innovation" has been defined as a character change that is substantial enough to trigger the diversification of a lineage (Galis, 2001). This theory assumes that some evolutionary changes are more "important" or more consequential than others (Hunter, 1998). The importance can be quantified by the appearance of higher taxa, the proliferation of species (Martin, McGirr, Richards, & St. John, 2019), and/or attainment of novel morphospace (Galis, 2001; Hunter, 1998). Because key innovations have been correlated with the newly emergent evolutionary success of a group of organisms, the term has often been used interchangeably with evolutionary novelty. This interpretation assumes, however, that novelty has certain connotations of success and proliferation.

One of the first to popularize a connection between the concept of novelty and the evolutionary diversification of a lineage was Ernst Mayr. According to Mayr, evolutionary novelties can be defined as a character that differs more than quantitatively from the character that gave rise to it (Mayr, 1960). Further, any character can qualify so long as the change permits an organism to perform a new function. This definition is virtually synonymous with key innovation, and importantly encompasses a significant change in character state rather than being restricted to the generation of a new character identity.

A representative example that fits Mayr's definition quite well is the bat wing. Bats are the only mammals capable of powered flight. This feat is possible because the forelimbs have become membranous wings in which the digits are extremely elongated, and the interdigital skin is retained and expanded (Swartz, Bennett, & Carrier, 1992; Thewissen & Babcock, 1992). Despite proportional differences, the bat wing is identifiable as a forelimb because of its bodily regionality and anatomical arrangements. First and foremost, the wing is in the region where a forelimb would be expected, affixed to the pectoral girdle. Second, the organization of the bony elements of the wing is identical to that seen in non-flying mammals, with a stylopod, zeugopod, and autopod. In fact, the styolopodial and zeugopodial elements are largely geometrically similar to other mammalian forelimbs, and it is the autopod that bears the most significant morphological divergence from the tetrapod common ancestor (Sears, Behringer, Rasweiler, & Niswander, 2006; Swartz et al., 1992).

These proportional changes in the forelimb allowed for a functional shift away from terrestrial to aerial locomotion. The bat wing is therefore a product of character state changes in homologous bony elements that together allow bats to perform a new function: flight. The unique morphology and function of the bat wing is undoubtedly a key innovation as it facilitated an explosive radiation in which bats adopted niches hitherto unexplored by mammals, and bats now constitute over 20% of all extant mammal species (Wilson & Reeder, 2005).

2.5 Camp 2: "Novelties are entirely new additions to the body plan"

Other researchers have suggested that evolutionary novelty should be saved for true additions to a body plan. For example, Müller and Wagner classify novelties as being neither homologous to a character in an ancestral lineage nor serially homologous to other characters within the same organism (Müller & Wagner, 1991).

According to this definition, novelty is a trait that cannot be traced to an ancestral state, which may be because no connection exists or none has yet been found. Therefore, a novel character would be a trait that is newly derived and therefore requires the evolution of a new character identity. This does not, however, bespeak the spontaneous generation of matter. Rather, the precursor of a novel character was present in the ancestor, perhaps in a developmental context that was permissive to change, but there was no fleshy outgrowth or other unique characteristic that might individuate it from its surroundings and connect it developmentally and phylogenetically to the derived character.

In this regard, let us return to the tetrapod limb and focus on how the autopod came to be. In the fish ancestors of tetrapods, each paired fin is terminated by a distal structure made up of lepidotrichia (Clack, 2009). Lepidotrichia are modified scales that are dermal in origin and develop from neural crest (Shubin, Daeschler, & Jenkins, 2006). A marked transition occurs in the limbs of early tetrapods, wherein the peripheral lepidotrichia are still present (Clack, 2009) but are accompanied by jointed bony elements of mesodermal (endochondral) origin that form digits (Nakamura, Gehrke, Lemberg, Szymaszek, & Shubin, 2016; Shubin et al., 2006). In more derived true tetrapods, the lepidotrichia are absent from the limb altogether.

The evolutionary loss of dermal appendages and origination of an array of endochondral skeletal elements distal to the zeugopod suggest the proximal elements of the limb originated prior to the autopod. Thus, the autopod is a novel addition to the limb ground plan present only in tetrapods. The novelty of the autopodial elements is further evidenced by a change in the terminal *Hox* code of the tetrapod limb that facilitates patterning of both mesopodial (carpals and tarsals) and autopodial (digit) elements (Andrey et al., 2013; Woltering & Duboule, 2010).

Finer scale analysis of the bat wing reveals something similar. Unlike the forelimb of terrestrial mammals, the membranous wing can generate lift. A key component to lift is a set of muscles unique to bats, the *Musculus plagiopatagiales proprii*, that are embedded within the membrane and act to tune the stiffness of the membrane during flight (Cheney et al., 2014). No other mammals have these muscles, making the plagiopatagiales a novel character nested among highly modified homologous bony and fleshy elements.

2.6 Distinguishing between morphological innovation and novelty

For the concept of evolutionary novelty to carry weight, we must distinguish between radically modified characters and ostensibly new ones. The debate between Camp 1 and Camp 2 comes down to whether we recognize a character as a novelty if we can trace its origin. Key innovations reflect the generation of new characters or character states that increase diversification of a clade, whereas true additions to the body plan represent the generation of novel character identities regardless of the degree of "evolutionary success." The former is problematic for multiple reasons: (1) it does not differentiate between exaggerated characters and new characters and (2) it couples novelty with diversification, a phenomenon that is not necessarily true for every addition to the body plan.

In an effort to avoid confusion, we would argue the following: If a pre-existing character is radically modified, it can be classified as a morphological innovation. If a new character is introduced, it is a novelty and by default it is accompanied by a change to the body plan. We previously cited the bat wing as an example of innovation, because it is a derivation of the tetrapod forelimb (Fig. 2). However, the bat wing is a unique example as it is also dependent upon the introduction of a novel character, the plagiopatagiales. Thus the bat wing is composed of both morphological innovations and novelties, exaggerated bones and skin and the plagiopatagiales

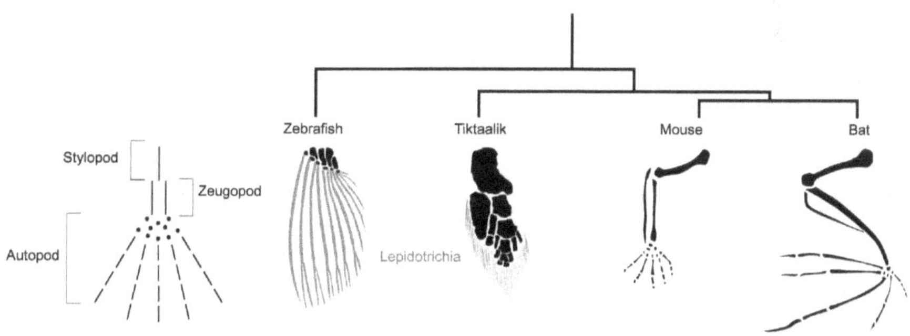

Fig. 2 Origination and elaboration of the tetrapod autopod. Phylogeny of vertebrates illustrating the appearance of a primitive tetrapod autopod in Tiktaalik. The autopod is absent in non-tetrapod vertebrates represented by modern teleost fish (zebrafish), and therefore is a tetrapod-specific novelty. Radical changes to autopod size and shape facilitated the evolution of the bat wing, an example of a morphological innovation.

respectively. Along these lines, the tetrapod autopod is also a novelty, having no morphological continuity with the distal elements of ray finned fishes and dependent upon the introduction of novel developmental patterning mechanisms.

It is important to keep in mind that these examples represent two extremes that may occur along a spectrum of all characters collectively. As we stated at the beginning of this section, it may be possible that continued quantitative change of a pre-existing character can alter the morphospace so dramatically that qualitatively the character no longer resembles its precursor. The linkage in a series of transition states, in contrast to the apparent de novo origination, is what defines an innovation versus a novelty. That said, since novelty is an absence of evidence for homology, new information can shift categorization of a novelty to an innovation.

Because homology is a morphological hypothesis based on phylogenetic connection to a common ancestor, the underlying assumption of homology is a continuity of genetic information. Since genes and the processes they act within can be homologized, developmental analysis can provide supporting evidence for phylogenetic hypotheses of homology (DiFrisco, 2019). Thus similarities in the development of a character can help validate apparent similarities or uniqueness. This requires a developmental model of multicellular character identity. In the following section, we propose a model for organ identity in multicellular development.

3. Part 2. Developmental and genetic determinants of likeness

"Who, if he has thought about embryological problems at all, has not felt the agonizing mental tension engendered by the difficulty of finding adequate verbal expression for something which has seemed to be tolerably clear in thought? And who, in such a predicament, has not eagerly welcomed the timely arrival of a suggestive metaphor or some convenient ellipsis? Such indeed is the relief, that the mind is lulled into complacency and no longer feels the urge to undertake the laborious analysis which is necessary if the makeshift metaphor is to be replaced by a direct statement in genuinely biological terms." (Woodger, 1945)

Historically, the concept of homology has represented the continuity of information (Van Valen, 1982). For many in evolutionary developmental biology, information is a genetic mechanism and the continuity of that mechanism explains its evolutionary traceability (DiFrisco, 2019; DiFrisco, Love, & Wagner, 2020). But what exactly is a traceable genetic mechanism? We use genetic mechanism to refer to a multitude of processes in developmental biology; gene regulatory networks, intracellular signaling cascades, patterning processes, and growth regulation have all been referred to as genetic mechanisms at one time or another. In many cases, each of the aforementioned "genetic mechanisms" acts pleiotropically (Dumont, Pécasse, & Maenhaut, 2001; Keller, 2000; Mayer, Blinov, & Loew, 2009). Given that many different processes are involved in the development of a character, and many are also used in other characters, we might therefore conjecture that a synthesis or integration of multiple networks and processes is involved in creating developmental uniqueness.

One of the first attempts to synthesize the developmental genetics of character identity was the hypothesis of Character Identity Networks (ChINs) to reconcile the many different molecular processes involved in making the development of a character unique (Wagner, 2007, 2014). In recent years, this conceptual model has undergone minor changes in order to recognize higher order processes that play a causal role in shaping character identity. DiFrisco et al., reformulated the hypothesis to Character Identity Mechanisms (ChIMs), combining the reductionist gene regulatory network (GRN) logic with cellular patterning processes (DiFrisco et al., 2020). This union better helps to clarify how dynamic interactions among molecular networks yield cellular behaviors that are not simply the consequence of any single gene or network. Ultimately the ChIM concept allows one to elucidate processes that are common among characters and those that are more specific (Fig. 3). In the sections that follow, we further refine the ChIM concept to represent the Character Identity Module (ChIMo), which defines multicellular organ/structure identity by more comprehensive developmental processes.

3.1 Patterning, GRNs, and specification

A central theme of ChIMs is that multiple molecular processes are involved in the developmental construction of a character, some of which may be pleiotropic and others that may be more modular or tissue-specific (DiFrisco et al., 2020). In the last decade or so, gene regulatory networks (GRNs) have held a prominent position in developmental biology research. GRNs are comprised of cis-regulatory elements (CREs) (i.e., regions in the vicinity of each gene that can affect its expression) plus the set of transcriptional regulators that share affinity for these regions of DNA (Davidson, 2001). GRNs were proposed as "modular subcircuits" (Davidson, 2009) whereupon each subcircuit is an information processing unit with a defined

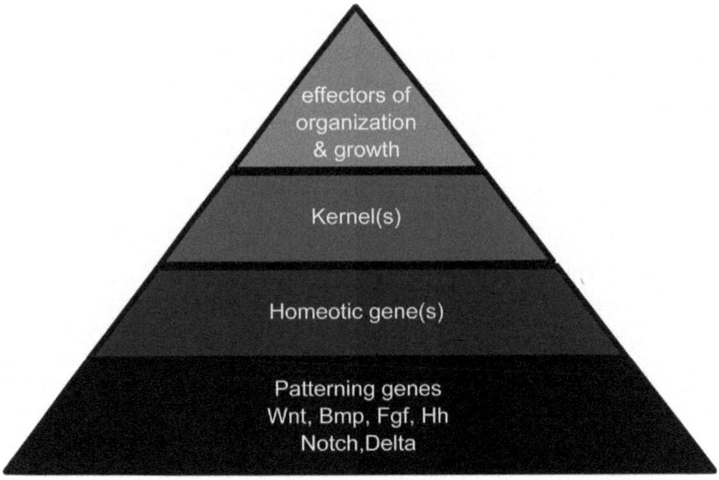

Fig. 3 Schematic representation of the relative uniqueness of molecular processes involved in character identity. The bottom most balck sector represents genes involved in patterning that are used throughout the organism. The next higher sector represents homeotic regulation common to a region of the organism. The sector above this represents kernels that are common to tissue types but that can be embedded within multiple regions. The top level of the pyramid represents networks that establish the "terminal details" common only to homologs.

output (Davidson, 2001; see (Hatleberg & Hinman, 2021; Stundl, Bertucci, Arendt, & Bronner, 2021)). This allure of determinism led to an explosion of research on how various GRNs lead to defined phenotypic outcomes with the expectation that all GRNs are modular and specific, and yet transcription factor loss of function often has massive pleiotropic phenotypic consequence.

One of the core recognitions of the initial concept of ChINs was the fact that GRNs come in various forms (Wagner, 2007, 2014). Some are widely used throughout the organism, while others are only found in a handful of locations. In fact, it is far more common for a GRN to play similar roles yet produce different outcomes in a wide variety of developing parts (Dumont et al., 2001; Mayer et al., 2009). A prime example is the GRN used in metazoans to pattern the anterior-posterior and proximo-distal axes of virtually every tissue in an organism. Patterning and organization throughout metazoa is facilitated by various interacting gene families including BMPs, WNTs, FGFs, and Hh (Ashe & Briscoe, 2006; Cotterell & Sharpe, 2010; Newman & Bhat, 2009; Rogers & Schier, 2011). Each of these genes acts as an extracellular signaling protein that produces a concentration gradient, and those concentrations are read and interpreted by their intracellular signaling cascades (Green & Sharpe, 2015).

In most cases where these factors are deployed, the regulation of other gene families occurs in a concentration dependent manner (Ashe & Briscoe, 2006; Cotterell & Sharpe, 2010; Rogers & Schier, 2011). In *Drosophila*, for example, the embryo is segmented by transverse stripes of WNT expression (Baker, 1988). WNT is expressed in the anterior compartment of each segment and diffuses to form a concentration gradient along the anterior-posterior axis (Kubota, Goto, & Hayashi, 2003). One of the primary roles of WNT during this stage is to induce expression of the homeobox gene *Distal-less* (*Dll*), which is required for limb outgrowth (Cohen et al., 1993).

Between the stripes of WNT lie dorsal expression domains of BMP and ventral expression domains of EGF (Kubota et al., 2003, Kubota, Goto, Eto, & Hayashi, 2000). BMP diffusion forms a concentration gradient that declines ventrally and EGF diffusion forms a concentration gradient that declines dorsally. Importantly, BMP and EGF signals are antagonistic of one another at high concentrations, and both are antagonistic of Dll (Kubota et al., 2000). Subsequently, *Dll* is induced in the region where BMP and EGF levels are low and WNT levels are high (Cohen et al., 1993; Kubota et al., 2003). Thus the interaction of multiple cell populations leads to the specification of limb primordial cells along the lateral line of the embryo.

This in effect establishes a unique transcriptional regulatory environment that allows for activation or repression of CREs. Early after differentiation, limb primordial cells themselves differentiate into a dorsal population that will become the wing and a ventral population that will form the legs (Cohen et al., 1993). An interesting requirement for the differentiation of the dorsal population is removal of WNT expression by T-box transcription factors (Hamaguchi, Yabe, Uchiyama, & Murakami, 2004; Reim, Lee, & Frasch, 2003). As T-box transcription factors are activated by BMP in the dorsal-most cells, WNT expression declines dorsally, facilitating the expression of a CRE (sna-DP) in the dorsal population of cells (Requena et al., 2017). Experimental inhibition of BMP signaling fails to induce T-box expression and sna-DP resulting in the loss of the dorsal population, and therefore of the wing (Requena et al., 2017).

This illustrates how dynamic interactions at the cellular level facilitate genetic interactions at the molecular level. In order to differentiate wing primordial cells from the population of leg primordial cells, a small anterior portion of the limb primordial trans environment must shift from WNT-control to BMP-control. This shift facilitates the expression of a new set of transcription factors that can activate a wing specific CRE that in turn changes the regulatory logic of these cells.

It is important to recognize that the redundancy and iterative use of the patterning GRN makes it unable to provide character identity specificity. This is evidenced by the fact that each morphogen is redeployed in virtually all developing regions of the body, not just appendages, to differentiate subdomains that vary among tissues (DiFrisco & Jaeger, 2020). The generic ability to organize cell populations within a tissue allows diverse use of the patterning GRN to yield different molecular responses and phenotypic outcomes in different regions of the body. For instance, during imaginal disc growth and differentiation in the larval stage, BMP induces the expression of different major transcriptional regulators in the eye (Curtiss & Mlodzik, 2000), wing (Nellen, Burke, Struhl, & Basler, 1996), and leg (Abu-shaar & Mann, 1998).

3.2 Kernels

A GRN can be character-specific if the genetic circuitry and phenotypic output is non-redundant (DiFrisco et al., 2020), and there are genetic circuits with highly specific phenotypic outcomes. At the genetic level these are often described as kernels because they act like a packaged deal (Davidson, 2001; Wagner, 2014). For instance, the induction of one transcription factor initiates a cascade effect, wherein downstream genes become sequentially activated (Arnone & Davidson, 1997; Davidson, 2001). Unlike patterning GRNs, kernels are made up of robust regulatory linkages that activate the same effectors when expressed in different regions of the body and can even produce similar phenotypes independently in distantly related organisms.

The outgrowth-induction kernel that identifies a field of cells as an appendage is a great example of how conserved regulatory linkages facilitate similar developmental and phenotypic outcomes in insects and vertebrates. Vertebrate and invertebrate limb bud formation requires the induction of *Dll* (Pueyo & Couso, 2005). *Dll* is specified in primordia of essentially any outgrowth of the body wall including limbs, genitals, and even teeth (Panganiban & Rubenstein, 2002).

Initiation of *Dll* ultimately drives expression of a core group of genes along the proximo-distal axis of the developing appendage. The gene *buttonhead* (*btd*) is involved in the initial activation of *Dll* in *Drosophila* (Estella, Rieckhof, Calleja, & Morata, 2003). Later in imaginal disc development, *btd* is expressed in the medial domain and is involved in further proximo-distal patterning of the limb (Estella et al., 2003). The orthologs of *Drosophila btd*, the *Sp8* genes, display a similar pattern and function in the ectoderm of the limb bud that is independently derived within and conserved among vertebrates (Kawakami et al., 2004). Knockdown of *btd/Sp8* dramatically perturbs limb outgrowth and patterning in both *Drosophila* and chick (Estella et al., 2003; Kawakami et al., 2004). It is important to note that these patterns are not due to the shared origin of invertebrate and vertebrate appendages, but rather a conserved set of genetic linkages that impose a conserved effect on cellular patterning.

The induction of *Dll* in early *Drosophila* imaginal discs is accompanied by induction of *homothorax* (*hth*) in a population of cells surrounding *Dll* cells (Abu-shaar & Mann, 1998; McKay, Estella, & Mann, 2009). As the disc grows, *dachshund* (*dac*) is expressed in the medial region of the disc, bounded by *Dll* cells distally and *hth* cells proximally (Abu-shaar & Mann, 1998; Dong, Chu, & Panganiban, 2001; Estella & Mann, 2008). A similar pattern is found in the vertebrate limb buds: *Meis1* (*hth* ortholog) is expressed proximally (Mercader et al., 1999), *Dach1* (*dac* ortholog) expressed medially (Hammond, Hanson, Brown, Lettice, & Hill, 1998), and *Dlx* family genes expressed in the distal ectoderm (Panganiban & Rubenstein, 2002). Knockdown of *Dll* ablates distal regions of legs, wings, and antenna in *Drosophila* (Gorfinkiel, Morata, & Guerrero, 1997), and double knockdown of *Dlx5* and *Dlx6* results in patterning defects in the vertebrate autopod (Panganiban & Rubenstein, 2002).

The outgrowth-induction kernel and others, such as those used in light sensing organs (Gehring, 2005; Kozmik, 2005) and cartilage (Tarazona, Slota, Lopez, Zhang, & Cohn, 2016), seem to have roots that can be traced back to the origin of metazoa (Shubin, Tabin, & Carroll, 2009). This conservation of genetic linkages and their associated phenotypic outcomes is commonly referred to as "deep homology" (Shubin et al., 2009). Deep homology reflects the conservation of genetic mechanisms in non-homologous characters that serve a similar function in distantly related species. For example, the last common ancestor of invertebrates and vertebrates likely lacked any appendages or had rudimentary appendages at best (Shubin, Tabin, & Carroll, 1997). The conservation of transcription factors and their linkages in insects and vertebrates to produce similar, but non-homologous limbs therefore represents deep homology of genetic and patterning mechanisms that likely arose for another purpose in a common ancestor.

Despite the astonishing conservation of kernel linkages and their patterning functions, kernels are not the sole proprietors of character identity. More likely, kernels merely specify the kind of cellular behaviors that will be available during development and adulthood. That is, kernels decide whether a cell, and the tissue as a whole, will sense light, be hard or soft, or be an appendage or not. Take for example the multitude of appendages in insects and vertebrates. Each uses the *Dll* kernel but produces entirely different structures with different functions (Fig. 4). Their substantial differences in form seem to be determined by the region of the organism they develop within. Thus, regional context in regulation of a kernel may be just as important as the deployment of a kernel in shaping character identity. In the following section, we detail how regional identity is determined by homeotic genes and how this may influence the patterning and phenotypic trajectory of kernels.

3.3 Hox genes establish regulatory environment

In the previous two sections we detailed how the patterning GRN facilitates the specification of a kernel and how a kernel confers a suite of conserved cellular patterning outcomes. Importantly, these two processes will have different outcomes in various regions of the body. In insects, early embryological actions of the patterning GRN will specify appendages in the head and thorax, but not in the abdomen (Angelini & Kaufman, 2005). Similarly, the outgrowth-induction kernel is activated in each of these regions, but produces morphologically distinct characters in the head, and thorax. The specification of character identity therefore requires regional information.

Regional information is conferred by homeobox genes. During embryonic development the rostrocaudal axis of the embryo is subdivided into semi-discrete domains of *Hox* expression (Hughes & Kaufman, 2002; Mallo, Wellik, & Deschamps, 2010). Note that appendages that share the outgrowth-induction kernel but are anatomically non-homologous develop in different homeotic environments (Fig. 4). These homeotic environments are key to setting up unique ChIMs for each character, evidenced by the fact that mutations or knockouts of homeotic genes result in the transformation of one character to another (Hughes & Kaufman, 2002).

The most characteristic examples of insect homeotic transformations include antennae to leg (Denell, Hummels, Wakimoto, & Kaufman, 1981;

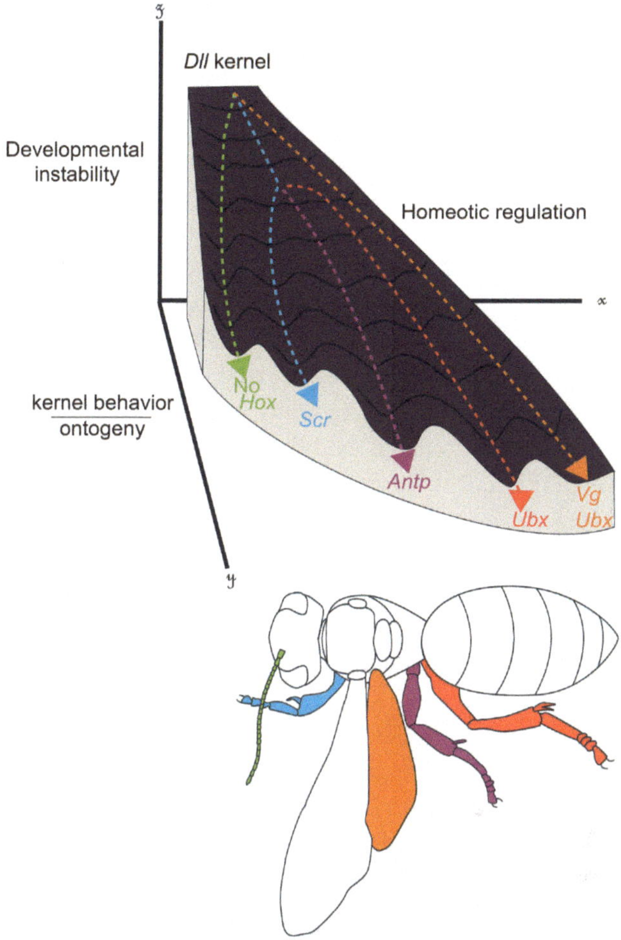

Fig. 4 An epigenetic landscape depiction of limb identity in the context of homeotic regulation and the "outgrowth-induction" (*Dll*) kernel. The landscape has three axes: the x-axis and width of the landscape represents different homeotic environments, the y-axis and length of the landscape represents the organizing principles imparted by the interaction of the *Dll* kernel with *Hox* genes during development. The z-axis represents developmental instability, where lower values represent more stable states. Different trajectories are available to the same *Dll* kernel because *Hox* genes regulate intra-, inter-, and extra-cellular physiology that causes differences in cellular patterning in different *Hox* environments. In the absence of *Hox* (green), the *Dll* kernel produces an antenna, a tubular and relatively evenly segmented structure. In *Scr* (blue), *Antp* (purple), and *Ubx* (red) environments, *Dll* produces tubular segmented legs but segments differ in proportion. When the *Vg* kernel is introduced in the *Ubx* environment (orange), the *Dll* kernel produces a wing, a planar sheet absent of segments.

Struhl, 1982), mouthpart to leg (Hughes & Kaufman, 2000; Pultz, Diederich, Cribbs, & Kaufman, 1988), and hindwings to forewings (Tomoyasu, Wheeler, & Denell, 2005; Weatherbee, Halder, Kim, Hudson, & Carroll, 1998). For simplicity however we will focus on the hindwings to forewings transformation because it has received the most experimental attention. In many insects, mutation or loss-of-function of the homeotic gene *Ultrabithorax* (*Ubx*) transforms the hindwings into a second set of forewings (Tomoyasu, 2017; Tomoyasu et al., 2005; Weatherbee et al., 1998, 1999). This simple genetic change suggests that *Hox* genes regulate a diverse set of genes involved in cellular patterning and growth. Indeed, *Ubx* regulates the expression of transcriptional cofactors (Hersh et al., 2007; Walsh & Carroll, 2007), the function of spatially patterned effectors (Simon & Guerrero, 2015), genes involved in cell-cell communication via juxtacrine (Pavlopoulos & Akam, 2011) and paracrine signaling (de Navas, Garaulet, & Sánchez-Herrero, 2006; Weatherbee et al., 1998), and genes that regulate the composition of the extracellular matrix (Diaz-de-la-Loza, Loker, Mann, & Thompson, 2020; José et al., 2018). Altogether *Ubx* studies hint at an unmatched regulatory power of *Hox* genes to regulate intra-, inter-, and extracellular environments to alter cellular patterning.

Regulation of virtually all aspects of the cellular environment affects how kernels and patterning GRNs behave, giving rise to unique patterns of differentiation and growth. *Hox* genes are crucial for establishing the regionalization of the body plan (Mallo et al., 2010), and this is probably crucial to the generation of unique ChIMs in different regions of the body. Indeed, the contribution of *Hox* genes to character identity has been appreciated for some time (Akam, 1995; Averof & Akam, 1995; Carroll, 1995; Ronshaugen, McGinnis, & McGinnis, 2002). *Hox* genes may be integral to the causal differences in the evolution of innovations and novelties, though not exclusively so.

3.4 Patterning GRN + kernel + Hox = character identity module

To summarize to this point, character identity cannot be attributed to any one molecular network, but is rather the cumulative interaction of networks and their influence on cellular behavior. In this light, we can slightly modify the ChIM hypothesis specifically for multicellular body parts, such that the combinatorial interactions of patterning genes, kernels, and *Hox* genes represent a Character Identity Module (ChIMo).

Based on the evidence provided above, a character identity module represents a delineated multicellular unit capable of some degree of self-organization via the utilization of the patterning GRN, and the rules governing that organization come from the molecular repertoire determined by kernels and *Hox* genes. The sequence of self-organization is roughly the following: at the cellular level, patterning GRNs and *Hox* genes work in unison to establish a unique regulatory environment within which a kernel(s) is specified. The combinatorial effect of the kernel(s) and *Hox* genes sets the initial molecular regulatory logic by establishing intra-, inter-, and extracellular composition. These cellular properties percolate up to influence the organ level by determining the expression domains of patterning GRN genes. The interactions among patterning genes feed back into cellular regulation by specifying distinct trans-environments within the organ, the exact identity of which are determined by the combinatorial interactions of kernel and *Hox* genes. Similarly, the location and size of trans-environments are determined by the composition of the ECM and setup of intracellular signaling cascades. These trans-environments then regulate effectors involved in hormonal control of growth and life history progression to shape the terminal details of the character through ontogeny.

Conceived in this way we can point toward distinct developmental processes underlying character identity and character state. Phylogenetic homology, and therefore continuity of character identity, reflects a conservation of regional specification and developmental patterning mechanism during character origination in the embryo (Roth, 1988). Thus the process of specifying a kernel in a given *Hox* environment may represent the base level of homology. Character state differences are based on conserved intrinsic patterning mechanisms and terminal differences in growth.

4. Part 3. Evolving away from likeness

4.1 Gene network co-option

To this point we have focused on the conceptual basis of homology and the mechanisms of character development. This is not to belabor the point but rather to build a model of character identity in multicellular development so that we may hypothesize the nature of dissimilarity. A central goal of this chapter is to determine if we can come to some level of appreciation regarding the mechanistic differences that underlie the morphological deviation of an existing character versus de novo origination of a character.

The presence of distinct ChIMos, which are manifest from combinatorial interactions of molecular pathways, suggests that various gene networks may be used in some characters and not others. We have suggested, as have many others (Davidson, 2001; DiFrisco et al., 2020; Shubin et al., 2009; Wagner, 2007, 2014), that kernels are an important component of establishing character identity because they determine the type of cellular behaviors accessible to the tissue. In addition we have highlighted a significant role of *Hox* genes in which they may *direct different outputs for a common kernel*. Taken together, this would suggest that characters within an organism differ in their cellular functions and patterning because of the differential usage of kernels and *Hox* genes, and therefore deployment of one or both of these factors in a different region of the body may have profound effects on character identity and character state.

The evolutionary event of expressing a gene and the resulting network in a different tissue is commonly referred to as "gene network co-option" (True & Carroll, 2002). Gene network co-option occurs via mutation in cis- or trans-regulatory elements to procure the transcription of a gene or set of genes in a new tissue (Monteiro & Podlaha, 2009; True & Carroll, 2002). Because GRNs, kernels, and even *Hox* genes behave as regulatory cascades, co-option of any transcriptional regulator of a given regulatory network is hypothesized to bring along downstream factors. This however relies upon the presence and/or absence of transcriptional cofactors, which as we described above likely vary among regions in the organism.

It is becoming clear that gene network co-option may play a key role in major morphological transitions. Over the last decade or so, the number of reported co-option events has continued to climb, each of which have been associated with major leaps in morphology. Examples include, but are not limited to, the evolution of the treehopper helmet (Fisher, Wegrzyn, & Jockusch, 2020; Prud'homme et al., 2011), beetle horns (Moczek & Rose, 2009; Moczek, Rose, Sewell, & Kesselring, 2006), bilaterian limbs (Lemons, Fritzenwanker, Gerhart, Lowe, & McGinnis, 2010), and wing color patterns in various insects (Carroll et al., 1994; Werner, Koshikawa, Williams, & Carroll, 2010).

Given that gene network co-option may have played a key role in the diversification of life it is interesting to ponder if different forms of co-option exist and how they might relate to innovation and novelty. For instance, if innovation builds upon an existing character in new ways and a novel character is defined by the absence of homology, meaning it is the first structure of its kind to develop in a given region of the body, then it follows that novelty might be mechanistically different from innovation in multiple ways. Morphological innovation may require co-opting a kernel that changes the intrinsic patterning and growth mechanisms after character specification. On the other hand, novelty might require introducing a kernel into a new region of the organism. Further, it is possible that novelty depends upon this region bearing a unique *Hox* expression environment that may direct the kernel to produce novel cellular patterning processes and the resulting identity of the character. Below, we dive into case studies of innovation and novelty associated with the co-option of a kernel to make these points.

4.2 Innovation and character state evolution: A case study in fly genitalia

One possible hypothesis for the evolution of morphological innovations, and therefore character states, is the co-option of a kernel into a character during later growth and morphological patterning phases. Perhaps the most elegant example of this comes from the laboratory of Mark Rebeiz. In *Drosophila melanogaster* and its closest relatives in the *melanogaster* clade, the male genitalia have a derived hook-shaped outgrowth called the posterior lobe (Jagadeeshan & Singh, 2006). The posterior lobe is highly diverse, both in size and in shape in the *melanogaster* clade, and is the only morphological way to faithfully identify each species (Glassford et al., 2015).

The unique shape of the posterior lobe has been traced back to a co-option event that brought in a kernel from an unlikely source: a *Hox*-related network involved in development of the larval posterior spiracle, a structure required for oxygen intake via the tracheal system (Glassford et al., 2015). The gene *Poxn* was found to be expressed in the posterior lobe of the male genital as well as the posterior spiracle (Boll & Noll, 2002). Its causal role in the morphogenesis of each structure is confirmed by *Poxn* mutants, which display phenotypic defects in each character (Boll & Noll, 2002; Glassford et al., 2015). Quite curiously, both the posterior spiracle and male genital are activated by the *Hox* gene *Abdominal-B* (*Abd-B*) (Estrada & Sánchez-Herrero, 2001; Hu & Castelli-Gair, 1999). This suggests that co-option of *Poxn* allocated a regulatory cascade molded from the spiracle ChIMo with previously evolved *Abd-B* regulation. The deployment of this regulatory network may therefore be quasi-predictable because of the *Abd-B* environment, but may produce unique cellular effects because of interactions with the outgrowth-initiation kernel involved in specification of the male genital.

During posterior spiracle development, *Abd-B* regulates multiple transcriptional regulators including *spalt*, *Poxn*, *ems*, *cut*, and *unpaired* (*upd*) (Lovegrove et al., 2006). Each of these genes (with the exception of *cut*) and several target genes are expressed in the posterior lobe, with a total of 10 genes shared between the posterior lobe and the spiracle kernel in *D. melanogaster* (Glassford et al., 2015). Most importantly, only two of these genes (*spalt* and *Abd-B*) are expressed in genitals of non-lobed species (Glassford et al., 2015). Enhancers for most of these genes drive reporter expression in both the spiracle and the posterior lobe in *D. melanogaster* (Glassford et al., 2015), consistent with the hypothesis that a cis-regulatory change associated with an upstream regulator of the ancestral spiracle network co-opted the downstream genes to the posterior lobe. A candidate for the upstream regulator is *upd*, as an extensive reporter screen of the *upd* locus in *D. melanogaster* found a region of regulatory DNA that is active only in the spiracle. This suggests that an as yet unidentified posterior lobe-specific enhancer may be responsible for *upd* expression and downstream network co-option in this organ. Altogether, these findings support a hypothesis that co-option of an upstream regulator of a kernel is sufficient to bring with it most downstream genes in the kernel.

The developmental effect of this co-option is realized as exaggerated growth of the posterior region of the genital. Compared to non-lobed drosophilids, the posterior lobe exhibits extreme changes in cell morphology owing to alterations in cytoskeletal arrangements and the composition of the ECM (Smith, Davidson, & Rebeiz, 2020). An apical ECM component, Dumpy, is expressed in the largest cells of the posterior lobe, and RNAi of *Dumpy* completely abolishes the posterior portion of the lobe corresponding to these cells (Smith et al., 2020). Thus, co-option of the spiracle kernel into the posterior region of the male genital modified the intra- and extracellular environment of the population of cells that changed the growth dynamics and ultimately expanded the existing morphospace.

The work detailed above demonstrates what is required to prove gene network co-option, and how co-option influences development of the new morphospace. Most tellingly, this work pushes away from the idea that co-option is and can be identified as correlations in gene expression. One needs to identify not only novel expression patterns, but also phenomena at the enhancer level. Once identified, it is essential to show how cellular-level processes are affected by this new developmental environment, compared to closely related species that do not share this co-option. Equivalent studies of this nature have been done in *Drosophila* wing color patterns (Werner et al., 2010), but many others have merely identified gene expression correlations absent of genomic, cellular, and comparative developmental data. These are fundamentally important if we are to truly appreciate and understand how significant changes in the form of a pre-existing character can occur in evolution.

4.3 Novelty and character identity evolution: A case study of the vertebrate pelvic appendages

We hypothesized above that the evolution of novelty requires the evolution of a new embryonic patterning event as well as new cellular patterning mechanism during later developmental stages. One way this can happen is via the co-option of a kernel into a region of the organism where previously no individuated character develops. An interesting example of this process can be found in the evolution of vertebrate pelvic appendages. However, their novelty is not immediately obvious. One of the major characteristics of jawed vertebrates (gnathostomes) is the presence of two sets of paired appendages: the pectoral appendages and the pelvic appendages (Carroll, 1988; Janvier, 1996; Owen, 1848). At first glance, the pectoral and pelvic appendages appear to be serial homologs, because they have a similar proximal-distal arrangement of parts (Coates & Cohn, 1998; Coates, Ruta, & Friedman, 2008; Diogo & Molnar, 2014; Miyashita & Diogo, 2016; Owen, 1848; Shubin et al., 1997), and the virtual identity of forelimb and hindlimb bud transcriptomes (Wagner, unpublished).

Despite convincing patterns of inter-limb similarity, the pectoral and pelvic appendages appear to have originated at different times during vertebrate evolution. The first vertebrates were jawless fish (agnathans, represented today by extant species of hagfish and lamprey) that lacked paired appendages altogether (Carroll, 1988; Janvier, 1996). The first muscular lateral pectoral outgrowths appear in the ostracoderms, yet neither in ostracoderms, nor pre-ostracoderms, is there any evidence of muscular pelvic outgrowths (Janvier, 2007; Janvier, Arsenault, & Desbiens, 2004). Much like the case of the insect wing, there are no extant species that bridge the gap between a single pair of pectoral appendages and two sets of paired appendages. Pelvic appendages appear much later in the first jawed vertebrates (Brazeau & Friedman, 2014; Zhu, Yu, Choo, Wang, & Jia, 2012), leaving a giant hole in vertebrate history and portraying a scenario where paired pectoral appendages evolved first in agnathans, followed by the origin of the pelvic appendages somewhere in the agnathan to gnathostome transition (Diogo, 2020).

As we have argued earlier in the chapter, morphological inference of homology requires developmental validation. The most obvious difference in pectoral and pelvic appendages is the region of the body where they develop. Pectoral appendage development is associated with the head and neck whereas the pelvic appendages are associated with development of the trunk (Diogo, 2020; Diogo, Linde-Medina, Abdala, & Ashley-Ross, 2013; Nagashima et al., 2016). Although they develop in anatomically discrete regions, their specification in the embryo involves a similar set of T-box genes, *Tbx5* and *Tbx4*, which are required for initiation of pectoral and pelvic limb bud development and identity, respectively (Don, Currie, & Cole, 2013; Gibson-Brown, Agulnik, Silver, Niswander, & Papaioannou, 1998; Gibson-Brown et al., 1996; Isaac et al., 1998; Logan, Simon, & Tabin, 1998; Ruvinsky & Gibson-Brown, 2000). The prevailing hypothesis suggests that upon gene duplication of the ancestral *Tbx4/5* gene, *Tbx5* was recruited first, facilitating emergence of the forelimb, followed by the co-option of *Tbx4* into the posterior flank mesoderm by *Pitx1* (Don et al., 2013; Logan & Tabin, 1999). Thus, their characterization as serial homologs from a developmental perspective relies upon the fact that they share the outgrowth-induction kernel (Willmer, 2003) in domains of the trunk that are permissive to form a limb (Ohuchi et al., 1997; Sekine et al., 1999).

The distinct evolutionary and developmental origins of the paired appendages, however, calls into question their serial homology. We have argued throughout this chapter that regional differences in the organism are constructed by homeotic genes, which may modify the molecular and cellular environments. The purpose in doing so is to break down an important problem: does the co-option of a kernel into a region of the organism in which a character did not previously exist represent serial homology or evolution of novelty?

Concordant with the ChIMo hypothesis, multiple interacting processes mold the developmental uniqueness of a character. Thus the presence of a similar kernel, in this case *Tbx* initiation of the outgrowth-induction kernel, is difficult to argue as serial homology for multiple reasons. For one, sexual organs and head barbels in vertebrates use some of the same genes in this kernel, and it would be difficult to argue for their serial homology (Archambeault, Taylor, & Crow, 2014; Freitas, Gómez-Marín, Wilson, Casares, & Gómez-Skarmeta, 2012; Freitas, Zhang, & Cohn, 2006). The penis, for example, is derived from the genital tubercle, which occurs in close proximity to the hindlimb buds in the embryo (Gredler et al., 2014; Herrera & Cohn, 2014; Tschopp et al., 2014). In both anolis and mouse, the transcriptomes of the genital tubercle and hindlimb bud are more similar early in development than they are at later stages (Tschopp et al., 2014), as might be expected from our landscape hypothesis (Fig. 4). Despite late stage differences, the penis and limb in mice both express key components of the outgrowth induction kernel, *Hoxa13* and *Hoxd13*, and even share many of the same enhancers (Amândio, Lopez-Delisle, Bolt, Mascrez, & Duboule, 2020; Gonzalez, Duboule, & Spitz, 2007; Lonfat, Montavon, Darbellay, Gitto, & Duboule, 2014). Thus, while the outgrowth-induction kernel facilitates the expression of a common set of genes, sometimes even using the same regulatory sequences, other aspects of developmental context differ to ultimately give rise to structures as anatomically and functionally distinct as the penis and limbs.

This leads into a key assumption of the pectoral-pelvic appendage homology: the idea that co-option of *Tbx4* equates to limb duplication. If this were true the fossil record should bear pelvic appendages that have the same anatomical arrangements as the pectoral appendages (Miyashita & Diogo, 2016; Reno, Horton Jr, & Lovejoy, 2013). The reasoning behind this assumption is twofold: (1) the idea that the outgrowth-induction kernel should produce an identical developmental pattern despite the different homeotic environments of the pectoral and pelvic appendages, and (2) this shared developmental regulation should produce a high degree of integration between the pectoral and pelvic appendages. Therefore, the pectoral and pelvic appendages should show a strong signal of morphological covariation dating back to the origin of the pelvic appendage.

When we look to the fossil record for answers, evidence puts the hypothesis of serial homology on even shakier ground. Several early vertebrate clades display pelvic appendages with significantly different arrangements of the bony elements compared to pectoral appendages (Brazeau & Friedman, 2014; Coates, 1994; Diogo, 2020). These findings seem to contradict the idea that the pelvic appendage is simply a duplication of the pectoral. This could support the idea that co-option of *Tbx* into the pelvic region facilitated novel cellular patterning processes due to the different *Hox* environment (Coates & Cohn, 1998). If then, the developmental output of a kernel is dependent upon the homeotic environment, can we equate gene network co-option to morphological duplication? For instance, if a kernel is co-opted to a different homeotic environment, then the segment or region of the organism from which the kernel originated was not duplicated. It is difficult then to accept the claim that co-option of *Tbx4* should duplicate pectoral anatomy in the new pelvic appendage.

It is therefore more likely that the vertebrate pelvic appendages represent a novel addition to the body plan and the appearance of serial homology in tetrapods may be in fact homoplasy or derived anatomical similarity (Don et al., 2013; Miyashita & Diogo, 2016; Roth, 1994). Thus, the independent co-option of the outgrowth-induction kernel to first the pectoral region and subsequently the pelvic region represents two distinct origination events. Vertebrate paired appendages therefore offer a unique system in which to study the problem of evolutionary novelty and how evolutionary continuity of form between characters relates to the continuity of their developmental mechanism.

A distinction can be drawn from the two case studies of gene network co-option presented here. Whereas both examples demonstrate the power of gene network co-option, only the latter can be said to have produced something new in the organism. This seems to highlight a potential difference in the relative timing of a co-option event. We hypothesize that the effect of gene network co-option as it relates to morphological evolution is contingent upon the developmental stage wherein co-option manifests. For instance, the posterior lobe is an example of co-option during growth and patterning that changes the shape of the male genital, whereas the pelvic appendage is an example of co-option that produces an entirely new structure. Thus, we hypothesize that if co-option occurs during periods of intrinsic growth and patterning, this can lead to the evolution of new character states. On the other hand, if co-option facilitates the early deployment of a network during specification this may have the power of producing a new character identity.

5. Conclusions and future research avenues

In this chapter, we have explored how homology and evolutionary novelty can be understood through the lens of development. Although homologies can be inferred by morphological criteria, the fact that homologous characters are defined by a shared common ancestry implies the genetic instructions for that character's development are also similar by descent. To make sense of genetic information as it relates to character uniqueness therefore requires a developmental model of multicellular organ identity. We have proposed a model for Character Identity Modules

(ChIMos), that combines patterning GRNs, kernels, and *Hox* genes, which interact to produce a multicellular unit capable of some degree of self-organization. The ChIMo model can be used to trace character identity over vast evolutionary time, which will be especially useful in cases where the fossil record is patchy to test hypotheses of novelty. Additionally, one may use it to formulate testable predictions with regard to the developmental evolution of characters, as we have done in the latter sections of this chapter.

A key to understanding morphological evolution at any scale comes from understanding how various molecular processes contribute to patterns of cell differentiation and proliferation that define organ identity versus those that regulate organ size, shape, and function. Perhaps most important to this endeavor is establishing some sort of multi-scale model of each character that predicts, or at the very least describes, how molecular level processes are translated to cellular states to influence tissue morphogenesis. This knowledge will not only provide much needed data for the ChIMo hypothesis, but may also allow biologists to make sense of the enormous genetic datasets procured from transcriptomic and genomic analyses. The ChIMo hypothesis with knowledge of the interaction between levels of organization can thus provide a common framework in which to design experiments that test the causal nature of genes that derive novelties and that open up new avenues for innovations.

Acknowledgments

We would like to thank Ceri Weber, Hannah Grunwald, and James DiFrisco for helpful comments on preliminary drafts of this manuscript. We would also like to thank Richard Gawne, Nafis Hasan, and Daniel Nicholsan for helpful discussions on the ideas presented in this chapter. We apologize to the authors whose work we were unable to include due to the breadth of this topic and space limitations.

References

Abu-shaar, M., & Mann, R. S. (1998). Generation of multiple antagonistic domains along the proximodistal axis during drosophila leg development. *Development, 125*, 3821–3830.

Akam, M. (1995). Hox genes and the evolution of diverse body plans. *Philosophical Transactions of the Royal Society of London. Series B, Biological Sciences, 349*, 313–319. https://doi.org/10.1098/rstb.1995.0119.

Amândio, A. R., Lopez-Delisle, L., Bolt, C. C., Mascrez, B., & Duboule, D. (2020). A complex regulatory landscape involved in the development of mammalian external genitals. *eLife, 9*, e52962. https://doi.org/10.7554/eLife.52962.

Andrey, G., Montavon, T., Mascrez, B., Gonzalez, F., Noordermeer, D., Leleu, M., et al. (2013). A switch between topological domains underlies HoxD genes collinearity in mouse limbs. *Science, 340*. https://doi.org/10.1126/science.1234167.

Angelini, D. R., & Kaufman, T. C. (2005). Insect appendages and comparative ontogenetics. *Developmental Biology, 286*, 57–77. https://doi.org/10.1016/j.ydbio.2005.07.006.

Archambeault, S., Taylor, J. A., & Crow, K. D. (2014). HoxA and HoxD expression in a variety of vertebrate body plan features reveals an ancient origin for the distal Hox program. *EvoDevo*, *5*, 1–10. https://doi.org/10.1186/2041-9139-5-44.

Arnone, M. I., & Davidson, E. H. (1997). The hardwiring of development: Organization and function of genomic regulatory systems. *Development*, *124*, 1851.

Ashe, H. L., & Briscoe, J. (2006). The interpretation of morphogen gradients. *Development*, *133*, 385–394. https://doi.org/10.1242/dev.02238.

Averof, M., & Akam, M. (1995). Hox genes and the diversification of insect and crustacean body plans. *Nature*, *376*, 420–423. https://doi.org/10.1038/376420a0.

Averof, M., & Cohen, S. M. (1997). Evolutionary origin of insect wings from ancestral gills. *Nature*, *385*, 627–630. https://doi.org/10.1038/385627a0.

Baker, N. E. (1988). Localization of transcripts from the wingless gene in whole drosophila embryos. *Development*, *103*, 289–298.

Boll, W., & Noll, M. (2002). The drosophila pox neuro gene: Control of male courtship behavior and fertility as revealed by a complete dissection of all enhancers. *Development*, *129*, 5667–5681. https://doi.org/10.1242/dev.00157.

Brazeau, M. D., & Friedman, M. (2014). The characters of Palaeozoic jawed vertebrates. *Zoological Journal of the Linnean Society*, *170*, 779–821. https://doi.org/10.1111/zoj.12111.

Bruce, H. S., & Patel, N. H. (2018). Insect wings and body wall evolved from ancient leg segments. *bioRxiv*, 244541. https://doi.org/10.1101/244541.

Carroll, R. (1988). *Vertebrate paleontology and evolution*. WH Freeman & Co (698 pp).

Carroll, S. B. (1995). Homeotic genes and the evolution of arthropods and chordates. *Nature*, *376*, 479–485. https://doi.org/10.1038/376479a0.

Carroll, S. B., Gates, J., Keys, D. N., Paddock, S. W., Panganiban, G., Selegue, J. E., et al. (1994). Pattern formation and eyespot determination in butterfly wings. *Science*, *265*, 109–114. https://doi.org/10.1126/science.7912449.

Cheney, J. A., Konow, N., Middleton, K. M., Breuer, K. S., Roberts, T. J., Giblin, E. L., et al. (2014). Membrane muscle function in the compliant wings of bats. *Bioinspiration & Biomimetics*, *9*. https://doi.org/10.1088/1748-3182/9/2/025007. https://doi.org/10.1088/1748-3182/9/2/025007.

Clack, J. A. (2009). The fin to limb transition: New data, interpretations, and hypotheses from paleontology and developmental biology. *Annual Review of Earth and Planetary Sciences*, *37*, 163–179. https://doi.org/10.1146/annurev.earth.36.031207.124146.

Clark-Hachtel, C. M., & Tomoyasu, Y. (2020). Two sets of candidate crustacean wing homologues and their implication for the origin of insect wings. *Nat. Ecol. Evol.*, *4*, 1694–1702. https://doi.org/10.1038/s41559-020-1257-8.

Coates, M. I. (1994). The origin of vertebrate limbs. *Development*, *1994*, 169–180.

Coates, M. I., & Cohn, M. J. (1998). Fins, limbs, and tails: Outgrowths and axial patterning in vertebrate evolution. *BioEssays*, *20*, 371–381. https://doi.org/10.1002/(SICI)1521-1878(199805)20:5<371::AID-BIES4>3.0.CO;2-R.

Coates, M. I., Ruta, M., & Friedman, M. (2008). Ever since Owen: Changing perspectives on the early evolution of tetrapods. *Annual Review of Ecology, Evolution, and Systematics*, *39*, 571–592. https://doi.org/10.1146/annurev.ecolsys.38.091206.095546.

Cohen, B., Simcox, A. A., & Cohen, S. M. (1993). Allocation of the thoracic imaginal primordia in the drosophila embryo. *Development*, *117*, 597–608.

Cotterell, J., & Sharpe, J. (2010). An atlas of gene regulatory networks reveals multiple three-gene mechanisms for interpreting morphogen gradients. *Molecular Systems Biology*, *6*, 425. https://doi.org/10.1038/msb.2010.74.

Curtiss, J., & Mlodzik, M. (2000). Morphogenetic furrow initiation and progression during eye development in drosophila: The roles of decapentaplegic, hedgehog and eyes absent. *Development*, *127*, 1325–1336.

Darwin, C. (1859). *On the origin of species by means of natural selection*. London: John Murray.
Davidson, E. H. (2001). *Genomic regulatory systems: In development and evolution*. Elsevier Science.
Davidson, E. H. (2009). Network design principles from the sea urchin embryo. *Genomes Evol.*, *19*, 535–540. https://doi.org/10.1016/j.gde.2009.10.007.
de Navas, L. F., Garaulet, D. L., & Sánchez-Herrero, E. (2006). The Ultrabithorax Hox gene of drosophila controls haltere size by regulating the Dpp pathway. *Development*, *133*, 4495. https://doi.org/10.1242/dev.02609.
Denell, R., Hummels, K., Wakimoto, B., & Kaufman, T. (1981). Developmental studies of lethality associated with the Antennapedia gene complex in Drosophila melanogaster. *Developmental Biology*, *81*, 43–50. https://doi.org/10.1016/0012-1606(81)90346-8.
Dequéant, M.-L., & Pourquié, O. (2008). Segmental patterning of the vertebrate embryonic axis. *Nature Reviews. Genetics*, *9*, 370–382. https://doi.org/10.1038/nrg2320.
Diaz-de-la-Loza, M.-C., Loker, R., Mann, R. S., & Thompson, B. J. (2020). Control of tissue morphogenesis by the HOX gene Ultrabithorax. *Development*, *147*, dev184564. https://doi.org/10.1242/dev.184564.
DiFrisco, J. (2019). Developmental homology. In L. Nuno de la Rosa, & G. Müller (Eds.), *Evolutionary developmental biology: A reference guide* (pp. 1–13). Springer International Publishing, Cham. https://doi.org/10.1007/978-3-319-33038-9_74-1.
DiFrisco, J., & Jaeger, J. (2020). Genetic causation in complex regulatory systems: An integrative dynamic perspective. *BioEssays*, *42*, 1900226. https://doi.org/10.1002/bies.201900226.
DiFrisco, J., Love, A. C., & Wagner, G. P. (2020). Character identity mechanisms: A conceptual model for comparative-mechanistic biology. *Biology and Philosophy*, *35*, 44. https://doi.org/10.1007/s10539-020-09762-2.
Diogo, R. (2020). Cranial or postcranial-dual origin of the pectoral appendage of vertebrates combining the fin-fold and gill-arch theories? *Developmental Dynamics*, *249*(10), 1182–1200. https://doi.org/10.1002/dvdy.192.
Diogo, R., Linde-Medina, M., Abdala, V., & Ashley-Ross, M. A. (2013). New, puzzling insights from comparative myological studies on the old and unsolved forelimb/hindlimb enigma. *Biological Reviews*, *88*, 196–214. https://doi.org/10.1111/j.1469-185X.2012.00247.x.
Diogo, R., & Molnar, J. (2014). Comparative anatomy, evolution, and homologies of tetrapod hindlimb muscles, comparison with forelimb muscles, and deconstruction of the forelimb-hindlimb serial homology hypothesis. *The Anatomical Record*, *297*, 1047–1075. https://doi.org/10.1002/ar.22919.
Don, E. K., Currie, P. D., & Cole, N. J. (2013). The evolutionary history of the development of the pelvic fin/hindlimb. *Journal of Anatomy*, *222*, 114–133. https://doi.org/10.1111/j.1469-7580.2012.01557.x.
Dong, P. S., Chu, J., & Panganiban, G. (2001). Proximodistal domain specification and interactions in developing drosophila appendages. *Development*, *128*, 2365–2372.
Dumont, J. E., Pécasse, F., & Maenhaut, C. (2001). Crosstalk and specificity in signalling: Are we crosstalking ourselves into general confusion? *Cellular Signalling*, *13*, 457–463. https://doi.org/10.1016/S0898-6568(01)00168-1.
Estella, C., & Mann, R. S. (2008). Logic of Wg and Dpp induction of distal and medial fates in the drosophila leg. *Development*, *135*, 627. https://doi.org/10.1242/dev.014670.
Estella, C., Rieckhof, G., Calleja, M., & Morata, G. (2003). The role of buttonhead and Sp1 in the development of the ventral imaginal discs of drosophila. *Development*, *130*, 5929–5941. https://doi.org/10.1242/dev.00832.
Estrada, B., & Sánchez-Herrero, E. (2001). The Hox gene abdominal-B antagonizes appendage development in the genital disc of drosophila. *Development*, *128*, 331–339.

Fisher, C. R., Wegrzyn, J. L., & Jockusch, E. L. (2020). Co-option of wing-patterning genes underlies the evolution of the treehopper helmet. *Nature Ecology and Evolution, 4*, 250–260. https://doi.org/10.1038/s41559-019-1054-4.

Freitas, R., Gómez-Marín, C., Wilson, J. M., Casares, F., & Gómez-Skarmeta, J. L. (2012). Hoxd13 contribution to the evolution of vertebrate appendages. *Developmental Cell, 23*, 1219–1229. https://doi.org/10.1016/j.devcel.2012.10.015.

Freitas, R., Zhang, G., & Cohn, M. J. (2006). Evidence that mechanisms of fin development evolved in the midline of early vertebrates. *Nature, 442*, 1033–1037. https://doi.org/10.1038/nature04984.

Galis, F. (2001). Key innovations and radiations. In *The character concept in evolutionary biology* Academic Press.

Gehring, W. J. (2005). New perspectives on eye development and the evolution of eyes and photoreceptors. *The Journal of Heredity, 96*, 171–184. https://doi.org/10.1093/jhered/esi027.

Gibson-Brown, J. J., Agulnik, S. I., Chapman, D. L., Alexiou, M., Garvey, N., Lee, S. M., et al. (1996). Evidence of a role for T-box genes in the evolution of limb morphogenesis and the specification of forelimb/hindlimb identity. *Mechanisms of Development, 56*, 93–101. https://doi.org/10.1016/0925-4773(96)00514-X.

Gibson-Brown, J. J., Agulnik, S. I., Silver, L. M., Niswander, L., & Papaioannou, V. E. (1998). Involvement of T-box genes Tbx2-Tbx5 in vertebrate limb specification and development. *Development, 125*, 2499–2509.

Glassford, W. J., Johnson, W. C., Dall, N. R., Smith, S. J., Liu, Y., Boll, W., et al. (2015). Co-option of an ancestral Hox-regulated network underlies a recently evolved morphological novelty. *Developmental Cell, 34*, 520–531. https://doi.org/10.1016/j.devcel.2015.08.005.

Gomez, C., Özbudak, E. M., Wunderlich, J., Baumann, D., Lewis, J., & Pourquié, O. (2008). Control of segment number in vertebrate embryos. *Nature, 454*, 335–339. https://doi.org/10.1038/nature07020.

Gonzalez, F., Duboule, D., & Spitz, F. (2007). Transgenic analysis of Hoxd gene regulation during digit development. *Developmental Biology, 306*, 847–859. https://doi.org/10.1016/j.ydbio.2007.03.020.

Gorfinkiel, N., Morata, G., & Guerrero, I. (1997). The homeobox gene distal-less induces ventral appendage development in drosophila. *Genes & Development, 11*, 2259–2271. https://doi.org/10.1101/gad.11.17.2259.

Gredler, M. L., Larkins, C. E., Leal, F., Lewis, A. K., Herrera, A. M., Perriton, C. L., et al. (2014). Evolution of external genitalia: Insights from reptilian development. *Sexual Development, 8*, 311–326. https://doi.org/10.1159/000365771.

Green, J. B., & Sharpe, J. (2015). Positional information and reaction-diffusion: Two big ideas in developmental biology combine. *Development, 142*, 1203–1211. https://doi.org/10.1242/dev.114991.

Hallgrímsson, B., Jamniczky, H. A., Young, N. M., Rolian, C., SCHMIDT-OTT, U., & Marcucio, R. S. (2012). The generation of variation and the developmental basis for evolutionary novelty. *Journal of Experimental Zoology. Part B, Molecular and Developmental Evolution, 318*, 501–517 (https://doi.org/10.1002/jez.b.22448).

Hamaguchi, T., Yabe, S., Uchiyama, H., & Murakami, R. (2004). Drosophila Tbx6-related gene, Dorsocross, mediates high levels of Dpp and Scw signal required for the development of amnioserosa and wing disc primordium. *Developmental Biology, 265*, 355–368. https://doi.org/10.1016/j.ydbio.2003.09.034.

Hammond, K. L., Hanson, I. M., Brown, A. G., Lettice, L. A., & Hill, R. E. (1998). Mammalian and drosophila dachshund genes are related to the ski proto-oncogene and are expressed in eye and limb. *Mechanisms of Development, 74*, 121–131. https://doi.org/10.1016/S0925-4773(98)00071-9.

Hatleberg, W. L., & Hinman, V. F. (2021). Modularity and hierarchy in biological systems: using gene regulatory networks to understand evolutionary change. *Current Topics in Developmental Biology, 141*, 39–73.

Herrera, A. M., & Cohn, M. J. (2014). Embryonic origin and compartmental organization of the external genitalia. *Scientific Reports, 4*, 6896. https://doi.org/10.1038/srep06896.

Hersh, B. M., Nelson, C. E., Stoll, S. J., Norton, J. E., Albert, T. J., & Carroll, S. B. (2007). The UBX-regulated network in the haltere imaginal disc of D. melanogaster. *Developmental Biology, 302*, 717–727. https://doi.org/10.1016/j.ydbio.2006.11.011.

Hu, N., & Castelli-Gair, J. (1999). Study of the posterior spiracles of drosophila as a model to understand the genetic and cellular mechanisms controlling morphogenesis. *Developmental Biology, 214*, 197–210. https://doi.org/10.1006/dbio.1999.9391.

Hughes, C. L., & Kaufman, T. C. (2000). RNAi analysis of deformed, proboscipedia and sex combs reduced in the milkweed bug Oncopeltus fasciatus: Novel roles for Hox genes in the hemipteran head. *Development, 127*, 3683–3694.

Hughes, C. L., & Kaufman, T. C. (2002). Hox genes and the evolution of the arthropod body plan1. *Evolution & Development, 4*, 459–499. https://doi.org/10.1046/j.1525-142X.2002.02034.x.

Hunter, J. P. (1998). Key innovations and the ecology of macroevolution. *Trends in Ecology & Evolution, 13*, 31–36. https://doi.org/10.1016/S0169-5347(97)01273-1.

Isaac, A., Rodriguez-Esteban, C., Ryan, A., Altabef, M., Tsukui, T., Patel, K., et al. (1998). Tbx genes and limb identity in chick embryo development. *Development, 125*, 1867–1875.

Jagadeeshan, S., & Singh, R. (2006). A time-sequence functional analysis of mating behaviour and genital coupling in drosophila: Role of cryptic female choice and male sex-drive in the evolution of male genitalia. *Journal of Evolutionary Biology, 19*, 1058–1070. https://doi.org/10.1111/j.1420-9101.2006.01099.x.

Janvier, P. (1996). *Early vertebrates and their extant relatives*. Oxford University Press.

Janvier, P. (2007). *Homologies and evolutionary transitions in early vertebrate history*. Bloomington, IN: Indiana University Press.

Janvier, P., Arsenault, M., & Desbiens, S. (2004). Calcified cartilage in the paired fins of the osteostracan Escuminaspis laticeps (Traquair 1880), from the late Devonian of Miguasha (Québec, Canada), with a consideration of the early evolution of the pectoral fin endoskeleton in vertebrates. *Journal of Vertebrate Paleontology, 24*, 773–779. https://doi.org/10.1671/0272-4634(2004)024[0773,CCITPF]2.0.CO;2.

Jardine, N. (1969). The observational and theoretical components of homology: A study based on the morphology of the dermal skull-roofs of rhipidistian fishes. *Biological Journal of the Linnean Society, 1*, 327–361. https://doi.org/10.1111/j.1095-8312.1969.tb00125.x.

Jones, K. E., Angielczyk, K. D., Polly, P. D., Head, J. J., Fernandez, V., Lungmus, J. K., et al. (2018). Fossils reveal the complex evolutionary history of the mammalian regionalized spine. *Science, 361*, 1249. https://doi.org/10.1126/science.aar3126.

José, M., García-Cortés, C., Foronda, D., Pastor-Pareja, J. C., Shashidhara, L., & Sánchez-Herrero, E. (2018). The drosophila Hox gene Ultrabithorax controls appendage shape by regulating extracellular matrix dynamics. *Development, 145*. https://doi.org/10.1242/dev.161844.

Kawakami, Y., Esteban, C. R., Matsui, T., Rodríguez-León, J., Kato, S., & Belmonte, J. C. I. (2004). Sp8 and Sp9, two closely related buttonhead-like transcription factors, regulate Fgf8 expression and limb outgrowth in vertebrate embryos. *Development, 131*, 4763–4774. https://doi.org/10.1242/dev.01331.

Keller, E. F. (2000). *Decoding the genetic program*. Cambridge: Cambridge University Press.

Keynes, R. J., & Stern, C. D. (1988). Mechanisms of vertebrate segmentation. *Development, 103*, 413–429.

Kozmik, Z. (2005). Pax genes in eye development and evolution. *Current Opinion in Genetics & Development, 15*, 430–438. https://doi.org/10.1016/j.gde.2005.05.001.
Kubota, K., Goto, S., Eto, K., & Hayashi, S. (2000). EGF receptor attenuates Dpp signaling and helps to distinguish the wing and leg cell fates in drosophila. *Development, 127*, 3769–3776.
Kubota, K., Goto, S., & Hayashi, S. (2003). The role of Wg signaling in the patterning of embryonic leg primordium in drosophila. *Developmental Biology, 257*, 117–126. https://doi.org/10.1016/S0012-1606(03)00062-9.
Kukalova-Peck, J. (1978). Origin and evolution of insect wings and their relation to metamorphosis, as documented by the fossil record. *Journal of Morphology, 156*, 53–125. https://doi.org/10.1002/jmor.1051560104.
Lemons, D., Fritzenwanker, J. H., Gerhart, J., Lowe, C. J., & McGinnis, W. (2010). Co-option of an anteroposterior head axis patterning system for proximodistal patterning of appendages in early bilaterian evolution. *Developmental Biology, 344*, 358–362. https://doi.org/10.1016/j.ydbio.2010.04.022.
Linz, D. M., & Tomoyasu, Y. (2018). Dual evolutionary origin of insect wings supported by an investigation of the abdominal wing serial homologs in Tribolium. *Proceedings of the National Academy of Sciences, 115*, E658. https://doi.org/10.1073/pnas.1711128115.
Logan, M., Simon, H.-G., & Tabin, C. (1998). Differential regulation of T-box and homeobox transcription factors suggests roles in controlling chick limb-type identity. *Development, 125*, 2825–2835.
Logan, M., & Tabin, C. J. (1999). Role of Pitx1 upstream of Tbx4 in specification of hindlimb identity. *Science, 283*, 1736–1739. https://doi.org/10.1126/science.283.5408.1736.
Lonfat, N., Montavon, T., Darbellay, F., Gitto, S., & Duboule, D. (2014). Convergent evolution of complex regulatory landscapes and pleiotropy at Hox loci. *Science, 346*, 1004–1006. https://doi.org/10.1126/science.1257493.
Lovegrove, B., Simões, S., Rivas, M. L., Sotillos, S., Johnson, K., Knust, E., et al. (2006). Coordinated control of cell adhesion, polarity, and cytoskeleton underlies Hox-induced organogenesis in drosophila. *Current Biology, 16*, 2206–2216. https://doi.org/10.1016/j.cub.2006.09.029.
Mallo, M., Wellik, D. M., & Deschamps, J. (2010). Hox genes and regional patterning of the vertebrate body plan. *Developmental Biology, 344*, 7–15. https://doi.org/10.1016/j.ydbio.2010.04.024.
Martin, C. H., McGirr, J. A., Richards, E. J., & St. John, M. (2019). How to investigate the origins of novelty: Insights gained from genetic, behavioral, and fitness perspectives. *Integrative Organismal Biology, 1*, obz018. https://doi.org/10.1093/iob/obz018.
Mayer, B. J., Blinov, M. L., & Loew, L. M. (2009). Molecular machines or pleiomorphic ensembles: Signaling complexes revisited. *Journal of Biology, 8*, 81. https://doi.org/10.1186/jbiol185.
Mayr, E. (1960). *The emergence of evolutionary novelties*. University of Chicago Press.
McKay, D. J., Estella, C., & Mann, R. S. (2009). The origins of the drosophila leg revealed by the cis-regulatory architecture of the Distalless gene. *Development, 136*, 61. https://doi.org/10.1242/dev.029975.
Mercader, N., Leonardo, E., Azpiazu, N., Serrano, A., Morata, G., Martínez-A, C., et al. (1999). Conserved regulation of proximodistal limb axis development by Meis1/Hth. *Nature, 402*, 425–429. https://doi.org/10.1038/46580.
Miyashita, T., & Diogo, R. (2016). Evolution of serial patterns in the vertebrate pharyngeal apparatus and paired appendages via assimilation of dissimilar units. *Frontiers in Ecology and Evolution, 4*, 71. https://doi.org/10.3389/fevo.2016.00071.
Moczek, A. P. (2008). On the origins of novelty in development and evolution. *BioEssays, 30*, 432–447. https://doi.org/10.1002/bies.20754.

Moczek, A. P., & Rose, D. J. (2009). Differential recruitment of limb patterning genes during development and diversification of beetle horns. *Proceedings of the National Academy of Sciences, 106*, 8992–8997. https://doi.org/10.1073/pnas.0809668106.

Moczek, A. P., Rose, D., Sewell, W., & Kesselring, B. R. (2006). Conservation, innovation, and the evolution of horned beetle diversity. *Development Genes and Evolution, 216*, 655–665. https://doi.org/10.1007/s00427-006-0087-2.

Monteiro, A., & Podlaha, O. (2009). Wings, horns, and butterfly eyespots: How do complex traits evolve? *PLoS Biology*, 7, e1000037. https://doi.org/10.1371/journal.pbio.1000037.

Müller, G. B., & Wagner, G. P. (1991). Novelty in evolution: Restructuring the concept. *Annual Review of Ecology and Systematics, 22*, 229–256. https://doi.org/10.1146/annurev.es.22.110191.001305.

Nagashima, H., Sugahara, F., Watanabe, K., Shibata, M., Chiba, A., & Sato, N. (2016). Developmental origin of the clavicle, and its implications for the evolution of the neck and the paired appendages in vertebrates. *Journal of Anatomy, 229*, 536–548. https://doi.org/10.1111/joa.12502.

Nakamura, T., Gehrke, A. R., Lemberg, J., Szymaszek, J., & Shubin, N. H. (2016). Digits and fin rays share common developmental histories. *Nature, 537*, 225–228. https://doi.org/10.1038/nature19322.

Nellen, D., Burke, R., Struhl, G., & Basler, K. (1996). Direct and long-range action of a DPP morphogen gradient. *Cell, 85*, 357–368. https://doi.org/10.1016/S0092-8674(00)81114-9.

Newman, S. A., & Bhat, R. (2009). Dynamical patterning modules: A "pattern language" for development and evolution of multicellular form. *The International Journal of Developmental Biology, 53*, 693–705. https://doi.org/10.1387/ijdb.072481sn.

Niwa, N., Akimoto-Kato, A., Niimi, T., Tojo, K., Machida, R., & Hayashi, S. (2010). Evolutionary origin of the insect wing via integration of two developmental modules. *Evolution & Development, 12*, 168–176. https://doi.org/10.1111/j.1525-142X.2010.00402.x.

Ohuchi, H., Nakagawa, T., Yamamoto, A., Araga, A., Ohata, T., Ishimaru, Y., et al. (1997). The mesenchymal factor, FGF10, initiates and maintains the outgrowth of the chick limb bud through interaction with FGF8, an apical ectodermal factor. *Development, 124*, 2235–2244.

Owen, R. (1848). *On the archetype and homologies of the vertebrate skeleton*. London: Richard and John Taylor.

Panganiban, G., & Rubenstein, J. L. (2002). Developmental functions of the Distal-less/Dlx homeobox genes. *Development, 129*, 4371–4386.

Pavlopoulos, A., & Akam, M. (2011). Hox gene Ultrabithorax regulates distinct sets of target genes at successive stages of drosophila haltere morphogenesis. *Proceedings of the National Academy of Sciences, 108*, 2855–2860. https://doi.org/10.1073/pnas.1015077108.

Peterson, T., & Müller, G. B. (2013). What is evolutionary novelty? Process versus character based definitions. *Journal of Experimental Zoology Part B: Molecular and Developmental Evolution, 320*, 345–350. https://doi.org/10.1002/jez.b.22508.

Prud'homme, B., Minervino, C., Hocine, M., Cande, J. D., Aouane, A., Dufour, H. D., et al. (2011). Body plan innovation in treehoppers through the evolution of an extra wing-like appendage. *Nature, 473*, 83–86. https://doi.org/10.1038/nature09977.

Pueyo, J. I., & Couso, J. P. (2005). Parallels between the proximal–distal development of vertebrate and arthropod appendages: Homology without an ancestor? *Current Opinion in Genetics & Development, 15*, 439–446. https://doi.org/10.1016/j.gde.2005.06.007.

Pultz, M. A., Diederich, R. J., Cribbs, D. L., & Kaufman, T. C. (1988). The proboscipedia locus of the Antennapedia complex: A molecular and genetic analysis. *Genes & Development, 2*, 901–920. https://doi.org/10.1101/gad.2.7.901.

Reim, I., Lee, H.-H., & Frasch, M. (2003). The T-box-encoding Dorsocross genes function in amnioserosa development and the patterning of the dorsolateral germ band downstream of Dpp. *Development, 130*, 3187–3204. https://doi.org/10.1242/dev.00548.

Remane, A. (1956). *Die Grundlagen des natürlichen Systems, der vergleichenden Anatomie und der Phylogenetik: theoretische Morphologie und Systematik*. Leipzig: Geest & Portig.

Reno, P. L., Horton, W. E., Jr., & Lovejoy, C. O. (2013). Metapodial or phalanx? An evolutionary and developmental perspective on the homology of the first ray's proximal segment. *Journal of Experimental Zoology Part B: Molecular and Developmental Evolution, 320*, 276–285. https://doi.org/10.1002/jez.b.22506.

Requena, D., Álvarez, J. A., Gabilondo, H., Loker, R., Mann, R. S., & Estella, C. (2017). Origins and specification of the drosophila wing. *Current Biology, 27*, 3826–3836.e5. https://doi.org/10.1016/j.cub.2017.11.023.

Rogers, K. W., & Schier, A. F. (2011). Morphogen gradients: From generation to interpretation. *Annual Review of Cell and Developmental Biology, 27*, 377–407. https://doi.org/10.1146/annurev-cellbio-092910-154148.

Ronshaugen, M., McGinnis, N., & McGinnis, W. (2002). Hox protein mutation and macroevolution of the insect body plan. *Nature, 415*, 914–917. https://doi.org/10.1038/nature716.

Roth, V. L. (1984). On homology. *Biological Journal of the Linnean Society, 22*, 13–29. https://doi.org/10.1111/j.1095-8312.1984.tb00796.x.

Roth, V. L. (1988). The biological basis of homology. In *Ontogeny and systematics* (pp. 1–26). Columbia University Press.

Roth, V. L. (1994). Within and between organisms: Replicators, lineages, and homologues. In *Homology* (pp. 301–337). Elsevier.

Roth, V. L. (2000). Character replication. In *The character concept in evolutionary biology* (pp. 81–107). Elsevier Science.

Ruvinsky, I., & Gibson-Brown, J. J. (2000). Genetic and developmental bases of serial homology in vertebrate limb evolution. *Development, 127*, 5233–5244.

Sears, K. E., Behringer, R. R., Rasweiler, J. J., & Niswander, L. A. (2006). Development of bat flight: Morphologic and molecular evolution of bat wing digits. *Proceedings of the National Academy of Sciences, 103*, 6581–6586. https://doi.org/10.1073/pnas.0509716103.

Sekine, K., Ohuchi, H., Fujiwara, M., Yamasaki, M., Yoshizawa, T., Sato, T., et al. (1999). Fgf10 is essential for limb and lung formation. *Nature Genetics, 21*, 138–141. https://doi.org/10.1038/5096.

Shubin, N., Daeschler, E. B., & Jenkins, F. A. (2006). The pectoral fin of Tiktaalik roseae and the origin of the tetrapod limb. *Nature, 440*, 764–771. https://doi.org/10.1038/nature04637.

Shubin, N., Tabin, C., & Carroll, S. (1997). Fossils, genes and the evolution of animal limbs. *Nature, 388*, 639–648. https://doi.org/10.1038/41710.

Shubin, N., Tabin, C., & Carroll, S. (2009). Deep homology and the origins of evolutionary novelty. *Nature, 457*, 818–823. https://doi.org/10.1038/nature07891.

Simon, E., & Guerrero, I. (2015). The transcription factor optomotor-blind antagonizes drosophila haltere growth by repressing decapentaplegic and hedgehog targets. *PLoS One, 10*, e0121239. https://doi.org/10.1371/journal.pone.0121239.

Slijper, E. (1946). *Comparative biologic anatomical investigations on the vertebral column and spinal musculature of mammals*. North-Holland Publishing Company.

Smith, S. J., Davidson, L. A., & Rebeiz, M. (2020). Evolutionary expansion of apical extracellular matrix is required for the elongation of cells in a novel structure. *eLife, 9*, e55965. https://doi.org/10.7554/eLife.55965.

Struhl, G. (1982). Genes controlling segmental specification in the drosophila thorax. *Proceedings of the National Academy of Sciences*, *79*, 7380–7384. https://doi.org/10.1073/pnas.79.23.7380.

Stundl, J., Bertucci, P. Y., Lauri, A., Arendt, D., & Bronner, M. E. (2021). Evolution of new cell types at the lateral neural border. *Current Topics in Developmental Biology*, *141*, 173–205.

Swartz, S. M., Bennett, M. B., & Carrier, D. R. (1992). Wing bone stresses in free flying bats and the evolution of skeletal design for flight. *Nature*, *359*, 726–729. https://doi.org/10.1038/359726a0.

Tarazona, O. A., Slota, L. A., Lopez, D. H., Zhang, G., & Cohn, M. J. (2016). The genetic program for cartilage development has deep homology within Bilateria. *Nature*, *533*, 86–89. https://doi.org/10.1038/nature17398.

Thewissen, J., & Babcock, S. (1992). The origin of flight in bats. *Bioscience*, *42*, 340–345. https://doi.org/10.2307/1311780.

Tomoyasu, Y. (2017). Ultrabithorax and the evolution of insect forewing/hindwing differentiation. *Current Opinion in Insect Science*, *19*, 8–15. https://doi.org/10.1016/j.cois.2016.10.007.

Tomoyasu, Y., Wheeler, S. R., & Denell, R. E. (2005). Ultrabithorax is required for membranous wing identity in the beetle Tribolium castaneum. *Nature*, *433*, 643–647. https://doi.org/10.1038/nature03272.

True, J. R., & Carroll, S. B. (2002). Gene co-option in physiological and morphological evolution. *Annual Review of Cell and Developmental Biology*, *18*, 53–80. https://doi.org/10.1146/annurev.cellbio.18.020402.140619.

Tschopp, P., Sherratt, E., Sanger, T. J., Groner, A. C., Aspiras, A. C., Hu, J. K., et al. (2014). A relative shift in cloacal location repositions external genitalia in amniote evolution. *Nature*, *516*, 391–394. https://doi.org/10.1038/nature13819.

Valentine, J. W., Erwin, D. H., & Jablonski, D. (1996). Developmental evolution of metazoan bodyplans: The fossil evidence. *Developmental Biology*, *173*, 373–381. https://doi.org/10.1006/dbio.1996.0033.

Van Valen, L. M. (1982). Homology and causes. *Journal of Morphology*, *173*, 305–312. https://doi.org/10.1002/jmor.1051730307.

Van Valen, L. M. (1993). Serial homology: The crests and cusps of mammalian teeth. *Acta Palaeontologica Polonica*, *38*, 145–158.

Wagner, G. P. (1989). The biological homology concept. *Annual Review of Ecology and Systematics*, *20*, 51–69.

Wagner, G. P. (2007). The developmental genetics of homology. *Nature Reviews. Genetics*, *8*, 473–479. https://doi.org/10.1038/nrg2099.

Wagner, G. P. (2014). *Homology, genes, and evolutionary innovation*. Princeton University Press.

Walsh, C. M., & Carroll, S. B. (2007). Collaboration between Smads and a Hox protein in target gene repression. *Development*, *134*, 3585. https://doi.org/10.1242/dev.009522.

Weatherbee, S. D., Halder, G., Kim, J., Hudson, A., & Carroll, S. (1998). Ultrabithorax regulates genes at several levels of the wing-patterning hierarchy to shape the development of the drosophila haltere. *Genes & Development*, *12*, 1474–1482. https://doi.org/10.1101/gad.12.10.1474.

Weatherbee, S. D., Nijhout, H. F., Grunert, L. W., Halder, G., Galant, R., Selegue, J., et al. (1999). Ultrabithorax function in butterfly wings and the evolution of insect wing patterns. *Current Biology*, *9*, 109–115.

Werner, T., Koshikawa, S., Williams, T. M., & Carroll, S. B. (2010). Generation of a novel wing colour pattern by the wingless morphogen. *Nature*, *464*, 1143–1148. https://doi.org/10.1038/nature08896.

Wigglesworth, V. (1973). Evolution of insect wings and flight. *Nature*, *246*, 127–129. https://doi.org/10.1038/246127a0.

Williams, J. A., Bell, J. B., & Carroll, S. B. (1991). Control of drosophila wing and haltere development by the nuclear vestigial gene product. *Genes & Development*, *5*, 2481–2495. https://doi.org/10.1101/gad.5.12b.2481.

Willmer, P. (2003). Convergence and homoplasy in the evolution of organismal form. In *The Vienna series of theoretical biology. Origination of organismal form* (pp. 33–50). MIT Press.

Wilson, D. E., & Reeder, D. M. (2005). *Mammal species of the world: A taxonomic and geographic reference*. JHU Press.

Woltering, J. M., & Duboule, D. (2010). The origin of digits: Expression patterns versus regulatory mechanisms. *Developmental Cell*, *18*, 526–532. https://doi.org/10.1016/j.devcel.2010.04.002.

Woodger, J. H. (1945). On biological transformations. In W. E. Le Gros Clark, & P. B. Medewar (Eds.), *Essays on growth and form presented to D'Arcy Wentworth Thompson* (pp. 95–120). Oxford: Clarendon Press.

Zhu, M., Yu, X., Choo, B., Wang, J., & Jia, L. (2012). An antiarch placoderm shows that pelvic girdles arose at the root of jawed vertebrates. *Biology Letters*, *8*, 453–456. https://doi.org/10.1098/rsbl.2011.1033.

CHAPTER TWO

Modularity and hierarchy in biological systems: Using gene regulatory networks to understand evolutionary change

William L. Hatleberg and Veronica F. Hinman*
Department of Biological Sciences, Carnegie Mellon University, Pittsburgh, PA, United States
*Corresponding author: e-mail address: veronica@cmu.edu

Contents

Abstract

Modularity and hierarchy are important theoretical concepts in biology, and both are useful frameworks to understand the evolution of complex systems. Gene regulatory networks (GRNs) provide a powerful mechanistic model for modularity in animal development, as they are made up of modular (or self-contained) circuits, which are deployed in a hierarchical manner over time. Over the years, studies in the sea urchin, *Strongylocentrotus purpuratus,* have provided an illustrative example of how these regulatory circuits are responsible for processes such as cell differentiation and cell state specificity. However, GRNs are themselves made up of a nested series of interactions, as each gene can be regulated by multiple *cis*-regulatory elements, which can be further

Current Topics in Developmental Biology, Volume 141
ISSN 0070-2153
https://doi.org/10.1016/bs.ctdb.2020.11.004

broken down into distinct transcription factor binding sites (TFBS). As a result, modularity can be applied to each "level" of this complex hierarchy. Throughout the literature, there is considerable discussion about the roles modular circuits, modular enhancers, and modular TFBS play in evolution, yet there is little discussion about how these nested interactions operate as a whole. In this chapter, we discuss how modular changes at different levels of the GRN hierarchy affect animal development and aim to provide a unified framework to understand the role of modularity in evolution.

1. Introduction

Over the last decade, gene regulatory networks (GRNs) have emerged as a prominent framework to understand both development and evolution (Bolouri, 2008; Davidson & Erwin, 2006; Erwin & Davidson, 2009; Longabaugh, Davidson, & Bolouri, 2005). At its most basic, a GRN is an experimentally-validated model for illustrating how regulatory genes interact with each other over space and time. These GRNs can be used as a mechanistic hypothesis for diverse developmental processes, such as cell type differentiation (e.g., Drapek, Sparks, & Benfey, 2017; Oliveri, Tu, & Davidson, 2008), cell migration (e.g., Christiaen et al., 2008; Martik & McClay, 2015; see chapter "Evolution of new cell types at the lateral neural border" by Stundl et al.), morphogenesis (e.g., Hill, Demarest, Gorsi, Smith, & Yost, 2017; Rafiq, Cheers, & Ettensohn, 2012), and regeneration (e.g., Gehrke et al., 2019; Smith et al., 2011).

Through the examination of these interactions, larger networks of relatively self-contained interactions emerge, which are known as modular circuits (Davidson & Erwin, 2006; Erwin & Davidson, 2009). These circuits are deployed sequentially over time and space, resulting in a broader hierarchical structure (Erwin & Davidson, 2009). However, the modular structure of GRN circuitry can be further broken down into smaller sub-modules, as the expression of each gene in the network relies on *cis*-regulatory elements, which are composed of multiple transcription factor binding sites (TFBS). Throughout the literature, the term "modular" is broadly used to describe different components of biological systems (circuits, enhancers, TFBSs), and is a useful conceptual framework to understand the evolution of complexity. Therefore, in this chapter, we aim to: (1) define modularity and hierarchy and discuss the relevance of these concepts to biological systems, (2) use the echinoderm GRN as a case study to illustrate the different ways in which GRNs are both modular and hierarchical, and (3) discuss the ways GRNs can be used as a mechanistic framework to understand the evolution of complex systems. Together, we hope to provide a unified framework to understand how biological systems evolve through modular changes at different hierarchical levels.

1.1 Modularity and hierarchy in biological systems

The concept of modularity is widespread across multiple academic disciplines, including engineering, computer science, mathematics, biology, and architecture (Simon, 1962). In all of these fields, modularity provides insight into the structure, and in some cases the evolution, of complex systems. At its most basic, modularity is the compartmentalization of a complex system into smaller, semi-autonomous parts or units that perform specific functions (Melo, Porto, Cheverud, & Marroig, 2016; Wagner, Pavlicev, & Cheverud, 2007). In biology, a straightforward example of a modular system is the eukaryotic cell, which is composed of distinct membrane bound organelles. Each organelle operates in relative autonomy to each other and provides a crucial function to the overall fitness of the cell as whole. Similarly, animal body plans are a single unit, which can be subdivided into discrete organ systems.

While ubiquitous, the use of the term "module" is highly varied throughout biology, and the concept has been applied to many subdisciplines, including comparative morphometrics, genetics, evolutionary biology, and genomics. A unifying definition of a module is a discrete trait (either phenotypic or genetic) that can be defined by a series of self-contained interactions (West-Eberhard, 2019). In many cases, these traits are physical, such as distinct morphometric phenotypes, which are closely tied to quantitatively linked genetic modules of covariant traits (Melo et al., 2016; Mitteroecker, 2009). However, in the age of genomics, the concept of modularity has been applied to broader system-wide interactions, and can also refer for suites of co-expressed genes (e.g., Xue et al., 2013), subnetworks of protein-protein interactions (e.g., Han et al., 2004), or metabolic pathways (e.g., Ravasz, Somera, Mongru, Oltvai, & Barabási, 2002). On a more macroscopic level, modules can also be made up of interactions of individual organisms that comprise a larger ecological food webs, trophic levels, community interactions, or ecosystems (Krause, Frank, Mason, Ulanowicz, & Taylor, 2003).

All of these examples and applications of modularity in biology are inherently tied to the concept of hierarchy, as modules can often be further broken down into a series of nested sub-modules. For instance, the earlier example of an organ can be further subdivided into distinct tissues, which are themselves composed of discrete cell types. Additionally, relationships between modules are reliant upon the underlying networks of interactions that are responsible for the formation and maintenance of each unit. These relationships can be temporal and spatial. For instance, smaller modules can be created and sustained within larger, parental modules (such as the preplacodal ectoderm becoming the neural crest and individual placodes; chapter "Evolution of new cell types at the lateral neural border" by Stundl et al.). Or, two smaller modules might work together to form a larger unit (as in the retina and lens primordia forming the eye). A simple depiction of hierarchical systems is a tree diagram, as used in phylogenetic analysis, cell lineage or differentiation diagrams, or community interaction networks. In this depiction, each node is a module that bifurcates into multiple "child" nodes, or clades (Fig. 1A). Often, tree diagrams such as these, illustrate increasing modularity over time. For instance, in phylogenetics, each node represents an ancestral state, and child modules indicate cladogenesis.

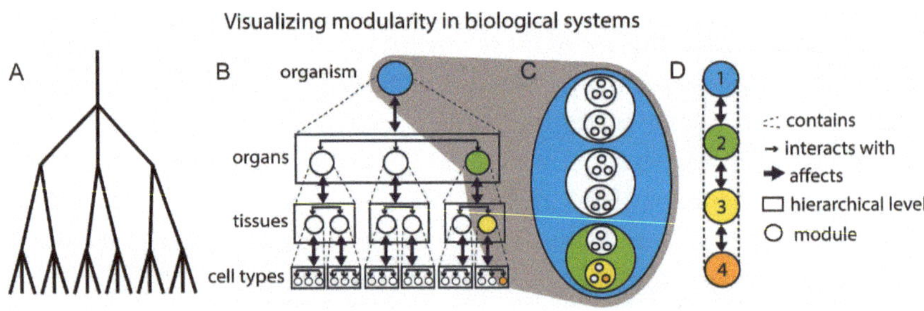

Fig. 1 Visualizing modularity and hierarchy in biological systems. Modularity is a ubiquitous phenomenon in biological systems, and distinct modules are often organized in a tiered, or hierarchical structure. (A) Tree-diagrams, as used in phylogenetics, cell-lineage mapping, and ecosystem dynamics, provide a simple depiction of modularity in hierarchical systems. Each node illustrates a module, which are subdivided into sub-modules. (B) The inclusion of network dynamics and hierarchical levels indicate how modules and sub-modules interact. For instance, an organism can be subdivided into multiple organs, which can be further subdivided into tissues and cell types. Color indicates the level of the hierarchical system, where blue is top tier and orange is bottom. (C) Alternative illustration of hierarchical modularity as a Venn diagram, where child modules are physically contained within their parent. (D) Reduced diagram depicting information flow between levels of the hierarchy, where interactions between sub-modules affect the overall network of the parent (i.e., cell-types affect tissues, which affect organs systems, which ultimately affect the viability of the organism as a whole).

We can embellish this model a little further (Fig. 1B) to see tiers emerging from the hierarchy, whereby interactions between modules of the same tier (i.e., the cells within a specific tissue, or the tissues within a specific organ) jointly affect the functioning of the parent module. This model, especially in systems occupying the same physical space, can also be depicted as a Venn diagram, where each module is like a nesting doll that contains multiple sub-modules (Fig. 1C). However, like phylogenies, complex clades can be visually collapsed to show overarching relationships between parts. Therefore, all hierarchical systems can be ultimately distilled down into flows of information, where parent modules can generate child modules, and the functions of child tiers can affect their parents. That the whole is based on the sum of its parts (high-level modules are made by low-level modules) seems obvious; but the low-level modules can be influenced by higher level modules, as well. The new advances in mechanotransduction, for instance, show that changes in tissue architecture can change cell shape and alter patterns of gene expression. Tissue-level forces sensed at the cell surface, for instance, can activate GRNs in the nucleus, leading to the differentiation of cells such as osteocytes or mammary epithelia (Furuta, Ren, Mao, & Bissell, 2018; Hamouda, Labouesse, & Chalut, 2020; Piccus & Brayson, 2020; Qin, Liu, Cao, & Xiao, 2020). In some instances, paracrine signals from healthy embryonic tissue can restore cancer cells to a non-oncogenic phenotype and integrate them into a normally functioning community (Aragona et al., 2020; Illmensee & Mintz, 1976; Kasemeier-Kulesa et al., 2008; Kenny & Bissell, 2003). Influences can even be from the environment to the gene level, as in predator-induced developmental defenses (Ritschar, Rabus, & Laforsch, 2020; Yin, Jin, Zhou, Wang, & Zhao, 2017) and larval metamorphosis cues (Freckelton, Nedved, & Hadfield, 2017; Tran & Hadfield, 2011). Child-tiers interact with one another to affect their parent system, and together, these iterative relationships make up the function of the system as a whole (Fig. 1D). As a result, the concept of modularity, as well as the applications of modular theory to biological systems, is relative to a broader hierarchical structure.

1.2 Development is a process of increasing modularity

von Baer (1828) noted that the organism became more complex as it developed, and animal development is a process of increasing modularity. On a morphological level, this modularity can be conceived of as the increasing compartmentalization of a single-celled zygote into distinct spatial territories over time, resulting in a multicellular body plan consisting of multiple discrete tissues and cell types.

An early model for this modularization process during development is Waddington's "epigenetic landscape," which likened cell differentiation to a marble rolling down a hill (Ladewig, Koch, & Brüstle, 2013; Waddington, 1957). Based on the topography of the hill, the marble is fated to roll down certain paths, ultimately leading to different end results. The ball makes a "decision" to become ectoderm or mesendoderm, followed by subsequent decisions (e.g., neural ectoderm or epidermal ectoderm), which is dependent upon earlier decisions. Eventually, it ends up in a basin of a particular cell type. This thought experiment is a metaphor for the unidirectional process of cell differentiation that occurs during development, which results in different cellular fates, or developmental outcomes. The branching patterns of potential marble paths form a tree structure, indicative of the spatio-temporal hierarchy of cellular differentiation. Thus, cell differentiation is a process of modularization that occurs during the physical compartmentalization of an embryo during development. In recent years, this model has been widely applied to cell fate mapping, and has been modernized through the addition of GRNs as a framework to conceptualize Waddington's landscape (Bhattacharya, Zhang, & Andersen, 2011; Cai et al., 2020). In this metaphor, changes to underlying gene regulatory networks (GRNs) are the topological "bumps" on the hill that influence whether the marble will become one cell fate versus another, and the basins are the stable GRNs that characterize a particular cell type. Moreover, unlike the inevitable gravitational pull on the marble (differentiation), it is possible to move the ball back up the hill (dedifferentiation) or to other positions at the bottom of the hill (transdifferentiation). The ability to induce differentiated cells to become pluripotent and then to re-differentiate into different specialized cell types shows that experimentation can turn the epigenetic landscape into a pinball game (Sareen & Svendsen, 2010; Takahashi, 2012).

Waddington's visual representation of cellular differentiation is perfectly corroborated by modern single cell technology (Flower, 2020; Wagner et al., 2018). When an embryo is dissociated into its individual cells and sequenced, it is possible to computationally identify cell type based on transcriptome similarity. This visual representation of a cell fate map illustrates how an embryo becomes increasingly modular as cells become terminally differentiated. Moreover, these single cell studies also allow us to investigate the specific changes that occur during development that lead to this cell differentiation. While changes in cell fate have been well characterized in stem cells (Graf & Enver, 2009; Novershtern et al., 2011), new single cell technologies allow researchers to uncover lineage tracing for all unique cell types, and computationally infer gene regulatory networks responsible for cell fate (e.g., Ding & Bar-Joseph, 2020; Friedman et al., 2018; Moignard et al., 2015).

2. Echinoderm development illustrates modularity and hierarchy at many levels

Echinoderms, including starfish and sea urchins, have been a developmental model system for over a hundred years, and provide an illustrative example of spatio-temporal modularization. Furthermore, the best characterized GRN is arguably that for the development of the sea urchin, *Strongylocentrotus purpuratus,* which has been used to model cell fate specification (e.g., Oliveri & Davidson, 2004). Through years of experimental validation, this network includes 72 genes and over 271 regulatory interactions, spanning the onset of fertilization through the formation of the late gastrula body plan (Cary et al., 2020). Therefore, this system serves as a good example to explain the concept of modularity in development (Peter & Davidson, 2009).

Cell specification begins during early cell cleavage (Arnone, Byrne, & Martinez, 2015; Gilbert, 2000). In Euechinoids, the subclass of Echinodermata containing the modern sea urchins, the first two cell cleavages are symmetrical, producing four evenly sized cells. However, the fourth cleavage is unequal, resulting in a 16-cell embryo with three different sized cells (macromere, mesomere, and micromeres). These asymmetrical cleavages begin the process of animal-vegetal axis formation, differentiating cells that will become the anterior part of the larva from the posterior. During subsequent cleavages, mesomere and macromere cells divide evenly, while the micromeres divide asymmetrically one more time to form both large and small micromeres. By 60-cell stage, each cell of the embryo has been largely specified into broader spatial territories, which are defined by unique gene expression patterns and result in discrete cell fates (Fig. 2).

These early territories become further subdivided into germ layers, whereby the mesomeres in the animal pole become the ectoderm, and the

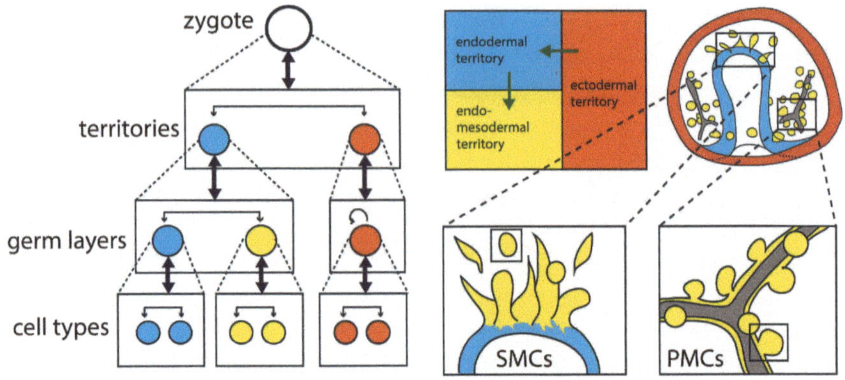

Fig. 2 Animal development is a process of increasing modularization. The compartmentalization of an embryo into distinct cell types forms a spatio-temporal hierarchy (A) the embryo is divided primarily into territories, such as ectoderm, and endomesoderm. In the sea urchin, initial territories are subdivided into the three primary germ layers, endo-, ecto-, and mesoderm. Territories are connected to each other via signaling pathways (Peter & Davidson, 2009). Germ layers are further divided into different cell types. Cell types perform a unique function and give rise to different cell lineages. For instance, while both endomesodermal in origin, during gastrulation, the primary and secondary mesenchyme cells (PMCs and SMCs) differentiate from one another. The PMCs are responsible for creating a structural biomineral skeleton, while SMCs go on to establish the immune system of the free-swimming larva (Ettensohn & Ruffins, 1993).

macromeres and micromeres predominantly become the endomesoderm (Fig. 2). While the germ layers are not irreversibly fated, this process of increasing cell differentiation establishes a spatio-temporal hierarchy of nested modular territories, which can be depicted as a cell lineage map, or cell fate tree. Mesomere cells become ectoderm, which largely form the outer layer of the echinoderm larvae, and are responsible for the formation of the larval nervous system. As is common among bilaterians, endoderm and mesoderm, which share common origin in the endomesoderm, are responsible for a wide range of cell types in the larva, including muscle cells, gut tissues, immune cells, and skeletogenic cells. The separation of the endomesoderm into discrete endo- and mesoderms result from a series of complex gene regulatory interactions (see Section 2.1), which are set in motion even before asymmetrical cleave begins.

2.1 GRN circuits are both modular and hierarchical, providing a mechanistic framework for development

At its most basic, a GRN is a set of interactions between regulatory genes and proteins over space and time. Regulatory genes, such as transcription factors, are able to activate or repress the expression of their downstream targets, which include other regulatory genes, as well as genes that produce the structural proteins resulting in phenotype. Tightly connected networks of regulatory interactions make up modular units of the GRN, known as subcircuits or motifs (Davidson, 2010; Sayed, Bocan, & Miskov-Zivanov, 2018; Smith, Theodoris, & Davidson, 2007). Broadly speaking, each subcircuit executes a specific job when deployed during a specific spatio-temporal context during development (Peter & Davidson, 2009). As the embryo develops, these circuits are responsible for sub-specialization within germ layers, and the differentiation of individual cell types. When a particular gene or circuit is solely responsible for a cell type, (such as a micromere cell lineage, or a neural precursor) it is known as a specification state (Ben-Tabou De-Leon & Davidson, 2007). Thus, when a specification state is deployed in a specific lineage, it is responsible for changing cell type identity.

A well-studied example of a temporal GRN activating a lineage-specific regulatory state is in primary mesenchyme cells (PMCs), which are derived from the large micromeres that form during the asymmetrical cleavage in the vegetal pole of the embryo (Fig. 2). During early stages of cleavage, the transcription factor *pmar1* is expressed transiently in the micromere lineage, due to activation by β-catenin and other maternal factors (Oliveri, Carrick, & Davidson, 2002; Oliveri et al., 2008). Pmar1 acts as a transcriptional repressor of *hesC*, which is itself a repressor of the downstream micromere-specific specification state circuit, which includes the genes *ets1*, *alx1*, *tbr* and *tel*. Therefore, while *pmar1* is expressed, the micromere-specific specification state genes are expressed, inducing micromere formation. Similarly, in cells where *pmar1* are not expressed, HesC represses micromere-specific specification state genes, thereby inhibiting micromere formation in those cells. This type of subcircuit, involving the repression of a repressor, is known as a double-negative gate, and is used in multiple developmental contexts to restrict spatial specification to specific cell types in a developing embryo (Davidson, 2010; Peter & Davidson, 2009). The tightly interconnected downstream regulatory network containing *ets1*, *alx1*, *erg*, *hex*, and *tgif*, forms another kind of subcircuit, known as a recursive positive feedback loop, which is responsible for maintaining cell type specificity (Oliveri et al., 2008). Each of these transcription factors positively regulates the others, thus expression of each one is necessary to maintain cell fate specificity.

Subcircuits such as the double negative gate and positive feedback loops are modular, as regulatory interactions between the genes acting in the subcircuit are largely self-contained (Peter & Davidson, 2009). Like development itself, these GRNs represent a temporal hierarchy, because expression of upstream regulators must happen in sequential order before the deployment of downstream modules, resulting in a regulatory cascade that can only unfold over time. In the example of micromere specification, for instance, upstream regulators such as *pmar1* are required for downstream lineage-specification, therefore the micromere-specific circuit including *ets1, alx1, tbr* and *tel,* will not be deployed without passing through the double negative gate. As a result, GRNs are seen as unidirectional and follow the temporal hierarchy of development.

2.2 GRN kernels and the evolutionary hierarchy of constraint

In addition to the temporal hierarchy, Davidson and Erwin (2006) argued that GRN subcircuits can be further categorized in a hierarchy of increasing evolutionary constraint. Through the comparison of GRN circuitry between echinoderm species, it is clear that certain subcircuits, known as "kernels," remain deeply conserved across evolutionary timescales. For example, the key players in the endomesoderm GRN in sea urchins are largely the same in sea stars, despite diverging from each other 500 million years ago (Hinman, Nguyen, Cameron, & Davidson, 2003). Alternatively, other subcircuits, known as "Plug-ins," can be readily co-opted for different purposes at different stages of development (more on this in Section 3.2) (Davidson & Erwin, 2006).

In the examples given so far, all genes described were transcription factors, and therefore affect the expression of downstream target genes. However, not all genes in a GRN are regulatory. Non-regulatory, or structural genes, are suites of genes that do not directly affect the expression of other genes. Instead of orchestrating the spatio-temporal expression of other regulators, the products of these "gene batteries" play important structural or metabolic roles in the cell, and occupy the terminal level of GRN hierarchy. These gene batteries play a key functional role of implementing cell-type specific processes.

For instance, the micromere cell lineage gives rise to skeletal mesoderm cells, or PMCs, which express batteries of genes responsible for the secretion of biomineral compounds and the formation of the complex larval skeletons in Euechinoids (Fig. 2). Therefore, the major fate changes during cellular differentiation leading to skeleton formation result from underlying changes to the temporally hierarchical GRN (i.e., double negative gate → micromere kernel specification genes → skeletogenic gene batteries). Additionally this follows the hypothetical evolutionary hierarchy, where well-conserved kernels are responsible for phylum-level traits, while downstream batteries are responsible for patterning more lineage-specific body plan characteristics (He & Deem, 2010).

The modularization of development, typified by cell differentiation and the formation of novel cell types over time, can be explained by underlying GRN circuitry, which is responsible for different specification states. These GRN circuits are both modular and hierarchical, and provide a powerful mechanistic framework to understand how changes in gene expression during development can result in different phenotypic outcomes. In addition, GRNs can illustrate another hypothetical form of hierarchy that consists of both deeply conserved kernels and terminal branches of the GRN network, which are responsible for orchestrating more lineage-specific traits (Davidson & Erwin, 2009; He & Deem, 2010).

2.3 GRNs as nesting dolls: Additional layers to the GRN hierarchy

While GRNs circuitry is a powerful framework to understand development, particularly the specification of cell types, it is important to note that GRN subcircuits reflect modularity at only one hierarchical level. Early GRN frameworks discussed circuitry as hierarchical in terms of temporal deployment, where modules are expressed in order over time and space. However, GRNs arguably need to be seen as hierarchical in an additional way: they can be broken down into a further series of nested interactions that affect the overall topology and dynamics of the network (Fig. 3).

In order to affect the expression of another gene, and thus to establish the GRN circuitry, *trans*-regulatory factors must interact with *cis*-regulatory DNA sequences, known as *cis*-regulatory modules or enhancers. These enhancers are non-coding sequences usually between 200 and 1000 nucleotides long, and can be 1000s of base pairs upstream or downstream of the gene they affect (Carroll, Grenier, & Weatherbee, 2013). Typically, *cis*-regulatory elements act either as an activator or a repressor, and directly influence expression of their associated gene. While *cis*-regulatory elements are non-coding, they function by interacting with proteins, such as transcription factors. Together, these nucleoprotein complexes interact with the basal promoter of the associated gene to regulate transcription.

Each gene is often associated with multiple *cis*-regulatory elements (CREs). It was traditionally believed that *cis*-regulatory elements were

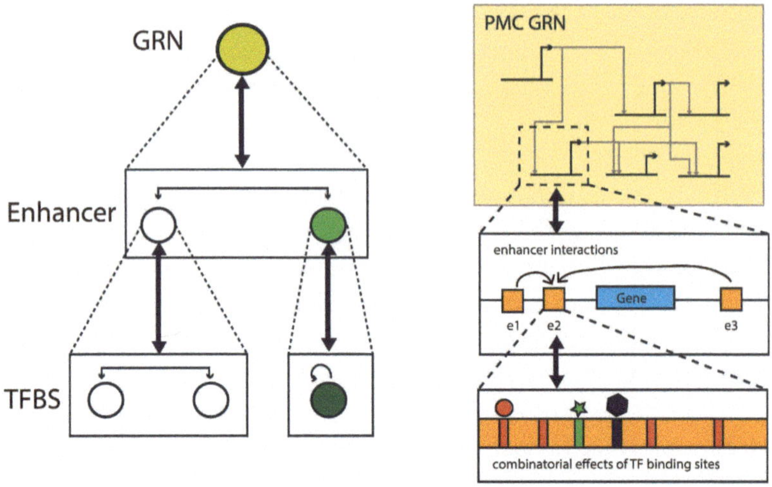

Fig. 3 Underlying gene regulatory networks are a mechanistic model for specification of cell type. *Cell type is defined by gene circuitry.* Cell type identity, such as the differentiation between PMC and SMC cells, is defined by the interactions of gene regulatory networks (GRN) within cells. This includes the regulation of gene expression via the interaction of *trans*-acting factors (such as TFs), which turn other regulatory genes "on" and "off." Such interactions create regulatory cascades and feedback loops, which make up complex network dynamics. For instance, the SMC-specific GRN acts as a specification state for pigment cell lineage, in which Pmar1 activity upstream of delta-notch signaling pathway activates the transcription factor, *gcm*. Gcm is the specification state of pigment cells, and is also responsible for repressing pivotal skeletogenic regulator, *alx1*, thereby preventing pigment cells from becoming skeletogenic PMCs (Ben-Tabou de-Leon & Davidson, 2007). *Gene circuity is regulated by enhancers.* Each *trans*-acting regulatory factor relies multiple non-coding *cis*-regulatory regions of the genome (i.e., enhancers), which usually lie within a few thousand base pairs from the transcription start site. Each gene has multiple enhancers, many of which operate in unique spatio-temporal contexts depending on *trans*-environment. This allows for different circuitry and GRN output in different developmental and environmental conditions. *Enhancers consist of transcription factor binding sites.* Each enhancer is composed of multiple transcription factor binding sites (TFBSs). Ultimately, the transcription of a gene (and thus the regulation of all downstream genes in the network) relies on multiple factors from the *trans*-environment.

modular, relying on a different enhancer for each spatio-temporal context (Kirchhamer, Yuh, & Davidson, 1996). However, recent studies have challenged this notion, suggesting that enhancers can also be pleiotropic (Sabarís, Laiker, Preger-Ben Noon, & Frankel, 2019). Additionally, while enhancer dynamics are often discussed in a Boolean way (causing genes to either be turned "on" or "off"), they are in fact more complex than that, including low affinity binding sites (Cheatle Jarvela et al., 2014; Nakagawa, Gisselbrecht, Rogers, Hartl, & Bulyk, 2013; Zhu et al., 2012) and different combinatorial effects (Reiter, Wienerroither, & Stark, 2017) depending on which proteins are bound, as well as the nuclear concentration of the protein. More specifically, there are competing models for how enhancers function, notably the enhanceosome versus billboard models (Arnosti & Kulkarni, 2005). The former model asserts that all TFBSs must be bound so that a stable protein complex is formed before transcription can be altered. Alternatively, the billboard model predicts that individually bound TFs and TF-combinations can have unique combinatorial effects. There is evidence that different genes rely on different strategies (Reiter et al., 2017), which can also be context specific (Yáñez-Cuna, Dinh, Kvon, Shlyueva, & Stark, 2012).

Enhancers themselves are made up of multiple transcription factor binding sites (TFBS), which are a non-coding nucleotide motif (usually between 10 and 15 base pairs), where a specific type of transcription factor protein can bind. Often, a single transcription factor protein may bind preferentially to multiple motifs, which can be visualized by different position weight matrices (PWM; Nakagawa et al., 2013; Zhu et al., 2012). Higher affinity (preferred) motifs are called primary, while low-affinity binding motifs are called secondary (Badis et al., 2009). Transcription factor binding motifs can be concentration specific, in which high concentrations result in the transcription factor binding to both primary and secondary motifs (Segal, Raveh-Sadka, Schroeder, Unnerstall, & Gaul, 2008). In contrast, when the transcription factor concentration is low, binding will only occur in the primary motif. For instance, in the sea star, *Patiria miniata,* the transcription factor T-brain (Tbr) is highly pleiotropic in early development, playing regulatory roles in all three germ layers (McCauley, Weideman, & Hinman, 2010), yet shows different binding affinities based on nuclear concentration (Cheatle Jarvela et al., 2014). Therefore, during sea star development, GRN output can change depending on concentration of the same *trans*-factor. While it was traditionally believed that low-affinity binding sites did not play a major role in development, it is becoming increasingly apparent that low affinity binding sites play a crucial role in gene regulation and the evolution of GRNs (Cary, Cheatle Jarvela, Francolini, & Hinman, 2017; Crocker, Noon, & Stern, 2016; Ramos & Barolo, 2013).

As GRNs are hierarchical, there are modular elements at different levels of organization. Modular circuits are orchestrated by modular enhancers, which are subsequently composed of modular TFBSs. Therefore, modularity in the context of GRN structure is a relative term, and position in the hierarchy has dramatic effects on the evolution of development. As illustrated in Fig. 1D, information flows from the lower tiers of a hierarchy, to parent tiers. Therefore, changes to modular TFBSs affect enhancer interactions, which can ultimately impact circuit dynamics. In some cases, these changes may not dramatically affect circuitry. For instance, the gain or loss of a single TFBS would not likely inhibit the proper function of a recursive positive feedback loop (Oliveri et al., 2008). However, as we will discuss in the next section, simple changes to *cis*-regulatory regions can lead to dramatic circuitry rewiring events, such as co-option (McQueen & Rebeiz, 2020), and alteration of TF binding affinities and/or number of binding sites can affect rates of transcription, which can ultimately affect circuit dynamics (Cheatle Jarvela et al., 2014). Thus, all levels of the GRN hierarchy can play important roles in causing phenotypic change.

3. Modularity, evolution, and homology

In many disciplines, discussions of modularity and hierarchy are closely tied to the emergence of complex structures. Herbert Simon's seminal work on the nature of hierarchical systems explored how modularity is both the product of complexity, yet also necessary for those complex systems to evolve (Simon, 1962). These ideas were eloquently summarized by Simon's thought experiment known as the watchmaker parable, which illustrated how structures that are composed of modular parts are much quicker to evolve than those that have to be built from scratch each time (Wu, 2013). The conceptual framework introduced in the watchmaker parable can be readily applied to biology. If a system is built using universal building blocks, these individual modules can be restructured, duplicated, and repurposed to make novel forms (Raff, 1996).

A central focus in evolutionary biology is to infer which modules are homologous, or share common ancestry. The alternative scenario is convergence, where seemingly similar structures arise independently of each other. To Simon, homology is closely akin to two watches made with the same modular parts, whereas his alternative scenario, in which each clock is made from scratch each time, illustrates convergence. Understanding homology is critical to piecing together a mechanism of evolution in modular systems. However, like many other aspects of complex biological systems, the concept of homology operates on its own unique hierarchy (Ochoterena, Vrijdaghs, Smets, & Claßen-Bockhoff, 2019). As a result, modular changes, including those in GRN structure, can be examined at the level of individuals (ontogeny), the population (tokogeny), or the species (phylogeny) (Ochoterena et al., 2019). This relativity of homology has been discussed since the beginning of embryology. Therefore, the comparisons of GRN structures both *within* and *between* species contain many parallels to early theoretical concepts of serial (ontogenetic) homology, versus historical, or phylogenetic homology. In the next section, we review the literature on historical concepts of homology and modularity in order to later recontextualize our understanding of GRN evolution.

3.1 Historical investigations of modular evolution

Early embryologists, such as Haeckel, believed that ontogenies could be broken down into temporal units or stages. For Haeckel, evolution of developmental processes evolved as new developmental stages were added on to the end of a life cycle. Other early embryologists similarly saw ontogenetic units as modules of evolutionary change. For instance, von Baer believed certain periods of development were characterized by increased evolutionary constraint, whereas other developmental stages were more amenable to evolutionary innovation, leading to hypotheses such as the hourglass model of evolution, or the phylotypic stage of development (Abzhanov, 2013). However, these early concepts of how development evolves, largely lacked a mechanistic framework. Instead, these theories explained differences in animal ontogenies through broader changes in developmental timing, such as the lengthening or shortening of specific developmental stages, known as heterochronies. Examples of such heterochronies were the retention of juvenile characteristics into adulthood, called neoteny or paedomorphosis. Similarly, embryologists studied the changes in relative morphological size throughout an ontogeny, known as allometry (Klingenberg, 1996). Together, these ideas formed a foundational understanding of how life-history strategies and morphological innovation were acquired throughout animal evolution. Recent permutations of these ideas have led to reinvestigations of embryological theories as genome-wide analyses that aim to link ontogeny to underlying genetics, such as theories surrounding the hourglass model of development (Levin et al., 2016), the phylotypic stage (Domazet-Lošo & Tautz, 2010), or the origins of larval body plans (Khalturin, 2020; Wang et al., 2020).

The concept of serial homology also dates back to Haeckel, who saw these repetitive phenotypic structures as a mechanism for evolutionary change. More recently, the concept of serial homology provides a powerful example of how modular structures can change within an ontogeny through the duplication/modification of modular parts during development. One example of this is the segmentation of the chordate vertebra. Together, the vertebral column is a defined structural unit; however, it can be broken down into individual modules. Each vertebrae is similar, yet has evolved morphological and functional variation depending on the position in the spinal cord. Similarly, there are indisputable morphological similarities (bone and muscle structures) between arms and legs, yet it is unlikely that these similarities evolved convergently. In contrast to serial homology, historical, or phylogenetic homology examines modular units *between* species. To continue upon the previous example, this would be the comparison of bat forelimbs (wings) and dolphin forelimbs (fins), both of which share similar underlying bone structures despite drastically different functions and exterior morphologies (Shubin, Tabin, & Carroll, 2009).

Both serial homology and phylogenetic homology represent shared ancestry. However, the two examples examine evolutionary change on different levels of the broader hierarchy. The same concept can be applied to other methodologies used to understand evolutionary change. For instance, in allometry, the comparative study of size of discrete morphometric traits, can be applied both *within* an ontogeny and *between* different levels of the evolutionary hierarchy, in individuals or species (Klingenberg, 1996). Additionally, nested underneath all of these comparisons are the genetic traits associated with each phenotypic trait.

With the rise of modern genetics, the modular unit of evolution shifted away from ontogeny toward the gene (Gilbert, Opitz, & Raff, 1996). In the eyes of a geneticist, evolution could only be explained by the random mutations that occurred in an organism's DNA and the subsequent selection through natural selection. In order to conceptualize development, one school thought turned inward, and tried to uncover the relationship between genotype and phenotype within an individual. Examples of these microevolutionary inquiries tried to pinpoint the underlying genetic traits, or modules of linked genes associated with morphologically distinct units, such as the bone morphometrics responsible for skull shape (Melo et al., 2016). Modern studies in that vein use quantitative trait loci to identify distinct modules of genes associated with a given phenotype

(Mitteroecker, 2009). In contrast, macroevolutionary geneticists saw the process of evolution through the broader differences and similarities between populations and species. These changes include differences in allelic frequencies across populations, which can ultimately lead to species level change, inline with Devgen-Popgen framework (Gilbert, 2003).

Starting in the 1980s, it became clear that the mechanism of evolution could not be fully summed up by genetics (Keller, 2002; Raff & Kaufman, 1991), as differences in genotype alone could not account for the conserved developmental stability seen during embryogenesis (Siegal & Bergman, 2002). As a result, pioneering evolutionary developmental biologists began a new synthesis of previous paradigms, or Evo-devo, which provided a framework to link early embryological theories of ontogeny with modern genetics. Central to this is the reinvigorated idea of epigenetics, which asserts that non-genetic (or developmental) processes can also be critical in the emergence of biological complexity during evolution. Examples of this include recontextualized historical embryological concepts, such as organizers and morphogenetic units, which were originally in vogue in the 1920s and 30s, to explain the origins of developmental complexity. This modern evo-devo framework places developmental processes in a broader hierarchy, describing the change at the level of the gene, the cell, and the tissue, and how changes at each level of this hierarchy can ultimately affect phenotype (Hall, 2003). Similar concepts, such as imaginal discs in drosophila further buoyed the idea that morphogenetic fields—groups of cells and their underlying cellular genetic interactions—can be seen as the evolutionary modules of an ontogeny (Gilbert et al., 1996).

GRNs, such as the early developmental GRN in the sea urchin (see Section 2), describe a mechanism for morphogenetic units (as described in Gilbert et al., 1996). In fact, the field of GRNs was largely born out of all of these historical frameworks. Therefore, much of the terminology used to describe modular evolutionary change in the past has been borrowed to describe modular changes to GRNs (Davidson & Erwin, 2009; Erwin & Davidson, 2009; Levine & Davidson, 2005). For instance, heterochronic shifts during development can now be contextualized as a change in GRN circuit deployment time (Israel et al., 2016). Using GRNs as a model for modularity, we can apply similar logic to the historical frameworks surrounding modular evolutionary change. Likewise, GRN circuits (as well as their underlying hierarchies) can be compared both *within* and *between* species (Fig. 4).

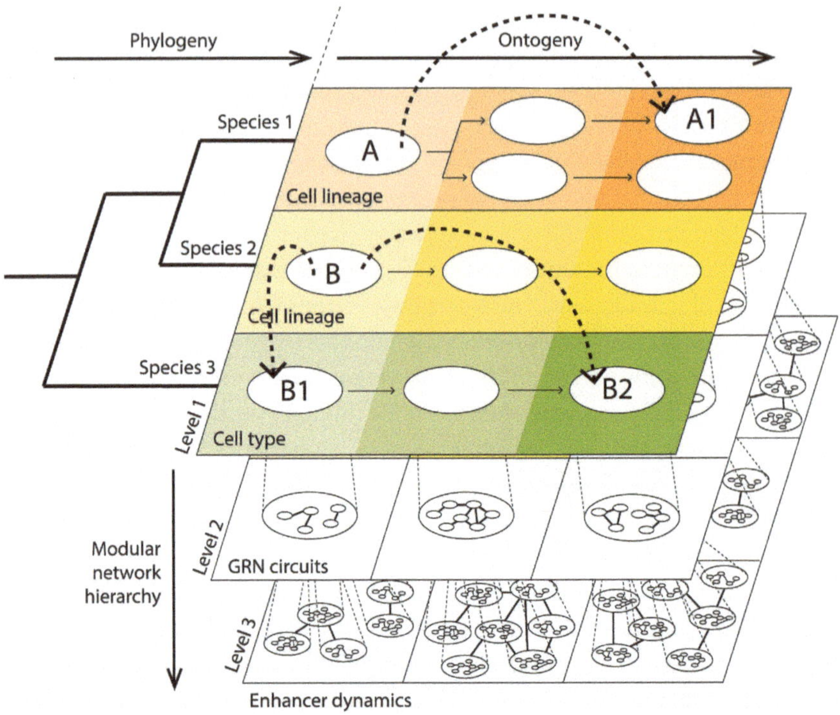

Fig. 4 Comparative framework to understand GRN changes in evolution. As with historical frameworks to discuss homology, the modularity of GRNs can be compared both within a species (A→A1), and between species (B→B1/B2). Together these comparisons encompass both phylogenetic as well as ontogenetic change. White ovals indicate a specific cell state, which are connected over space and time in an ontogeny (Level 1). Underlying these cell types are unique GRN circuits (Level 2), which can be further characterized by underlying enhancer dynamics (Level 3). These planes (Levels 1–3) make up nested layers of modular network hierarchy. The concept for this figure was adapted to describe GRN hierarchies from Klingenberg's (1996) discussion on multivariate allometry.

3.2 Co-option and the comparison of GRN homology *within* species

As noted in the watchmaker parable, it is theoretically quicker to duplicate and repurpose the same modules, than to rebuild from scratch each time. In the context of GRNs, this could involve re-deploying the same circuits in different developmental scenarios, and subsequently adapting those pre-existing modules to a new purpose. Conceptually, this process is known as co-option, and there are many examples of circuits being deployed in different spatio-temporal contexts to perform novel tasks (reviewed in

McQueen & Rebeiz, 2020). For instance, Hox genes are famous for their role in anterior-posterior axis formation, yet they also play a role in hand development, which similarly requires complex spatial patterning (Kmita, Fraudeau, Hérault, & Duboule, 2002). The effects of co-opting circuits can vary greatly depending where in the GRN hierarchy the co-option occurs.

3.2.1 Evolution at the level of the circuit

Due to the high level of modularity of GRNs, there is significant evidence that existing circuitry can be co-opted, or expressed within a different spatio-temporal context during development. In the literature, there are a number of examples of such "wholesale co-option events," where entire circuits are reused in different tissues or stages of development (McQueen & Rebeiz, 2020). For instance, in echinoderms, it is hypothesized the skeletogenic GRN subcircuit used in the larval stage represents a lineage-specific co-option event, which allowed the modular subcircuit to be deployed at a different time during the animal's ontogeny and the evolution of a novel phenotype (Gao & Davidson, 2008). Such co-option events can also explain duplicated structures in development, such as serial homologs (Akam, 1998; Carroll, Weatherbee, & Langeland, 1995; Monteiro, 2008), as well as allometric changes to morphological size (Amore & Casares, 2010).

In all developmental systems, there are examples of similar circuits being expressed in different tissues or at different times. While it is tempting to infer co-option from similar genes that are co-expressed in different tissues, it is necessary to functionally validate that those genes are in fact in the same network through perturbation experiments. From these results, if the exact same genes and regulatory connections are present in two different tissues, it is likely that the whole circuit has been co-opted (Fig. 5A.I), such as in the GRN subcircuit responsible for biomineralization in echinoderm larva and adults (Gao & Davidson, 2008). However, in many circumstances, the regulatory wiring may be different (Fig. 5A.II), which could either be the result of divergence over time, or the genes were independently recruited via convergence. Additionally, if only a single gene is pleiotropic, it is possible that only that gene was co-opted to a previously existing network (Fig. 5A.III). As a result, it can be very difficult to definitively distinguish between these evolutionary scenarios looking at gene circuits alone. Therefore, in order to tease apart these evolutionary hypotheses, it is necessary to delve deeper into the network hierarchy.

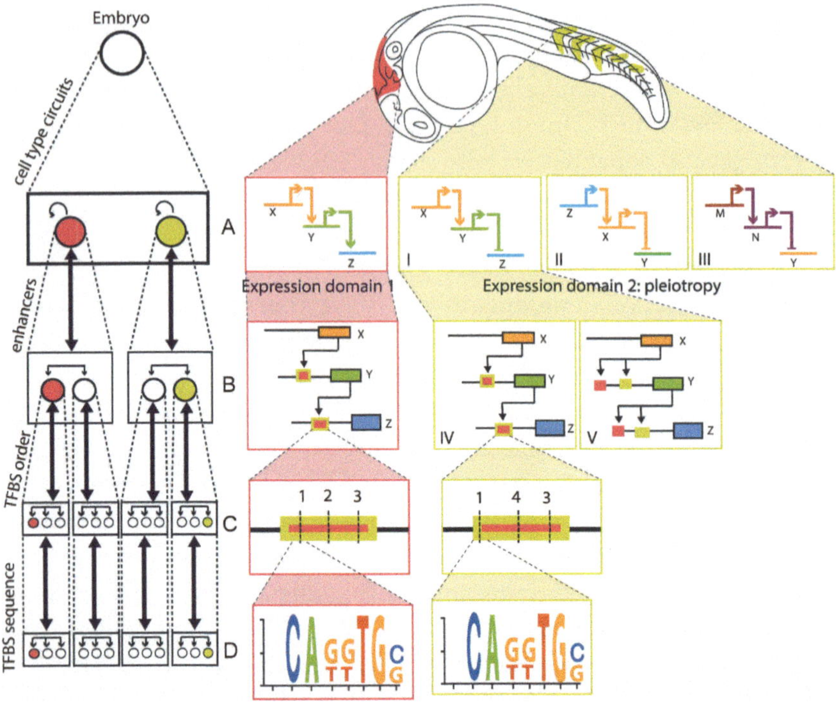

Fig. 5 Co-option events during development. There are many examples of pleiotropic genes that are expressed in two distinct spatial and/or temporal domains (example: domain #1 in magenta, and domain #2 in lime). Pleiotropy can be the result of multiple evolutionary scenarios, including the co-option of an entire GRN circuit, *de novo* GRN assembly, or the individual recruitment of the gene to an existing network (McQueen & Rebeiz, 2020; Monteiro & Gupta, 2016; Monteiro & Podlaha, 2009). (A) The first step to tease apart these evolutionary scenarios is to examine the GRN circuitry associated with each expression domain. (A.I) If both networks contain the same genes, and share conserved circuit wiring, it is likely that the pleiotropy is the result of a whole-circuit co-option event. (A.II) If both expression domains consist of the same genes, yet the specific regulatory information between them is different, it is either possible that the circuit was co-opted and then diverged, or the same genes were independently recruited to a similar network. (A.III) Lastly if the gene is the only pleiotropic gene in the two networks, it was likely to be independently recruited instead of the result of whole circuit co-option. (B) To conclusively distinguish between co-option and *de novo* assembly, one must examine the underlying enhancers in the network. (B.IV) If a network has been co-opted, one would expect to find that pleiotropic enhancers drive both spatio-temporal domains. (B.V) However, if similar circuits were the result of independently derived networks (from separate *de novo* events), one would expect to find modular enhancers for each domain (Monteiro & Podlaha, 2009). (C) As *cis*-regulatory regions are made up of modular transcription factor binding sites (TFBS), further investigation of potential co-option events can examine whether the enhancer contains the same TFBS, and whether or not they are in the same order and confirmation. (D) Lastly, transcription factors themselves often have different binding affinities, therefore it is possible to break down each TFBS into nucleotide sequence, as well as analyze the position weight matrix (PWM) of each binding motif (Nakagawa et al., 2013; Zhu et al., 2012).

3.2.2 Evolution at the level of the enhancer

The mechanism for such a wholesale co-option event would either be a mutation in the *cis*-regulatory or the coding sequence of a high-level regulator, which would enable it to be expressed in a novel tissue (McQueen & Rebeiz, 2020). However, due to the highly pleiotropic nature of *trans*-acting regulators, it is believed that these regulatory changes leading to co-option events primarily occur in CREs. This is because a transcription factor protein must be able to bind many *cis*-regulatory elements; therefore, mutations to these pleiotropic regulators would prevent all of its disparate functions, which could be highly deleterious for the organism. Thus, due to this evolutionary "burden," pleiotropic enhancers would be highly constrained, thereby increasing the selective pressure for the redundancy and/or modularity of enhancers (Stern, 2000).

If that single gene is responsible for the specification state of a cell type, then novel expression in a different tissue would be enough to deploy all downstream circuitry. Therefore, an entire circuit can be co-opted, and all downstream regulatory elements should be the same in both spatiotemporal contexts. It was previously assumed that enhancers were modular, where each enhancer was responsible for a single expression domain; however, there is mounting evidence that enhancers are often pleiotropic (Sabarís et al., 2019). Therefore, instead of modular enhancers, a circuit that resulted from a co-option event would likely rely on pleiotropic enhancers (Fig. 5B; Monteiro & Podlaha, 2009). In contrast, if those genes were independently recruited to be expressed in a new cell type, we would expect to find different enhancers for each expression domain. Thus, the presence of modular enhancers would provide evidence for *de novo* GRN assembly.

Real world scenarios appear to be more complex than this, as enhancer dynamics typically rely on the presence of many *cis*-regulatory elements for each gene. The concept of duplication/divergence is widespread when thinking of protein-coding gene evolution (Force et al., 1999), and can be as prominent as examples of subfunctionalization or neofunctionalization following whole genome duplication (Sandve, Rohlfs, & Hvidsten, 2018). However, as with protein-coding sequences, *cis*-regulatory elements can also be readily duplicated, creating multiple functional versions of the same enhancer. As a result, duplication of regulatory elements reduces evolutionary constraint, allowing sub- or neofunctionalization of enhancers. For instance, duplicate functional copies of *cis*-regulatory sequences, known as shadow enhancers, have been hypothesized to play a role in phenotypic innovation, particularly by promoting new regulatory connections between genes

(Hong, Hendrix, & Levine, 2008). Therefore, when circuit is co-opted, multiple functional copies of enhancer sequences are redeployed, providing additional sequence for evolution to act on. Resulting from this observation, Monteiro and Gupta (2016) proposed a new model for evolution following a co-option event, known as *Cis*-Regulatory Element-Duplication, Degeneration, and Complementation Model (CRE-DDC; Monteiro & Gupta, 2016). This model suggests that circuit co-option events can be the source of novel phenotypes, as re-deployment of duplicated CREs are free to diverge, forming new domain specific regulatory adaptations.

It is difficult to computationally infer enhancers; however, open chromatin in developing tissues suggests that as much as half of the enhancers involved in development are pleiotropic (Sabarís et al., 2019). Thus, these pleiotropic enhancers are critical to our understanding of co-option during evolution. While duplication/divergence appears to be a common way to form new regulatory connections, it remains a possibility that *cis*-regulatory elements evolve *de novo*. As enhancers themselves are made up of modular TFBSs, a deeper understanding of TFBS order and specific binding affinities could shed more light on the evolutionary mechanism (Fig. 5C and D).

3.2.3 Evolution at the level of the TFBS

As there have only been a few studies that closely examine the structure of pleiotropic enhancers, we primarily have theoretical models for how they could be structured (Sabarís et al., 2019). As enhancers can be broken down into discrete TFBS, it is possible that all TFBS can be used in both expression domains. Alternatively, some TFBS within a pleiotropic enhancer could be only used during one of the two spatio-temporal contexts, which is known as "site pleiotropy" (Sabarís et al., 2019). Additionally, there could be any combination in between, whereby some TFBS are pleiotropic while other are context specific, or modular (e.g., Crocker et al., 2016; Preger-Ben Noon et al., 2018; Sabarís et al., 2019).

The CRE-DDC model (Monteiro & Gupta, 2016) explains how co-option at the circuit level could provide new material for *cis*-regulatory evolution. At the level of the TFBS, there is a lot of evidence for how relatively few mutations can impair binding activity. Both insertion and deletions have been shown to modify regulatory interactions, by either loss or gain of repressor/enhancer function (Wittkopp & Kalay, 2011). Additionally, new TFBSs can evolve from non-regulatory sequences (Tuğrul, Paixão, Barton, & Tkačik, 2015), cell-specific enhancers have been shown to evolve independently via retrotransposon activity (Franchini et al., 2011; Lynch et al., 2015), and changes in spacing between TFBS can also affect regulatory function (Wittkopp & Kalay, 2011). Thus, duplicated CREs can readily diverge at the level of the TFBS, affecting both pleiotropy and modularity of enhancer activity. In fact, it has been shown that there are high rates of TFBS turnover, suggesting that the gain of novel TFBSs are as common as losses (Ludwig & Kreitman, 1995), and this turnover appears to be under positive selection (He, Holloway, Maerkl, & Kreitman, 2011).

It is becoming increasingly clear of the dizzying numbers of possibilities for ways CREs can evolve. Therefore over time, it can be difficult to tease apart the exact mechanism of evolution. However, as highlighted in previous studies (Crocker et al., 2016; Preger-Ben Noon et al., 2018; Sabaris et al., 2019), the complex regulatory implications of pleiotropic TFBSs calls our previously tidy framework for enhancer modularity into question.

3.2.4 A summary of GRN co-option

Comparing GRN deployment within a species over different spatial and temporal domains provides key insight into how modular changes occur in an ontogeny. Specifically, this provides a mechanistic understanding of how GRNs can be co-opted and redeployed during development. Looking at different levels of the GRN hierarchy provides different insights into how modular circuits evolved in a species. In theory, when a GRN has been co-opted, it inherits all downstream modules. Therefore, we would expect to find pleiotropic enhancers. On the other hand, *de novo* GRN evolution builds new networks from the bottom-up. However, the prevalence of duplication and divergence events in CREs makes it difficult to fully distinguish between co-option and *de novo* assembly. Thus, the presence of pleiotropic enhancers does not necessarily mean that it was part of a co-option event, as lots of dynamics occur at the level of the TFBS, including site pleiotropy and independent convergences of binding sites.

3.3 Implications of co-option: GRN comparisons *between* species

Similar problems are faced when comparing GRNs across large scales of evolutionary time. GRNs are often used in a comparative framework between species. A common motivating question is whether similar traits are the result of common ancestry (homology) or have independently evolved (convergence), and it has been widely accepted that similar networks between species are indicative of common origin. However, like examination of co-opted networks within species, it can be difficult to differentiate between common origin (homology), homology with divergence, and independent recruitment of similar gene players. For example, there are many examples of well-conserved GRNs between phyla including examples in eye development (retinal determination network), Hox genes, apical sensory organ in marine larva (Marlow et al., 2014), Heart evolution (Olson, 2006), Pharyngeal cluster/gill slits (Olson, 2006), and gut development (Arendt, Technau, & Wittbrodt, 2001).

In addition to the literature that hypothesizes that these similarities result from shared common ancestry, there are a number of papers that assert that similarities in GRNs are the result of convergence, suggesting that the same players have been recruited independently in each lineage. As with potentially co-opted circuits, comparisons of GRN hierarchies can help differentiate between different evolutionary scenarios (Fig. 6).

3.3.1 Evolution at the level of the circuit

In many instances, we find the same key players interacting with one another in different species, which could be the result of three primary scenarios: (1) it represents a homologous network, sharing the same common ancestor with sister taxa, (2) it represents a lineage-specific co-option event, or (3) the network evolved convergently in different lineages (Fig. 6).

If the same genes are expressed in homologous tissues in different species, it is likely that the similarities in GRN composition results from shared ancestry. To test this hypothesis, closer examination of the specific regulatory networks, as well as interactions with signaling pathways are required to conclude homology. For instance, the development of the pharyngeal pouches in vertebrates (Graham & Richardson, 2012) has many similarities to the expression of gill slits in basal chordates (Holland & Holland, 1996; Kozmik et al., 2007) and even basal deuterostomes (Gillis, Fritzenwanker, & Lowe, 2012; Ogasawara, Wada, Peters, & Satoh, 1999), suggesting common ancestry between gill slits and pharyngeal pouches, as well as the underlying GRNs.

However, in many instances, similar GRNs are present in two very different tissues in different species. For instance, the retinal determination circuit is expressed in bilaterian eye development as well as in vertebrate muscle formation (Relaix & Buckingham, 1999), or Hox gene expression in both vertebrate AP-axis formation and *Drosophila* leg bristles (Rozowski & Akam, 2002). In these examples, direct GRN homology cannot be inferred. Instead, it is likely that the network is the result of an ancient co-option event in a previous last common ancestor, or even a more recent lineage-specific adaptation. In classic examples of commonly co-opted circuits, such as the retinal determination network, there is considerable body

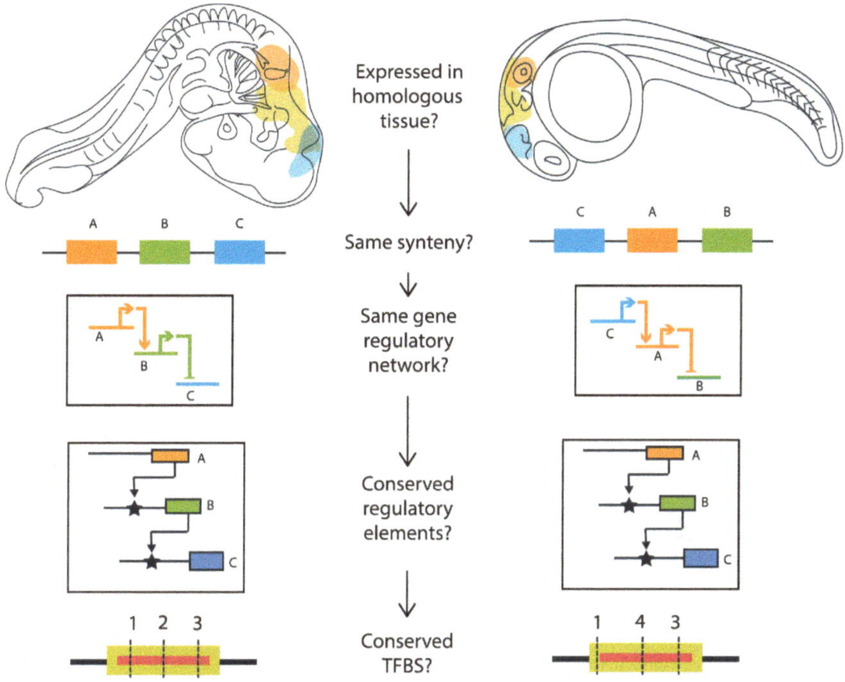

Fig. 6 GRN comparisons between species. As with the investigation of co-option events within species, it is possible to use the GRN hierarchy to infer homology versus convergence. There are many examples of the same genes being co-expressed in similar tissues in divergent species, yet it is difficult to assess whether those similarities are the result of shared ancestry. A first attempt to infer homology is to see if co-expressed genes are being deployed in homologous tissues in the two organisms. If so, it is possible that they represent a homologous GRN. If not, it could be possible that they are the result of either a lineage-specific co-option event, or the result of convergent GRN assembly. If the regulatory connections between genes are conserved between species, there is increasing evidence that they are homologous circuits. Similarly, if conserved non-coding elements (CNEs) can be identified in both species, it is likely that the genes shared a common GRN in their last common ancestor. Over large evolutionary distances, it can be difficult to find conserved transcription factor binding sites (TFBS) due to the high turnover of *cis*-regulatory elements over time; however, examination of TBFS, motifs, and binding affinities can shed light on the evolutionary mechanisms of GRN evolution.

of literature discussing the extent to which these networks can be considered homologous (Wagner, 2007). However, if the regulatory connections are dramatically different, and/or the circuits are expressed in highly distinct spatio-temporal contexts between species, it is likely that the genes were independently recruited to a similar network via convergence.

Based on the prevalence of co-option, duplication, and divergence events outlined in Section 3.2, it is likely that similar circuits between species merely appear to be homologous, while in fact originated independently (Conant & Wagner, 2003). As with our discussion of co-opted circuits, GRN hierarchy and modularity can provide further insight in determining which GRNs are truly homologous between species. In order to further understand the mechanisms by which these similarities arose, it is necessary to look at lower levels of the GRN.

3.3.2 Evolution in the levels below: Conserved non-coding elements (CNEs) and TFBS

If similar GRNs in different species share common ancestry, then they should have had the same *cis*-regulatory elements in their last common ancestor. As a result, extant species could possibly share the same *cis*-regulatory elements (CREs), which are known as conserved non-coding elements (CNEs; Nelson & Wardle, 2013). In closely related species, there are many examples of CNEs driving similar networks (McEwen et al., 2009; Navratilova et al., 2009; Ritter et al., 2010). However, there are also putative examples that extend across metazoans (Maeso, Irimia, Tena, Casares, & Gómez-Skarmeta, 2013).

As discussed in the previous section, CREs display rapid turnover (He et al., 2011); therefore, it is common for homologous circuits to not maintain CNEs to an identifiable level. Furthermore, due to divergence, some CREs will be unlikely to continue to operate in the same regulatory context in each species (Nelson & Wardle, 2013). To make things even harder, many CNEs do not show high degrees of broad sequence similarity (or any at all), making them exceptionally difficult to identify over deep evolutionary time (Nelson & Wardle, 2013). That said, confirmation that the same CNE drives the same gene expression pattern in two species is highly indicative of common ancestry. Therefore, one way to validate GRN homology is through the identification of a CNE between species. Interestingly, some studies have shown that similar phenotypes can independently arise from similar changes of *cis*-regulatory elements (Frankel, Wang, & Stern, 2012; Partha et al., 2017), in which case a close examination of TFBS order and sequence may be necessary to rule out convergence.

Additionally, evolution of regulatory networks can result from specific binding affinities to certain TFBSs (Cheatle Jarvela et al., 2014). For instance, in the previously mentioned study in echinoderms, the authors found evolutionary variability between primary and secondary binding sites for T-brain. While both the sea urchin and sea star had highly conserved primary binding motifs, only the sea star was able to respond to secondary binding sites under high T-brain concentrations. Thus, it is possible that novel regulatory interaction can also evolve via differences in transcription factor binding affinities.

3.4 Future horizons: Genome organization and 4d synteny

Some examples of regulatory networks conserved between phyla also share conserved genome organization (4d chromatin structure), such as well described syntenic regions. In these well-studied examples, such as Hox, Parahox, and pharyngeal clusters, the same suites of genes are found in a conserved order across phyla (Garcia-Fernàndez, 2005; Simakov et al., 2015). In some cases, this order is completely intact, whereas in other examples, certain genes have been lost, or their direction has been reversed (Cameron et al., 2006). A common hypothesis for the maintenance of synteny is to preserve the function of tightly intertwined regulatory information between genes. By maintaining co-expressed genes in physical proximity, it may be easier to ensure that necessary *cis*-regulatory elements in between the protein-coding regions are maintained (Darbellay et al., 2019). In many cases, syntenic blocks are co-expressed together as a module. However, the same syntenic blocks do not always perform the same function in different species (Deschamps & van Nes, 2005; Rozowski & Akam, 2002). This indicates that syntenic blocks can be readily co-opted to perform different tasks, yet remain together as a regulatory unit (Lonfat, Montavon, Darbellay, Gitto, & Duboule, 2014). This provides a perfect example of genetic modularity, where syntenic regions are building blocks within a genetic toolkit. However, the complexity of chromatin markers suggests that there is a lot more than we currently understand when it comes to functional genomics. Therefore, it is likely that 4d chromatin architecture is an example of another level in the hierarchy of biological complexity that we still do not know enough about.

4. Conclusions

Complex biological systems are both modular and hierarchical, and GRNs provide a mechanistic framework to understand evolutionary change. Research often examines concepts such as pleiotropy, co-option, modularity, and homology; however, it is impossible to fully understand these ideas without examining the entire hierarchical system. Through our awareness of this hierarchical system, we are better able to understand how it evolved.

Investigation of circuit co-option/*de novo* assembly looks at modular evolutionary changes at the level of the individual (ontogeny). Whereas homology/convergence examines similarly modular changes on the level of the species. Due to the hierarchy of ontogeny and phylogeny, these are parallel evolutionary processes. When a circuit is duplicated or co-opted during development, the GRNs share common ancestry. However, when similar circuits are independently formed through *de novo* GRN assembly, the two circuits are convergent. The former are top-down, while the latter are bottom-up.

In order to understand how GRNs evolved, we must look at network changes both within and between species. Recent single-cell datasets allow us to recreate lineage trajectories during development. By doing this in closely related species, it will be possible to link evolutionary studies of both ontogeny and phylogeny.

Acknowledgments

We would like to acknowledge Cheryl Telmer, Anne Meyer, and Olga Zueva for their feedback on the manuscript. V.F.H. acknowledges funding from NSF. IOS 1557431, BSF (Binational Science Foundation) 2015031 and NSF MCB 1715721.

References

Abzhanov, A. (2013). von Baer's law for the ages: Lost and found principles of developmental evolution. *Trends in Genetics*, *29*(12), 712–722.

Akam, M. (1998). Hox genes, homeosis and the evolution of segment identity: No need for hopeless monsters. *The International Journal of Developmental Biology*, *42*(3), 445–451.

Amore, G., & Casares, F. (2010). Size matters: The contribution of cell proliferation to the progression of the specification Drosophila eye gene regulatory network. *Developmental Biology*, *344*(2), 569–577.

Aragona, M., Sifrim, A., Malfait, M., Song, Y., Van Herck, J., Dekoninck, S., et al. (2020). Mechanisms of stretch-mediated skin expansion at single-cell resolution. *Nature*, *584*(7820), 268–273.

Arendt, D., Technau, U., & Wittbrodt, J. (2001). Evolution of the bilaterian larval foregut. *Nature*, *409*(6816), 81–85.

Arnone, M. I., Byrne, M., & Martinez, P. (2015). Echinodermata. In A. Wanninger (Ed.), *Evolutionary developmental biology of invertebrates* vol. 6. (pp. 1–58). Vienna: Springer.

Arnosti, D. N., & Kulkarni, M. M. (2005). Transcriptional enhancers: Intelligent enhanceosomes or flexible billboards? *Journal of Cellular Biochemistry*, *94*(5), 890–898.

Badis, G., Berger, M. F., Philippakis, A. A., Talukder, S., Gehrke, A. R., Jaeger, S. A., et al. (2009). Diversity and complexity in DNA recognition by transcription factors. *Science*, *324*(5935), 1720–1723.

Ben-Tabou De-Leon, S., & Davidson, E. H. (2007). Gene regulation: Gene control network in development. *Annual Review of Biophysics and Biomolecular Structure*, *36*, 191.

Bhattacharya, S., Zhang, Q., & Andersen, M. E. (2011). A deterministic map of Waddington's epigenetic landscape for cell fate specification. *BMC Systems Biology*, *5*, 85.
Bolouri, H. (2008). *Computational modeling of gene regulatory networks—A primer*. World Scientific Publishing Company.
Cai, W., Huang, J., Zhu, Q., Li, B. E., Seruggia, D., Zhou, P., et al. (2020). Enhancer dependence of cell-type-specific gene expression increases with developmental age. *Proceedings of the National Academy of Sciences of the United States of America*, *117*(35), 21450–21458.
Cameron, R. A., Rowen, L., Nesbitt, R., Bloom, S., Rast, J. P., Berney, K., et al. (2006). Unusual gene order and organization of the sea urchin hox cluster. *Journal of Experimental Zoology. Part B, Molecular and Developmental Evolution*, *306*(1), 45–58.
Carroll, S. B., Grenier, J. K., & Weatherbee, S. D. (2013). *From DNA to diversity: Molecular genetics and the evolution of animal design*. John Wiley & Sons.
Carroll, S. B., Weatherbee, S. D., & Langeland, J. A. (1995). Homeotic genes and the regulation and evolution of insect wing number. *Nature*, *375*(6526), 58–61.
Cary, G. A., Cheatle Jarvela, A. M., Francolini, R. D., & Hinman, V. F. (2017). Genome-wide use of high- and low-affinity Tbrain transcription factor binding sites during echinoderm development. *Proceedings of the National Academy of Sciences of the United States of America*, *114*(23), 5854–5861.
Cary, G. A., McCauley, B. S., Zueva, O., Pattinato, J., Longabaugh, W., & Hinman, V. (2020). *Systematic comparison of developmental GRNs explains how novelty is incorporated in early development*. BioRxiv.
Cheatle Jarvela, A. M., Brubaker, L., Vedenko, A., Gupta, A., Armitage, B. A., Bulyk, M. L., et al. (2014). Modular evolution of DNA-binding preference of a Tbrain transcription factor provides a mechanism for modifying gene regulatory networks. *Molecular Biology and Evolution*, *31*(10), 2672–2688.
Christiaen, L., Davidson, B., Kawashima, T., Powell, W., Nolla, H., Vranizan, K., et al. (2008). The transcription/migration interface in heart precursors of Ciona intestinalis. *Science*, *320*(5881), 1349–1352.
Conant, G. C., & Wagner, A. (2003). Convergent evolution of gene circuits. *Nature Genetics*, *34*(3), 264–266.
Crocker, J., Noon, E. P.-B., & Stern, D. L. (2016). The soft touch: Low-affinity transcription factor binding sites in development and evolution. *Current Topics in Developmental Biology*, *117*, 455–469.
Darbellay, F., Bochaton, C., Lopez-Delisle, L., Mascrez, B., Tschopp, P., Delpretti, S., et al. (2019). The constrained architecture of mammalian Hox gene clusters. *Proceedings of the National Academy of Sciences of the United States of America*, *116*(27), 13424–13433.
Davidson, E. H. (2010). Emerging properties of animal gene regulatory networks. *Nature*, *468*(7326), 911–920.
Davidson, E. H., & Erwin, D. H. (2006). Gene regulatory networks and the evolution of animal body plans. *Science*, *311*(5762), 796–800.
Davidson, E. H., & Erwin, D. H. (2009). An integrated view of precambrian eumetazoan evolution. *Cold Spring Harbor Symposia on Quantitative Biology*, *74*, 65–80.
Deschamps, J., & van Nes, J. (2005). Developmental regulation of the Hox genes during axial morphogenesis in the mouse. *Development*, *132*(13), 2931–2942.
Ding, J., & Bar-Joseph, Z. (2020). Analysis of time series regulatory networks. *Current Opinion in Systems Biology*, *21*, 16–24.
Domazet-Lošo, T., & Tautz, D. (2010). A phylogenetically based transcriptome age index mirrors ontogenetic divergence patterns. *Nature*, *468*(7325), 815–818.
Drapek, C., Sparks, E. E., & Benfey, P. N. (2017). Uncovering gene regulatory networks controlling plant cell differentiation. *Trends in Genetics*, *33*(8), 529–539.
Erwin, D. H., & Davidson, E. H. (2009). The evolution of hierarchical gene regulatory networks. *Nature Reviews. Genetics*, *10*(2), 141–148.

Ettensohn, C. A., & Ruffins, S. W. (1993). Mesodermal cell interactions in the sea urchin embryo: properties of skeletogenic secondary mesenchyme cells. *Development, 117*(4), 1275–1285.

Flower, M. (2020). How to visualize cells as overlapping trajectories of profiles. In B. Latour, & P. Weibel (Eds.), *Critical zones: The science and politics of landing on earth* (pp. 356–363). Cambridge: MIT Press.

Force, A., Lynch, M., Pickett, F. B., Amores, A., Yan, Y. L., & Postlethwait, J. (1999). Preservation of duplicate genes by complementary, degenerative mutations. *Genetics, 151*(4), 1531–1545.

Franchini, L. F., López-Leal, R., Nasif, S., Beati, P., Gelman, D. M., Low, M. J., et al. (2011). Convergent evolution of two mammalian neuronal enhancers by sequential exaptation of unrelated retroposons. *Proceedings of the National Academy of Sciences of the United States of America, 108*(37), 15270–15275.

Frankel, N., Wang, S., & Stern, D. L. (2012). Conserved regulatory architecture underlies parallel genetic changes and convergent phenotypic evolution. *Proceedings of the National Academy of Sciences of the United States of America, 109*(51), 20975–20979.

Freckelton, M. L., Nedved, B. T., & Hadfield, M. G. (2017). Induction of invertebrate larval settlement; different bacteria, different mechanisms? *Scientific Reports*, 7, 42557.

Friedman, C. E., Nguyen, Q., Lukowski, S. W., Helfer, A., Chiu, H. S., Miklas, J., et al. (2018). Single-cell transcriptomic analysis of cardiac differentiation from human PSCs reveals HOPX-dependent cardiomyocyte maturation. *Cell Stem Cell, 23*(4), 586–598.e8.

Furuta, S., Ren, G., Mao, J. H., & Bissell, M. J. (2018). Laminin signals initiate the reciprocal loop that informs breast-specific gene expression and homeostasis by activating NO, p53 and microRNAs. *elife*, 7. https://doi.org/10.7554/eLife.26148.

Gao, F., & Davidson, E. H. (2008). Transfer of a large gene regulatory apparatus to a new developmental address in echinoid evolution. *Proceedings of the National Academy of Sciences of the United States of America, 105*(16), 6091–6096.

Garcia-Fernàndez, J. (2005). Hox, ParaHox, ProtoHox: Facts and guesses. *Heredity, 94*(2), 145–152.

Gehrke, A. R., Neverett, E., Luo, Y.-J., Brandt, A., Ricci, L., Hulett, R. E., et al. (2019). Acoel genome reveals the regulatory landscape of whole-body regeneration. *Science, 363*(6432), eaau6173. https://doi.org/10.1126/science.aau6173.

Gilbert, S. F. (2000). *The early development of sea urchins—Developmental biology* (9th ed.). Sinuauer Associates, Inc.

Gilbert, S. F. (2003). Evo-devo, devo-evo, and devgen-popgen. *Biology and Philosophy, 18*(2), 347–352.

Gilbert, S. F., Opitz, J. M., & Raff, R. A. (1996). Resynthesizing evolutionary and developmental biology. *Developmental Biology, 173*(2), 357–372.

Gillis, J. A., Fritzenwanker, J. H., & Lowe, C. J. (2012). A stem-deuterostome origin of the vertebrate pharyngeal transcriptional network. *Proceedings. Biological Sciences/the Royal Society, 279*(1727), 237–246.

Graf, T., & Enver, T. (2009). Forcing cells to change lineages. *Nature, 462*(7273), 587–594.

Graham, A., & Richardson, J. (2012). Developmental and evolutionary origins of the pharyngeal apparatus. *EvoDevo, 3*(1), 24.

Hall, B. K. (2003). Unlocking the black box between genotype and phenotype: Cell condensations as morphogenetic (modular) units. *Biology and Philosophy, 18*(2), 219–247.

Hamouda, M. S., Labouesse, C., & Chalut, K. J. (2020). Nuclear mechanotransduction in stem cells. *Current Opinion in Cell Biology, 64*, 97–104. https://doi.org/10.1016/j.ceb.2020.05.005.

Han, J.-D. J., Bertin, N., Hao, T., Goldberg, D. S., Berriz, G. F., Zhang, L. V., et al. (2004). Evidence for dynamically organized modularity in the yeast protein-protein interaction network. *Nature, 430*(6995), 88–93.

He, J., & Deem, M. W. (2010). Hierarchical evolution of animal body plans. *Developmental Biology*, *337*(1), 157–161.

He, B. Z., Holloway, A. K., Maerkl, S. J., & Kreitman, M. (2011). Does positive selection drive transcription factor binding site turnover? A test with Drosophila cis-regulatory modules. *PLoS Genetics*, 7(4), e1002053.

Hill, J. T., Demarest, B., Gorsi, B., Smith, M., & Yost, H. J. (2017). Heart morphogenesis gene regulatory networks revealed by temporal expression analysis. *Development*, *144*(19), 3487–3498.

Hinman, V. F., Nguyen, A. T., Cameron, R. A., & Davidson, E. H. (2003). Developmental gene regulatory network architecture across 500 million years of echinoderm evolution. *Proceedings of the National Academy of Sciences of the United States of America*, *100*(23), 13356–13361.

Holland, L. Z., & Holland, N. D. (1996). Expression of AmphiHox-1 and AmphiPax-1 in amphioxus embryos treated with retinoic acid: Insights into evolution and patterning of the chordate nerve cord and pharynx. *Development*, *122*(6), 1829–1838.

Hong, J.-W., Hendrix, D. A., & Levine, M. S. (2008). Shadow enhancers as a source of evolutionary novelty. *Science*, *321*(5894), 1314.

Illmensee, K., & Mintz, B. (1976). Totipotency and normal differentiation of single teratocarcinoma cells cloned by injection into blastocysts. *Proceedings of the National Academy of Sciences of the United States of America*, *73*(2), 549–553.

Israel, J. W., Martik, M. L., Byrne, M., Raff, E. C., Raff, R. A., McClay, D. R., et al. (2016). Comparative developmental transcriptomics reveals rewiring of a highly conserved gene regulatory network during a major life history switch in the sea urchin genus heliocidaris. *PLoS Biology*, *14*(3), e1002391.

Kasemeier-Kulesa, J. C., Teddy, J. M., Postovit, L.-M., Seftor, E. A., Seftor, R. E. B., Hendrix, M. J. C., et al. (2008). Reprogramming multipotent tumor cells with the embryonic neural crest microenvironment. *Developmental Dynamics*, *237*(10), 2657–2666.

Keller, E. F. (2002). *The century of the gene*. Cambridge: Harvard University Press.

Kenny, P. A., & Bissell, M. J. (2003). Tumor reversion: Correction of malignant behavior by microenvironmental cues. *International Journal of Cancer*, *107*(5), 688–695.

Khalturin, K. (2020). The origin of metazoan larvae. *Nature Ecology & Evolution*, *4*(5), 674–675.

Kirchhamer, C. V., Yuh, C. H., & Davidson, E. H. (1996). Modular cis-regulatory organization of developmentally expressed genes: Two genes transcribed territorially in the sea urchin embryo, and additional examples. *Proceedings of the National Academy of Sciences of the United States of America*, *93*(18), 9322–9328.

Klingenberg, C. P. (1996). Multivariate allometry. In L. F. Marcus, M. Corti, A. Loy, G. J. P. Naylor, & D. E. Slice (Eds.), *Advances in morphometrics* (pp. 23–49). Boston, MA: Springer US.

Kmita, M., Fraudeau, N., Hérault, Y., & Duboule, D. (2002). Serial deletions and duplications suggest a mechanism for the collinearity of Hoxd genes in limbs. *Nature*, *420*(6912), 145–150.

Kozmik, Z., Holland, N. D., Kreslova, J., Oliveri, D., Schubert, M., Jonasova, K., et al. (2007). Pax-Six-Eya-Dach network during amphioxus development: Conservation in vitro but context specificity in vivo. *Developmental Biology*, *306*(1), 143–159.

Krause, A. E., Frank, K. A., Mason, D. M., Ulanowicz, R. E., & Taylor, W. W. (2003). Compartments revealed in food-web structure. *Nature*, *426*(6964), 282–285.

Ladewig, J., Koch, P., & Brüstle, O. (2013). Leveling Waddington: The emergence of direct programming and the loss of cell fate hierarchies. *Nature Reviews. Molecular Cell Biology*, *14*(4), 225–236.

Levin, M., Anavy, L., Cole, A. G., Winter, E., Mostov, N., Khair, S., et al. (2016). The mid-developmental transition and the evolution of animal body plans. *Nature, 531*(7596), 637–641.

Levine, M., & Davidson, E. H. (2005). Gene regulatory networks for development. *Proceedings of the National Academy of Sciences of the United States of America, 102*(14), 4936–4942.

Lonfat, N., Montavon, T., Darbellay, F., Gitto, S., & Duboule, D. (2014). Convergent evolution of complex regulatory landscapes and pleiotropy at Hox loci. *Science, 346*(6212), 1004–1006.

Longabaugh, W. J. R., Davidson, E. H., & Bolouri, H. (2005). Computational representation of developmental genetic regulatory networks. *Developmental Biology, 283*(1), 1–16.

Ludwig, M. Z., & Kreitman, M. (1995). Evolutionary dynamics of the enhancer region of even-skipped in Drosophila. *Molecular Biology and Evolution, 12*(6), 1002–1011.

Lynch, V. J., Nnamani, M. C., Kapusta, A., Brayer, K., Plaza, S. L., Mazur, E. C., et al. (2015). Ancient transposable elements transformed the uterine regulatory landscape and transcriptome during the evolution of mammalian pregnancy. *Cell Reports, 10*(4), 551–561.

Maeso, I., Irimia, M., Tena, J. J., Casares, F., & Gómez-Skarmeta, J. L. (2013). Deep conservation of cis-regulatory elements in metazoans. *Philosophical Transactions of the Royal Society of London. Series B, Biological Sciences, 368*(1632), 20130020.

Marlow, H., Tosches, M. A., Tomer, R., Steinmetz, P. R., Lauri, A., Larsson, T., et al. (2014). Larval body patterning and apical organs are conserved in animal evolution. *BMC Biology, 12*, 7.

Martik, M. L., & McClay, D. R. (2015). Deployment of a retinal determination gene network drives directed cell migration in the sea urchin embryo. *eLife, 4*, e08827.

McCauley, B. S., Weideman, E. P., & Hinman, V. F. (2010). A conserved gene regulatory network subcircuit drives different developmental fates in the vegetal pole of highly divergent echinoderm embryos. *Developmental Biology, 340*(2), 200–208.

McEwen, G. K., Goode, D. K., Parker, H. J., Woolfe, A., Callaway, H., & Elgar, G. (2009). Early evolution of conserved regulatory sequences associated with development in vertebrates. *PLoS Genetics, 5*(12), e1000762.

McQueen, E., & Rebeiz, M. (2020). On the specificity of gene regulatory networks: How does network co-option affect subsequent evolution? *Current Topics in Developmental Biology, 139*, 375–405.

Melo, D., Porto, A., Cheverud, J. M., & Marroig, G. (2016). Modularity: Genes, development and evolution. *Annual Review of Ecology, Evolution, and Systematics, 47*, 463–486.

Mitteroecker, P. (2009). The developmental basis of variational modularity: Insights from quantitative genetics, morphometrics, and developmental biology. *Evolutionary Biology, 36*(4), 377–385.

Moignard, V., Woodhouse, S., Haghverdi, L., Lilly, A. J., Tanaka, Y., Wilkinson, A. C., et al. (2015). Decoding the regulatory network of early blood development from single-cell gene expression measurements. *Nature Biotechnology, 33*(3), 269–276.

Monteiro, A. (2008). Alternative models for the evolution of eyespots and of serial homology on lepidopteran wings. *BioEssays: News and Reviews in Molecular, Cellular and Developmental Biology, 30*(4), 358–366.

Monteiro, A., & Gupta, M. D. (2016). Identifying coopted networks and causative mutations in the origin of novel complex traits. *Current Topics in Developmental Biology, 119*, 205–226.

Monteiro, A., & Podlaha, O. (2009). Wings, horns, and butterfly eyespots: How do complex traits evolve? *PLoS Biology*, 7(2), e37.

Nakagawa, S., Gisselbrecht, S. S., Rogers, J. M., Hartl, D. L., & Bulyk, M. L. (2013). DNA-binding specificity changes in the evolution of forkhead transcription factors. *Proceedings of the National Academy of Sciences of the United States of America, 110*(30), 12349–12354.

Navratilova, P., Fredman, D., Hawkins, T. A., Turner, K., Lenhard, B., & Becker, T. S. (2009). Systematic human/zebrafish comparative identification of cis-regulatory activity around vertebrate developmental transcription factor genes. *Developmental Biology, 327*(2), 526–540.

Nelson, A. C., & Wardle, F. C. (2013). Conserved non-coding elements and cis regulation: Actions speak louder than words. *Development, 140*(7), 1385–1395.

Novershtern, N., Subramanian, A., Lawton, L. N., Mak, R. H., Haining, W. N., McConkey, M. E., et al. (2011). Densely interconnected transcriptional circuits control cell states in human hematopoiesis. *Cell, 144*(2), 296–309.

Ochoterena, H., Vrijdaghs, A., Smets, E., & Claßen-Bockhoff, R. (2019). The search for common origin: Homology revisited. *Systematic Biology, 68*(5), 767–780.

Ogasawara, M., Wada, H., Peters, H., & Satoh, N. (1999). Developmental expression of Pax1/9 genes in urochordate and hemichordate gills: Insight into function and evolution of the pharyngeal epithelium. *Development, 126*(11), 2539–2550.

Oliveri, P., Carrick, D. M., & Davidson, E. H. (2002). A regulatory gene network that directs micromere specification in the sea urchin embryo. *Developmental Biology, 246*(1), 209–228.

Oliveri, P., & Davidson, E. H. (2004). Gene regulatory network controlling embryonic specification in the sea urchin. *Current Opinion in Genetics & Development, 14*(4), 351–360.

Oliveri, P., Tu, Q., & Davidson, E. H. (2008). Global regulatory logic for specification of an embryonic cell lineage. *Proceedings of the National Academy of Sciences of the United States of America, 105*(16), 5955–5962.

Olson, E. N. (2006). Gene regulatory networks in the evolution and development of the heart. *Science, 313*(5795), 1922–1927.

Partha, R., Chauhan, B. K., Ferreira, Z., Robinson, J. D., Lathrop, K., Nischal, K. K., et al. (2017). Subterranean mammals show convergent regression in ocular genes and enhancers, along with adaptation to tunneling. *eLife, 6*, e25884.

Peter, I. S., & Davidson, E. H. (2009). Modularity and design principles in the sea urchin embryo gene regulatory network. *FEBS Letters, 583*(24), 3948–3958.

Piccus, R., & Brayson, D. (2020). The nuclear envelope: LINCing tissue mechanics to genome regulation in cardiac and skeletal muscle. *Biology Letters, 16*. https://doi.org/10.1098/rsbl.2020.0302.

Preger-Ben Noon, E., Sabarís, G., Ortiz, D. M., Sager, J., Liebowitz, A., Stern, D. L., et al. (2018). Comprehensive analysis of a cis-regulatory region reveals pleiotropy in enhancer function. *Cell Reports, 22*(11), 3021–3031.

Qin, L., Liu, W., Cao, H., & Xiao, G. (2020). Molecular mechanosensors in osteocytes. *Bone Research, 8*(23). https://doi.org/10.1038/s41413-020-0099-y.

Raff, R. A. (1996). *The shape of life: Genes, development, and the evolution of animal form*. University of Chicago Press.

Raff, R. A., & Kaufman, T. C. (1991). *Embryos, genes, and evolution: The developmental-genetic basis of evolutionary change*. Indiana University Press.

Rafiq, K., Cheers, M. S., & Ettensohn, C. A. (2012). The genomic regulatory control of skeletal morphogenesis in the sea urchin. *Development, 139*(3), 579–590.

Ramos, A. I., & Barolo, S. (2013). Low-affinity transcription factor binding sites shape morphogen responses and enhancer evolution. *Philosophical Transactions of the Royal Society of London. Series B, Biological Sciences, 368*(1632), 20130018.

Ravasz, E., Somera, A. L., Mongru, D. A., Oltvai, Z. N., & Barabási, A. L. (2002). Hierarchical organization of modularity in metabolic networks. *Science, 297*(5586), 1551–1555.

Reiter, F., Wienerroither, S., & Stark, A. (2017). Combinatorial function of transcription factors and cofactors. *Current Opinion in Genetics & Development, 43*, 73–81.

Relaix, F., & Buckingham, M. (1999). From insect eye to vertebrate muscle: Redeployment of a regulatory network. *Genes & Development, 13*(24), 3171–3178.

Ritschar, S., Rabus, M., & Laforsch, C. (2020). Predator-specific inducible morphological defenses of a water flea against two freshwater predators. *Journal of Morphology*, *281*(6), 653–661.
Ritter, D. I., Li, Q., Kostka, D., Pollard, K. S., Guo, S., & Chuang, J. H. (2010). The importance of being cis: Evolution of orthologous fish and mammalian enhancer activity. *Molecular Biology and Evolution*, *27*(10), 2322–2332.
Rozowski, M., & Akam, M. (2002). Hox gene control of segment-specific bristle patterns in Drosophila. *Genes & Development*, *16*(9), 1150–1162.
Sabarís, G., Laiker, I., Preger-Ben Noon, E., & Frankel, N. (2019). Actors with multiple roles: Pleiotropic enhancers and the paradigm of enhancer modularity. *Trends in Genetics*, *35*(6), 423–433.
Sandve, S. R., Rohlfs, R. V., & Hvidsten, T. R. (2018). Subfunctionalization versus neofunctionalization after whole-genome duplication. *Nature Genetics*, *50*(7), 908–909.
Sareen, D., & Svendsen, C. N. (2010). Stem cell biologists sure play a mean pinball. *Nature Biotechnology*, *28*(4), 333–335.
Sayed, K., Bocan, K. N., & Miskov-Zivanov, N. (2018). Automated extension of cell signaling models with genetic algorithm. *Annual International Conference of the IEEE Engineering in Medicine and Biology Society*, *2018*, 5030–5033.
Segal, E., Raveh-Sadka, T., Schroeder, M., Unnerstall, U., & Gaul, U. (2008). Predicting expression patterns from regulatory sequence in Drosophila segmentation. *Nature*, *451*(7178), 535–540.
Shubin, N., Tabin, C., & Carroll, S. (2009). Deep homology and the origins of evolutionary novelty. *Nature*, *457*(7231), 818–823.
Siegal, M. L., & Bergman, A. (2002). Waddington's canalization revisited: Developmental stability and evolution. *Proceedings of the National Academy of Sciences of the United States of America*, *99*(16), 10528–10532.
Simakov, O., Kawashima, T., Marlétaz, F., Jenkins, J., Koyanagi, R., Mitros, T., et al. (2015). Hemichordate genomes and deuterostome origins. *Nature*, *527*(7579), 459–465.
Simon, H. A. (1962). The architecture of complexity. *Proceedings of the American Philosophical Society*, *106*(6), 467–482. Retrieved August 18, 2020, from http://www.jstor.org/stable/985254.
Smith, J., Morgan, J. R., Zottoli, S. J., Smith, P. J., Buxbaum, J. D., & Bloom, O. E. (2011). Regeneration in the era of functional genomics and gene network analysis. *The Biological Bulletin*, *221*(1), 18–34.
Smith, J., Theodoris, C., & Davidson, E. H. (2007). A gene regulatory network subcircuit drives a dynamic pattern of gene expression. *Science*, *318*(5851), 794–797.
Stern, D. L. (2000). Evolutionary developmental biology and the problem of variation. *Evolution*, *54*(4), 1079–1091.
Takahashi, K. (2012). Cellular reprogramming—lowering gravity on Waddington's epigenetic landscape. *Journal of Cell Science*, *125*(Pt. 11), 2553–2560.
Tran, C., & Hadfield, M. G. (2011). Larvae of Pocillopora damicornis (Anthozoa) settle and metamorphose in response to surface-biofilm bacteria. *Marine Ecology Progress Series*, *433*, 85–96.
Tuğrul, M., Paixão, T., Barton, N. H., & Tkačik, G. (2015). Dynamics of transcription factor binding site evolution. *PLoS Genetics*, *11*(11), e1005639.
von Baer, K. E. (1828). *Über Entwickelungsgeschichte der Thiere: Beobachtung und Reflexion*. Königsberg: Bornträger.
Waddington, C. H. (1957). *The strategy of the genes*. London: Allen and Unwin.
Wagner, G. P. (2007). The developmental genetics of homology. *Nature Reviews. Genetics*, *8*(6), 473–479.
Wagner, G. P., Pavlicev, M., & Cheverud, J. M. (2007). The road to modularity. *Nature Reviews. Genetics*, *8*(12), 921–931.

Wagner, D. E., Weinreb, C., Collins, Z. M., Briggs, J. A., Megason, S. G., & Klein, A. M. (2018). Single-cell mapping of gene expression landscapes and lineage in the zebrafish embryo. *Science*, *360*(6392), 981–987.

Wang, J., Zhang, L., Lian, S., Qin, Z., Zhu, X., Dai, X., et al. (2020). Evolutionary transcriptomics of metazoan biphasic life cycle supports a single intercalation origin of metazoan larvae. *Nature Ecology & Evolution*, *4*(5), 725–736.

West-Eberhard, M. J. (2019). Modularity as a universal emergent property of biological traits. *Journal of Experimental Zoology. Part B, Molecular and Developmental Evolution*, *332*(8), 356–364.

Wittkopp, P. J., & Kalay, G. (2011). Cis-regulatory elements: Molecular mechanisms and evolutionary processes underlying divergence. *Nature Reviews. Genetics*, *13*(1), 59–69.

Wu, J. (2013). Hierarchy theory: An overview. In R. Rozzi, S. T. A. Pickett, C. Palmer, J. J. Armesto, & J. B. Callicott (Eds.), *Linking ecology and ethics for a changing world: Values, philosophy, and action* (pp. 281–301). Dordrecht: Springer Netherlands.

Xue, Z., Huang, K., Cai, C., Cai, L., Jiang, C., Feng, Y., et al. (2013). Genetic programs in human and mouse early embryos revealed by single-cell RNA sequencing. *Nature*, *500*(7464), 593–597.

Yáñez-Cuna, J. O., Dinh, H. Q., Kvon, E. Z., Shlyueva, D., & Stark, A. (2012). Uncovering cis-regulatory sequence requirements for context-specific transcription factor binding. *Genome Research*, *22*(10), 2018–2030.

Yin, X., Jin, W., Zhou, Y., Wang, P., & Zhao, W. (2017). Hidden defensive morphology in rotifers: Benefits, costs, and fitness consequences. *Scientific Reports*, 7(1), 4488.

Zhu, X., Ahmad, S. M., Aboukhalil, A., Busser, B. W., Kim, Y., Tansey, T. R., et al. (2012). Differential regulation of mesodermal gene expression by Drosophila cell type-specific Forkhead transcription factors. *Development*, *139*(8), 1457–1466.

CHAPTER THREE

Molecular insights into deuterostome evolution from hemichordate developmental biology

Christopher J. Lowe*
Hopkins Marine Station, Department of Biology, Stanford University, Pacific Grove, CA, United States
*Corresponding author: e-mail address: clowe@stanford.edu

Contents

Abstract

Hemichordates, along with echinoderms and chordates, belong to the lineage of bilaterians called the deuterostomes. Their phylogenetic position as an outgroup to chordates provides an opportunity to investigate the evolutionary origins of the chordate body plan and reconstruct ancestral deuterostome characters. The body plans of the hemichordates and chordates are organizationally divergent making anatomical comparisons very challenging. The developmental underpinnings of animal body plans

Current Topics in Developmental Biology, Volume 141
ISSN 0070-2153
https://doi.org/10.1016/bs.ctdb.2020.12.002

are often more conservative than the body plans they regulate, and offer a novel data set for making comparisons between morphologically divergent body architectures. Here I review the hemichordate developmental data generated over the past 20 years that further test hypotheses of proposed morphological affinities between the two taxa, but also compare the conserved anteroposterior, dorsoventral axial patterning programs and germ layer specification programs. These data provide an opportunity to determine which developmental programs are ancestral deuterostome or bilaterian innovations, and which ones occurred in stem chordates or vertebrates representing developmental novelties of the chordate body plan.

1. Introduction

Hemichordates have long occupied an important place in the study of chordate origins. The link between hemichordates and chordates was recognized by early zoologists and embryologists: Kowalevsky recognized the chordate like gill slits of acorn worms (Kowalevsky, 1866) and Bateson further developed hypotheses on morphological affinities between the two groups (Bateson, 1886). The close relationship between hemichordates and echinoderms was first established by Metchnikoff who recognized the basic organizational similarities in their larval forms (Metschnikoff, 1881), despite the organizational disparity in their adult body plans. Interest in hemichordate biology waned with the rise of genetics and the decline in the influence of embryology (Hall, 2005), and during the next 50 years the phylum received little research attention, and only in the past few decades has there been some headway in resolving some of the promising hypotheses raised by the early investigators.

Hemichordates are divided into 2 classes; the small largely colonial pterobranchs, and the solitary enteropneust worms (Hyman, 1955). Most recent attention has focused on enteropneusts, and this review will focus attention on contemporary developmental biology research from a range of enteropneust species, representing contrasting early life history strategies. There are key anatomical characters that were identified by early investigators as potential morphological homologs of some of the defining features of the chordate body plan. However, the body plans of hemichordates and chordates are sufficiently divergent that establishing unambiguous anatomical homologies has been very challenging, and as I will discuss, molecular work has helped, but not entirely resolved this problem. Hemichordates also offer insights into the evolution of the unique conserved features of the early developmental programs of chordates and, along with echinoderms, offer an outgroup perspective into the evolution of some of the key features of the vertebrate developmental program, including the organizer (Gerhart, 2001), which plays a central role in the early establishment of the vertebrate body plan (Harland & Gerhart, 1997).

Some of the most important work on the development of these marine animals was carried out by embryologists in the late 1800s (Bateson, 1885, 1886; Morgan, 1891, 1894) but with a few exceptions, work on these animals dwindled until late in the next century. Interest in hemichordates, along with other less well known phyla, was reignited with the advent of molecular developmental biology and the emerging field of evo-devo: the availability of new tools for addressing classical zoological and developmental hypotheses, which had languished due to the ambiguous conclusions reached from anatomical comparisons, provided a jolt of new data and reinvigorated interest in animal origins. This has been especially useful in animals with divergent body plans, which have been challenging to make even basic axial comparisons with other bilaterian groups, due to the lack of plausible homologous characters. The molecular underpinnings of embryological development involved in the early establishment of animal body plans can be viewed as another anatomical character—a cryptic one that can only be revealed by molecular manipulations (Slack, Holland, & Graham, 1993). As we move through the fourth decade of contemporary evo-devo, the broader sampling of the molecular underpinnings of animal body plans has taught us that molecular comparisons can be just as fraught with homoplasy as anatomical characters (Abouheif et al., 1997; Hejnol & Lowe, 2015; Martin-Duran et al., 2018; Wagner, 2007). Most comparative developmental data have focused on arthropods and chordates, two groups with highly derived and sophisticated body plans, yet most major bilaterian phyla are characterized by far simpler body plans (Brusca, Moore, & Shuster, 2016). Hemichordates represent one of these simple body plans and have revealed that the relationship between conserved gene regulatory networks and morphological evolution is not nearly as tight as comparisons between arthropods and chordates have suggested (Lowe et al., 2003, 2006; Pani et al., 2012). In the past 20 years, new data have been generated from a range of hemichordate species, representing a broad sampling of the morphological diversity and life history strategies within the group. However, data from Pterobranchs are conspicuously lacking for rounding out a broader picture of evolution within the group. Developmental data have been mostly generated from two species, *Saccoglossus kowalevskii*, a direct-developing species, and *Ptychodera flava*, an indirect-developing species with a long-lived larva that metamorphoses into the adult body plan (Kaul-Strehlow & Röttinger, 2015; Rottinger & Lowe, 2012). Molecular studies from additional species have broadened sampling within enteropneusts, and we are beginning to be able to distinguish between taxon-specific features and ancestral developmental characters for the group. Studies have included research focusing on hypotheses of morphological homology between hemichordates and chordates, as well as others focusing on more classical developmental questions on axis patterning and germ layer specification. Great progress has been made in comparing the hemichordate body plan to that of other bilaterians, and particularly to chordates.

2. The animals

The Hemichordata consists of two major classes, the pterobranchs and the enteropneusts. Enteropneusts are solitary marine worms that burrow in sandy or muddy sediment and feed by a combination of filter and deposit feeding ranging in size from a few millimeters to a meter in length (Cannon, Rychel, Eccleston, Halanych, & Swalla, 2009; Kaul-Strehlow & Röttinger, 2015). Pterobranchs are often small colonial animals connected to each other by a common stalk living as a series of inter-connected zooids (Sato, Bishop, & Holland, 2008).

As someone who has worked on these animals now for 20 years, I can attest to the many challenges of using hemichordates as developmental model species, both at the level of finding, collecting and working with the adult animals, but also challenges around working with their embryos. Research progress on pterobranchs has been slow largely due to their scarcity, with fewer global sites described, but there may be a diversity of these animals still waiting to be discovered (Tassia et al., 2016). Enteropneusts are broadly distributed globally, but more easily found on reef flats in tropical regions and intertidally in more temperate zones. The majority of their diversity and biomass may be found subtidally, as intertidal populations generally require protected bays or inlets. Recent studies have found a surprising deep sea diversity, again suggesting further exploration will uncover additional diversity (Osborn et al., 2012).

The phylogenetic position of hemichordates as sister group to echinoderms within the Ambulacraria is well established (Bromham & Degnan, 1999; Cameron, Garey, & Swalla, 2000; Cannon et al., 2014; Furlong & Holland, 2002) forming the Ambulacraria. Also, their close relationship with chordates mostly well supported by phylogenomic analyses (Fig. 1A).

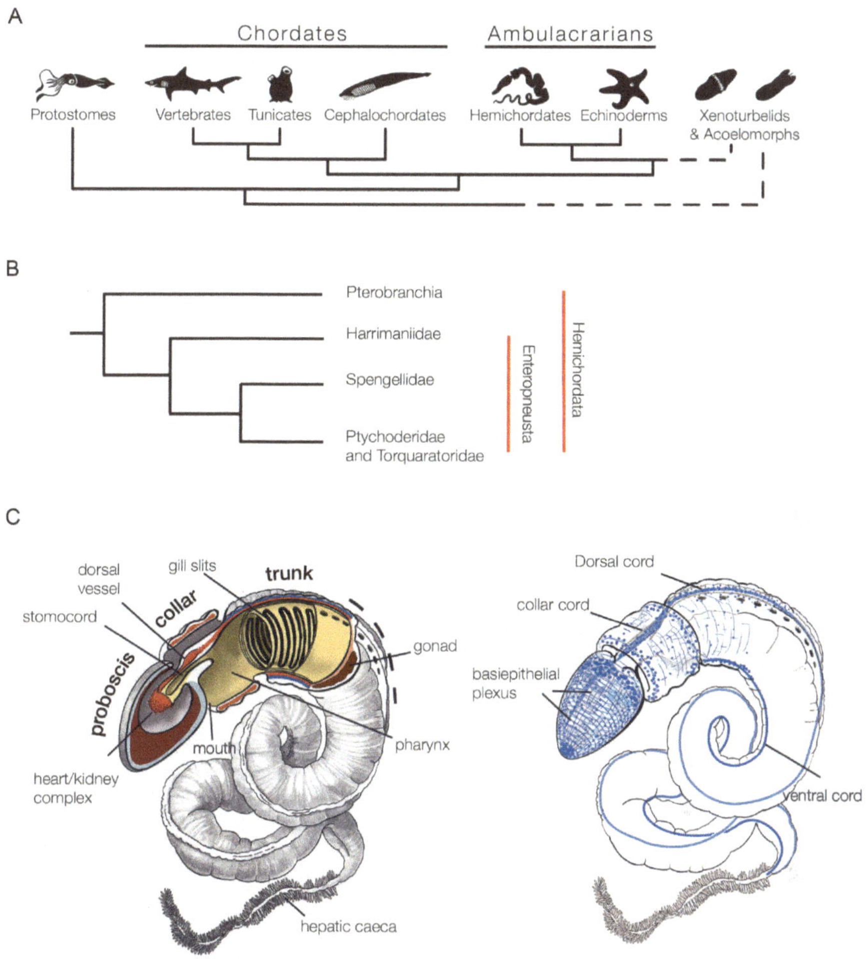

Fig. 1 Hemichordate body plan and phylogenetic relationships. (A) Summary of phylogenetic relationships of deuterostome taxa based on Philippe et al. (2011) and Hejnol et al. (2009). (B) Phylogenetic relationships within hemichordates showing the relationships between pterobranchs and enteropneusts, and the interrelationships between enteropneust families. Based on Cannon et al. (2014). (C) Diagrams of a generalized enteropneust showing the organization of the adult body plan and key anatomical features including the nervous system showing broad distribution of a neural plexus and two nerve cords. *Panel A and C: figure modified from Lowe, C.J., Clarke, D.N., Medeiros, D.M., Rokhsar, D.S., & Gerhart, J. (2015). The deuterostome context of chordate origins.* Nature, 520, *456–465.*

However, their relationship to acoels and Xenoturbellids remains controversial with some studies grouping Xenoturbellids into the deuterostomes, and some grouping acoels and Xenoturbellids as a monophyletic group (Xenacoelomorpha) into deuterostomes (Bourlat et al., 2006; Dunn et al., 2008; Mwinyi et al., 2010; Philippe et al., 2011, 2019) or as basally branching bilaterians (Hejnol et al., 2009; Rouse, Wilson, Carvajal, & Vrijenhoek, 2016; Simakov et al., 2015; Srivastava, Mazza-Curll, van Wolfswinkel, & Reddien, 2014). Nevertheless the core grouping of deuterostome taxa composed of chordates, hemichordates and echinoderms has been stable for some time. However, a recent preprint (Kapli et al., 2020) proposes that support for the classical deuterostome clade is equivocal, and an alternative radical departure to this classical grouping is equally likely; namely a possible closer affinity of chordates to protostomes, leaving the Ambulacraria as a clade separate from chordates and paraphyletic deuterostomes. Such a reorganization would require a reconsideration of how we think about chordate origins, and merits a thorough investigation, but is outside of the scope of this review.

Our understanding of the relationships within hemichordates is now greatly improved with better sampling, and while earlier molecular studies suggested that pterobranchs may nest within enteropneusts (Cameron et al., 2000), a more recent study organizes pterobranchs and enteropneusts into two monophyletic groups (Cannon et al., 2014). Enteropneusts are organized into four families; the Harrimanidae, Spengelidae, Ptychoderidae, with the deep sea Torquaratoridae nested within the Ptychoderidae (Fig. 1B) (Cannon et al., 2014).

3. Body plan and key anatomical features

The body plan of enteropneusts is well adapted to their marine life history as burrowing particulate and filter feeders. The following succinct description of their anatomy is largely to highlight key anatomies that have been important in the debate on the importance of hemichordates in understanding chordate origins. These features are highlighted in Fig. 1C. Other reviews provide a more comprehensive consideration of anatomy and historical background (Gee, 2018; Kaul-Strehlow & Röttinger, 2015; Satoh, 2016). Enteropneusts have a tripartite organization with an anterior proboscis/prosome, middle collar/mesosome, and posterior trunk/metasome (Hyman, 1955). The proboscis is strongly muscular and is involved in both burrowing and feeding. The proboscis, pharynx and gill slits in the anterior trunk are capable of creating feeding currents through ciliary action, passing phytoplankton and diatoms through the ventral mouth into the pharynx, then trapping them in paired pharyngeal gill slits before transporting them ventrally and posteriorly by ciliary action (Cameron, 2002; Gonzalez & Cameron, 2009). Worms also utilize direct particle capture in mucus secreted from the proboscis ectoderm and moved down through the mouth by ciliary action. These ropes of mucus and particles are transferred from the pharynx and into the posterior gut. In the anterior gut, the stomochord (Bateson, 1886; Ruppert, 2005), a supportive structure for the heart/kidney complex (Balser & Ruppert, 1990), projects into the proboscis. The trunk can be very long with the gut running the entire length into a terminal anus. In Harrimaniids, newly hatched juveniles are characterized by a post-anal tail (Bateson, 1885), which is used by the juvenile to move around in the burrow and as a hold fast, but is resorbed as the animals grow, and is not present in other hemichordate groups. The most anterior trunk is defined by the posterior pharynx and paired cartilaginous gill slits (Gillis, Fritzenwanker, & Lowe, 2011; Rychel, Smith, Shimamoto, & Swalla, 2006; Rychel & Swalla, 2007), which give way to the paired gonad, and in some species, a prominent region of hepatic saccules. The nervous system is defined by a basiepithelial plexus throughout the ectoderm and pharynx, but most conspicuous anteriorly in the proboscis and collar (Bullock, 1940; Dilly, Welsch, & Storch, 1970; Hess, 1937; Kaul-Strehlow, Urata, Minokawa, Stach, & Wanninger, 2015; Knight-Jones, 1952; Silèn, 1950). The plexus is most prominent at the base of the proboscis, forming a thick, dense layer of neurites, thickest in the dorsal region. There are regions of potentially centralized elements: two cords running along the dorsal and ventral midlines (Knight-Jones, 1952; Nomaksteinsky et al., 2009; Pickens, 1970). The dorsal cord runs from the most anterior region of the proboscis to the anus. In both the proboscis and trunk, this cord represents a thickening of the pervasive plexus into a superficial cord, with an associated cell bodies or soma, but in the most anterior extent of the dorsal cord that runs the length of the collar, it is internalized by a process that resembles chordate neurulation (Bateson, 1885; Brown, Prendergast, & Swalla, 2008; Bullock, 1945; Kaul & Stach, 2010; Knight-Jones, 1952; Luttrell, Konikoff, Byrne, Bengtsson, & Swalla, 2012; Morgan, 1894). The ventral cord begins at the anterior trunk, connecting to the dorsal cord just posterior to the collar and extends down the entirety of the trunk. Both cords, shown by EM to contain closely packed axons, have associated neural cell bodies, suggestive of an integrative function

(Kaul-Strehlow, Urata, Praher, & Wanninger, 2017; Nomaksteinsky et al., 2009), but their potential role in coordinating neural function remains poorly characterized (Cameron & Mackie, 1996; Pickens, 1970).

4. Life history strategies of species used for developmental studies

There have been a range of species of hemichordate utilized for developmental studies that span much of the phylogenetic diversity of the group (Bateson, 1884, 1885; Brown et al., 2008; Gonzalez, Jiang, & Lowe, 2018; Henry, Tagawa, & Martindale, 2001; Lin, Tung, Yu, & Su, 2016; Miyamoto & Saito, 2007; Morgan, 1894; Sato et al., 2008; Urata & Yamaguchi, 2004) although there are no developmental studies available for any torquaratorid species so far, and molecular studies on pterobranch development are still very limited (Sato et al., 2008; Sato & Holland, 2008). Enteropneusts are split into two developmental categories; direct developers and indirect developers (Fig. 2). Developmental mode falls squarely into a phylogenetic framework with the harrimanid lineage all direct-developers whereas the rest of the class are indirect-developers. The majority of developmental data has been generated from two species; *P. flava* and *S. kowalevskii*. *P. flava* is from the Ptychoderidae, which has a broad Indopacific distribution in sandy reef flats, and is characterized by indirect development (Rottinger & Lowe, 2012). *S. kowalevskii*, from the Harrimaniidae, is characterized by direct development and found intertidally along the majority of the Eastern Seaboard of the USA (Lowe, Tagawa, Humphreys, Kirschner, & Gerhart, 2004). Several other indirect-developing species have been utilized in developmental comparisons; two other ptychoderid species; *Balanoglossus simodensis*, and *Balanoglossus misakiensis* from Japan, and the spengellid *Schizocardium californicum* from the west coast of the USA.

5. Developmental contributions to testing hypotheses of morphological homology

Much of the initial motivation for examining developmental mechanisms in hemichordates was driven largely by the promise of applying new data and techniques to address classical morphological-based hypotheses raised over 100 years ago. Hypotheses of chordate origins have been particularly challenging to test due the large morphological disparity between

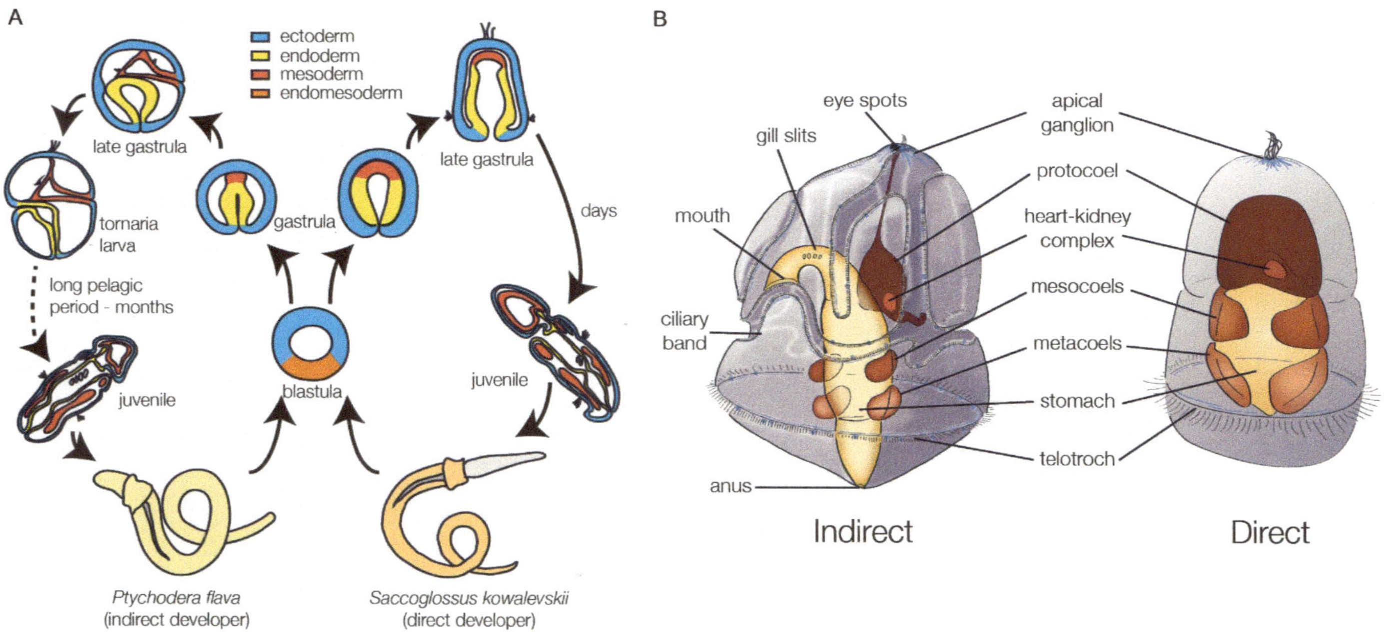

Fig. 2 See figure legend on next page.

chordates and their closest invertebrate outgroups, echinoderms and hemichordates (Lowe, Clarke, Medeiros, Rokhsar, & Gerhart, 2015; Tagawa, 2016). Comparative studies based on morphological comparisons had reached an impasse, and molecular data has the potential to help resolve some of these long-standing hypotheses.

6. Pharyngeal gill slits

The new molecular data has not provided unambiguous support to either accept or refute the proposed anatomical homologies between hemichordate and chordate structures. However, the pharyngeal gill slits are an exception (Fig. 1). Molecular work strongly supports the homology between hemichordate and chordate gill slits (Fritzenwanker, Gerhart, Freeman, & Lowe, 2014; Gillis et al., 2011; Ogasawara, Wada, Peters, & Satoh, 1999; Okai, Tagawa, Humphreys, Satoh, & Ogasawara, 2000; Rychel et al., 2006; Rychel & Swalla, 2007; Simakov et al., 2015). Data supporting homology on morphological characters was already strong: enteropneust gill slits resemble those in amphioxus, both from an anatomical

Fig. 2 Developmental mode of hemichordate species. (A) model of the two contrasting life histories of enteropneust hemichordates. Indirect developers develop into planktonic larvae spending months in the plankton before metamorphosis into benthic juveniles. Direct developers have abbreviated development forming the adult body plan from the embryo. They are defined by a very short pelagic period, and hatch as competent juveniles that can swim or immediately begin to burrow and feed. (B) Detailed models of the two contrasting life histories. The generalized larva is representative of all indirect developing species, late in larval development, many weeks into planktonic development, close to metamorphosis. The model of the direct developer represents 48 h. of development, after gastrulation during embryo elongation, following the formation of the coelomic pouches. Both models are shown as a ventro-lateral tilted view with anterior at the top and posterior at the bottom. *Figure modified from Lowe, C.J., Clarke, D.N., Medeiros, D.M., Rokhsar, D.S., Gerhart, J. (2015). The deuterostome context of chordate origins.* Nature, 520, *456–465. and Simakov, O., Kawashima, T., Marletaz, F., Jenkins, J., Koyanagi, R., Mitros, T., Hisata, K., Bredeson, J., Shoguchi, E., Gyoja, F., Yue, J.X., Chen, Y.C., Freeman, R.M., Jr., Sasaki, A., Hikosaka-Katayama, T., Sato, A., Fujie, M., Baughman, K.W., Levine, J., Gonzalez, P., Cameron, C., Fritzenwanker, J.H., Pani, A.M., Goto, H., Kanda, M., Arakaki, N., Yamasaki, S., Qu, J., Cree, A., Ding, Y., Dinh, H.H., Dugan, S., Holder, M., Jhangiani, S.N., Kovar, C.L., Lee, S.L., Lewis, L.R., Morton, D., Nazareth, L.V., Okwuonu, G., Santibanez, J., Chen, R., Richards, S., Muzny, D.M., Gillis, A., Peshkin, L., Wu, M., Humphreys, T., Su, Y.H., Putnam, N.H., Schmutz, J., Fujiyama, A., Yu, J.K., Tagawa, K., Worley, K.C., Gibbs, R.A., Kirschner, M.W., Lowe, C.J., Satoh, N., Rokhsar, D.S., Gerhart, J. (2015). Hemichordate genomes and deuterostome origins.* Nature, 527, *459–465.*

and functional perspective (Gonzalez & Cameron, 2009; Rychel & Swalla, 2007). In the sister group to hemichordates, the echinoderms, there is no equivalent structure in extant species, yet in certain stem echinoderm fossils there is evidence of gill slits that were subsequently lost in crown groups (Dominguez, Jacobson, & Jefferies, 2002). Genes expression studies of transcription factors with reported roles in chordate pharyngeal and gill pouch patterning including *pax1/9* and *six1/2*, *eya* and three Fox genes *Fox I*, *FoxC* and *FoxL1* (Fritzenwanker et al., 2014; Gillis et al., 2011; Ogasawara et al., 1999) showed conserved expression domains in *S. kowalevskii* during gill pouch development. Genomic data has now provided added data further strengthening homology: Hemichordates share the same pharyngeal gene cluster that was originally described in chordates (Simakov et al., 2015). This cluster contains 6 genes, 4 of which are transcription factors (*pax1/9, nk2–1, nk2–2 and foxA*) with expression domains in the pharyngeal slits or in the pharyngeal endoderm surrounding them. The genes are in the same order as in the vertebrate clusters, and an ordered cluster is even detected in the genome of an asteroid, making this genomic feature clearly part of a conserved ancestral developmental program for anterior gut patterning in deuterostomes (Baughman et al., 2014; Simakov et al., 2015).

7. Nerve cords and stomochord

Perhaps the most intense interest in hemichordates concerns their potential for providing insights into evolution of the chordate nervous system. The organizational principles of the enteropneust nervous system have long been debated, as there are elements of a pervasive nerve plexus suggestive of local processing (Bullock, 1940; Lowe et al., 2003), but also centralized elements including the collar cord (Kaul & Stach, 2010; Knight-Jones, 1952; Luttrell et al., 2012), which has long been a structure that has fascinated zoologists because of its morphogenesis is similar to the vertebrate nerve cord (Fig. 1). How the nervous system processes information and the role of cord's integration of sensory information will require further neurophysiology experiments (Cameron & Mackie, 1996; Pickens, 1970). Further, if the cords do represent centralized processing centers, are either of them homologous to any region of the chordate dorsal central nervous system? Several groups have either explicitly addressed these questions or have produced data relevant to the question (Formery, Schubert, & Croce, 2019). In Lowe et al. (2003), described in more detail below, classical transcriptional regulators of anteroposterior and mediolateral patterning of the vertebrate central nervous system (CNS) were largely expressed in circumferential domains throughout the ectoderm of *S. kowalevskii*, rather than being expressed coincident with the more centralized elements of the nervous system. They proposed that this may reflect the fundamental organizational principle of hemichordates nervous system as a plexus (Lowe et al., 2003). Later work further tested this hypothesis and examined the expression of a range of marker genes conserved in fruit flies and/or vertebrates for neural progenitor cells (*soxB1* and *msi*), differentiating neurons (*elav, prox, bruA* and *dclk*), and mature neurons (*syt1, gad, th, tph, vmat* and 5-HT) (Cunningham & Casey, 2014). The initial distribution of neurons is broadly spread throughout the ectoderm but this was transitory, and before hatching the developing nerve cells of the trunk become more aligned with the midlines (Cunningham & Casey, 2014) as the cords begin to form. Neural markers such as Gaba, 5HT, VaChT rather than exclusively patterning genes were also examined in adult *P. flava* (Nomaksteinsky et al., 2009). They reveal a clear aggregation of neurons at the base of the proboscis, concentrated dorsally, and also along the dorsal cord and ventral cord, with far fewer neurons in the general ectoderm. This pattern was also observed in the post-metamorphic juveniles in another indirect-developing species, *B. misakiensis* (Kaul-Strehlow et al., 2017). The cords are organized with a concentration of cell bodies and a clear neuropil. Cholinergic and GABA-ergic cell bodies were identified in the cords and proboscis stem, with only sparse expression of the cholinergic marker VaChT outside of these domains (Nomaksteinsky et al., 2009). Nomaksteinsky and colleagues propose that the cords and proboscis stem represent a bonafide CNS, with the remaining sparser distribution of neurons throughout the ectoderm representing a peripheral nervous system. The authors remark on the close similarities between the internalized collar cord and the dorsal cord of vertebrates, but in conclusion do not make a strong statement of homology of any part of the enteropneust CNS to chordates, and raise the possibility of deep homology of the patterning module rather than specific neuroanatomies between hemichordates and chordates.

Further studies have focused more specifically on the potential homology of the collar cord to the chordate dorsal nervous system (Kaul-Strehlow et al., 2017; Luttrell et al., 2012; Miyamoto & Wada, 2013). The striking morphological similarities between hemichordate collar cord and vertebrate neural tube, and the process of neurulation were first noted by classical zoologists (Bateson, 1884; Bullock, 1945; Dawydoff, 1948; Knight-Jones, 1952; Morgan, 1894) and have been recently reinvestigated by several groups

(Kaul & Stach, 2010; Luttrell et al., 2012; Miyamoto & Wada, 2013). Miyamoto and Wada (2013) investigated the developmental basis of neurulation in *B. misakiensis* and demonstrate that the stomochord and dorsal endoderm that underlie the region of collar ectoderm that gives rise to collar cord, are a broad source of hedgehog (hh) ligand, which has been shown to be a key determinant in the D/V dorsoventral patterning of the vertebrate neural tube (Echelard et al., 1993). They also demonstrate that there is some spatial pattern in the D/V aspect of the collar cord using conserved markers of neural tube D/V patterning, such as *bmp2/4*, *dlx* and *pax3/7*. However, this D/V patterning is not nearly as extensive as that found in vertebrates, representing more dorsal than ventral markers. In *S. kowalevskii*, *bmp2/4* and *dlx* are similarly localized along the midline, with an additional dorsal neural tube marker, *tbx2/3*, but their expression extends along the entire length of the dorsal midline rather than restricted to the collar cord. In addition, classical markers of the vertebrate floor plate and ventral neural tube, such as *netrin*, *sim*, and the motor neuron marker *mnx*, are expressed ventrally, making comparisons quite complex as markers are split across both cords (Lowe et al., 2006). Kaul-Strehlow et al. (2017) again investigated the expression of mediolateral patterning genes in post metamorphic juveniles of *B. simodensis*, which showed little evidence of conserved spatial domains in the collar cord, and they concluded that the collar cord may represent an independent acquisition in the enteropneust lineage.

So far, the evidence in support of homology of any region of the hemichordate nervous system to the CNS of vertebrates is decidedly mixed, and while molecular patterning data do not explicitly rule out homology, it is also equally plausible that cords evolved independently in enteropneusts. Similar data from a variety of protostome groups also show a poor correspondence of mediolateral patterning data and cord formation, and suggest patterning similarities between model species may represent convergence rather than homology, with neuroanatomies evolving independently (Martin-Duran et al., 2018). Despite the extensive developmental patterning data available, clearly more work is needed to decipher the neural organization of hemichordates. There are few contemporary studies of the neural organization of hemichordates with a limited array of cross reactive antisera (Gonzalez et al., 2018; Kaul-Strehlow et al., 2015; Miyamoto, Nakajima, Wada, & Saito, 2010), and ultrastructure (Kaul & Stach, 2010; Kaul-Strehlow et al., 2015; Kaul-Strehlow & Stach, 2013), but a better understanding of neural organization, distribution, cell type composition and how it is regulated by conserved networks of regulatory genes is badly needed.

The origin of the chordate notochord has a long history in the classical comparative literature (Annona, Holland, & D'Aniello, 2015) and the hemichordate stomochord (Fig. 1) has long been proposed as a potential homolog of this key feature of the chordate body plan. The idea was originally proposed by Bateson (Bateson, 1886) and supported by other investigators (Balser & Ruppert, 1990; Miyamoto & Wada, 2013). However, an analysis and survey of structural and patterning genes during stomochord formation fail to provide support for the homology of the two structures (Satoh et al., 2014). Rather, expression of FoxE quite specifically in the stomochord (Fritzenwanker et al., 2014) provides evidence that this structure may have more affinities with the chordate endostyle. FoxE (TTF2), a thyroid specific transcription factor involved in vertebrate thyroid gland development (Damante, Tell, & Di Lauro, 2001) and is also expressed in the club-shaped gland of amphioxus larvae that eventually gives rise to the endostyle in adults (Yu, Holland, Jamrich, Blitz, & Hollan, 2002), as well as in the endostyle of ascidians (Hiruta, Mazet, Yasui, Zhang, & Ogasawara, 2005).

8. Insights from comparative axial patterning

While most interest in the developmental biology of hemichordates has focused on testing morphological homology, extensive patterning data now provide insights into body plan evolution at the level of axial patterning, not only for providing insights into the origin and evolution of chordates, but also to our understanding of how conserved axial patterning relates to the evolution of morphology. While it remains challenging to use developmental data to reconstruct ancestral deuterostome anatomical characters (Lowe et al., 2015), patterning data provides a suite of characters often more conserved than the morphologies they regulate, so offering an opportunity to infer ancestral deuterostome developmental modules. Using this approach, we can begin to test what aspects of the vertebrate axial programs evolved in stem chordates or stem vertebrates, and are therefore associated with the assembly of the features of the chordate body plan, and which ones are deep deuterostome or bilaterian innovations. There are now both descriptive and functional developmental data on early axial patterning for both the larval and adult body plans. Below I discuss the evolutionary implications of these studies for both body plan evolution and chordate origins.

9. Anteroposterior patterning

For the most part, comparative anteroposterior (A/P) and dorsoventral (D/V) patterning has been examined across phyla during ectodermal patterning, largely focused on central nervous system development. Particularly striking has been the close patterning similarities between arthropods, the errant annelid, *Platynereis dumerilii* and vertebrates (Arendt & Nubler-Jung, 1996; De Robertis, 2008; Reichert & Simeone, 2001). Suites of transcription factors with conserved relative spatial register and similar patterning roles have been described in all three groups during CNS development leading to the hypothesis of an ancestral complex CNS in the common ancestor of bilaterians (Denes et al., 2007; Hirth et al., 2003; Tomer, Denes, Tessmar-Raible, & Arendt, 2010; Vergara et al., 2017). A comprehensive examination of the expression of conserved orthologous transcription factors with conserved roles in CNS patterning during early development in *S. kowalevskii* shows the same exquisite conservation of pattern along the A/P axis, with the same relative spatial register as we find in the other models with complex central nervous system organization (Fig. 3A). However, the major difference is that in most cases, the expression domains wrap around entire ectoderm rather than being restricted to either the dorsal or ventral side (Lowe et al., 2003), as is the case for the many expression domains restricted to the neural plate of vertebrates. A subset of these genes has also been examined in two other hemichordate species with expression domains that are very similar (Gonzalez, Uhlinger, & Lowe, 2017; Kaul-Strehlow et al., 2017). For example, vertebrate forebrain patterning genes such as *six3*, *foxG*, *retinal homeobox* and *nk2–1* are restricted to the developing proboscis ectoderm, those involved in midbrain patterning such as *otx*, *pax6* and *barH* down into the collar/mesosome, and hindbrain, and trunk patterning genes such as *engrailed (en)*, *gbx*, and Hox genes are expressed in the metasome/trunk. Hox genes show some divergence from this pattern, with certain members of the cluster associating more with the midlines and nerve cord development than the entire ectoderm (Aronowicz & Lowe, 2006; Gonzalez et al., 2017). This general pattern has now been shown in three difference species: most comprehensively in the direct developer *S.kowalevskii (*Aronowicz & Lowe, 2006*;* Lowe et al., *2003)*, but also in representative species from the two remaining enteropneust families; the ptychoderid *B.simodensis* (Kaul-Strehlow et al., 2017) and the Spengellid, *S.californicum (*Gonzalez et al., 2017*)* and thus

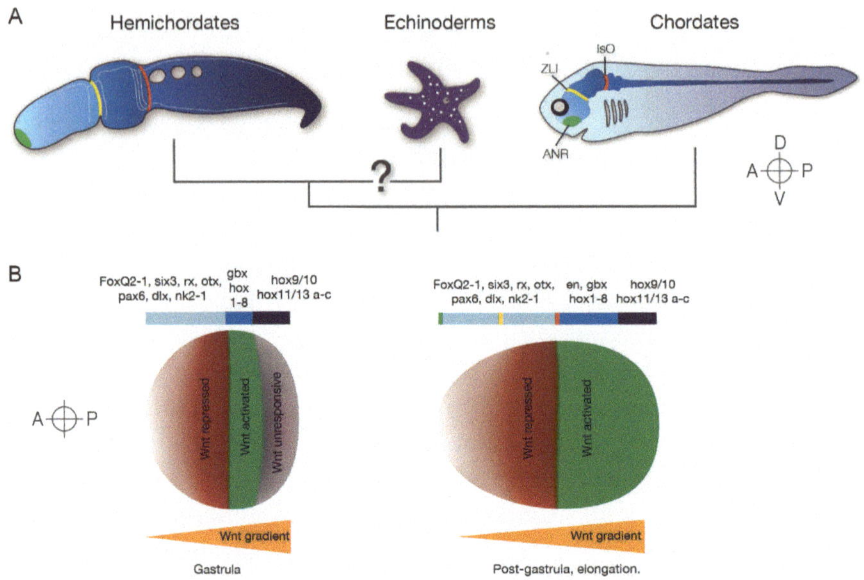

Fig. 3 Summary of ectodermal anteroposterior axial patterning of enteropneusts and chordates. (A) Comparison of A/P patterning in a generalized enteropneust and vertebrate. The blue gradient represents the similarities in gene expression of homologous transcription factors involved in specifying the main regional compartments of the vertebrate CNS and enteropneust ectoderm. The green yellow and red stripes represent homologous local signaling centers positioned at the same relative position in the CNS of vertebrates and general ectoderm of hemichordates. (B) Summary of the role of cWnt signaling during early development of *S. kowalevskii*. The blue rectangles represent the regional expression of groups of conserved transcription factors along the A/P axis of the embryos. Two developmental stages are shown and represent a surface view of the ectoderm at late gastrula and post gastrula stage. The red territory represents the anterior embryonic territories that are repressed by the action of cWnt. The green territory represents the territory dependent on cWnt signaling for gene activation. The gray blastoporal territory in the posterior of the gastrula represents a region not responsive to cWnt during blastula and gastrulation. *Panel A is modified from Lowe, C.J., Clarke, D.N., Medeiros, D.M., Rokhsar, D.S., Gerhart, J. (2015). The deuterostome context of chordate origins.* Nature, 520, *456–465.*

represents the ground state of enteropneusts (Kaul-Strehlow et al., 2017). Expression domains of the majority of these genes are in the general ectoderm rather than uniquely associated with the neural cords or regions of neural concentration at the base of the proboscis. This suggests a general role in providing spatial patterning information throughout the ectoderm, to both nervous system and epidermal tissue. A general role in ectodermal patterning for many of the A/P patterning genes may be a conserved feature of this suite of genes. Holland et al. (2013) point out that many of the transcription factors with distinct expression territories in *S. kowalevskii* that were investigated in Lowe et al. (2003) also have expression domains in the ectoderm outside of the CNS of chordates.

Interest has largely focused on the adult body plan of enteropneusts, but the same patterning network is also responsible for patterning the larval A/P axis and has been examined explicitly in *S. californicum* and to some extent also in *P. flava* (Gonzalez et al., 2017; Rottinger, DuBuc, Amiel, & Martindale, 2015; Rottinger & Martindale, 2011; Su et al., 2019). The metamorphosis from larval to adult body plans in enteropneusts is dramatic, resulting in a rapid change in body plan from a planktonic feeding larva, constructed of a gel-filled body with thin epithelia, using ciliary bands for feeding and swimming, to a benthic, muscular burrowing worm over the course of 48h. (Gonzalez et al., 2018; Miyamoto et al., 2010; Morgan, 1891). Despite these major morphological transformations of the body regions between the two life history stages, the underlying A/P patterning network seems largely conserved (Gonzalez et al., 2017). The transcription factors with conserved roles in anterior patterning, markers of the proboscis and collar in the adult, are deployed in similar relative expression domains to their distribution in the adult as the larval body plan begins to form during embryogenesis and beyond. The most terminal markers, the posterior class Hox genes, are also expressed around the blastopore of embryos and eventually, the anus, following embryogenesis. However, the anterior/mid trunk markers, the medial and anterior class Hox genes, are not activated during the establishment of the larval body plans, and are only transcribed late in larval development when the adult body plan is beginning to form, closer to metamorphosis (Gonzalez et al., 2017). A similar distribution of patterning genes is also described in echinoderm larvae (Yankura, Martik, Jennings, & Hinman, 2010) suggesting that ambulacrarian larvae are a trunkless body plan and are best described as swimming heads (Lacalli, 2005). I will further discuss the importance of considering life history in comparative studies in a later section.

10. Origin of CNS signaling centers

In *S. kowalevskii* embryos the A/P patterning similarities with vertebrates can be extended even further. The vertebrate brain and central nervous system develop by successive refinement of the neurectoderm into specialized territories and domains. Signaling centers act as secondary organizers that mediate regional patterning in the CNS, and are defined by the local secretion of specific ligands (Echevarria, Vieira, Gimeno, & Martinez, 2003; Kiecker & Lumsden, 2012). The three main centers are the anterior neural ridge (ANR), the zona limitans intrathalamica (ZLI) and the midbrain hindbrain organizer or isthmus (Iso), which are conserved throughout vertebrates (Fig. 3A) but are either absent or divergent in invertebrate chordates (Albuixech-Crespo et al., 2017; Scholpp & Lumsden, 2010). However, some investigators (Holland et al., 2013) still contest the absence of local neural organizers in cephalochordates. This would suggest that local neural signaling centers may have been a stem vertebrate innovation that was associated with the evolution of a more complex CNS. Surprisingly, in *S. kowalevskii*, these same signaling center regulatory programs are established at equivalent regions of the A/P transcriptional network following gastrulation, as the embryo begins to elongate and establish the major divisions of the body plan; an ANR-like domain at the anterior tip of the proboscis/prosome expressing *fgf8/17/18* and *secreted frizzled 1/2/5*, a ZLI-like domain at the boundary between the developing proboscis and the collar expressing *hh*, and an Iso-like domain expressing *fgf8/17/18/Wnt1* (Fig. 3A). Functional evidence supports similarities in the regulatory program both upstream and downstream of these localized domains of ligand expression, suggesting a conservation of the regulatory program in the absence of any morphological homology (Pani et al., 2012). Although this interpretation has been challenged (Holland et al., 2013), suggesting that differences in gene expression domains between hemichordates and vertebrates are more consequential than acknowledged. For example, *irx* and *fezF*, two important transcription factors with abutting domains in the anterior neural plate of chordates, including amphioxus (Irimia et al., 2010), are important in the establishment of the ZLI in vertebrates, are not expressed in the same pattern in *S. kowalevskii*, leading the authors to question the ZLI comparison. However, a follow up study of cis-regulation of the defining ligand of the ZLI, *hh*, strengthens the evidence of the homology of the ZLI regulatory program between hemichordates and vertebrates (Yao et al., 2016). Comparative sequence analysis is generally used to test for conservation of regulatory sequences between taxa, yet it is clear that functionally conserved enhancers do not always display strong sequence conservation (Fisher, Grice, Vinton, Bessling, & McCallion, 2006; Hare, Peterson, Iyer,

Meier, & Eisen, 2008), and genes with similar spatiotemporal expression domains may share conserved regulatory logic. The defining ligand of the ZLI, Sonic Hedgehog (shh), has many distinct expression territories within the vertebrate CNS under the regulatory control of distinct enhancers (Epstein, McMahon, & Joyner, 1999; Jeong et al., 2008; Jeong, El-Jaick, Roessler, Muenke, & Epstein, 2006; Jeong & Epstein, 2003). Two enhancers have been identified that are involved in expression of Shh in mouse at the ZLI—SBE1 and SBE5. SBE 1 is an intronic enhancer in intron 2 of the Shh locus and involved in expression in the ventral forebrain and ZLI (Epstein et al., 1999). Surprisingly, an analysis of intron 2 of Hh in *S. kowalevskii* revealed a putative homologous enhancer with similar transcription factor binding motifs to SBE1, but in a scrambled order relative to the mouse enhancer. Recent progress in the development of transgenic approaches in *S. kowalevskii* (Minor et al., 2019) facilitated testing this enhancer for activity *in vivo*. Injection of a GFP transgene of this hemichordate intronic enhancer demonstrated activity in the putative hemichordate ZLI-like domain, at the boundary between the proboscis and collar, supportive of a conserved function. Most surprising was that cross species enhancer swaps demonstrated that the regulatory logic of the hemichordate SBE1-like enhancer was sufficient to drive reporter expression in mouse in largely the same spatiotemporal domains as the endogenous mouse SBE1 element. Conversely, the mouse enhancer drove identical expression to the endogenous hemichordate SBE1 enhancer during early hemichordate development. So, despite the fundamental body plan differences and 600 million years of independent evolution that separates vertebrates and hemichordates, the cis-regulatory logic of an enhancer, along with its trans regulators for a key developmental ligand, is conserved and involved in patterning very different ectodermal structures. This enhancer is not present in either amphioxus or tunicates indicating secondary loss from an ancestral state, and outlines the importance of outgroup comparisons for determining which gene regulatory innovations are deuterostome-specific, predating the origins of the chordate body plan, and which ones are stem chordate, or stem vertebrate innovations, and potentially instrumental in the evolution of specific features of the chordate body plan. This work clearly demonstrates the challenges of using gene regulatory networks for reconstructing ancestral morphological characters. Given how diverse the body plans are that these homologous programs regulate, our ability to reconstruct morphological characters from these data is limited.

11. Early establishment of A/P pattern

From a wide range of vertebrate studies, the early establishment of A/P pattern in the embryo is now well characterized. A number of signaling pathways are implicated, with minor differences between animal species, yet largely conserved roles throughout the groups are largely conserved (Arnold & Robertson, 2009; Schier, 2001; Schoenwolf, 2000). One of the most important signaling pathways for A/P specification is the canonical Wnt (cWnt) cascade. It has the strongest effect on head patterning, repressing the most anterior neural fates of forebrain and midbrain, but also promoting more posterior, hindbrain fates in a dose-dependent manner (Glinka et al., 1998; Kiecker & Niehrs, 2001; Leyns, Bouwmeester, Kim, Piccolo, & De Robertis, 1997; McGrew, Hoppler, & Moon, 1997; Nordstrom, Jessell, & Edlund, 2002; Piccolo et al., 1999). Wnt antagonists, such as Dkk and secreted frizzled, are localized anteriorly in the neural plate, prechordal plate and organizer, to protect the anterior neural territory from the posteriorizing effects of Wnt ligands secreted from the posterior neural plate and underlying mesoderm (Glinka et al., 1998; Houart et al., 2002; Kiecker & Niehrs, 2001; Piccolo et al., 1999). This has led to the hypothesis that a simple cWnt gradient, low in the anterior and high in the posterior, is largely responsible for the patterning of the ectoderm, and has been proposed as a fundamental conserved mechanism of A/P patterning (Kiecker & Niehrs, 2001; Niehrs, 2010; Nordstrom et al., 2002). This model is strongly supported from comparative vertebrate studies focused on forebrain, midbrain and hindbrain, but surprisingly little data has been generated from the most posterior regions of the neuraxis, and recent work has challenged the importance of Wnt in posterior specification (Polevoy et al., 2019).

In the invertebrate chordates, the role of cWnt in A/P patterning is likely broadly conserved for the most part. However, in both amphioxus and ascidians there are distinct differences in details, with cWnt largely involved in the specification of the far anterior and posterior of the A/P axis (Feinberg, Roure, Piron, & Darras, 2019; Onai et al., 2009) suggesting that their neuraxes are not fully patterned by cWnt. In echinoderms, nothing is known about the role of Wnt in adult patterning; but in larvae, cWnt also plays a conserved role in anterior patterning (Range, 2018; Range, Angerer, & Angerer, 2013; Range & Wei, 2016), even though the lack of a larval trunk territory (Yankura et al., 2010) makes a full comparison of axial patterning with vertebrates difficult. The role of cWnt patterning in protostomes has been most comprehensively examined in planarians where it is involved in the establishment of the A/P axis during homeostasis and regeneration, and regulates the developmental decision between head and tail (Gurley, Rink, & Sanchez Alvarado, 2008; Petersen & Reddien, 2008). In the polychaete *Platynereis dumerilli*, cWnt also plays a key role in head suppression, and in arthropods there are several examples of cWnt involved in posterior growth (Angelini & Kaufman, 2005; Marlow et al., 2014; Martin & Kimelman, 2009; Miyawaki et al., 2004), and some evidence from the beetle Tribolium that Wnt may also be involved in head patterning (Fu et al., 2012). It is therefore broadly accepted that cWnt is a conserved element of A/P patterning (Petersen & Reddien, 2009).

The close transcriptional similarities between vertebrates and hemichordates during ectodermal A/P patterning raise the possibility that the upstream regulation of this gene regulatory network is also conserved. A recent study of cWnt signaling in *S. kowalevskii* again exhibits some striking similarities with vertebrate patterning (Darras et al., 2018). The 13 Wnt ligands, four frizzled receptors, and a range of Wnt antagonists show strongly localized patterns along the animal/vegetal and A/P axes during embryogenesis. The ligands are mostly concentrated in the posterior ectoderm from blastula through gastrula stages, and throughout almost all ectoderm with the exclusion of the most anterior territory. Like the expression of transcription factors along the A/P axis, ligands, receptors and antagonists are expressed in ectodermal rings. Unlike sea urchins, we do not observe any ligand expression in the endomesoderm precursors (Robert, Lhomond, Schubert, & Croce, 2014). Wnt antagonists Sfrps and Dkk1/2/4 are localized anteriorly in the ectoderm, in a manner very similar to their disposition in other bilaterian species, including vertebrates (Glinka et al., 1998; Range, 2018). Manipulation of the levels of cWnt signaling during the most critical stage of A/P patterning, mid-blastula through gastrulation, results in major transformations of the fate of the A/P axis. Over-expression of Wnt3 by RNA injection, or constitutive activation of cWnt signaling by the application of a Gsk3ß antagonist, resulted in the truncation of anterior fates in a dose dependent manner: both proboscis and collar fates were lost, with the anterior trunk marker engrailed (en) expressed at the far anterior tip of experimental embryos, and a remaining trunk territory that seemed relatively normal. The reciprocal experiment of lowering cWnt activity by over-expression of Wnt antagonists resulted in the complementary phenotype of a greatly enlarged proboscis with reduced trunk. However, in neither experiment was the axis completely anteriorized or posteriorized, very similar to the phenotypes in vertebrate experiments (Kiecker & Niehrs, 2001). A QPCR analysis following GSK3ß inhibitor treatments through gastrulation, resulting in the constitutive activation of cWnt signaling, demonstrated a clear molecular read out that cWnt repressed anterior fates in a dose dependent manner down to the posterior collar, and promoted anterior and mid trunk fates. Notably, GSK3ß inhibitor treatments through gastrula stage had no effect on the most posterior fates defined by the expression of posterior Hox at this early stage of A/P specification. It was only after gastrulation, as the embryo started to elongate during an extended period of posterior growth, that posterior Hox became entrained into cWnt regulation in a regulatory loop with Brachyury that bares a strong resemblance to the mechanisms of posterior growth in vertebrates (Fritzenwanker, Uhlinger, Gerhart, Silva, & Lowe, 2019). The targets of cWnt signaling during embryogenesis in *S. kowalevskii* are strikingly similar to those described in vertebrates; repression of markers of forebrain-midbrain/proboscis-collar markers and promotion of hindbrain/anterior trunk markers (Fig. 3B). The lack of early cWnt sensitivity of posterior Hox during embryogenesis during the initial establishment of posterior fates challenges the simple model of cWnt function in early A/P patterning (Niehrs, 2010). The possibility that this may be a conserved feature of deuterostomes is supported by work in Xenopus reporting similar findings and deserves further investigation in other species (Polevoy et al., 2019).

The exquisite conservation of a cWnt regulated transcriptional network involved in ectodermal A/P regionalization between such contrasting body plans again reveals the rather loose coupling of gene regulatory networks with morphological evolution, and demonstrates how tight regulatory conservation seems to be permissive rather than restrictive in morphological output over macroevolutionary time frames. This type of data is extremely valuable for making broad comparisons across groups and establishing regional rather than anatomical homologies. This is particularly useful in comparisons where morphological disparity is large, making even basic regional comparisons challenging. From both the transcriptional and cWnt patterning data, the proboscis of hemichordates and the forebrain of vertebrates share fundamental patterning similarities as does the midbrain and collar, and the hindbrain-spinal cord/trunk. This does not imply any type of morphological homology, but rather reflects a common ancestry from a homologous ancestral territory; a deep homology (Shubin, Tabin, & Carroll, 2009). However, these data establish a good regional framework for investigating anatomical homology. For example, if the collar cord is homologous to any region of the vertebrate CNS, then it would not be the hindbrain or spinal cord, but rather the midbrain, as the collar transcriptional profile most closely resembles this region of the vertebrate CNS (Kaul-Strehlow et al., 2017). Also, the first gill slit in both vertebrates and hemichordates develops in a similar transcriptional regional territory along the A/P axis, further supporting homology.

12. Dorsoventral patterning

The D/V axis has featured prominently in comparative body patterning (Arendt & Nubler-Jung, 1994; De Robertis, 2008; De Robertis & Sasai, 1996). Details of axis patterning have been well characterized in both vertebrates and arthropods, and it is very well established that the CNS of arthropods and chordates are established by a homologous signaling system. However, how these close patterning similarities relate to the homology of anatomies patterned by this conservative regulatory system is less clear (Arendt, 2018; Lowe et al., 2006; Martin-Duran et al., 2018). The CNS of both arthropods and chordates forms in a region of the ectoderm that is protected from the neural-repressive effects of Decapentaplegic/Bmp by Short gastrulation(sog)/Chordin (Holley & Ferguson, 1997). In arthropods, the CNS forms on the ventral side, and in chordates on the dorsal side. The distribution of the Bmp/Chordin axis defines the position of the CNS in both of these distantly related groups, so this molecular axis is inverted in chordates in relation to arthropods, and gives support to an old classical hypothesis that chordates evolved by D/V axis inversion of an ancestor with a CNS orientated on the ventral side (Geoffroy-St. Hilaire, 1822). The phylogenetic position of hemichordates, as a group closely related to chordates, with two nerve cords; one on the dorsal and the other on the ventral side makes this group particularly interesting for the investigation of the role of a Bmp/Chordin axis during early development. This axis has now been investigated in both indirect and direct-developing species with some common and contrasting findings (Lowe et al., 2006; Su et al., 2019). In *S. kowalevskii* (Lowe et al., *2006)*, *bmp2/4* is expressed throughout embryogenesis and into juvenile development on the dorsal side in the ectoderm, overlapping the region where the dorsal cord forms. Chordin is expressed early in gastrulation very broadly in the ectoderm, but excluded from the dorsal side, and becomes increasingly localized to the ventral side at later stages. This orientation of a Bmp/Chordin axis resembles that of protostomes rather than chordates, where ventral is defined as the side of the body apposed to the substratum and the side on which the mouth opens. Manipulation of Bmp signaling through embryogenesis has a profound effect on the D/V axis in all three germ layers. Knockdown of *bmp2/4* results in ventralization of the body plan. There is an expansion of the mouth territory dorsally to the extent that at later stages the proboscis becomes detached from the rest of the body. Ventral molecular markers expand and dorsal markers are repressed. Following *bmp2/4* over expression, treated embryos exhibit a reciprocal dorsalized phenotype. Interestingly, A/P organization seems largely unaffected without any major changes suggesting that D/V and A/P patterning may be mostly uncoupled in this species. Strikingly, following over-expression of *bmp2/4*, neural fates are not repressed as they are in both arthropods and chordates. The role of a Bmp/Chordin axis in *S. kowalevskii* is largely to provide broad D/V patterning information across all three germ layers, including the ectoderm, similar to other groups. The inversion of the molecular axis between protostomes plus hemichordates and chordates suggests that some sort of axis inversion occurred in stem chordates. There are some clear anatomical patterns related to this axis that seem well conserved; on the Bmp side, there are gill slits, a contractile heart-like dorsal vessel and coelomic mesoderm; on the chordin side there is the majority of the striated muscle. This anatomical relationship to the Bmp/Chordin axis is conserved between hemichordates and chordates and was inherited from a common ancestor (Lowe et al., 2015). However, there are some striking differences in two of the main morphological features polarized along the D/V axis. First, the functional data clearly demonstrated that the hemichordate mouth is sensitive to bmp levels—it expanded in the absence of Bmp and was repressed in excess Bmp. This is in contrast to chordates where the mouth forms in the Bmp territory and suggests a novel relationship of the chordate mouth to Bmp signaling. Second, the nervous system shows a very different relationship to the Bmp/Chordin axis. *Bmp2/4* in *S. kowalevskii* is expressed along the dorsal midline, coincident with the forming dorsal cord Thus, if either of the neural cords is homologous to the chordate dorsal cord, then an inversion scenario would suggest that this would be the ventral cord rather that the dorsal cord. However, as previously discussed, we do not see strong molecular support for homology of either cord. In addition, manipulation of the levels of Bmp in *S. kowalevskii* seemed to play more of a patterning role in the ectoderm rather than a role in segregation of a discrete neurectoderm. Clearly this critical component of bilaterian D/V patterning can support a range of neural architectures while maintaining a primary role in D/V patterning. D/V axis inversion can therefore be considered independent from issues of nervous system centralization and the complexities of making direct comparisons between hemichordate and chordate nervous systems.

Data from indirect-developing species share some patterning similarities, but also some distinctive differences with *S. kowalevskii* (Su et al., 2019). First, it is important to note that embryogenesis gives rise to a larval body plan rather than an adult body plan, so the morphological output of the embryonic patterning is very distinct between indirect and direct developers. Nothing is known about how the adult nervous system is patterned during metamorphosis, as the larval nervous system is replaced by the adult nervous system. The larval nervous system is organized with an apical territory and a more posterior territory tracking the ciliary bands (Dautov & Nezlin, 1992; Gonzalez et al., 2018; Miyamoto et al., 2010; Nakajima, Humphreys, Kaneko, & Tagawa, 2004). Manipulating Bmp signaling during early embryogenesis has a similar patterning effect on the D/V as was observed in *S. kowalevskii*; Bmp protein treatments dorsalized embryos, and Bmp inhibition with an inhibitor resulted in ventralization. Several of the same markers were used in this study, and the response was remarkably similar between these contrasting life histories. The response of the mouth to Bmp manipulations was also the same; Bmp inhibition resulted in a dorsal expansion of the mouth, and Bmp protein treatments blocked mouth formation. However, unlike in *S. kowalevskii*, the authors argue that Bmp represses neural fates; the apical neural territory that gives rise to the ciliated apical organ and the larval eye spots is lost when Bmp is over expressed, as is the more posterior ciliary band territory and associated neurons. The authors argue that their data supports a conserved role of BMP in neural suppression, in line with other bilaterian model species. However, transitory treatments with Bmp protein through gastrulation results in the repositioning of the ciliary band territory more ventrally rather than suppression of neural fates altogether, raising the possibility that loss of neural territories is perhaps due to a patterning effect. Nevertheless, in both cases Bmp activity during embryogenesis is on the dorsal and not the ventral side. Data from both life history strategies strongly support a clear molecular inversion between hemichordates and chordates in D/V organization, yet the large differences in D/V composition of the two groups does not allow for a clear step by step reconstruction of how this inversion may have occured.

13. Endomesoderm specification

The specification of endomesoderm and the formation of a separate endoderm and mesoderm has been studied quite extensively in deuterostome taxa. The canonical Wnt pathway again emerges with a critical and conserved developmental role in the specification of the animal/vegetal axis. In this case, β-catenin plays a key role in the specification of the animal vegetal axis and in endomesoderm induction in a wide variety of animal taxa, from cnidarians (Wikramanayake et al., 2003), to nemerteans (Henry, Perry, Wever, Seaver, & Martindale, 2008), echinoderms (Logan, Miller, Ferkowicz, & McClay, 1999; Miyawaki et al., 2003; Wikramanayake, Huang, & Klein, 1998) and chordates (Imai, Takada, Satoh, & Satou, 2000). This role is proposed to have deep ancestry in metazoan development (Wikramanayake et al., 2003). Work in *S. kowalevskii* has furthered supported this hypothesis and established the importance of β-catenin in endomesoderm specification (Darras, Gerhart, Terasaki, Kirschner, & Lowe, 2011). Embryological experiments in the 1950s on *S. kowalevskii* (Colwin & Colwin, 1950, 1953) established the timing and importance of specification of the endomesoderm in the vegetal pole for normal embryonic patterning. The presence of the vegetal pole is critical for the correct patterning of the overlying animal territory that gives rise of the embryonic ectoderm. Removal of the vegetal pole at the 4 or 8 cell stage, and culture of animal explants results in the failure of any posterior fates to form, and the explants take on apical fates, with long stiff cilia typical of the apical tuft (Darras et al., 2011). This also indicates that the endomesoderm is a source of early posteriorizing factors starting at blastula stage. Based on data from indirect-developing echinoderms (Logan et al., 1999; Wikramanayake et al., 1998) and direct-developing hemichordates, the role of β-catenin in endomesoderm specification can be confidently reconstructed as an ancestral feature of deuterostomes.

14. Mesoderm specification

The mesoderm of adult enteropneusts is divided into three major territories; the most anterior territory, called the protocoel, forms all the muscle and mesoderm of the proboscis/prosome and forms as a single anterior coelom. The mesoderm of the collar/mesosome and trunk/metasome derive from two pairs of coeloms called the mesocoels and metacoels, respectively

(Hyman, 1955) (Fig. 2B). This tripartite coelomic organization is shared with other invertebrate deuterostomes including cephalochordates and echinoderms (Bateson, 1884; Conklin, 1932; Remane, 1963), and a tripartite coelomic arrangement is proposed to be ancestral for deuterostomes. In both indirect and direct-developing species, the protocoel forms at the end of gastrulation from the tip of the archenteron. However, the timing of mesocoel and metacoel formation is different between life history strategies (Fig. 2). In *S. kowalevskii*, these two pairs of posterior coeloms form immediately following gastrulation by enterocoely, as the embryo begins to elongate (Bateson, 1884; Green, Norris, Terasaki, & Lowe, 2013), whereas their formation in indirect-developing species occurs long after embryogenesis, very late in larval development, close to the onset of metamorphosis (Gonzalez et al., 2018; Lin et al., 2016; Morgan, 1891, 1894; Urata & Yamaguchi, 2004).

A molecular perspective into the evolution of mesoderm in deuterostomes is becoming much clearer. Work in a variety of vertebrate model species has determined the importance of both Nodal and FGF signaling in this process (Kimelman, 2006), with FGF signaling being important both for induction and maintenance of mesodermal fate (Fletcher, Baker, & Harland, 2006; Schulte-Merker & Smith, 1995). In amphioxus, FGF signaling has been demonstrated to play an important role in the formation of primary somites (Bertrand et al., 2011). In ascidians it is involved in the specification of the notochord, mesenchyme, and secondary muscle development (Darras & Nishida, 2001; Kim & Nishida, 2001; Yasuo & Hudson, 2007). Outside of chordates, in echinoderms, FGF signaling has been implicated in myogenesis and skeletal morphogenesis (Andrikou, Pai, Su, & Arnone, 2015), so it is clear that FGF is a promising ligand for understanding the evolution of deuterostome mesoderm.

Like vertebrates, hemichordate mesoderm is formed, not by the inheritance of a molecular determinant, but through induction by extrinsic factors. The nature of the molecular signals for mesoderm specification in enteropneusts has now been established in two species with contrasting developmental modes. This was first characterized in *S. kowalevskii* (Green et al., *2013)*. Fgf8/17/18, the ancestral ligand that gave rise to FGF 8, 17 and 18 in vertebrates, is expressed in the ectoderm overlying the regions of the endomesoderm that form the coeloms, in the right spatiotemporal pattern to be a candidate for an inducing signal. In addition, one of the two FGF receptors, FGFRB, is expressed in the archenteron in the regions where mesoderm is forming. Both ligand and receptor localization suggested that the ectoderm plays a role in signaling to the underlying endomesoderm to trigger the formation of mesoderm. Functional analysis by siRNA knockdown or RNA over expression supported this: knockdown of either the *fgf8/17/18* ligand or the receptor *fgfRB* resulted in the loss of mesodermal compartments. Over expression of *fgf8/17/18* had the opposite effect, with an over proliferation of muscle and loss of endoderm. Embryological explant experiments combined with siRNA knockdown established that the formation of mesoderm from endomesoderm requires direct contact from the ectoderm, the expression of ligand in the ectoderm, and FGFRB in the endomesoderm.

Studies in *P. flava* also establishes the importance of FGF signaling the early formation of the larval anterior mesoderm and muscle differentiation, with results very similar to those described in *S. kowalevskii* using FGF signaling inhibitors (Fan, Ting, Yu, & Su, 2018). Posterior coelomic compartments form late in larval development, close to metamorphosis, and although the details of the molecular determinants of their development are unknown, they may also be under the regulation of FGF. The formation of the adult body plan through metamorphosis of *P. flava* is accompanied by a burst of muscle differentiation, and Fan et al. (2018) report that FGF signaling may play a role in this transition. The process of metamorphosis is thought to be partly triggered by the settlement of larvae onto sand. Autoclaved sand is less effective at triggering metamorphosis than untreated sand suggesting that some bioactive material in the sand is also a critical component (Lin et al., 2016). The addition of bFGF protein resulted in modest increase in the rate of metamorphosis but only in the presence of sand; bFGF alone was not effective at promoting the transition to adult body plan. Inhibition of metamorphosis by the use of FGF signaling inhibitors showed variable results depending on the inhibitor used, so a definitive role of FGF remains to be further tested. However, very little is known about the process of metamorphosis and the complex integration of environmental and molecular cues, so this is an exciting new set of observations.

A separate study in the role of Nodal-like signaling in *P. flava* has argued for a role of Nodal signaling in mesoderm induction (Rottinger et al., 2015). There is a single Nodal gene and it has two expression domains; in the vegetal pole at the boundary of ectoderm and endomesoderm at early blastula, and in the ventral ectoderm following gastrulation. Early embryonic treatment with a Nodal-like signaling inhibitor resulted in the loss of anterior mesoderm (Rottinger et al., 2015). Further experiments are required to investigate this observation to determine whether this is a direct or indirect effect, and comparisons in *S. kowalevskii* will also be important to determine whether this is a conserved feature of enteropneusts or unique to indirect developers.

In summary, data from a wide range of species now suggests that the role of β-catenin and FGF signaling in the specification of endomesoderm and mesoderm are largely conserved in deuterostomes.

15. Life history considerations

In the grand hypotheses of chordate origins, the significance of life history has loomed large (Gee, 1996). Many hypotheses have focused on how modifications to an ancestral adult body plan gave rise to chordates, whereas probably the best known of classical chordate origins hypotheses have derived chordates by transformation of an ancestral larval body plan (Gee, 1996). One of the most influential figures in chordate origins is Walter Garstang who developed the Auricularian hypothesis, which was refined over a series of publications starting in 1894 through 1928 (Garstang, 1894, 1928). This hypothesis argued that the earliest chordates were simple and resembled a tunicate tadpole that evolved by transformation of the larval body plan of an ancestral deuterostome animal. This ancestral animal would have had a sessile adult and a larva that resembled the dipleurula larva of echinoderms and hemichordates. There are many elements to this elegant hypothesis that were refined over a series of papers (Holland, 2011), but the most significant element of the hypothesis was the derivation of the chordate central nervous system from a dorsal migration and fusion of the larval ciliary bands, and eventual transformation into the CNS of a chordate tadpole. This hypothesis was later championed and modified by Romer and Berrill (Berrill, 1955; Romer, 1972), and an inversion scenario was incorporated to fit with contemporary developmental data (Nielsen, 1999).

One major challenge to the hypothesis came from molecular phylogenetics: a key aspect of the hypotheses was that the tunicates represent the most basal branch of chordates. We now know that tunicates are in fact the sister group to vertebrates, and cephalochordates are the most basally branching chordate group (Delsuc, Brinkmann, Chourrout, & Philippe, 2006), making some key elements of the hypothesis untenable. Observations from paleontology have also challenged the hypothesis: a key component of the Auricularian hypothesis was the presence of a sessile deuterostome ancestral adult, which in later iterations by Berrill proposed to be similar to hemichordate pterobranchs. The latest fossil data suggest that the ancestral body plan of hemichordates was likely to be motile, combining elements of both enteropneust and pterobranch body plans (Nanglu, Caron, & Cameron, 2020; Nanglu, Caron, Conway Morris, & Cameron, 2016), weakening one of the elements of the hypothesis. Similar conclusions have been reached from the synthesis of molecular genetics and phylogenetics (Brown et al., 2008). Molecular genetic data have generally not been very supportive of the idea of larval origins of the chordate body plan. First, work from several species (as previously described), has revealed that it is the adult body plan with the closest patterning similarities to the chordate body plan along the A/P axis, not the larva. Patterning data from ambulacrarian larvae reveal that the larval body plan represents a truncated A/P axis, and essentially a "swimming head" from a patterning perspective (Gonzalez et al., 2017; Lacalli, 2005). A larval origins scenario would require a wholesale heterochronic shift of most of the adult patterning program earlier into the larval body plan. Based on current data, it seems more plausible that the chordate body plan evolved from an adult rather than larval life history stage. However, recent work in *P. flava* (Su et al., 2019) has revived this idea based on a modified version of the Garstang/Romer hypothesis as proposed by Nielsen (Nielsen, 1999), with ciliary bands of a dipleurula-like ancestor converging on the ventral side of the larva before reorienting to the dorsal side through a D/V axis inversion. As described in the previous section on D/V patterning, the authors experimentally manipulated the levels of Bmp signaling in both enteropneust and echinoid larvae. These data demonstrate that the position of the ciliary band responds to the levels of Bmp, and transitory Bmp treatments result in ventral relocation of the band and the formation of a ventral territory of neurogenic ectoderm. They also show that genes involved in mediolateral patterning of the vertebrate neural plate also line up in a similar fashion in the ectoderm of ambulacrarian larvae and respond similarly to Bmp signaling. Based on these similarities, they argue that tinkering with Bmp signaling in an ancestral larval body plan may have resulted in the initiation of a Garstang-like transformation of an ancestral larval body plan, which could have been the first phase in the path toward a chordate body plan. However, the discussion of the data fails to consider the evidence in favor of the alternative scenario, adult origins of the chordate body plan. Haag (2005) lays out similar problems of larval origins from the perspective of echinoderm body plans, and argues that it is the adult rather than the larva that merits close comparisons with chordates. Currently, when considering the entirety of the patterning data from enteropneusts, it is strongly supportive of adult origins rather than larval origins of the chordate body.

The comparisons of embryogenesis across phyla rarely consider the implications of comparisons of non-equivalent life history stages, but this could clearly have a significant impact on evolutionary comparisons. Where larval and adult body plans have significant differences in their basic organization, we may expect marked differences between the early embryogenesis of indirect and direct developers. For example, the nervous system organization of larval and adult enteropneusts is completely different. Most of the larval nervous system is lost at metamorphosis, and the formation and patterning of the adult nervous system occurs *de novo*, late in larval development and through metamorphosis (Gonzalez et al., 2018; Miyamoto et al., 2010). We have little understanding of the morphogenesis of adult structures by remodeling of an existing larval body plan, but the developmental programs responsible for adult nerve cord morphogenesis are likely to be distinct from those involved in the morphogenesis of the larval nervous system in the apical organ and ciliary bands. In another example, work from *S. kowalevskii* demonstrates that the early action of Wnt signaling through embryogenesis results in the activation of anterior and medial Hox genes during gastrulation that are involved in regulating trunk morphogenesis (Darras et al., 2018). However, gene expression analysis in the indirect developing species *S. californicum* clearly shows no expression of the orthologous Hox genes until closer to metamorphosis as the adult body plan is beginning to form, indicating a lack of trunk in larvae (Gonzalez et al., 2017). The difference in Hox activity between life history strategies may be a result of differences in Wnt activity during embryogenesis, and we should certainly expect to find significant differences in embryonic programs between indorevt and direct developers related to such striking patterning differences. These examples demonstrate that when species are chosen to represent developmental mechanisms of a clade of animals, it is important to consider life history strategy as a significant source of within phylum developmental variation.

16. Summary and future directions

The past 20 years has seen much progress in hemichordate biology. There has been progress on many fronts; from molecular phylogenetics, genomics, developmental biology to paleontology and anatomy. A synthesis of insights from these disparate fields allows a tentative glimpse into the early evolution of deuterostomes. The ability to reconstruct detailed morphological characters remains challenging, as the large organizational and morphological disparity between hemichordates and chordates that was daunting for comparative morphologists remains a major issue, even with the new types of data generated over the last two decades. However, there is building consensus that early deuterostomes are likely to have been benthic worms that used a set of pharyngeal gills for feeding. Molecular, genomic and paleontological data have strongly supported the homology of pharyngeal gill slits (Cannon et al., 2014; Caron, Morris, & Cameron, 2013; Gillis et al., 2011; Nanglu et al., 2016, 2020; Ogasawara et al., 1999; Rychel et al., 2006; Simakov et al., 2015). The details of the morphological characters of this benthic worm are still difficult to reconstruct. However, it is clear from the strong axial conservation between hemichordates and chordates that this worm had a well patterned A/P axis (Aronowicz & Lowe, 2006; Darras et al., 2018; Kaul-Strehlow et al., 2017; Lowe et al., 2003; Pani et al., 2012), defined by an anterior ectodermal region that likely utilized a conserved transcription factor gene regulatory network to partition the entire ectoderm into major divisions of the body plan. The entire extent of the A/P axis would have been regulated by a cWnt gradient, with a well-defined "head" territory negatively regulated by cWnt, and trunk fates promoted by cWnt, and a posteriorly-situated Wnt/brachyury loop involved in trunk elongation. The loose coupling of A/P patterning to any specific neural architecture means that the details of the nervous system are difficult to reconstruct, but it is likely that there was some degree of centralized processing (Nomaksteinsky et al., 2009). However, given the likely feeding mode of mucociliary suspension/particle feeding, this ancestral nervous system was unlikely complex. The D/V axis, like all bilaterians, was patterned by a Bmp/Chordin axis; Bmp expressed dorsally and chordin on the ventral side. Some sort of axis inversion would have occurred in stem chordates. Gill slits, a dorsal heart-like contractile dorsal vessel and coelomic mesoderm would have been associated with the Bmp side, and striated muscle more associated with the ventral side (Lowe et al., 2006). The mouth was ventral and negatively regulated by Bmp. The relationship of the nervous system to the Bmp/chordin axis also suggests that this patterning mechanism can support a range of neural conformations over macroevolutionary time periods. So again, reconstruction of ancestral deuterostome neural characters along this axis is challenging. The similarities of early embryogenesis between hemichordates, cephalochordates and echinoderms suggest that we can confidently reconstruct several aspects of the early specification of germ layers. β-catenin is central in patterning the animal vegetal axis and the early specification of the endomesoderm (Darras et al., 2011; Wikramanayake et al., 1998, 2003), and mesoderm would have formed by enterocoely and was regulated at least in part by FGF signaling (Andrikou et al., 2015; Bertrand et al., 2011; Fan et al., 2018; Green et al., 2013).

Future work in hemichordates must carry out some basic characterizations of pterobranch development and develop some genomic resources. There are few contemporary studies on these animals and they are necessary to provide a broader understanding of the development of the clade (Sato et al., 2008; Sato & Holland, 2008; Stach, Gruhl, & Kaul-Strehlow, 2012). Recent work on regeneration in hemichordates demonstrates their enormous potential for providing insights into the limited regenerative capacities of vertebrates. Many hemichordate species are capable of anterior regeneration as well as posterior regeneration of the trunk and tail, rapidly regrowing the anterior region of the body plan that shares so many patterning similarities with vertebrate brains (Arimoto & Tagawa, 2015, 2018; Luttrell, Gotting, Ross, Alvarado, & Swalla, 2016; Rychel & Swalla, 2009; Yoshimura, Morino, & Wada, 2019). A molecular characterization of these processes will provide some valuable comparative information to deuterostome regenerative biology. The proliferation of available hemichordates genomes will make a genomics approach to understanding hemichordate biology increasingly important. Improvements in genome sequencing and assembly are facilitating the rapid production of high quality, chromosomal level assemblies. A high-quality assembly of the genome of *S. californicum* is now available, and improvements to the original *P. flava* genome are in the works. Single cell technologies have enormous potential to provide insights into cell type characterizations in the nervous system and provide a basis for more detailed comparisons with other bilaterians. Several groups have ongoing projects using this technology. Better genomes are now facilitating improved functional approaches to understanding the biology of hemichordates. So far, these have been limited to *S. kowalevskii*, but have been technically challenging in indirect-developing species. Therefore, generating new experimental approaches by adopting and modifying emerging technologies from model species will be critical to allow more mechanistic insights into the development of hemichordates and the origins of vertebrates.

Acknowledgments

I would like to thank Laurent Formery, John Gerhart and Paul Bump for providing feedback on earlier drafts of the manuscript, and the diverse range of colleagues and collaborators that have contributed to this body of work in hemichordate developmental biology. This work was supported by NSF grant IOS 1656628.

References

Abouheif, E., Akam, M., Dickinson, W. J., Holland, P. W., Meyer, A., Patel, N. H., et al. (1997). Homology and developmental genes. *Trends in Genetics*, *13*, 432–433.

Albuixech-Crespo, B., Lopez-Blanch, L., Burguera, D., Maeso, I., Sanchez-Arrones, L., Moreno-Bravo, J. A., et al. (2017). Molecular regionalization of the developing amphioxus neural tube challenges major partitions of the vertebrate brain. *PLoS Biology*, *15*, e2001573.

Andrikou, C., Pai, C. Y., Su, Y. H., & Arnone, M. I. (2015). Logics and properties of a genetic regulatory program that drives embryonic muscle development in an echinoderm. *eLife*, *4*.

Angelini, D. R., & Kaufman, T. C. (2005). Functional analyses in the milkweed bug Oncopeltus fasciatus (Hemiptera) support a role for Wnt signaling in body segmentation but not appendage development. *Developmental Biology*, *283*, 409–423.

Annona, G., Holland, N. D., & D'Aniello, S. (2015). Evolution of the notochord. *EvoDevo*, *6*, 30.

Arendt, D. (2018). Animal evolution: Convergent nerve cords? *Current Biology: CB*, *28*, R225–R227.

Arendt, D., & Nubler-Jung, K. (1994). Inversion of dorsoventral axis? *Nature*, *371*, 26.

Arendt, D., & Nubler-Jung, K. (1996). Common ground plans in early brain development in mice and flies. *BioEssays*, *18*, 255–259.

Arimoto, A., & Tagawa, K. (2015). Hedgehog expression during development and regeneration in the hemichordate, Ptychodera flava. *Zoological Science*, *32*, 33–37.

Arimoto, A., & Tagawa, K. (2018). Regeneration in the enteropneust hemichordate, Ptychodera flava, and its evolutionary implications. *Development, Growth & Differentiation*, *60*, 400–408.

Arnold, S. J., & Robertson, E. J. (2009). Making a commitment: Cell lineage allocation and axis patterning in the early mouse embryo. *Nature Reviews. Molecular Cell Biology*, *10*, 91–103.

Aronowicz, J., & Lowe, C. J. (2006). Hox gene expression in the hemichordate Saccoglossus kowalevskii and the evolution of deuterostome nervous systems. *Integrative and Comparative Biology*, *46*, 890–901.

Balser, E. J., & Ruppert, E. E. (1990). Structure, ultrastructure, and function of the Preoral heart-kidney in Saccoglossus kowalevskii (Hemichordata, Enteropneusta) including new data on the Stomochord. *Acta Zoologica*, *71*, 235–249.

Bateson, W. (1884). Early stages in the development of Balanoglossus (sp. incert.). *Quarterly Journal of Microscopical Science*, *24*, 208–236.

Bateson, W. (1885). Later stages in the development of Balanoglossus Kowalevskii with a suggestion as to the affinities of the Enteropneusta. *Quarterly Journal of Microscopical Science*, *25*, 81–128.

Bateson, W. (1886). The ancestry of the chordata. *Quarterly Journal of Microscopical Science*, *26*, 535–571.

Baughman, K. W., McDougall, C., Cummins, S. F., Hall, M., Degnan, B. M., Satoh, N., et al. (2014). Genomic organization of Hox and ParaHox clusters in the echinoderm, Acanthaster planci. *Genesis*, *52*, 952–958.

Berrill, N. J. (1955). *The origin of the vertebrates*. Oxford: Clarendon Press.

Bertrand, S., Camasses, A., Somorjai, I., Belgacem, M. R., Chabrol, O., Escande, M. L., et al. (2011). Amphioxus FGF signaling predicts the acquisition of vertebrate morphological traits. *Proceedings of the National Academy of Sciences of the United States of America*, *108*, 9160–9165.

Bourlat, S. J., Juliusdottir, T., Lowe, C. J., Freeman, R., Aronowicz, J., Kirschner, M., et al. (2006). Deuterostome phylogeny reveals monophyletic chordates and the new phylum Xenoturbellida. *Nature*, *444*, 85–88.

Bromham, L. D., & Degnan, B. M. (1999). Hemichordates and deuterostome evolution: Robust molecular phylogenetic support for a hemichordate + echinoderm clade. *Evolution & Development, 1*, 166–171.
Brown, F. D., Prendergast, A., & Swalla, B. J. (2008). Man is but a worm: Chordate origins. *Genesis, 46*, 605–613.
Brusca, R. C., Moore, W., & Shuster, S. M. (2016). *Invertebrates* (3rd edition). Sunderland, Massachusetts U.S.A: Sinauer Associates, Inc.
Bullock, T. H. (1940). The functional organisation of the nervous system of the Enteropneusta. *Biological Bulletin, 79*, 91.
Bullock, T. H. (1945). The anatomical organization of the nervous system of enteropneusta. *The Quarterly Journal of Microscopical Science, 86*, 55–112.
Cameron, C. B. (2002). Particle retention and flow in the pharynx of the enteropneust worm Harrimania planktophilus: The filter-feeding pharynx may have evolved before the chordates. *The Biological Bulletin, 202*, 192–200.
Cameron, C. B., Garey, J. R., & Swalla, B. J. (2000). Evolution of the chordate body plan: New insights from phylogenetic analyses of deuterostome phyla. *Proceedings of the National Academy of Sciences of the United States of America, 97*, 4469–4474.
Cameron, C. B., & Mackie, G. O. (1996). Conduction pathways in the nervous system of Saccoglossus sp. (Enteropneusta). *Canadian Journal of Zoology, 74*, 15–19.
Cannon, J. T., Kocot, K. M., Waits, D. S., Weese, D. A., Swalla, B. J., Santos, S. R., et al. (2014). Phylogenomic resolution of the hemichordate and echinoderm clade. *Current Biology: CB, 24*, 2827–2832.
Cannon, J. T., Rychel, A. L., Eccleston, H., Halanych, K. M., & Swalla, B. J. (2009). Molecular phylogeny of hemichordata, with updated status of deep-sea enteropneusts. *Molecular Phylogenetics and Evolution, 52*, 17–24.
Caron, J. B., Morris, S. C., & Cameron, C. B. (2013). Tubicolous enteropneusts from the Cambrian period. *Nature, 495*, 503–506.
Colwin, A. L., & Colwin, L. H. (1950). The developmental capacities of separated early blastomeres of an enteropneust, Saccoglossus kowalevskii. *Journal of Experimental Zoology, 115*, 263–295.
Colwin, A. L., & Colwin, L. H. (1953). The normal embryology of Saccoglossus kowalevskii. *Journal of Morphology, 92*, 401–453.
Conklin, E. G. (1932). The embryology of amphioxus. *Journal of Morphology, 54*, 69–151.
Cunningham, D., & Casey, E. S. (2014). Spatiotemporal development of the embryonic nervous system of Saccoglossus kowalevskii. *Developmental Biology, 386*, 252–263.
Damante, G., Tell, G., & Di Lauro, R. (2001). A unique combination of transcription factors controls differentiation of thyroid cells. *Progress in Nucleic Acid Research and Molecular Biology, 66*, 307–356.
Darras, S., Fritzenwanker, J. H., Uhlinger, K. R., Farrelly, E., Pani, A. M., Hurley, I. A., et al. (2018). Anteroposterior axis patterning by early canonical Wnt signaling during hemichordate development. *PLoS Biology, 16*, e2003698.
Darras, S., Gerhart, J., Terasaki, M., Kirschner, M., & Lowe, C. J. (2011). Beta-catenin specifies the endomesoderm and defines the posterior organizer of the hemichordate Saccoglossus kowalevskii. *Development, 138*, 959–970.
Darras, S., & Nishida, H. (2001). The BMP signaling pathway is required together with the FGF pathway for notochord induction in the ascidian embryo. *Development, 128*, 2629–2638.
Dautov, S. S., & Nezlin, L. P. (1992). Nervous system of the Tornaria larva (Hemichordata: Enteropneusta). A histochemical and ultrastructural study. *The Biological Bulletin, 183*, 463–475.
Dawydoff, C. (1948). Classe des Entéropneustes. In P. P. Grassé (Ed.), *Traité de Zoologie* (pp. 369–453). Paris: Libraries de l'Academie de Médecine.

De Robertis, E. M. (2008). Evo-devo: Variations on ancestral themes. *Cell, 132*, 185–195.
De Robertis, E. M., & Sasai, Y. (1996). A common plan for dorsoventral patterning in Bilateria. *Nature, 380*, 37–40.
Delsuc, F., Brinkmann, H., Chourrout, D., & Philippe, H. (2006). Tunicates and not cephalochordates are the closest living relatives of vertebrates. *Nature, 439*, 965–968.
Denes, A. S., Jekely, G., Steinmetz, P. R., Raible, F., Snyman, H., Prud'homme, B., et al. (2007). Molecular architecture of annelid nerve cord supports common origin of nervous system centralization in bilateria. *Cell, 129*, 277–288.
Dilly, P. N., Welsch, U., & Storch, V. (1970). The structure of the nerve fiber layer and neurocord in the enteropneusts. *Zeitschrift für Zellforschung und Mikroskopische Anatomie, 103*, 129–148.
Dominguez, P., Jacobson, A. G., & Jefferies, R. P. (2002). Paired gill slits in a fossil with a calcite skeleton. *Nature, 417*, 841–844.
Dunn, C. W., Hejnol, A., Matus, D. Q., Pang, K., Browne, W. E., Smith, S. A., et al. (2008). Broad phylogenomic sampling improves resolution of the animal tree of life. *Nature, 452*, 745–749.
Echelard, Y., Epstein, D. J., St-Jacques, B., Shen, L., Mohler, J., McMahon, J. A., et al. (1993). Sonic hedgehog, a member of a family of putative signaling molecules, is implicated in the regulation of CNS polarity. *Cell, 75*, 1417–1430.
Echevarria, D., Vieira, C., Gimeno, L., & Martinez, S. (2003). Neuroepithelial secondary organizers and cell fate specification in the developing brain. *Brain Research. Brain Research Reviews, 43*, 179–191.
Epstein, D. J., McMahon, A. P., & Joyner, A. L. (1999). Regionalization of sonic hedgehog transcription along the anteroposterior axis of the mouse central nervous system is regulated by Hnf3-dependent and -independent mechanisms. *Development, 126*, 281–292.
Fan, T. P., Ting, H. C., Yu, J. K., & Su, Y. H. (2018). Reiterative use of FGF signaling in mesoderm development during embryogenesis and metamorphosis in the hemichordate Ptychodera flava. *BMC Evolutionary Biology, 18*, 120.
Feinberg, S., Roure, A., Piron, J., & Darras, S. (2019). Antero-posterior ectoderm patterning by canonical Wnt signaling during ascidian development. *PLoS Genetics, 15*, e1008054.
Fisher, S., Grice, E. A., Vinton, R. M., Bessling, S. L., & McCallion, A. S. (2006). Conservation of RET regulatory function from human to zebrafish without sequence similarity. *Science, 312*, 276–279.
Fletcher, R. B., Baker, J. C., & Harland, R. M. (2006). FGF8 spliceforms mediate early mesoderm and posterior neural tissue formation in Xenopus. *Development, 133*, 1703–1714.
Formery, L., Schubert, M., & Croce, J. C. (2019). Ambulacrarians and the ancestry of deuterostome nervous systems. *Results and Problems in Cell Differentiation, 68*, 31–59.
Fritzenwanker, J. H., Gerhart, J., Freeman, R. M., Jr., & Lowe, C. J. (2014). The fox/Forkhead transcription factor family of the hemichordate Saccoglossus kowalevskii. *EvoDevo, 5*, 17.
Fritzenwanker, J. H., Uhlinger, K. R., Gerhart, J., Silva, E., & Lowe, C. J. (2019). Untangling posterior growth and segmentation by analyzing mechanisms of axis elongation in hemichordates. *Proceedings of the National Academy of Sciences of the United States of America, 116*, 8403–8408.
Fu, J., Posnien, N., Bolognesi, R., Fischer, T. D., Rayl, P., Oberhofer, G., et al. (2012). Asymmetrically expressed axin required for anterior development in Tribolium. *Proceedings of the National Academy of Sciences of the United States of America, 109*, 7782–7786.
Furlong, R. F., & Holland, P. W. (2002). Bayesian phylogenetic analysis supports monophyly of ambulacraria and of cyclostomes. *Zoological Science, 19*, 593–599.
Garstang, W. (1894). Preliminary note on a new theory of the phylogeny of the chordata. *Zoologischer Anzeiger, 22*, 122–125.

Garstang, W. (1928). The morphology of the Tunicata. *The Quarterly Journal of Microscopical Science*, *72*, 51–189.
Gee, H. (1996). *Before the backbone*. London: Chapman & Hall.
Geoffroy-St. Hilaire, E. (1822). Considérations générales sur les vertebrés. *Mémoires du Muséum National d'Histoire Naturelle*, *9*, 89–119.
Gerhart, J. (2001). Evolution of the organizer and the chordate body plan. *The International Journal of Developmental Biology*, *45*, 133–153.
Gillis, J. A., Fritzenwanker, J. H., & Lowe, C. J. (2011). *A stem-deuterostome origin of the vertebrate pharyngeal transcriptional network*. (Proceedings. Biological sciences/The Royal Society).
Glinka, A., Wu, W., Delius, H., Monaghan, A. P., Blumenstock, C., & Niehrs, C. (1998). Dickkopf-1 is a member of a new family of secreted proteins and functions in head induction. *Nature*, *391*, 357–362.
Gonzalez, P., & Cameron, C. B. (2009). The gill slits and pre-oral ciliary organ of Protoglossus (Hemichordata: Enteropneusta) are filter-feeding structures. *Biological Journal of the Linnean Society*, *98*, 898–906.
Gonzalez, P., Jiang, J. Z., & Lowe, C. J. (2018). The development and metamorphosis of the indirect developing acorn worm Schizocardium californicum (Enteropneusta: Spengelidae). *Frontiers in Zoology*, *15*, 26.
Gonzalez, P., Uhlinger, K. R., & Lowe, C. J. (2017). The adult body plan of indirect developing hemichordates develops by adding a Hox-patterned trunk to an anterior larval territory. *Current Biology*, *27*, 87–95.
Green, S. A., Norris, R. P., Terasaki, M., & Lowe, C. J. (2013). FGF signaling induces mesoderm in the hemichordate Saccoglossus kowalevskii. *Development*, *140*, 1024–1033.
Gurley, K. A., Rink, J. C., & Sanchez Alvarado, A. (2008). Beta-catenin defines head versus tail identity during planarian regeneration and homeostasis. *Science*, *319*, 323–327.
Gee, H. (2018). *Across the bridge: Understanding the origin of the vertebrates*. Chicago: The University of Chicago Press.
Haag, E. S. (2005). Echinoderm rudiments, rudimentary bilaterians, and the origin of the chordate CNS. *Evolution & Development*, *7*, 280–281.
Hall, B. K. (2005). Betrayed by Balanoglossus: William Bateson's rejection of evolutionary embryology as the basis for understanding evolution. *Journal of Experimental Zoology. Part B, Molecular and Developmental Evolution*, *304*, 1–17.
Hare, E. E., Peterson, B. K., Iyer, V. N., Meier, R., & Eisen, M. B. (2008). Sepsid even-skipped enhancers are functionally conserved in drosophila despite lack of sequence conservation. *PLoS Genetics*, *4*, e1000106.
Harland, R., & Gerhart, J. (1997). Formation and function of Spemann's organizer. *Annual Review of Cell and Developmental Biology*, *13*, 611–667.
Hejnol, A., & Lowe, C. J. (2015). Embracing the comparative approach: how robust phylogenies and broader developmental sampling impacts the understanding of nervous system evolution. *Philosophical transactions of the Royal Society of London. Series B, Biological sciences*, *370*, 1–16.
Hejnol, A., Obst, M., Stamatakis, A., Ott, M., Rouse, G. W., Edgecombe, G. D., et al. (2009). Assessing the root of bilaterian animals with scalable phylogenomic methods. *Proceedings Biological Sciences/The Royal Society*, *276*, 4261–4270.
Henry, J. Q., Perry, K. J., Wever, J., Seaver, E., & Martindale, M. Q. (2008). Beta-catenin is required for the establishment of vegetal embryonic fates in the nemertean. *Cerebratulus lacteus. Developmental biology*, *317*, 368–379.
Henry, J. Q., Tagawa, K., & Martindale, M. Q. (2001). Deuterostome evolution: Early development in the enteropneust hemichordate, Ptychodera flava. *Evolution & Development*, *3*, 375–390.
Hess, W. N. (1937). The nervous system of Dolichoglossus kowalevskyi. *Journal of Experimental Zoology*, *79*, 1–11.

Hirth, F., Kammermeier, L., Frei, E., Walldorf, U., Noll, M., & Reichert, H. (2003). An urbilaterian origin of the tripartite brain: Developmental genetic insights from drosophila. *Development, 130*, 2365–2373.

Hiruta, J., Mazet, F., Yasui, K., Zhang, P., & Ogasawara, M. (2005). Comparative expression analysis of transcription factor genes in the endostyle of invertebrate chordates. *Developmental Dynamics, 233*, 1031–1037.

Holland, N. D. (2011). Walter Garstang: A retrospective. *Theory in Biosciences, 130*, 247–258.

Holland, L. Z., Carvalho, J. E., Escriva, H., Laudet, V., Schubert, M., Shimeld, S. M., et al. (2013). Evolution of bilaterian central nervous systems: A single origin? *EvoDevo, 4*, 27.

Holley, S. A., & Ferguson, E. L. (1997). Fish are like flies are like frogs: Conservation of dorsal-ventral patterning mechanisms. *BioEssays, 19*, 281–284.

Houart, C., Caneparo, L., Heisenberg, C., Barth, K., Take-Uchi, M., & Wilson, S. (2002). Establishment of the telencephalon during gastrulation by local antagonism of Wnt signaling. *Neuron, 35*, 255–265.

Hyman, L. H. (1955). *The invertebrates* (1st ed.). New York: McGraw-Hill.

Imai, K., Takada, N., Satoh, N., & Satou, Y. (2000). (Beta)-catenin mediates the specification of endoderm cells in ascidian embryos. *Development, 127*, 3009–3020.

Irimia, M., Pineiro, C., Maeso, I., Gomez-Skarmeta, J. L., Casares, F., & Garcia-Fernandez, J. (2010). Conserved developmental expression of Fezf in chordates and drosophila and the origin of the zona Limitans Intrathalamica (ZLI) brain organizer. *EvoDevo, 1*, 7.

Jeong, Y., El-Jaick, K., Roessler, E., Muenke, M., & Epstein, D. J. (2006). A functional screen for sonic hedgehog regulatory elements across a 1 Mb interval identifies long-range ventral forebrain enhancers. *Development, 133*, 761–772.

Jeong, Y., & Epstein, D. J. (2003). Distinct regulators of Shh transcription in the floor plate and notochord indicate separate origins for these tissues in the mouse node. *Development, 130*, 3891–3902.

Jeong, Y., Leskow, F. C., El-Jaick, K., Roessler, E., Muenke, M., Yocum, A., et al. (2008). Regulation of a remote Shh forebrain enhancer by the Six3 homeoprotein. *Nature Genetics, 40*, 1348–1353.

Kapli, P., Natsidis, P., Leite, D. J., Fursman, M., Jeffrie, N., Rahman, I. A., et al. (2020). Lack of support for Deuterostomia prompts reinterpretation of the first Bilateria. *bioRxiv*. 2020.2007.2001.182915.

Kaul, S., & Stach, T. (2010). Ontogeny of the collar cord: Neurulation in the hemichordate Saccoglossus kowalevskii. *Journal of Morphology, 271*, 1240–1259.

Kaul-Strehlow, S., & Röttinger, E. (2015). Hemichordata. In A. Wanninger (Ed.), *Evolutionary developmental biology of invertebrates 6: Deuterostomia* (pp. 59–89). Vienna: Springer Vienna.

Kaul-Strehlow, S., & Stach, T. (2013). A detailed description of the development of the hemichordate Saccoglossus kowalevskii using SEM, TEM, histology and 3D-reconstructions. *Frontiers in Zoology, 10*, 1–31.

Kaul-Strehlow, S., Urata, M., Minokawa, T., Stach, T., & Wanninger, A. (2015). Neurogenesis in directly and indirectly developing enteropneusts: Of nets and cords. *Organisms, Diversity and Evolution, 15*, 405–422.

Kaul-Strehlow, S., Urata, M., Praher, D., & Wanninger, A. (2017). Neuronal patterning of the tubular collar cord is highly conserved among enteropneusts but dissimilar to the chordate neural tube. *Scientific Reports, 7*, 7003.

Kiecker, C., & Lumsden, A. (2012). The role of organizers in patterning the nervous system. *Annual Review of Neuroscience, 35*, 347–367.

Kiecker, C., & Niehrs, C. (2001). A morphogen gradient of Wnt/beta-catenin signalling regulates anteroposterior neural patterning in Xenopus. *Development, 128*, 4189–4201.

Kim, G. J., & Nishida, H. (2001). Role of the FGF and MEK signaling pathway in the ascidian embryo. *Development, Growth & Differentiation, 43*, 521–533.

Kimelman, D. (2006). Mesoderm induction: From caps to chips. *Nature Reviews. Genetics*, *7*, 360–372.
Knight-Jones, E. (1952). On the nervous system of Saccoglossus cambriensis (Enteropneusta). *Philosophical Transactions of the Royal Society of London. Series B, Biological Sciences*, *236*, 315–354.
Kowalevsky, A. (1866). Anatomie des Balanoglossus. *Mem Acad Imp Sci St Petersburg*, *7*, 16.
Lacalli, T. C. (2005). Protochordate body plan and the evolutionary role of larvae: Old controversies resolved? *Canadian Journal of Zoology*, *83*, 216–224.
Leyns, L., Bouwmeester, T., Kim, S. H., Piccolo, S., & De Robertis, E. M. (1997). Frzb-1 is a secreted antagonist of Wnt signaling expressed in the Spemann organizer. *Cell*, *88*, 747–756.
Lin, C. Y., Tung, C. H., Yu, J. K., & Su, Y. H. (2016). Reproductive periodicity, spawning induction, and larval metamorphosis of the hemichordate acorn worm Ptychodera flava. *Journal of Experimental Zoology. Part B, Molecular and Developmental Evolution*, *326*, 47–60.
Logan, C. Y., Miller, J. R., Ferkowicz, M. J., & McClay, D. R. (1999). Nuclear beta-catenin is required to specify vegetal cell fates in the sea urchin embryo. *Development*, *126*, 345–357.
Lowe, C. J., Clarke, D. N., Medeiros, D. M., Rokhsar, D. S., & Gerhart, J. (2015). The deuterostome context of chordate origins. *Nature*, *520*, 456–465.
Lowe, C. J., Tagawa, K., Humphreys, T., Kirschner, M., & Gerhart, J. (2004). Hemichordate embryos: procurement, culture. *and basic methods. Methods Cell Biol*, *74*, 171–194.
Lowe, C. J., Terasaki, M., Wu, M., Freeman, R. M., Jr., Runft, L., Kwan, K., et al. (2006). Dorsoventral patterning in hemichordates: Insights into early chordate evolution. *PLoS Biology*, *4*, e291.
Lowe, C. J., Wu, M., Salic, A., Evans, L., Lander, E., Stange-Thomann, N., et al. (2003). Anteroposterior patterning in hemichordates and the origins of the chordate nervous system. *Cell*, *113*, 853–865.
Luttrell, S. M., Gotting, K., Ross, E., Alvarado, A. S., & Swalla, B. J. (2016). Head regeneration in hemichordates is not a strict recapitulation of development. *Developmental Dynamics*, *245*, 1159–1175.
Luttrell, S., Konikoff, C., Byrne, A., Bengtsson, B., & Swalla, B. J. (2012). Ptychoderid hemichordate neurulation without a notochord. *Integrative and Comparative Biology*, *52*, 829–834.
Marlow, H., Tosches, M. A., Tomer, R., Steinmetz, P. R., Lauri, A., Larsson, T., et al. (2014). Larval body patterning and apical organs are conserved in animal evolution. *BMC Biology*, *12*, 7.
Martin, B. L., & Kimelman, D. (2009). Wnt signaling and the evolution of embryonic posterior development. *Current Biology: CB*, *19*, R215–R219.
Martin-Duran, J. M., Pang, K., Borve, A., Le, H. S., Furu, A., Cannon, J. T., et al. (2018). Convergent evolution of bilaterian nerve cords. *Nature*, *553*, 45–50.
McGrew, L. L., Hoppler, S., & Moon, R. T. (1997). Wnt and FGF pathways cooperatively pattern anteroposterior neural ectoderm in Xenopus. *Mechanisms of Development*, *69*, 105–114.
Metschnikoff, V. E. (1881). Über die systematische Stellung von Balanoglossus. *Zoologischer Anzeiger*, *4*, 139–157.
Minor, P. J., Clarke, D. N., Andrade Lopez, J. M., Fritzenwanker, J. H., Gray, J., & Lowe, C. J. (2019). I-SceI Meganuclease-mediated transgenesis in the acorn worm, Saccoglossus kowalevskii. *Developmental Biology*, *445*, 8–15.
Miyamoto, N., Nakajima, Y., Wada, H., & Saito, Y. (2010). Development of the nervous system in the acorn worm Balanoglossus simodensis: Insights into nervous system evolution. *Evolution & Development*, *12*, 416–424.

Miyamoto, N., & Saito, Y. (2007). Morphology and development of a new species of Balanoglossus (Hemichordata: Enteropneusta: Ptychoderidae) from Shimoda, Japan. *Zoological Science, 24*, 1278–1285.
Miyamoto, N., & Wada, H. (2013). Hemichordate neurulation and the origin of the neural tube. *Nature Communications, 4*, 2713.
Miyawaki, K., Mito, T., Sarashina, I., Zhang, H., Shinmyo, Y., Ohuchi, H., et al. (2004). Involvement of wingless/armadillo signaling in the posterior sequential segmentation in the cricket, Gryllus bimaculatus (Orthoptera), as revealed by RNAi analysis. *Mechanisms of Development, 121*, 119–130.
Miyawaki, K., Yamamoto, M., Saito, K., Saito, S., Kobayashi, N., & Matsuda, S. (2003). Nuclear localization of beta-catenin in vegetal pole cells during early embryogenesis of the starfish Asterina pectinifera. *Development, Growth & Differentiation, 45*, 121–128.
Morgan, T. (1891). The growth and metamorphosis of tornaria. *Journal of Morphology, 5*, 407–458.
Morgan, T. (1894). Development of Balanoglossus. *Journal of Morphology, 9*, 1–86.
Mwinyi, A., Bailly, X., Bourlat, S. J., Jondelius, U., Littlewood, D. T., & Podsiadlowski, L. (2010). The phylogenetic position of Acoela as revealed by the complete mitochondrial genome of Symsagittifera roscoffensis. *BMC Evolutionary Biology, 10*, 309.
Nakajima, Y., Humphreys, T., Kaneko, H., & Tagawa, K. (2004). Development and neural organization of the tornaria larva of the Hawaiian hemichordate, Ptychodera flava. *Zoological Science, 21*, 69–78.
Nanglu, K., Caron, J. B., & Cameron, C. B. (2020). *Cambrian Tentaculate Worms and the origin of the hemichordate body plan.* CB: Current biology.
Nanglu, K., Caron, J. B., Conway Morris, S., & Cameron, C. B. (2016). Cambrian suspension-feeding tubicolous hemichordates. *BMC Biology, 14*, 56.
Niehrs, C. (2010). On growth and form: A Cartesian coordinate system of Wnt and BMP signaling specifies bilaterian body axes. *Development, 137*, 845–857.
Nielsen, C. (1999). Origin of the chordate central nervous system - and the origin of chordates. *Development Genes and Evolution, 209*, 198–205.
Nomaksteinsky, M., Rottinger, E., Dufour, H. D., Chettouh, Z., Lowe, C. J., Martindale, M. Q., et al. (2009). Centralization of the deuterostome nervous system predates chordates. *Current Biology: CB, 19*, 1264–1269.
Nordstrom, U., Jessell, T. M., & Edlund, T. (2002). Progressive induction of caudal neural character by graded Wnt signaling. *Nature Neuroscience, 5*, 525–532.
Ogasawara, M., Wada, H., Peters, H., & Satoh, N. (1999). Developmental expression of Pax1/9 genes in urochordate and hemichordate gills: Insight into function and evolution of the pharyngeal epithelium. *Development, 126*, 2539–2550.
Okai, N., Tagawa, K., Humphreys, T., Satoh, N., & Ogasawara, M. (2000). Characterization of gill-specific genes of the acorn worm Ptychodera flava. *Developmental Dynamics, 217*, 309–319.
Onai, T., Lin, H. C., Schubert, M., Koop, D., Osborne, P. W., Alvarez, S., et al. (2009). Retinoic acid and Wnt/beta-catenin have complementary roles in anterior/posterior patterning embryos of the basal chordate amphioxus. *Developmental Biology, 332*, 223–233.
Osborn, K. J., Kuhnz, L. A., Priede, I. G., Urata, M., Gebruk, A. V., & Holland, N. D. (2012). Diversification of acorn worms (Hemichordata, Enteropneusta) revealed in the deep sea. *Proceedings Biological Sciences/The Royal Society, 279*, 1646–1654.
Pani, A. M., Mullarkey, E. E., Aronowicz, J., Assimacopoulos, S., Grove, E. A., & Lowe, C. J. (2012). Ancient deuterostome origins of vertebrate brain signalling centres. *Nature, 483*, 289–294.

Petersen, C. P., & Reddien, P. W. (2008). Smed-betacatenin-1 is required for anteroposterior blastema polarity in planarian regeneration. *Science*, *319*, 327–330.

Petersen, C. P., & Reddien, P. W. (2009). Wnt signaling and the polarity of the primary body axis. *Cell*, *139*, 1056–1068.

Philippe, H., Brinkmann, H., Copley, R. R., Moroz, L. L., Nakano, H., Poustka, A. J., et al. (2011). Acoelomorph flatworms are deuterostomes related to Xenoturbella. *Nature*, *470*, 255–258.

Philippe, H., Poustka, A. J., Chiodin, M., Hoff, K. J., Dessimoz, C., Tomiczek, B., et al. (2019). Mitigating anticipated effects of systematic errors supports sister-group relationship between Xenacoelomorpha and Ambulacraria. *Current Biology: CB*, *29*(1818–1826), e1816.

Piccolo, S., Agius, E., Leyns, L., Bhattacharyya, S., Grunz, H., Bouwmeester, T., et al. (1999). The head inducer Cerberus is a multifunctional antagonist of nodal. *BMP and Wnt signals*. *Nature*, *397*, 707–710.

Pickens, P. E. (1970). Conduction along the ventral cord of a hemichordate worm. *Journal of Experimental Zoology*, *53*, 515–528.

Polevoy, H., Gutkovich, Y. E., Michaelov, A., Volovik, Y., Elkouby, Y. M., & Frank, D. (2019). New roles for Wnt and BMP signaling in neural anteroposterior patterning. *EMBO Reports*, *20*.

Range, R. C. (2018). Canonical and non-canonical Wnt signaling pathways define the expression domains of frizzled 5/8 and frizzled 1/2/7 along the early anterior-posterior axis in sea urchin embryos. *Developmental Biology*, *444*, 83–92.

Range, R. C., Angerer, R. C., & Angerer, L. M. (2013). Integration of canonical and non-canonical Wnt signaling pathways patterns the neuroectoderm along the anterior-posterior axis of sea urchin embryos. *PLoS Biology*, *11*, e1001467.

Range, R. C., & Wei, Z. (2016). An anterior signaling center patterns and sizes the anterior neuroectoderm of the sea urchin embryo. *Development*, *143*, 1523–1533.

Reichert, H., & Simeone, A. (2001). Developmental genetic evidence for a monophyletic origin of the bilaterian brain. Philosophical transactions of the Royal Society of London. *Series B, Biological sciences*, *356*, 1533–1544.

Remane. (1963). The enterocoelic origin of the coelom. In T. L. Metazoa (Ed.), *E., D.* Berkeley, CA: University of California Press.

Robert, N., Lhomond, G., Schubert, M., & Croce, J. C. (2014). A comprehensive survey of wnt and frizzled expression in the sea urchin Paracentrotus lividus. *Genesis*, *52*, 235–250.

Romer, A. S. (1972). The vertebrate as a dual animal-somatic and visceral. *Evolutionary Biology*, *6*, 121–156.

Rottinger, E., DuBuc, T. Q., Amiel, A. R., & Martindale, M. Q. (2015). Nodal signaling is required for mesodermal and ventral but not for dorsal fates in the indirect developing hemichordate, Ptychodera flava. *Biol Open*, *4*, 830–842.

Rottinger, E., & Lowe, C. J. (2012). Evolutionary crossroads in developmental biology: Hemichordates. *Development*, *139*, 2463–2475.

Rottinger, E., & Martindale, M. Q. (2011). Ventralization of an indirect developing hemichordate by NiCl(2) suggests a conserved mechanism of dorso-ventral (D/V) patterning in Ambulacraria (hemichordates and echinoderms). *Developmental Biology*, *354*, 173–190.

Rouse, G. W., Wilson, N. G., Carvajal, J. I., & Vrijenhoek, R. C. (2016). New deep-sea species of Xenoturbella and the position of Xenacoelomorpha. *Nature*, *530*, 94–97.

Ruppert, E. E. (2005). Key characters uniting hemichordates and chordates: Homologies or homoplasies? *Canadian Journal of Zoology*, *83*, 8–23.

Rychel, A. L., Smith, S. E., Shimamoto, H. T., & Swalla, B. J. (2006). Evolution and development of the chordates: Collagen and pharyngeal cartilage. *Molecular Biology and Evolution*, *23*, 541–549.

Rychel, A. L., & Swalla, B. J. (2007). Development and evolution of chordate cartilage. *Journal of Experimental Zoology. Part B, Molecular and Developmental Evolution, 308*, 325–335.
Rychel, A. L., & Swalla, B. J. (2009). Regeneration in hemichordates and echinoderms. In B. Rinkevich, & V. Matranga (Eds.), *Stem cells in marine organisms* (pp. 245–265). Heidelberg etc.: Springer, Dordrecht.
Sato, A., Bishop, J. D., & Holland, P. W. (2008). Developmental biology of pterobranch hemichordates: History and perspectives. *Genesis, 46*, 587–591.
Sato, A., & Holland, P. W. (2008). Asymmetry in a pterobranch hemichordate and the evolution of left-right patterning. *Developmental Dynamics, 237*, 3634–3639.
Satoh, N., Tagawa, K., Lowe, C. J., Yu, J. K., Kawashima, T., Takahashi, H., et al. (2014). On a possible evolutionary link of the stomochord of hemichordates to pharyngeal organs of chordates. *Genesis*.
Schier, A. F. (2001). Axis formation and patterning in zebrafish. *Current Opinion in Genetics & Development, 11*, 393–404.
Schoenwolf, G. C. (2000). Molecular genetic control of axis patterning during early embryogenesis of vertebrates. *Annals of the New York Academy of Sciences, 919*, 246–260.
Scholpp, S., & Lumsden, A. (2010). Building a bridal chamber: Development of the thalamus. *Trends in Neurosciences, 33*, 373–380.
Schulte-Merker, S., & Smith, J. C. (1995). Mesoderm formation in response to Brachyury requires FGF signalling. *Current Biology: CB, 5*, 62–67.
Shubin, N., Tabin, C., & Carroll, S. (2009). Deep homology and the origins of evolutionary novelty. *Nature, 457*, 818–823.
Silèn, L. (1950). On the nervous system of Glossibalanus marginatus meek (Enteropneusta). *Acta Zoologica, 31*, 149–175.
Simakov, O., Kawashima, T., Marletaz, F., Jenkins, J., Koyanagi, R., Mitros, T., et al. (2015). Hemichordate genomes and deuterostome origins. *Nature, 527*, 459–465.
Slack, J. M., Holland, P. W., & Graham, C. F. (1993). The zootype and the phylotypic stage. *Nature, 361*, 490–492.
Srivastava, M., Mazza-Curll, K. L., van Wolfswinkel, J. C., & Reddien, P. W. (2014). Whole-body acoel regeneration is controlled by Wnt and bmp-Admp signaling. *Current Biology: CB, 24*, 1107–1113.
Stach, T., Gruhl, A., & Kaul-Strehlow, S. (2012). The central and peripheral nervous system of Cephalodiscus gracilis (Pterobranchia, Deuterostomia). *Zoomorphology (Berlin), 131*, 11–24.
Su, Y. H., Chen, Y. C., Ting, H. C., Fan, T. P., Lin, C. Y., Wang, K. T., et al. (2019). BMP controls dorsoventral and neural patterning in indirect-developing hemichordates providing insight into a possible origin of chordates. *Proceedings of the National Academy of Sciences of the United States of America, 116*, 12925–12932.
Satoh, N. (2016). *Chordate origins and evolution: The molecular evolutionary road to vertebrates*. Amsterdam: Academic Press.
Tagawa, K. (2016). Hemichordate models. *Current Opinion in Genetics & Development, 39*, 71–78.
Tassia, M. G., Cannon, J. T., Konikoff, C. E., Shenkar, N., Halanych, K. M., & Swalla, B. J. (2016). The global diversity of Hemichordata. *PLoS One, 11*, e0162564.
Tomer, R., Denes, A. S., Tessmar-Raible, K., & Arendt, D. (2010). Profiling by image registration reveals common origin of annelid mushroom bodies and vertebrate pallium. *Cell, 142*, 800–809.
Urata, M., & Yamaguchi, M. (2004). The development of the enteropneust hemichordate Balanoglossus misakiensis KUWANO. *Zoological Science, 21*, 533–540.

Vergara, H. M., Bertucci, P. Y., Hantz, P., Tosches, M. A., Achim, K., Vopalensky, P., et al. (2017). Whole-organism cellular gene-expression atlas reveals conserved cell types in the ventral nerve cord of Platynereis dumerilii. *Proceedings of the National Academy of Sciences of the United States of America*, *114*, 5878–5885.

Wagner, G. P. (2007). The developmental genetics of homology. *Nature Reviews. Genetics*, *8*, 473–479.

Wikramanayake, A. H., Hong, M., Lee, P. N., Pang, K., Byrum, C. A., Bince, J. M., et al. (2003). An ancient role for nuclear beta-catenin in the evolution of axial polarity and germ layer segregation. *Nature*, *426*, 446–450.

Wikramanayake, A. H., Huang, L., & Klein, W. H. (1998). Beta-catenin is essential for patterning the maternally specified animal-vegetal axis in the sea urchin embryo. *Proceedings of the National Academy of Sciences of the United States of America*, *95*, 9343–9348.

Yankura, K. A., Martik, M. L., Jennings, C. K., & Hinman, V. F. (2010). Uncoupling of complex regulatory patterning during evolution of larval development in echinoderms. *BMC Biology*, *8*, 143.

Yao, Y., Minor, P. J., Zhao, Y. T., Jeong, Y., Pani, A. M., King, A. N., et al. (2016). Cis-regulatory architecture of a brain signaling center predates the origin of chordates. *Nature Genetics*, *48*, 575–580.

Yasuo, H., & Hudson, C. (2007). FGF8/17/18 functions together with FGF9/16/20 during formation of the notochord in Ciona embryos. *Developmental Biology*, *302*, 92–103.

Yoshimura, K., Morino, Y., & Wada, H. (2019). Regeneration of the acorn worm pygochord with the implication for its convergent evolution with the notochord. *Development, Growth & Differentiation*, *61*, 158–165.

Yu, J. K., Holland, L. Z., Jamrich, M., Blitz, I. L., & Hollan, N. D. (2002). AmphiFoxE4, an amphioxus winged helix/forkhead gene encoding a protein closely related to vertebrate thyroid transcription factor-2: Expression during pharyngeal development. *Evolution & Development*, *4*, 9–15.

CHAPTER FOUR

Cephalochordates: A window into vertebrate origins

Linda Z. Holland* and Nicholas D. Holland
Marine Biology Research Division, Scripps Institution of Oceanography, University of California San Diego, La Jolla, CA, United States
*Corresponding author: e-mail address: lzholland@ucsd.edu

Contents

Abstract

How vertebrates evolved from their invertebrate ancestors has long been a central topic of discussion in biology. Evolutionary developmental biology (evodevo) has provided a new tool—using gene expression patterns as phenotypic characters to infer homologies between body parts in distantly related organisms—to address this question. Combined with micro-anatomy and genomics, evodevo has provided convincing evidence that vertebrates evolved from an ancestral invertebrate chordate, in many respects resembling a modern amphioxus. The present review focuses on the role of evodevo in addressing two major questions of chordate evolution: (1) how the vertebrate brain evolved from the much simpler central nervous system (CNS) in of this ancestral chordate and (2) whether or not the head mesoderm of this ancestor was segmented.

Current Topics in Developmental Biology, Volume 141
ISSN 0070-2153
https://doi.org/10.1016/bs.ctdb.2020.07.001

1. Introduction

Cephalochordates (amphioxus, lancelets) were initially thought to be mollusks (Pallas, 1774). Sixty years later, Costa (1834), unaware of Pallas' description, described amphioxus as a fish. Amphioxus continued to be regarded as a primitive or perhaps degenerate fish until the turn of the 20th century when Herdman (1904) authoritatively reclassified it as a cephalochordate. For the next century, amphioxus was considered the closest relative of the vertebrates, and tunicates, because of their simplified body plans, were considered to be the most basal chordates. These relationships were later shown to be consistent with molecular phylogenetic analyses with rDNA sequences (Mallatt & Sullivan, 1998; Wada & Satoh, 1994).

Then, about 15 years ago, Delsuc and colleagues performed phylogenetic analyses with large concatenated sets of nuclear genes (Delsuc, Brinkmann, Chourrout, & Philippe, 2006; Delsuc, Tsagkogeorga, Lartillot, & Philippe, 2008). Surprisingly, these analyses reversed the positions of tunicates and amphioxus, placing tunicates as sister group of the vertebrates and amphioxus basally in the chordates (Fig. 1). In addition, comparisons of genome sequences showed that tunicates are evolving especially rapidly compared to vertebrates and that amphioxus species are evolving more slowly than the slowest-evolving vertebrate known, the elephant shark (Yue et al., 2014). In light of their very different evolutionary rates, it is not surprising that tunicates are quite diverse while the three genera of amphioxus have similar body plans and closely resemble Cambrian chordates such as *Pikaia*, generally thought to be an ancestral cephalochordate, as well as *Myllokunmingia*, a small jawless fish. As noted by Bone (1960) and others, modern amphioxus also bears strong resemblance to ammocoete larvae of lampreys (Fig. 2). The most notable difference between amphioxus and these fossil jawless fish and ammocoete larvae is the presence of paired eyes in the vertebrates, and their lack in amphioxus and *Pikaia*. In addition, amphioxus lacks vertebrae, thus differing markedly from the jawed vertebrates (gnathostomes) as well as from jawless vertebrates (agnathans), which have reduced cartilaginous relics of vertebrae (Janvier, 2011; Ota, 2018; Ota, Fujimoto, Oisi, & Kuratani, 2011). Because modern cephalochordates are simple but vertebrate-like, it is not surprising that they have been at the center of the debate on how the vertebrates evolved from their invertebrate ancestors.

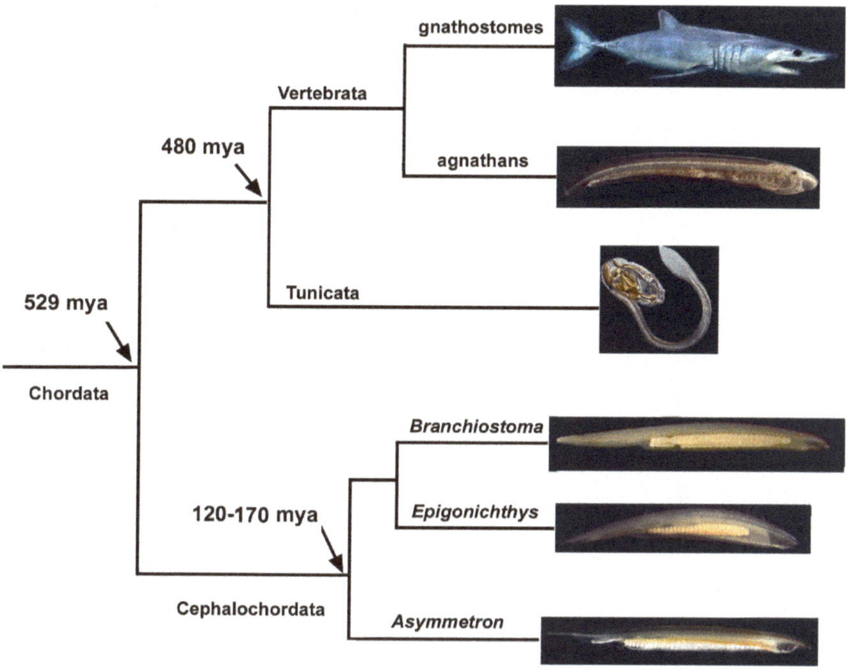

Fig. 1 Phylogenetic tree of the Phylum Chordata. Animals shown are not to scale. All three genera of cephalochordates (amphioxus) are shown with anterior to the right. *Branchiostoma* is the largest genus with about 30 species with the largest individuals (maximum length ~6 cm) and has rows of gonads ventrally on both right and left sides. In contrast, both *Epigonichthys* and *Asymmetron* have a row of gonads only on the right side. They are smaller than *Branchiostoma*; *Asymmetron* has a maximum length of ~2 cm. The tunicate shown is *Oikopleura dioica*, in the Class Appendicularia. The other classes of tunicates are Ascidiacea and Thaliacea. Tunicates are evolving quite rapidly and are very diverse. There are over 3000 species of ascidians. For vertebrates, there are two superclasses: agnathans, the jawless fish and gnathostomes, jawed vertebrates. Divergence times (mya = millions of years ago) from (Delsuc et al., 2018; Kon et al., 2007; Nohara, Nishida, Miya, & Nishikawa, 2005; Subirana, Farstey, Bertrand, & Escriva, 2020; Yue, Yu, Putnam, & Holland, 2014). Photo of *Epigonichthys* from (Igawa et al., 2017). Photograph of the tunicate *Oikopleura dioica* courtesy of Cristian Canestro; photograph of larval agnathan, the lamprey *Petromyzon marinus* courtesy of Jeramiah Smith.

2. The debate on vertebrate origins

By the mid-20th century, it was generally agreed, based on the similarity of the amphioxus and vertebrate body plans with dorsal, hollow nerve cords, notochord, segmental muscles and gill slits, that amphioxus and

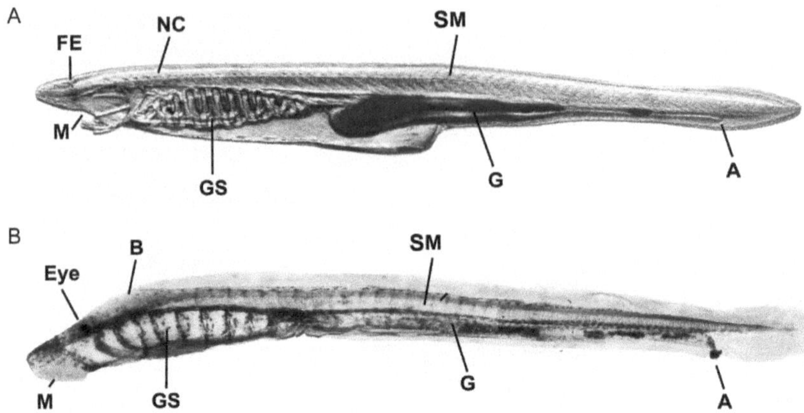

Fig. 2 Comparison of (A) juvenile amphioxus (*Branchiostoma floridae*) with metamorphosis nearly completed and (B) an ammocoete larva of a lamprey (*Lampetra*). Anterior to the left. Major features are indicated. A = anus, B = brain, Eye = lamprey eye, FE = frontal eye (a non-image-forming ciliary photoreceptor), G = gut, GS = gll slits, M = mouth, NC = nerve cord, SM = segmental muscles. These organisms are both about 5 mm long.

vertebrates shared a common ancestor that was probably very amphioxus-like (Romer, 1956). Although homologies of some structures such as the gill slits, notochord and the endostyle in amphioxus and vertebrates have been generally accepted (Dohrn, 1885; Kingsbury, 1920; Kluge, Renault, & Rohr, 2005; Koltzoff, 1902; Mallatt, 1996), possible homologies of other structures have been questioned. For example, Goodrich (1902, 1909) felt that the amphioxus kidneys were most closely related to those of annelids. In contrast, Ruppert (1994) argued "the nephridia of amphioxus are an anatomically, histologically, and cytologically reasonable model for the plesiomorphic design of the vertebrate nephron," while electron microscopy revealed similarity between the amphioxus kidneys (cyrtopodocytes) and the podocytes of larval vertebrate kidneys (Gans, 1993; Holland, 2017).

The most controversial homologies of amphioxus and vertebrates have concerned their central nervous systems (CNS) (Ariëns Kappers, Huber, & Crosby, 1960; Balfour, 1885; Gans & Northcutt, 1983; Willey, 1894) and the evolutionary relationship, if any, between the unsegmented head muscles of vertebrates and the anterior paraxial somites of amphioxus (reviewed in Olsson, Ericsson, & Cerny, 2005). The problem is that while there are some anatomical similarities between the amphioxus and vertebrate nerve cords and between the head muscles in amphioxus and vertebrates, the over 500 million years separating the two lineages has been ample for both divergence and convergence of structures. However, in recent years, comparisons of extensive genomic data have indicated that the similarities between the two subphyla are more likely due to evolutionary conservation rather than to convergence. That is because, as noted above, both vertebrates and amphioxus are evolving exceptionally slowly (Venkatesh et al., 2014; Yue et al., 2014, 2016). Moreover, while the early development of amphioxus resembles that of other invertebrates such as sea urchins, later development is much more like that of vertebrates. In fact, Kowalevsky (1867), who first described amphioxus development, noted the affinity of amphioxus and vertebrate development. Thus, both comparative genomics and comparative morphology point to modern amphioxus is the best available proxy for the ancestral vertebrate and the best organism to compare with vertebrates to ask how vertebrates evolved from their invertebrate ancestors. Moreover, the use of gene expression patterns in embryos and larvae as characters to infer homologies—an evodevo tool that was developed in the latter part of the 20th century—provided an important new approach to address questions of homologies. Interpretation of such gene expression patterns is limited by three factors—(1) the morphological divergence of homologous structures from their common ancestors, (2) co-option of parts of gene networks for new functions, and (3) the convergence of non-homologous structures. Although the last two of these limitations are less likely in slowly evolving organisms, there are on-going controversies as to whether expression of a particular gene(s) in similar structures in two organisms reflects true homologies or convergence involving the independent co-option of genes to form similar structures.

3. The dawn of evodevo

The surprising discovery in the late 20th century that developmental genes are broadly conserved across phyla provided the impetus for evodevo. Before then, in spite of years of research on genetics of organisms such as *Drosophila* and yeast, it was widely thought that genes were not conserved except in very close relatives (Mayr, 1963). However, at the same time as Ernst Mayr was articulating this concept as part of the "new synthesis," Linus Pauling and Emile Zuckerkandl were demonstrating that, to the contrary, the amino acid sequences of house-keeping proteins are sufficiently conserved across modern organisms that they can be compared to reconstruct the ancestral protein sequence (Pauling & Zuckerkandl, 1963; Zuckerkandl & Pauling, 1965).

A second important development for evodevo was in situ hybridization. Techniques for DNA/DNA and DNA/RNA hybridization with radioactive probes were developed in the 1960s (reviewed in Hennig, 1973). The radioactive signals were detected by placing photographic emulsion on the slides (autoradiography). Non-radioactive probes, including biotinylated ones, were developed in the early 1980s (Gomes et al., 1985), and whole mount in situ hybridization with non-radioactive probes soon followed (Tautz & Pfeifle, 1989), just at the time it was found that not only house-keeping, but also developmental genes are broadly conserved across phyla. Combined with in situ hybridization, this realization led to the discoveries in the 1980s that Hox genes are linearly expressed along the body axes in both developing flies and vertebrates. This marked the dawn of the field of evolutionary developmental biology (De Robertis, Oliver, & Wright, 1990; McGinnis, Garber, Wirz, Kuroiwa, & Gehring, 1984; Scott & Weiner, 1984).

4. How did the vertebrate brain evolve from that of an amphioxus-like ancestor?

4.1 The problem

To address how vertebrates evolved from their invertebrate ancestors, we adopted an evo/devo approach using amphioxus as a stand-in for the proximate ancestor of the vertebrates. The amphioxus nerve cord, which has an enlarged anterior end (the cerebral vesicle) and a trunk/tail nerve cord (sometimes termed the "spinal cord") has only an estimated 20,000 neurons compared to millions or billions for vertebrates (Nicol & Meinertzhagen, 1991). Bone (1960) in his histochemical study of the amphioxus spinal cord concluded that the many similarities between the amphioxus and vertebrate central nervous systems indicate that the simplicity of the amphioxus CNS is "more probably truly primitive than the result of secondary simplification" and that the "resemblances between the two are in fact true homologies." One similarity is the presence of Reissner's fiber, made of SCO-spondin and secreted by cells located ventrally within the cerebral vesicle in amphioxus and initially by both the floor plate and the dorsal subcommisural organ in the vertebrate diencephalon. Reissner's fiber extends posteriorly down the full length of the nerve cord (Olsson, 1955; Olsson, Yulis, & Rodriguez, 1994; Sterba, Fredriksson, & Olsson, 1983). However, other anatomical landmarks are scarce in the amphioxus CNS; the most obvious are the four kinds of photoreceptors (Fig. 3). At the anterior tip of the amphioxus CNS is a photoreceptor composed of two rows of ciliated cells

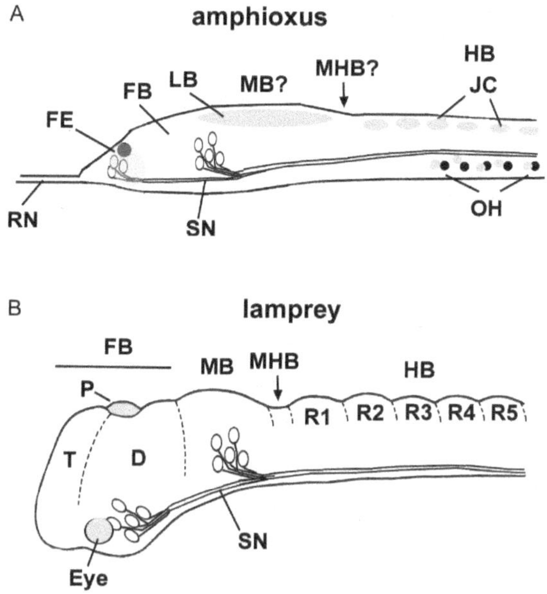

Fig. 3 Diagram of the developing brains of (A) amphioxus and (B) lamprey. Anterior to left. Indicated in gray are the photoreceptors (FE = frontal eye; LB = lamellar body (homologous to the vertebrate pineal), JC = Joseph cells, OH = organs of Hesse, P = pineal, Eye = the paired eyes of the lamprey). Amphioxus and lamprey have comparable groups of serotonergic neurons (SN). In amphioxus, the anterior most group comprises row 2 of the photoreceptive neurons of the frontal eye. RN = paired rostral nerves.

plus a pigment cell. This has been termed the "frontal eye," although it is not image-forming (Lacalli, 1996). In larvae, the cilia protrude slightly from the anterior neuropore, which remains after the neural plate has curled up into a tube. The larvae use this photoreceptor to orient to the light as they hang in the water column to filter feed (Stokes & Holland, 1995a). More posteriorly in the cerebral vesicle is a dorsal putative photoreceptor, the lamellar body, composed of whorls of long cilia of a number of cells. The lamellar body is structurally very like the pineal in a larval lamprey (Lacalli, 2004; Pu & Dowling, 1981). Just behind the lamellar body are bilateral arrays of large microvillar photoreceptors termed Joseph cells. The microvilli of these cells are on the side of the cell facing the neural canal. A fourth type of photoreceptor, the organs of Hesse, occurs ventrally in the CNS posterior to the cerebral vesicle. These are also microvillar and each associated with a pigment cell (Ruiz & Anadon, 1991). Other than these photoreceptor cells, most of the cells in the amphioxus CNS are small with 5–10 μm cell bodies. The large

Rhode cells are an exception (Bone, 1959, 1960). The relationship of these cells with the Rohan-Beard cells of larval vertebrate brains has been debatable.

Because there are so few anatomical similarities between the amphioxus and vertebrate nerve cords, ideas for exactly how the two are evolutionarily related have run the gamut from the vertebrate brain being an elaboration of the small cerebral vesicle of amphioxus (Hatschek, 1892), to the anterior forebrain being a vertebrate innovation (Huxley, 1875), to both the midbrain and forebrain being vertebrate innovations, with only the hindbrain evolving from the preexisting anterior region of the ancestral nerve cord (Gans & Northcutt, 1983). The advent of evodevo in the 1980s–90s provided a new approach for addressing how the vertebrate brain may have evolved from an amphioxus-like CNS in the ancestral chordate. Today, this question is largely resolved, but whether the head muscles of the ancestral vertebrate were segmented like those of modern amphioxus is still controversial.

4.2 How evodevo addressed the evolution of the CNS

Research on amphioxus embryos had largely ceased in the early 1900s due to the decline of major populations of *Branchiostoma lanceolatum* in Italy and reaction against Haeckel's recapitulation theory (Rasmussen, 1991). However, in the late 1980s, an accessible population of amphioxus (*Branchiostoma floridae*) in Old Tampa Bay, Florida, USA, attracted our attention. We initially discovered how to spawn the animals in the laboratory during the breeding season on days when they would normally have spawned in the field and examined fertilization and early development. The dawn of the evodevo era changed our focus. Inspired by a review emphasizing that Hox genes are expressed co-linearly in the vertebrate nerve cord as they are along the body axis in the developing fruitfly (De Robertis et al., 1990), we addressed the question of how the vertebrate brain may have evolved from a much simpler nerve cord in an amphioxus-like ancestor by determining if Hox genes are also expressed collinearly in the amphioxus nerve cord. Peter Holland in the U.K. had the same idea at the same time. The three Hollands collaborated and published the first genes and development paper on amphioxus in 1992 (Holland, Holland, Williams, & Holland, 1992). This paper showed that the anterior limit of expression of *Hox3* in the amphioxus CNS was at the level of the boundary between somites 4 and 5 in the paraxial mesoderm—comparable to the anterior limit of vertebrate *HoxB3* at the boundary between rhombomeres 4 and 5 in the vertebrate hindbrain (Fig. 4). The conclusion was that amphioxus probably

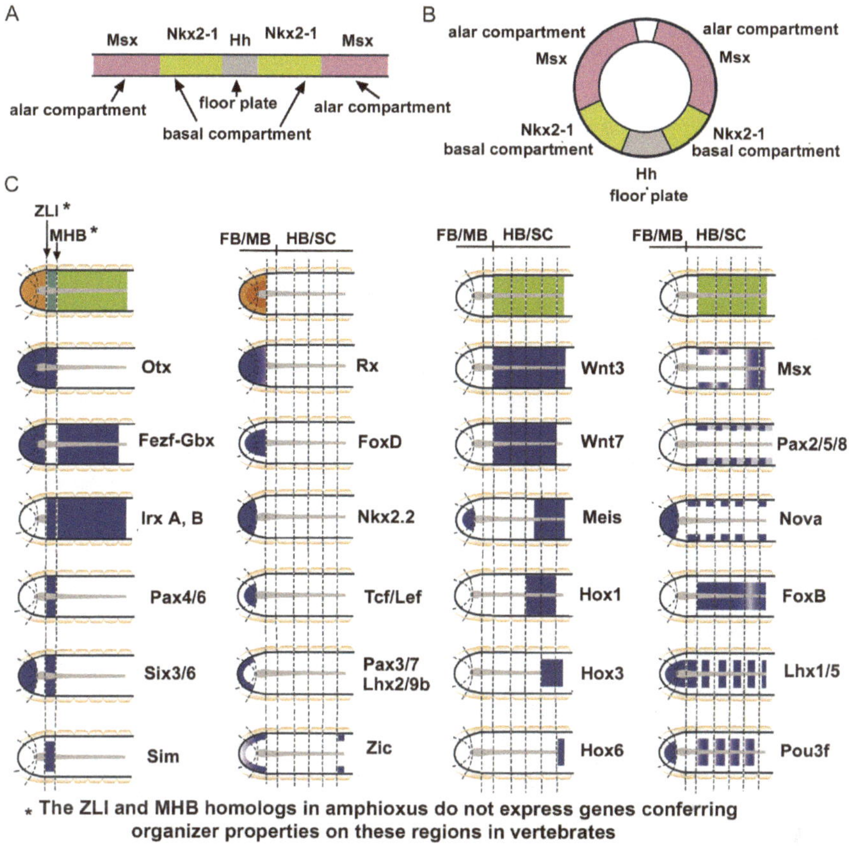

Fig. 4 Gene expression divides the developing neural tube of amphioxus into mediolateral and anterior-posterior domains that correspond to homologous domains in the vertebrate nerve cord. (A) and (B) Cross section through neural plate (A) and neural tube (B) showing Msx expressed laterally (dorsally), Nkx2-1 expressed medially (laterally) and Hh expressed ventrally in the floor plate. (C) Gene expression domains demarcate anterior-posterior compartments in the amphioxus neural tube. ZLI = zona limitans intrathalamica, MHB = midbrain/hindbrain boundary, FB = forebrain, HB = hindbrain, SC = Spinal cord. *Panel (C) adapted from Albuixech-Crespo, B., Lopez-Blanch, L., Burguera, D., Maeso, I., Sanchez-Arrones, L., Moreo-Bravo, J. A. et al. (2017). Molecular regionalization of the developing amphioxus neural tube challenges major partitions of the vertebrate brain.* PLoS Biology, 15*(4), e2001573. doi:10.1371/journal.pbio.2001573. Data from Candiani, S., Castagnola, P., Oliveri, D., & Pestarino, M. (2002). Cloning and developmental expression of* AmphiBrn1/3/4, *a* POU III *gene in amphioxus.* Mechanisms of Development, 116*, 231–234, Castro, L. F. C., Rasmussen, S. L. K., Holland, P. W. H., Holland, N. D., & Holland, L. Z. (2006). A* Gbx *homeobox gene in amphioxus: Insights into ancestry of the ANTP class and evolution of the midbrain/hindbrain boundary.* Developmental Biology, 295*, 40–51, Glardon, S., Holland, L. Z., Gehring, W. J., & Holland, N. D. (1998). Isolation and developmental expression of the amphioxus* Pax-6 *gene* (AmphiPax-6*): Insights into eye and photoreceptor evolution.* Development, 125*, 2701–2710, Gostling, N. J., & Shimeld, S. M. (2003). Protochordate* Zic *genes define primitive*

(Continued)

Fig. 4—Cont'd somite compartments and highlight molecular changes underlying neural crest evolution. *Evolution and Development, 5,* 136–144, Holland, L. Z., Schubert, M., Kozmik, Z., & Holland, N. D. (1999). Characterization of *AmphiPax3/7*, an amphioxus paired box gene: Insights into chordate myogenesis, neurogenesis, and the possible evolutionary precursor of definitive vertebrate neural crest. *Evolution and Development, 1,* 153–165, Holland, L. Z., Venkatesh, T. V., Gorlin, A., Bodmer, R., & Holland, N. D. (1998). Characterization and developmental expression of *AmphiNk2-2*, an NK2 class homeobox gene from amphioxus (Phylum Chordata; Subphylum Cephalochordata). *Development, Genes and Evoution, 208,* 100–105, Irimia, M., Pineiro, C., Maeso, I., Gomez-Skarmeta, J. L., Casares, F., & Garcia-Fernandez, J. (2010). Conserved developmental expression of *Fezf* in chordates and *Drosophila* and the origin of the zona limitans intrathalamica (ZLI) brain organizer. *EvoDevo, 1*(1), 7. doi:10.1186/2041-9139-1-7, Irimia, M., Denuc, A., Burguera, D., Somorjai, I., Martín-Durán, J. M., Genikhovich, G., et al. (2011). Stepwise assembly of the *Nova*-regulated alternative splicing network in the vertebrate brain. *Proceedings of the National Academy of Sciences of the United States of America, 108,* 5319–5324, Kaltenbach, S. L., Holland, L. Z., Holland, N. D, & Koop D. (2009). Developmental expression of the three iroquois genes of amphioxus (*BflrxA*, *BflrxB*, and *BflrxC*) with special attention to the gastrula organizer and anteroposterior boundaries in the central nervous system. *Gene Expression Patterns, 9,* 329–334, Kozmik Z, Holland N. D., Kalousova A., Paces J., Schubert M., Holland L. Z. (1999). Characterization of an amphioxus paired box gene, *AmphiPax2/5/8*: Developmental expression patterns in optic support cells, nephridium, thyroid-like structures and pharyngeal gill slits, but not in the midbrain-hindbrain boundary region. *Development, 126,* 1295–1304, Kozmik, Z., Holland, N. D., Kreslova, J., Oliveri, D., Schubert, M., & Jonasova K. et al. (2007). Pax-Six-Eya-Dach network during amphioxus development: Conservation in vitro but context specificity in vivo. *Developmental Biology, 306,* 143–159, Lin, H. C., Holland, L. Z., & Holland, N. D. (2006). Expression of the *AmphiTcf* gene in amphioxus: Insights into the evolution of the *TCF/LEF* gene family during vertebrate evolution. *Developmental Dynamics, 235,* 3396–3403, Mazet, F., & Shimeld, S. M. (2002). The evolution of chordate neural segmentation. *Developmental Biology, 251,* 258–270, Schubert, M., Holland, L. Z., & Holland, N. D. (2000). Characterization of two amphioxus *Wnt* genes (*AmphiWnt4* and *AmphiWnt7b*) with early expression in the developing central nervous system. *Developmental Dynamics, 217,* 205–215, Schubert, M., Holland, L. Z., Stokes, M. D., & Holland, N. D. (2001). Three amphioxus *Wnt* genes (*AmphiWnt3*, *AmphiWnt5*, and *AmphiWnt6*) associated with the tail bud: The evolution of somitogenesis in chordates. *Developmental Biology, 240,* 262–273, Schubert, M., Holland, N. D., Laudet, V., & Holland, L. Z. (2006). A retinoic acid-Hox hierarchy controls both anterior/posterior patterning and neuronal specification in the developing central nervous system of the cephalochordate amphioxus. *Developmental Biology, 296,* 190–202, Sharman, A. C., Shimeld, S. M., & Holland, P. W. H. (1999). An amphioxus *Msx* gene expressed predominantly in the dorsal neural tube. *Development Genes and Evolution, 209,* 260–263, Shimeld, S. M. (1999). The evolution of the hedgehog gene family in chordates: Insights from amphioxus hedgehog. *Development Genes and Evoution, 209,* 40–47, Venkatesh, T. V., Holland, N. D., Holland, L. Z., Su, M. T., & Bodmer, R. (1999). Sequence and developmental expression of amphioxus *AmphiNk2-1*: Insights into the evolutionary origin of the vertebrate thyroid gland and forebrain. *Development Genes and Evolution, 209,* 254–259, Williams, N. A., & Holland, P. W. H. (1996). Old head on young shoulders. *Nature, 383,* 490, Yu, J. -K., Holland, N. D., & Holland, L. Z. (2002). An amphioxus winged helix/forkhead gene, *AmphiFoxD*: Insights into vertebrate neural crest evolution. *Developmental Dynamics, 225,* 289–297.

has a homolog of the vertebrate hindbrain as well as more anterior regions. Subsequently, in situ hybridization for five more of the 15 Hox genes in amphioxus showed that like their vertebrate counterparts, they are co-linearly expressed in the CNS of early embryos (Schubert et al., 2006).

Additional comparisons of genes expressed in the central nervous systems (CNS) of amphioxus and vertebrates complemented and added to the initial conclusions from Hox genes. Expression of *Pax4/6* and *Otx* in the anterior and posterior regions of the amphioxus cerebral vesicle suggested homologies of the anterior photoreceptor of amphioxus and the vertebrate paired eyes (Glardon, Holland, Gehring, & Holland, 1998; Williams & Holland, 1996). In addition, gene expression patterns have suggested that amphioxus has homologs of two CNS organizers—the *Zona Limitans Intrathalamica* (ZLI) and the midbrain/hindbrain boundary (MHB) (Fig. 4). The ZLI resides within the diencephalon. In both vertebrates and amphioxus, it is characterized by an anterior domain of *Fezf* that abuts a posterior domain of *IrxB* (Irimia et al., 2010; Kaltenbach, Holland, Holland, & Koop, 2009). However, in vertebrates, but not amphioxus, the *Shh* domain coincides where those of *FezF* and *IrxB* abut, suggesting that while this region in vertebrates functions as an organizer, in amphioxus, it does not. Similarly, in both vertebrates and amphioxus, the domains of *Otx* and *Gbx* abut at the MHB (Castro, Rasmussen, Holland, Holland, & Holland, 2006). In vertebrates, this region also has stripes of *Pax2/5/8* and *En* expression and secretes Wnt1 and Fgf8 signaling proteins. In contrast, in amphioxus, none of the other MHB markers is expressed in a stripe at the *Otx/Gbx* boundary. *En* is expressed in a few cells in the CNS both more anteriorly and more posteriorly to this (Holland, Kene, Williams, & Holland, 1997), while *Pax2/5/8* is expressed in the anterior photoreceptor and throughout the entire CNS posterior to the cerebral vesicle (Kozmik et al., 1999), *Fgf8/17/18* is broadly expressed in the cerebral vesicle (Bertrand et al., 2011), and *Wnt1* is not expressed in the CNS (although *Wnt8* is expressed more anteriorly in the cerebral vesicle) (Holland & Holland, 2000; Schubert, Holland, Panopoulou, Lehrach, & Holland, 2000). These expression patterns, summarized in Albuixech-Crespo et al. (2017) strongly suggest that the amphioxus CNS likely has the territories that evolved into the ZLI and MHB in vertebrates, with the genes conferring organizer properties being recruited into these territories early in the vertebrate lineage (Fig. 4C).

In amphioxus and vertebrates, not only are expression patterns of anterior/posterior patterning genes similar, but those of medio-lateral (dorsal-ventral) patterning genes are also conserved (Fig. 4A and B). Within the CNS, *Nkx2* is expressed dorsally, *Gsh* a little more ventrally and *Msx* even more ventrally (Sharman, Shimeld, & Holland, 1999). Taken together, these data indicate that the amphioxus and vertebrate nerve cords have conserved anterior/posterior and dorsal/ventral patterning and, therefore, both likely evolved from the nerve cord of an ancestral chordate with the vertebrates taking advantage of duplicate genes deriving from whole genome duplications to invent new structures such as neural crest and a telencephalon.

4.3 Micro-anatomy

The conclusions about evolution of the CNS from gene expression patterns have been bolstered by detailed microanatomical studies, particularly the serial 3D reconstructions of neural cell populations and their wiring diagrams done by Thurston Lacalli. For example, he showed that the anterior photoreceptor (frontal eye) is composed of a pigment cell and 16 photoreceptor cells that extend their cilia out the neuropore and send axons posteriorly into ventro-lateral tracts of nerve fibers that end just behind the lamellar body (Lacalli, 1996). He postulated that this region, which includes large dorsal rhabdomeric photoreceptors (the Joseph cells) that may be used to detect shadows, evolved into the tectum in vertebrates as the Joseph cells were lost. Because the photoreceptor cells of the vertebrate retina are ciliary, Lacalli (2018) also proposed that the vertebrate eyes and the amphioxus frontal photoreceptor are homologous.

Comparisons of brains of amphioxus and the lamprey have been particularly informative. For example, the patterns of serotonergic nerves are very much the same in amphioxus and the lamprey (Fig. 3). Cells associated with the amphioxus frontal eye are serotonergic and there is a large cluster of such neurons at the posterior end of the cerebral vesicle just behind the lamellar body. The pattern in the larval lamprey (ammocoete) is similar. The axons of these neurons extend posteriorly in the ventral neuropile. Taken together, patterns of gene expression and micro-anatomy indicated that the ancestral chordate nerve cord was divided into spinal cord, midbrain, diencephalic forebrain but lacked a telencephalon and neural crest. Moreover, the frontal photoreceptor of this ancestral brain likely evolved into the paired eyes of vertebrates.

5. The origins of neural crest and placodes

In vertebrates, neural crest cells migrate from the dorsal part of the developing CNS and wander ventrally, giving rise to pigment cells, bones and cartilage of the head, connective tissue, parts of the cranial ganglia, sensory neurons, and cells of the adrenal medulla (Bronner, 2015; Bronner & LeDourin, 2012; Knecht & Bronner-Fraser, 2002). The first animals in evolution to have definitive neural crest are lampreys and hagfish (Ota, Kuraku, & Kuratani, 2007). However, lampreys lack parts of the gene network common to head neural crest in amniotes, suggesting that during vertebrate evolution, the neural crest network for the head was gradually elaborated upon (Martik et al., 2019). Although amphioxus lacks neural crest, gene expression suggests that the ectoderm adjacent the neural plate sets the evolutionary stage for neural crest. First, genes that specify the neural plate as well as those that specify the neural plate borders are similarly expressed in amphioxus and vertebrates (Holland, 2013; Meulemans & Bronner-Fraser, 2004) (Fig. 5). In addition, amphioxus *Dlx*, homologous to vertebrate *Dlx5*, a marker of both premigratory and migratory neural crest in vertebrates, is expressed in the ectoderm adjacent the amphioxus neural plate (Holland, Panganiban, Henyey, & Holland, 1996). Moreover, this amphioxus tissue migrates, not as single cells, but as a sheet, over the neural plate. The evolution of neural crest not only provided a source of new structures but of new cell-cell interactions allowing old cells to alter their fate. For example, in the mouse, *Dlx5*-expressing neural crest interacts with pharyngeal mesoderm, and null mutations eliminate several muscles derived from the 4th pharyngeal arch (Sugii et al., 2017).

5.1 Are ectodermal sensory cells in amphioxus related to neurogenic placodes in vertebrates?

Although amphioxus does not have mechanosensory cells organized into discrete structures like vertebrate placodes and their derivatives, amphioxus has several types of ectodermal sensory cells, differing in the length of the cilium and the number and size of microvilli (Holland & Holland, 2001; Holland & Yu, 2002; Lacalli & Hou, 1999; Stokes & Holland, 1995b; Zieger et al., 2018). These are scattered down the sides of larvae and adults. In adults, there are also ventral ectodermal sensory cells. Many of the ectodermal sensory cells in amphioxus are primary neurons, sending axons into the CNS. In the early amphioxus embryo, some ectodermal cells in the ventral midline that express *Delta* and *Tlx* lose their cilia, withdraw to the base of the ectoderm and migrate dorsally along the basal lamina (Holland, 2020; Kaltenbach, Yu, & Holland, 2009; Rasmussen, Holland, Schubert, Beaster-Jones, & Holland, 2007). They then extend to the surface of the ectoderm, regrow cilia, and extend axons that go between the somites into the CNS (Kaltenbach, Holland, et al., 2009). As more is learned about the genesis and migration of neurons from neurogenic placodes in the vertebrate

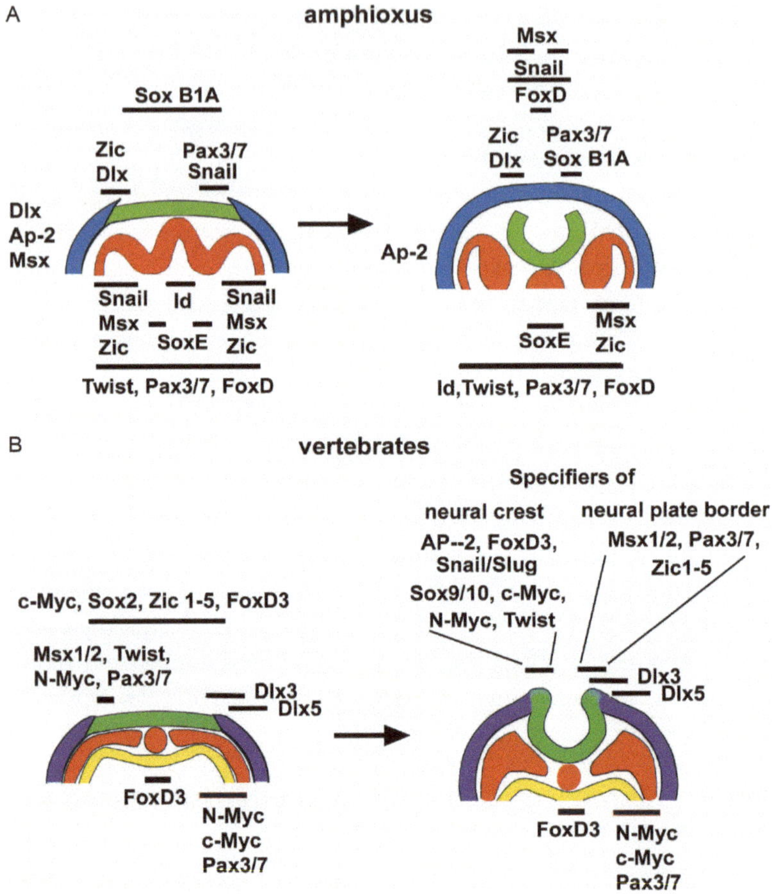

Fig. 5 Comparison of expression patterns of genes specifying the neural plate and the neural plate border in (A) amphioxus with those specifying the neural plate, neural plate border and neural crest in (B) vertebrates. Cross sections. Blue = ectoderm, green = neural plate, red = mesendoderm/mesoderm, yellow = endoderm. *Adapted from Holland, L. Z. (2013). Evolution of new characters after whole genome duplications: Insights from amphioxus.* Seminars in Cell and Developmental Biology, 24, *101–109. Data from Aybar, M. J., Nieto, M. A., & Mayor, R. (2003).* Snail *precedes* Slug *in the genetic cascade required for the specification and migration of the* Xenopus *neural crest.* Development, 130, *483–494, Holland, L. Z., Schubert, M., Holland, N. D., & Neuman, T. (2000). Evolutionary conservation of the presumptive neural plate markers* AmphiSox1/2/3 *and* AmphiNeurogenin *in the invertebrate chordate amphioxus.* Developmental Biology, 226, *18–33, Khudyakov, J., & Bronner-Fraser, M. (2009). Comprehensive spatiotemporal analysis of early chick neural crest network genes.* Developmental Dynamics, 238, *716–723, Meulemans, D., & Bronner-Fraser, M. (2004). Gene-regulatory interactions in neural crest evolution and development.* Developmental Cell, 7, *291–299, Sauka-Spengler, T., & Bronner-Fraser, M. (2008). A gene regulatory network orchestrates neural crest formation.* Nature Reviews Molecular Cell Biology, 9, *557–568, Yu, J. -K., Meulemans, D., McKeown, S. J., & Bronner-Fraser, M. (2008). Insights from the amphioxus genome on the origin of vertebrate neural crest.* Genome Research, 18, *1127–1132.*

head, clear parallels with the ectodermal sensory cells of amphioxus have come to light suggesting an evolutionary relationship. These placodal cells, like the amphioxus ectodermal sensory cells, migrate dorsally under the placodal epithelium. A subset of cells is bipolar with a long leading axon and a short posterior one. Some of these leading axons make contact with the developing neural tube and extend anteriorly or posteriorly within it (Smith, Fleenor, & Begbie, 2015), similar to the amphioxus sensory cells (Holland & Yu, 2002). Moreover, both these vertebrate and amphioxus sensory neurons express many of the same genes including *Delta, hu/elav, tlx, soxb1c, coe, islet,* and *eya* (Lara-Ramírez, Poncelet, Patthey, & Shimeld, 2017). Taken together, the data indicate that in evolving neurogenic placodes, vertebrates elaborated upon the ectodermal sensory cells in an amphioxus-like ancestor. The idea that the pan-placodal ectoderm of vertebrates evolved from the larval ectoderm of an amphioxus-like common ancestor (Holland, 2009) has been questioned (Schlosser, 2017). Even so, Schlosser (2017) does not dispute that the genetic mechanisms for distinguishing neural from non-neural ectoderm and for specifying ectodermal sensory cells are conserved to a large extent in amphioxus, tunicates and vertebrates. Vertebrates simply elaborated upon and added to these mechanisms to invent placodes.

6. The head-segmentation conundrum

There is a long-standing question whether or not the head mesoderm of the ancestral vertebrate contained segmental paraxial muscles like that of modern amphioxus. The answer to this question has been elusive for several reasons.

(1) A major problem is that most of the work on the development of vertebrate head muscles has been on amniotes, whose heads are quite derived compared to those of jawless vertebrates and even chondrichthyes and teleost fishes. Even so, although modern vertebrates lack segmental muscles in the head, the vertebrate head is not entirely unsegmented. In vertebrates, as in amphioxus, the pharynx has a series of gill slits; in reptiles, birds and mammals, the embryonic pharynx produces endoderm-lined pouches on either side, but these never penetrate to form patent gill slits.

(2) The muscular somites in all vertebrates extend only from the level of the posterior hindbrain to the tip of the tail. Anterior to the somites in embryos of jawless vertebrates and sharks is a series of mesodermal segments with lumens called head cavities (premandibular, mandibular, hyoid, glossopharyngeal). These head cavities extend projections ventrally similar to the ventral outgrowth of lateral plate mesoderm from the somites.

(3) In shark embryos, scanning electron microscopy shows very clear condensations of paraxial mesoderm anterior to the somites and posterior to the head cavities.

(4) In mammals, stereo pairs of scanning electron micrographs have indicated to some the presence of a series of mesodermal condensations called somitomeres anterior to the true somites. However, because they are so difficult to see, many researchers have denied their existence (reviewed in Noden & Trainor, 2005).

(5) In vertebrates, neural crest (i.e., pluripotent cells that migrate away from the neural tube) gives rise to the cartilage, bone and connective tissue of the head including the gill bars. However, amphioxus lacks neural crest. Consequently, genes such as *Dlx* that are expressed in neural crest derivatives of the vertebrate pharyngeal arches are not expressed in the pharyngeal arches in amphioxus, while those such as *Tbx1/10* that are expressed in endoderm and head mesoderm not derived from neural crest are expressed similarly in the pharyngeal arches of both amphioxus ad vertebrates (Fig. 6).

6.1 Competing scenarios

One scenario proposes that the lateral mesoderm of the common ancestor of chordates was segmented along the full length of the animal and that during vertebrate evolution segmentation of the head was extensively modified and then lost. This scenario, which takes into account both gene expression and developmental anatomy, argues that the ventral extensions of the anteriormost 8–10 somites of an amphioxus-like ancestral chordate evolved into most of the head mesoderm of vertebrates, including the branchial muscles expressing *Tbx1/10* and jaw/velar muscles, while the somites themselves gave rise to somitomeres in gnathostomes and some of the extra-ocular muscles (Aldea et al., 2019; Holland, Holland, & Gilland, 2008). This scenario relies heavily upon similarities between amphioxus and the jawless agnathans as well as sharks.

The opposing idea, that head mesoderm of the ancestral chordate was unsegmented is chiefly advocated by Kuratani and colleagues. They have consistently maintained that the vertebrate head mesoderm "evolved, not

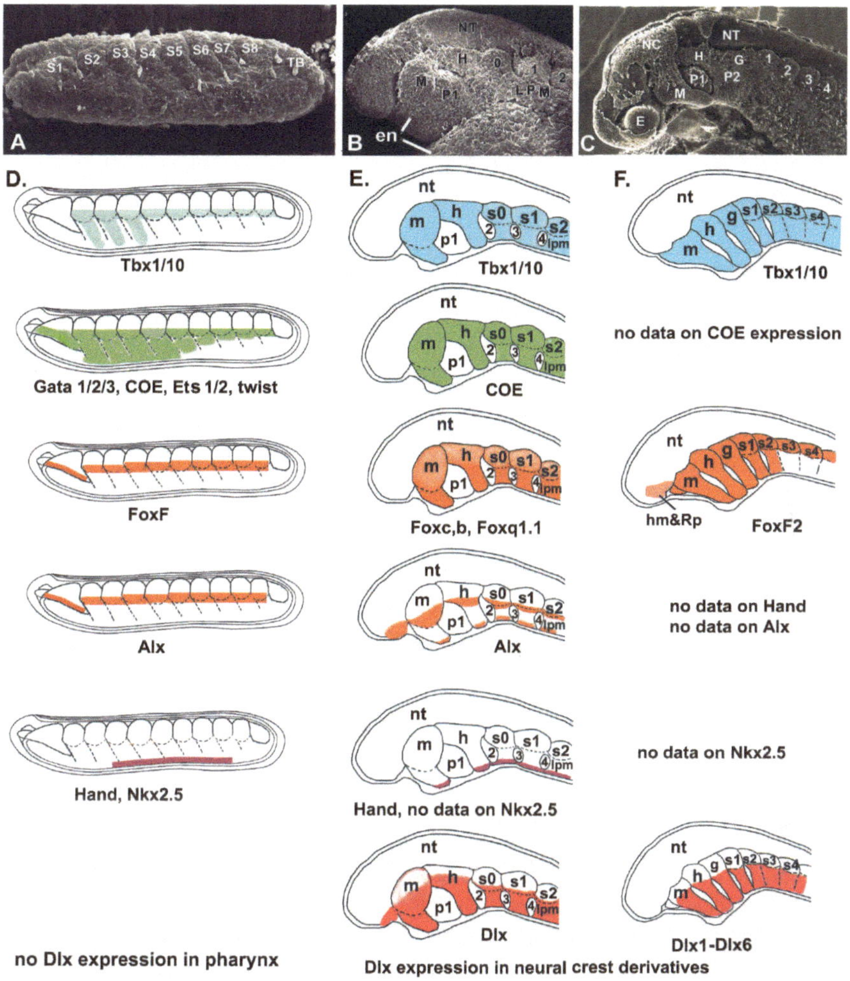

Fig. 6 Embryos of amphioxus-mid-neurula (A, D), lamprey (B, E) and shark (C, F), anterior to the left. (A–C) Scanning electron micrographs of embryos with the ectoderm removed. (D–F) Diagrams of gene expression in the developing paraxial and ventrolateral mesoderm in amphioxus (D), lamprey (E), and (F) shark. Side views of the head, anterior to left. (A–C) Abbreviations: S0–S8, 0–4 = somites, E = eye, en = endoderm, G = glossopharyngeal mesoderm, H = hyoid mesoderm, M = mandibular mesoderm, NC = neural crest, NT = neural tube, P1–P4 = pharyngeal pouches, TB = tailbud. *Panel (A) reprinted with permission from Kaltenbach, S. L., Yu, J. -K., & Holland, N. D. (2009). The origin and migration of the earliest-developing sensory neurons in the peripheral nervous system of amphioxus.* Evolution and Development, 11, *142–151, Panel (B) reprinted with permission from Kuratani, S., Horigome, N., & Hirano, S. (1999). Developmental morphology of the head mesoderm and reevaluation of segmental theories of the vertebrate head: Evidence from embryos of an agnathan vertebrate, Lampetra japonica.* Developmental Biology, 210, *381–400, Panel (C) courtesy of E. H. Gilland, Dept. Anatomy, Howard, Univ. from Gilland, E. H. (1992).* Morphogenesis of segmental units

(Continued)

by the modification of rostral somites of an amphioxus-like ancestor, but through the polarization of unspecified paraxial mesoderm into head mesoderm anteriorly and trunk somites posteriorly" (Kuratani & Adachi, 2016). They have since used this argument as the basis, in part, for proposing that vertebrates should be a separate phylum and not included with tunicates and amphioxus in the Phylum Chordata (Irie, Satoh, & Kuratani, 2018).

Fig. 4—Cont'd in the chordamesoderm and neuroepithelium of Squalus acanthias. *Ph. D. dissertation. (pp. 306). Harvard University, Panel (D) data from Aldea, D., Subirana, L., Keime, C., Meister, L., Maeso, I., Marcellini, S. et al. (2019). Genetic regulation of amphioxus somitogenesis informs the evolution of the vertebrate head mesoderm.* Nature Ecology and Evolution, 3, *1233–1240, Cattell, M., Lai, S., Cerny, R., & Medeiros D. M. (2011). A new mechanistic scenario for the origin and evolution of vertebrate cartilage.* PLoS One, 6, *e22474, Cerny, R., Cattell, M., Sauka-Spengler, T., Bronner-Fraser, M., Yu, F. Q., & Medeiros, D. M. (2010). Evidence for the prepattern/cooption model of vertebrate jaw evolution.* Proceedings of the National Academy of Sciences of the United States of America, 107, *17262–17267, Debiais-Thibaud, M., Metcalfe, C. J., Pollack, J., Germon, I., Ekker, M., Depew, M. et al. (2013). Heterogeneous conservation of* Dlx *paralog co-expression in jawed vertebrates.* PLoS One, 8, *e68182, Holland, N. D., Venkatesh, T. V., Holland, L. Z., Jacobs, D. K., & Bodmer, R. (2003).* AmphiNk2-tin, *an amphioxus homeobox gene expressed in myocardial progenitors: Insights into evolution of the vertebrate heart.* Developmental Biology, 255, *128–137, Lara-Ramírez, R., Poncelet, G., Patthey, C., & Shimeld, S. M. (2017). The structure, splicing, synteny and expression of lamprey* COE *genes and the evolution of the* COE *gene family in chordates.* Development Genes and Evolution, 227, *319–338, Mahadevan, N. R., Horton, A. C., & Gibson-Brown, J. J. (2004). Developmental expression of the amphioxus* Tbx1/10 *gene illuminates the evolution of vertebrate branchial arches and sclerotome.* Development Genes and Evolution, 214, *559–566. doi:10.1007/s00427-004-0433-1, Mazet, F., Masood, S., Luke, G. N., Holland, N. D., & Shimeld, S. M. (2004). Expression of* AmphiCoe, *an amphioxus* COE/EFB *gene, in the developing central nervous system and epidermal sensory neurons.* Genesis, 38, *58–65, Meulemans, D., & Bronner-Fraser, M. (2007). Insights from amphioxus into the evolution of vertebrate cartilage.* PLoS One, 2, *e787, Pascual-Anaya, J., Albuixech-Crespo, B., Somorjai, I. M., Carmona, R., Oisi, Y., Alvarez, S. et al. (2013). The evolutionary origins of chordate hematopoiesis and vertebrate endothelia.* Developmental Biology, 375, *182–192, Sauka-Spengler, T., Le Mentec, C., Lepage, M., & Mazan, S. (2002). Embryonic expression of* Tbx1, *a DiGeorge syndrome candidate gene, in the lamprey* Lampetra fluviatilis. Gene Expression Patterns, 2, *99–103, Wotton, K. R., Mazet, F., & Shimeld, S. M. (2008). Expression of* FoxC, FoxF, FoxL1, *and* FoxQ1 *genes in the dogfish* Scyliorhinus canicula *defines ancient and derived roles for* Fox *genes in vertebrate development.* Developmental Dynamics, 237, *1590–1603, Wotton, K. R., & Shimeld, S. M. (2011). Analysis of lamprey clustered* Fox *genes: Insight into* Fox *gene evolution and expression in vertebrates.* Gene, 489, *30–40, Yasui, K., Zhang, S. C., Uemura, M., Aizawa, S., & Ueki, A. (1998). Expression of a twist-related gene,* Bbtwist, *during the development of a lancelet species and its relation to cephalochordate anterior structures.* Developmental Biology, 195, *49–59.*

More or less in agreement with this idea, Vyas, Nandkishore, and Sambasivan (2019) proposed that cranial skeletal muscles in vertebrates evolved, not from somite-derived muscles, but from cardiac progenitors in an ancestral chordate. However, given the many similarities in vertebrate and amphioxus development, placing vertebrates into a separate phylum does not seem justified.

6.2 The evidence

In amphioxus, muscular somites extend the full length of the body. The anterior 8–10 somites (the head somites) are segmented from the dorsolateral mesoderm of the gastrula and early neurula, while the more posterior ones are pinched off from the tailbud. During development, the somites give rise to ventral extensions that fuse with one another anterior/posteriorly to form continuous sheets of lateral mesoderm. Anteriorly, these sheets are penetrated by the gill slits. The region remaining between adjacent gill slits develops into the mesoderm of the pharyngeal arches. Other derivatives of these mesodermal sheets include the pterygial muscle that contracts the atrium and some coelomic epithelia.

The most appropriate vertebrates to compare to amphioxus are lampreys as their ammocoete larvae most closely resemble modern amphioxus (Fig. 2). Not only are lampreys an early branch off the vertebrate lineage, but, like the earliest fossil fish and amphioxus, lampreys are jawless and have a series of pharyngeal gill slits. In lampreys as in other vertebrates, the true somites are segmented from bands of mesoderm deriving from the tailbud and extend from the level of the posterior hindbrain to the tip of the tail. Ventral to the somites is the lateral plate mesoderm which gives rise anteriorly to the heart and blood. In addition, anterior to the somites, lampreys have three mesodermal head segments (premandibular, mandibular, and hyoid). These are formed enterocoelically like the anterior somites of amphioxus and, like amphioxus somites, they send processes ventrally into the pharyngeal arches (Koltzoff, 1902).

Sharks, jawed cartilaginous fish, also have mesodermal head segments anterior to the somites. Balfour (1876, 1877) initially described four anterior head segments, each with an internal space (head cavity) in the shark—the premandibular, mandibular, hyoid and glossopharyngeal (Fig. 6). The cavities in these segments, unlike those in the lamprey, are formed by schizocoely like the posterior somites of an amphioxus. Scanning electron microscopy has shown that these "head cavities" are the anteriormost of a series of mesodermal condensations or somites located anterior to the true somites that give rise to paraxial muscle (Gilland, 1992). Moreover, each of these mesodermal head somites extends ventrally into the pharyngeal arches of the shark. In addition, they give rise to the jaw, eye muscles and other muscles of the head (Neal, 1918).

More problematic are gnathostome somitomeres, first observed in chick embryos anterior to the true somites (Meier, 1979). Subsequently, up to seven on each side were described in teleosts, amphibians and in several amniotes (reviewed in Jacobson, 1988). Even though no ventral extensions of the somitomeres are visible by scanning electron microscopy, it has been suggested that they may represent vestiges of somites inherited from an ancestor like amphioxus that had somites extending the full length of the body (Holland et al., 2008). Alternatively, they have been considered to represent an evolutionarily arrested state of somite formation (Noden & Trainor, 2005).

6.3 Gene expression

Gene expression in head mesoderm of amphioxus, lamprey and shark is consistent with homology of the head mesoderm in the three groups, exclusive of neural crest derivatives in the latter two. First, in amphioxus, *engrailed*, a transcriptional repressor, is expressed in the posterior portion of each of the anterior, enterocoelically-formed, somites, but not in the posterior ones that derive from the tailbud (Holland et al., 1997). Similarly, in both lamprey and shark, the posterior walls of the mandibular head cavity, but none of the true somites, express *engrailed* (Adachi, Takechi, Hirai, & Kuratani, 2012; Holland, Holland, Honma, & Fujii, 1993). These patterns suggest that the head cavities in lampreys and shark may have evolved from the anterior somites of an amphioxus-like ancestral vertebrate (Holland et al., 2008). Even so, since the expression of *engrailed* in the mandibular mesoderm of the shark is weak, Adachi and colleagues argued against homology of the head cavities of vertebrates with the anterior somites of amphioxus (Adachi et al., 2012).

Fig. 6 diagrams conserved expression of a suite of genes coding for transcription factors in the anterior somites or head cavities and their ventro/lateral extensions in amphioxus, shark and lamprey. Most notable is *Tbx1/10* (Adachi & Kuratani, 2012; Adachi et al., 2012; Gilland, 1992; Holland et al., 1993, 1997; Koop et al., 2014; Mahadevan, Horton, & Gibson-Brown, 2004) (Fig. 6). In amphioxus, knockdown of *Tbx1/10* impairs pharyngeal segmentation (Koop et al., 2014). In the shark, the *Tbx1* domains include the mandibular arch, hyoid arch and branchial arches, as well as the epipharyngeal mesenchyme of the hyoid cavity and the ventral part of the otic vesicle (Adachi et al., 2012). The *Tbx1* domains in the lamprey are similar, including the lateral plate mesoderm, hyoid, mandibular and pharyngeal arch mesenchyme and the otic vesicle (Sauka-Spengler, Le Mentec, Lepage, & Mazan, 2002; Tiecke et al., 2007). In bony vertebrates such as the zebrafish, *Tbx1* is likewise expressed in the lateral cranial mesoderm, which gives rise to cardiopharyngeal progenitors including the outflow tract, and muscles of the head and pharyngeal arches (Guner-Ataman et al., 2018). Pharyngeal arch development is impaired in *Tbx1* null mutants (Piotrowski et al., 2003). In the frog *Xenopus*, the *Tbx1* domains include lateral plate and head mesoderm and later the pharyngeal arches, ventral part of the otic vesicle and the mesenchyme near the outflow tract of the heart and head mesenchyme (Ataliotis, Ivins, Mohun, & Scambler, 2005).

In amphioxus and vertebrates, other genes with conserved expression in the ventro-lateral mesoderm, which gives rise to the contractile ventral haemal vessel of amphioxus and the heart field in vertebrates (Prummel et al., 2019), include *Tbx4/5, Tbx20, Hand* and *Nkx2.5* (tinman) (Kokubo et al., 2010) (Fig. 6). In addition, Wang, Holland, and Takahashi (2019) suggested an evolutionary link between zebrafish head mesoderm and the anterior somites of amphioxus since embryos of both organisms express *Tbx20* in the lateral/ventral mesoderm and *FoxC* medial to *Tbx1/10*. Thus, they argued that, in the ancestral chordate, paraxial and lateral plate mesoderm were already established and, therefore that the anterior somites of amphioxus and the cranial mesoderm of vertebrates are homologous; the vertebrates simply lost segmentation in the head mesoderm. This conclusion agrees with that of Aldea et al. (2019), who studied the gene network mediating somite formation and proposed that the vertebrate head mesoderm evolved from the ventral portion of the segmented anterior paraxial mesoderm of an amphioxus-like ancestor. Their evolutionary scenario has three phases—first, the paraxial mesoderm segregated from the lateral plate mesoderm which then lost segmentation and, second, was regionalized along the anterior/posterior axis. Third, the anterior paraxial mesoderm was lost, allowing the lateral plate mesoderm to form the head muscles. In agreement with this theory, Prummel et al. (2019) found that an enhancer of the *draculin* (*drl*) gene from zebrafish directs expression throughout the entire lateral plate mesoderm in the zebrafish and the chick as well as to the lateral mesendoderm in the lamprey, to cardiac and pharyngeal muscle precursors in the ascidian tunicate *Ciona*, and to the ventral portion of the somites and their ventrolateral extensions in amphioxus.

Moreover, genes including *FoxL* have conserved domains in the mandibular and eye mesenchyme in the shark. Notable exceptions include *Dlx* and *Hox* genes that are expressed in neural crest derived components of the pharynx in vertebrates, but not in the amphioxus pharynx, evidently due to the lack of neural crest in the latter (Parker, Bronner, & Krumlauf, 2019).

7. Conclusions

Comparisons of gene expression patterns and developmental mechanisms between amphioxus and vertebrates lend considerable weight to the thesis that vertebrates evolved from an organism that resembled a modern amphioxus. Homologies inferred from evodevo studies are strengthened by comparisons from paleontology, comparative anatomy and genomics. First, early chordate fossils have close resemblance to modern amphioxus. These include the eyeless *Pikaia* from the middle Cambrian that is usually considered an early chordate, (Mallatt & Holland, 2013) as well as similar chordates from the lower Cambrian with apparent eyes such as *Haikouella* and *Yunannozoan* together with *Myllokunmingia* and *Metaspriggina* from the middle Cambrian (Conway Morris & Caron, 2014). Second, anatomically, amphioxus resembles the ammocoete larva of a lamprey with a jawless mouth, segmental muscles, pharynx with gill slits, notochord and dorsal nerve cord. Third, amphioxus genomes are exceptionally slowly evolving (Yue et al., 2014, 2016). Recently, a chromosome-level genome assembly of *Branchiostoma floridae* has confirmed that the ancestral chordate likely had 17 genomic linkage groups (chromosomes) and demonstrated a very high level of synteny at the chromosome level with vertebrate genomes. This study concluded that during chordate evolution, one whole genome duplication occurred by autotetraploidization before lampreys and jawed vertebrates diverged, and a second whole genome duplication occurred before the split of cartilaginous fish and bony fish by allotetraploidization—that is a whole genome duplication after two closely-related species hybridized (Simakov et al., 2020). Importantly, these authors found that the lamprey chromosomes can be grouped into the same 17 chromosome-linkage groups, underscoring that in spite of the evolution of such vertebrate-specific characters as neural crest after the initial whole genome duplication, the similarities between amphioxus and larval lampreys are most likely due to evolutionary conservation and not convergent evolution. Taken together, all the evidence indicates that amphioxus is an excellent stand-in for the ancestral chordate for giving insights into how the vertebrates evolved from their invertebrate ancestors.

Acknowledgment

This study was supported by NSF grant IOS 1952567.

References

Adachi, N., & Kuratani, S. (2012). Development of head and trunk mesoderm in the dogfish, *Scyliorhinus torazame*: I. Embryology and morphology of the head cavities and related structures. *Evolution and Development*, *14*, 234–256.

Adachi, N., Takechi, M., Hirai, T., & Kuratani, S. (2012). Development of the head and trunk mesoderm in the dogfish, *Scyliorhinus torazame*: II. Comparison of gene expression between the head mesoderm and somites with reference to the origin of the vertebrate head. *Evolution and Development*, *14*, 257–276.

Albuixech-Crespo, B., Lopez-Blanch, L., Burguera, D., Maeso, I., Sanchez-Arrones, L., Moreo-Bravo, J. A., et al. (2017). Molecular regionalization of the developing amphioxus neural tube challenges major partitions of the vertebrate brain. *PLoS Biology*, *15*(4). e2001573. https://doi.org/10.1371/journal.pbio.2001573.

Aldea, D., Subirana, L., Keime, C., Meister, L., Maeso, I., Marcellini, S., et al. (2019). Genetic regulation of amphioxus somitogenesis informs the evolution of the vertebrate head mesoderm. *Nature Ecology and Evolution*, *3*, 1233–1240.

Ariëns Kappers, C. U., Huber, G. C., & Crosby, E. C. (1960). The comparative anatomy of the nervous system of vertebrates, including man. *Vol. 1,* New York: Hafner Publishing Company, pp. 135–148, Vol 2. Pp. 868–869.

Ataliotis, P., Ivins, S., Mohun, T. J., & Scambler, P. J. (2005). *XTbx1* is a transcriptional activator involved in head and pharyngeal arch development in *Xenopus laevis*. *Developmental Dynamics*, *232*, 979–991.

Balfour, F. M. (1876). On the development of elasmobranch fishes. From stages B to G. *Journal of Anatomy and Physiology*, *10*, 672–688 + pl. XXIX).

Balfour, F. M. (1877). The development of elasmobranch fishes. *Journal of Anatomy and Physiology*, *11*, 406–490.

Balfour, F. M. (1885). A treatise on comparative embryology. *Vols. 1&2,* London: Macmillan, 591.

Bertrand, S., Camasses, A., Somorjai, I. M. L., Belgacem, M. R., Chabrol, O., Escande, M. L., et al. (2011). Amphioxus FGF signaling predicts the acquisition of vertebrate morphological traits. *Proceedings of the National Academy of Sciences of the United States of America*, *108*, 9160–9165.

Bone, Q. (1959). The central nervous system in larval acraniates. *Quarterly Journal of Microscopical Science*, *100*, 509–527.

Bone, Q. (1960). The central nervous system in amphioxus. *Journal of Comparative Nerurology*, *115*, 27–64.

Bronner, M. (2015). On the crest of becoming vertebrate. *Nature*, *527*, 311–312.

Bronner, M., & LeDourin, N. (2012). Development and evolution of the neural crest: An overview. *Developmental Biology*, *366*, 2–9.

Castro, L. F. C., Rasmussen, S. L. K., Holland, P. W. H., Holland, N. D., & Holland, L. Z. (2006). A *Gbx* homeobox gene in amphioxus: Insights into ancestry of the ANTP class and evolution of the midbrain/hindbrain boundary. *Developmental Biology*, *295*, 40–51.

Conway Morris, S., & Caron, J.-B. (2014). A primitive fish from the Cambrian of North America. *Nature*, *512*, 419.

Costa, O. G. (1834). *Cenni zoologici ossia descrizione sommaria delle specie nuove di animali discoperti in diverse contrade del regno nell'anno 1834*. Napoli: Azzolino, 90.

De Robertis, E. M., Oliver, G., & Wright, C. V. E. (1990). Homeobox genes and the vertebrate body plan. *Scientific American*, *263*, 46–53.

Delsuc, F., Brinkmann, H., Chourrout, D., & Philippe, H. (2006). Tunicates and not cephalochordates are the closest living relatives of the vertebrates. *Nature*, *439*, 965–968.

Delsuc, F., Philippe, H., Tsagkogeorga, G., Simion, P., Tilak, M.-K., Turon, S., et al. (2018). A phylogenomic framework and timescale for comparative studies of tunicates. *BMC Biology*, *16*, 39. https://doi.org/10.1186/s12915-018-0499-2.

Delsuc, F., Tsagkogeorga, G., Lartillot, N., & Philippe, H. (2008). Additional molecular support for the new chordate phylogeny. *Genesis*, *46*, 592–604.

Dohrn, A. (1885). Studien zur Urgeschichte des Wirbelthierkorpers. VIII. Die Thyroidea bei Petromyzon, Amphioxus und Tunicaten. In *Vol. 6. Mittheilungen aus der Zoologischen Station zu Neapel* (pp. 1–92).

Gans, C. (1993). Evolutionary origin of the vertebrate skull. In J. Hanken & B. K. Hall (Eds.), *The skull, Vol. 2: Patterns of structural and systematic diversity* (pp. 1–35). Chicago: University Chicago Press.

Gans, C., & Northcutt, R. G. (1983). Neural crest and the origin of vertebrates: A new head. *Science*, *220*, 268–274.

Gilland, E. H. (1992). *Morphogenesis of segmental units in the chordamesoderm and neuroepithelium of* Squalus acanthias. Ph. D. dissertation Harvard University, 306.

Glardon, S., Holland, L. Z., Gehring, W. J., & Holland, N. D. (1998). Isolation and developmental expression of the amphioxus *Pax-6* gene (*AmphiPax-6*): Insights into eye and photoreceptor evolution. *Development*, *125*, 2701–2710.

Gomes, S. A., Nascimento, J. P., Siqueira, M. M., Krawczuk, M. M., Pereira, H. G., & Russell, W. C. (1985). In situ hybridization with biotinylated DNA probes: A rapid diagnostic test for adenovirus upper respiratory infections. *Journal of Virological Methods*, *12*, 105–110.

Goodrich, E. S. (1902). On the structure of the excretory organs of amphioxus. *Quarterly Journal of Microscopical Science*, *45*, 493–501 + pl. XXVII).

Goodrich, E. S. (1909). On the structure of the excretory organs of amphioxus. Parts 2–4. *Quarterly Journal of Microscopical Science*, *54*, 185–205 + pl. XI-XVI).

Guner-Ataman, B., González-Rosa, J. M., Shah, H. N., Butty, V. L., Jeffrey, S., Abrial, M., et al. (2018). Failed progenitor specification underlies the cardiopharyngeal phenotypes in a zebrafish model of 22q11.2 deletion syndrome. *Cell Reports*, *24*, 1342–1354.e5.

Hatschek, B. (1892). Die Metamerie des Amphioxus und des Ammocoetes. *Verhandlungen der Anatomischen Gesellschaft*, *6*, 136–161.

Hennig, W. (1973). Molecular hybridization of DNA and RNA in situ. In G. H. Bourne, J. F. Danielli, & K. W. Jeon (Eds.), *International review of cytology* (pp. 1–44). Academic Press.

Herdman, W. A. (1904). Ascidians and amphioxus. In S. F. Harmer & H. E. Shipley (Eds.), *The Cambridge natural history* (pp. 35–138). London: Macmillan.

Holland, L. Z. (2009). Chordate roots of the vertebrate nervous system: Expanding the molecular toolkit. *Nature Reviews Neruroscience*, *10*, 736–746.

Holland, L. Z. (2013). Evolution of new characters after whole genome duplications: Insights from amphioxus. *Seminars in Cell and Developmental Biology*, *24*, 101–109.

Holland, N. D. (2017). The long and winding path to understanding kidney structure in amphioxus—A review. *International Journal of Developmental Biology*, *61*, 683–688.

Holland, N. D. (2020). Hunting needles in a haystack: Migrating precursors of epidermal sensory neurons in amphioxus found with serial blockface scanning electron microscopy

(SBSEM). *Acta Zoologica Stockholm*, e-ublished ahead of print. https://doi.org/10.1111/azo.12324.
Holland, L. Z., & Holland, N. D. (2000). Developmental expression of *AmphiWnt1*, an amphioxus gene in the *Wnt1/wingless* subfamily. *Development Genes and Evolution, 210*, 522–524.
Holland, L. Z., & Holland, N. D. (2001). Evolution of neural crest and placodes: Amphioxus as a model for the ancestral chordate. *Journal of Anatomy, 199*, 85–98.
Holland, L. Z., Holland, N. D., & Gilland, E. (2008). The amphioxus and the evolution of head segmentation. *Integrative and Comparative Biology, 48*, 630–646.
Holland, N. D., Holland, L. Z., Honma, Y., & Fujii, T. (1993). *Engrailed* expression during development of a lamprey, Lampetra japonica—A possible clue to homologies between agnathan and gnathostome muscles of the mandibular arch. *Development, Growth & Differentiation, 35*, 153–160.
Holland, P. W. H., Holland, L. Z., Williams, N. A., & Holland, N. D. (1992). An amphioxus homeobox gene: Sequence conservation, spatial expression during development and insights into vertebrate evolution. *Development, 116*, 653–661.
Holland, L. Z., Kene, M., Williams, N. A., & Holland, N. D. (1997). Sequence and embryonic expression of the amphioxus *engrailed* gene (*AmphiEn*): The metameric pattern of transcription resembles that of its segment-polarity homolog in *Drosophila*. *Development, 124*, 1723–1732.
Holland, N. D., Panganiban, G., Henyey, E. L., & Holland, L. Z. (1996). Sequence and developmental expression of *AmphiDll*, an amphioxus *Distal-less* gene transcribed in the ectoderm, epidermis and nervous system: Insights into evolution of craniate forebrain and neural crest. *Development, 122*, 2911–2920.
Holland, N. D., & Yu, J.-K. (2002). Epidermal receptor development and sensory pathways in vitally stained amphioxus (*Branchiostoma floridae*). *Acta Zooligica Stockholm, 83*, 309–319.
Huxley, T. H. (1875). Preliminary note upon the brain and skull of *Amphioxus lanceolatus*. *Proceedings of the Royal Society of London, 23*, 127–132.
Igawa, T., Nozawa, M., Suzuki, D. G., Reimer, J. D., Morov, A. R., Wang, Y., et al. (2017). Evolutionary history of the extant amphioxus lineage with shallow-branching diversification. *Scientific Reports, 7*, 1157. https://doi.org/10.1038/s41598-017-00786-5.
Irie, N., Satoh, N., & Kuratani, S. (2018). The phylum vertebrata: A case for zoological recognition. *Zoological Letters, 4*, 32. https://doi.org/10.1186/s40851-018-0114-y.
Irimia, M., Pineiro, C., Maeso, I., Gomez-Skarmeta, J. L., Casares, F., & Garcia-Fernandez, J. (2010). Conserved developmental expression of *Fezf* in chordates and *Drosophila* and the origin of the zona limitans intrathalamica (ZLI) brain organizer. *EvoDevo, 1*(1), 7. https://doi.org/10.1186/2041-9139-1-7.
Jacobson, A. G. (1988). Somitomeres: Mesodermal segments of vertebrate embryos. *Development, 104*(Suppl), 209–220.
Janvier, P. (2011). Comparative anatomy: All vertebrates do have vertebrae. *Current Biology, 21*, R661–R663.
Kaltenbach, S. L., Holland, L. Z., Holland, N. D., & Koop, D. (2009). Developmental expression of the three *iroquois* genes of amphioxus (*BfIrxA, BfIrxB*, and *BfIrxC*) with special attention to the gastrula organizer and anteroposterior boundaries in the central nervous system. *Gene Expression Patterns, 9*, 329–334.
Kaltenbach, S. L., Yu, J.-K., & Holland, N. D. (2009). The origin and migration of the earliest-developing sensory neurons in the peripheral nervous system of amphioxus. *Evolution and Development, 11*, 142–151.
Kingsbury, B. F. (1920). The developmental origin of the notochord. *Science, 51*, 190–193.
Kluge, B., Renault, N., & Rohr, K. B. (2005). Anatomical and molecular reinvestigation of lamprey endostyle development provides new insight into thyroid gland evolution. *Development Genes and Evolution, 215*, 32–40.

Knecht, A. K., & Bronner-Fraser, M. (2002). Induction of the neural crest: A multigene process. *Nature Reviews Genetics*, *3*, 453–461.

Kokubo, N., Matsuura, M., Onimaru, K., Tiecke, E., Kuraku, S., Kuratani, S., et al. (2010). Mechanisms of heart development in the Japanese lamprey, *Lethenteron japonicum*. *Evolution and Development*, *12*, 34–44.

Koltzoff, N. K. (1902). Entwickelungsgeschichte des Kofes von *Petromyzon planeri*. *Bulletin de la Société Impériale des Naturalistes de Moscou N S*, *15*, 259–589 + pl. I-VII).

Kon, T., Nohara, M., Yamanoue, Y., Fujiwara, Y., Nishida, M., & Nishikawa, T. (2007). Phylogenetic position of a whale-fall lancelet (Cephalochordata) inferred from whole mitochondrial genome sequences. *BMC Evolutionary Biology*, *7*, 1–12.

Koop, D., Chen, J., Theodosiou, M., Carvalho, J. E., Alvarez, S., de Lera, A. R., et al. (2014). Roles of retinoic acid and *Tbx1/10* in pharyngeal segmentation: Amphioxus and the ancestral chordate condition. *EvoDevo*, *5*(1), 36. https://doi.org/10.1186/2041-9139-5-36.

Kowalevsky, A. (1867). Entwickelungsgeschichte des Amphioxus lanceolatus. In *Mémoires de l'Académie impériale des sciences de St Pétersbourg 5e série (Ser VII)* . *11*, 1–17 + pl. I-III).

Kozmik, Z., Holland, N. D., Kalousova, A., Paces, J., Schubert, M., & Holland, L. Z. (1999). Characterization of an amphioxus paired box gene, *AmphiPax2/5/8*: Developmental expression patterns in optic support cells, nephridium, thyroid-like structures and pharyngeal gill slits, but not in the midbrain-hindbrain boundary region. *Development*, *126*, 1295–1304.

Kuratani, S., & Adachi, N. (2016). What are head cavities?—A history of studies on vertebrate head segmentation. *Zoological Science*, *33*, 213–228.

Lacalli, T. C. (1996). Frontal eye circuitry, rostral sensory pathways, and brain organization in amphioxus larvae: Evidence from 3D reconstructions. *Philosophical Transactions of the Royal Society B*, *351*, 243–263.

Lacalli, T. C. (2004). Sensory systems in amphioxus: A window on the ancestral chordate condition. *Brain, Behavior and Evolution*, *64*, 148–162.

Lacalli, T. (2018). Amphioxus, motion detection, and the evolutionary origin of the vertebrate retinotectal map. *EvoDevo*, *9*, 6. https://doi.org/10.1186/s13227-018-0093-2.

Lacalli, T. C., & Hou, S. F. (1999). A reexamination of the epithelial sensory cells of amphioxus (*Branchiostoma*). *Acta Zoologica Stockholm*, *80*, 125–134.

Lara-Ramírez, R., Poncelet, G., Patthey, C., & Shimeld, S. M. (2017). The structure, splicing, synteny and expression of lamprey *COE* genes and the evolution of the *COE* gene family in chordates. *Development Genes and Evolution*, *227*, 319–338.

Mahadevan, N. R., Horton, A. C., & Gibson-Brown, J. J. (2004). Developmental expression of the amphioxus *Tbx1/10* gene illuminates the evolution of vertebrate branchial arches and sclerotome. *Development Genes and Evolution*, *214*, 559–566. https://doi.org/10.1007/s00427-004-0433-1.

Mallatt, J. (1996). Ventilation and the origin of jawed vertebrates: A new mouth. *Zoological Journal of the Linnean Society*, *117*, 329–404.

Mallatt, J., & Holland, N. (2013). *Pikaia gracilens* Walcott: Stem chordate, or already specialized in the Cambrian? *Journal of Experimental Zoology Part B: Molecular and Developmental Evolution*, *320*, 247–271.

Mallatt, J., & Sullivan, J. (1998). 28S and 18S rDNA sequences support the monophyly of lampreys and hagfishes. *Molecular Biology and Evolution*, *15*, 1706–1718.

Martik, M. L., Gandhi, S., Uy, B. R., Gillis, J. A., Green, S. A., Simoes-Costa, M., et al. (2019). Evolution of the new head by gradual acquisition of neural crest regulatory circuits. *Nature*, *574*, 675–678.

Mayr, E. (1963). *Animal species and evolution*. Boston, MA: Harvard University Press.

McGinnis, W., Garber, R. L., Wirz, J., Kuroiwa, A., & Gehring, W. J. (1984). A homologous protein-coding sequence in drosophila homeotic genes and its conservation in other metazoans. *Cell*, *37*, 403–408.

Meier, S. (1979). Development of the chick embryo mesoblast: Formation of the embryonic axis and establishment of the metameric pattern. *Developmental Biology*, *73*, 25–45.

Meulemans, D., & Bronner-Fraser, M. (2004). Gene-regulatory interactions in neural crest evolution and development. *Developmental Cell*, 7, 291–299.

Neal, H. V. (1918). Neuromeres and metamers. *Journal of Morphology*, *31*, 292–315.

Nicol, D., & Meinertzhagen, I. A. (1991). Cell counts and maps in the larval central nervous system of the ascidian *Ciona intestinalis* (L.). *The Journal of Comparative Neurology*, *309*, 415–429.

Noden, D. M., & Trainor, P. A. (2005). Relations and interactions between cranial mesoderm and neural crest populations. *Journal of Anatomy*, *207*, 575–601.

Nohara, M., Nishida, H., Miya, M., & Nishikawa, T. (2005). Evolution of the mitochondrial genome in Cephalochordata as inferred from complete nucleotide sequences from two *Epigonichthys* species. *Journal of Molecular Evolution*, *60*, 526–537.

Olsson, R. (1955). Structure and development of Reissner's fibre in the caudal end of amphioxus and some lower vertebrates. *Acta Zoologica Stockholm*, *36*, 167–198.

Olsson, L., Ericsson, R., & Cerny, R. (2005). Vertebrate head development: Segmentation, novelties, and homology. *Theory in Biosciences*, *124*, 145–163.

Olsson, R., Yulis, R., & Rodriguez, E. M. (1994). The infundibular organ of the lancelet (*Branchiostoma lanceolatum*, Acrania): An immunocytochemical study. *Cell and Tissue Research*, *277*, 107–114.

Ota, K. G. (2018). Recent advances in hagfish developmental biology in a historical context: Implications for understanding the evolution of the vertebral elements. In K. Kobayashi, T. Kitano, Y. Iwao, & M. Kondo (Eds.), *Reproductive and developmental strategies: The continuity of life* (pp. 615–634). Tokyo, Japan: Springer.

Ota, K. G., Fujimoto, S., Oisi, Y., & Kuratani, S. (2011). Identification of vertebra-like elements and their possible differentiation from sclerotomes in the hagfish. *Nature Communications*, *2*, 1–6.

Ota, K. G., Kuraku, S., & Kuratani, S. (2007). Hagfish embryology with reference to the evolution of the neural crest. *Nature*, *446*, 672–675.

Pallas, P. S. (1774). *Spicilegia zoologica, Vol 1, fascicle 10*. Berlin: G. A. Lange.

Parker, H. J., Bronner, M. E., & Krumlauf, R. (2019). An atlas of anterior *Hox* gene expression in the embryonic sea lamprey head: *Hox*-code evolution in vertebrates. *Developmental Biology*, *453*, 19–33.

Pauling, L., & Zuckerkandl, E. (1963). Chemical paleogenetics. Molecular "restoration studies" of extinct forms of life. *Acta Chemica Scandavica*, *17*, S9–S16.

Piotrowski, T., Ahn, D.-g., Schilling, T. F., Nair, S., Ruvinsky, I., Geisler, R., et al. (2003). The zebrafish *van gogh* mutation disrupts *tbx1*, which is involved in the DiGeorge deletion syndrome in humans. *Development*, *130*, 5043–5052.

Prummel, K. D., Hess, C., Nieuwenhuize, S., Parker, H. J., Rogers, K. W., Kozmikova, I., et al. (2019). A conserved regulatory program initiates lateral plate mesoderm emergence across chordates. *Nature Communications*, *10*, 3857.

Pu, G. A., & Dowling, J. E. (1981). Anatomical and physiological characteristics of pineal photoreceptor cell in the larval lamprey, *Petromyzon marinus*. *Journal of Neurophysiology*, *46*, 1018–1038.

Rasmussen, N. (1991). The decline of recapitulationism in early twentieth-century biology: Disciplinary conflict and consensus of the battleground of theory. *Journal of the History of Biology*, *24*, 51–89.

Rasmussen, S. L. K., Holland, L. Z., Schubert, M., Beaster-Jones, L., & Holland, N. D. (2007). Amphioxus *AmphiDelta*: Evolution of Delta protein structure, segmentation, and neurogenesis. *Genesis*, *45*, 113–122.

Romer, A. S. (1956). *A shorter version of the second edition of The Vertebrate Body*. Philadelphia, London: W.B. Saunders, Co, 486.

Ruiz, M. S., & Anadon, R. (1991). Some considerations on the fine structure of rhabdomeric photoreceptors in the amphioxus, *Branchiostoma lanceolatum* (Cephalochordata). *Journal für Hirnforschung*, *32*, 159–164.

Ruppert, E. E. (1994). Evolutionary origin of the vertebrate nephron. *American Zoologist*, *34*, 542–553.

Sauka-Spengler, T., Le Mentec, C., Lepage, M., & Mazan, S. (2002). Embryonic expression of *Tbx1*, a DiGeorge syndrome candidate gene, in the lamprey *Lampetra fluviatilis*. *Gene Expression Patterns*, *2*, 99–103.

Schlosser, G. (2017). From so simple a beginning—What amphioxus can teach us about placode evolution. *International Journal of Developmental Biology*, *61*, 633–648.

Schubert, M., Holland, N. D., Laudet, V., & Holland, L. Z. (2006). A retinoic acid-Hox hierarchy controls both anterior/posterior patterning and neuronal specification in the developing central nervous system of the cephalochordate amphioxus. *Developmental Biology*, *296*, 190–202.

Schubert, M., Holland, L. Z., Panopoulou, G. D., Lehrach, H., & Holland, N. D. (2000). Characterization of amphioxus *AmphiWnt8*: Insights into the evolution of patterning of the embryonic dorsoventral axis. *Evolution and Development*, *2*, 85–92.

Scott, M. P., & Weiner, A. J. (1984). Structural relationships among genes that control development: Sequence homology between the *Antennapedia*, *Ultrabithorax*, and *fushi tarazu* loci of *Drosophila*. *Proceedings of the National Academy of Sciences of the United States of America*, *81*, 4115–4119.

Sharman, A. C., Shimeld, S. M., & Holland, P. W. H. (1999). An amphioxus *Msx* gene expressed predominantly in the dorsal neural tube. *Development Genes and Evolution*, *209*, 260–263.

Simakov, O., Marletaz, F., Yue, J.-X., O'Connell, B. O., Jenkins, J., Brandt, A., et al. (2020). Deeply conserved synteny resolves early events in vertebrate evolution. *Nature Ecology and Evolution*, *4*, 820–830. https://doi.org/10.1038/s41559-020-1156-z.

Smith, A. C., Fleenor, S. J., & Begbie, J. (2015). Changes in gene expression and cell shape characterise stages of epibranchial placode-derived neuron maturation in the chick. *Journal of Anatomy*, *227*, 89–102.

Sterba, G., Fredriksson, G., & Olsson, R. (1983). Immunocytochemical investigations of the infundibular organ in amphioxus (*Branchiostoma lanceolatum*; Cephalochordata). *Acta Zoologica Stockholm*, *64*, 149–153.

Stokes, M. D., & Holland, N. D. (1995a). Ciliary hovering in larval lancelets (= amphioxus). *Biological Bulletin*, *188*, 231–233.

Stokes, M. D., & Holland, N. D. (1995b). Embryos and larvae of a lancelet, *Branchiostoma floridae*, from hatching through metamorphosis—Growth in the laboratory and external morphology. *Acta Zoolologica Stockholm*, *76*, 105–120.

Subirana, L., Farstey, V., Bertrand, S., & Escriva, H. (2020). *Asymmetron lucayanum*: How many species are valid? *PLoS One*, *15*, e0229119.

Sugii, H., Grimaldi, A., Li, J., Parada, C., Vu-Ho, T., Feng, J., et al. (2017). The *Dlx5-FGF10* signaling cascade controls cranial neural crest and myoblast interaction during oropharyngeal patterning and development. *Development*, *144*, 4037–4045.

Tautz, D., & Pfeifle, C. (1989). A non-radioactive in situ hybridization method for the localization of specific RNAs in *Drosophila* embryos reveals translational control of the segmentation gene hunchback. *Chromosoma*, *98*, 81–85.

Tiecke, E., Matsuura, M., Kokubo, N., Kuraku, S., Kusakabe, R., Kuratani, S., et al. (2007). Identification and developmental expression of two *Tbx1/10*-related genes in the agnathan *Lethenteron japonicum*. *Development Genes and Evolution*, *217*, 691–697.

Venkatesh, B., Lee, A. P., Ravi, V., Maurya, A. K., Lian, M. M., Swann, J. B., et al. (2014). Elephant shark genome provides unique insights into gnathostome evolution. *Nature*, *505*, 174–179.

Vyas, B., Nandkishore, N., & Sambasivan, R. (2019). Vertebrate cranial mesoderm: Developmental trajectory and evolutionary origin. *Cellular and Molecular Life Sciences*, *77*, 1933–1945. https://doi.org/10.1007/s00018-019-03373-1.

Wada, H., & Satoh, N. (1994). Details of the evolutionary history from invertebrates to vertebrates, as deduced from the sequences of 18S rDNA. *Proceedings of the National Academy of Sciences of the United States of America*, *91*, 1801–1804.

Wang, H., Holland, P. W. H., & Takahashi, T. (2019). Gene profiling of head mesoderm in early zebrafish development: Insights into the evolution of cranial mesoderm. *EvoDevo*, *10*, 14. https://doi.org/10.1186/s13227-019-0128-3.

Willey, A. (1894). *Amphioxus and the ancestry of the vertebrates*. New York: Macmillan, 316.

Williams, N. A., & Holland, P. W. H. (1996). Old head on young shoulders. *Nature*, *383*, 490.

Yue, J.-X., Kozmikova, I., Ono, H., Nossa, C. W., Kozmik, Z., Putnam, N. H., et al. (2016). Conserved noncoding elements in the most distant genera of cephalochordates: The Goldilocks principle. *Genome Biology and Evolution*, *8*, 2387–2405.

Yue, J.-X., Yu, J.-K., Putnam, N. H., & Holland, L. Z. (2014). The transcriptome of an amphioxus, *Asymmetron lucayanum*, from the Bahamas: A window into chordate evolution. *Genome Biolology and Evolution*, *6*, 2681–2696.

Zieger, E., Garbarino, G., Robert, N. S. M., Yu, J.-K., Croce, J. C., Candiani, S., et al. (2018). Retinoic acid signaling and neurogenic niche regulation in the developing peripheral nervous system of the cephalochordate amphioxus. *Cellular and Molecular Life Sciences*, *75*, 2407–2429.

Zuckerkandl, E., & Pauling, L. (1965). Molecules as documents of evolutionary history. *Journal of Theoretical Biology*, *8*, 357–866.

CHAPTER FIVE

Transitional chordates and vertebrate origins: Tunicates

Alexander Fodor[a,b], Jiatai Liu[a,b], Lindsay Turner[a,b], and Billie J. Swalla[a,b,*]
[a]Biology Department, University of Washington, Seattle, WA, United States
[b]Friday Harbor Laboratories, Friday Harbor, WA, United States
*Corresponding author: e-mail address: bjswalla@uw.edu

Contents

Abstract

The Origin of Chordates has fascinated scientists from the time of Charles Darwin's publication "Descent of Man" in 1871. For over 100 years, it was accepted that chordates evolved from tunicates, our sessile invertebrate sister group. However, genomic and embryonic analyses have shown that lancelets have a body plan and genome much more like vertebrates than do tunicates. In 2000, we proposed a worm-like hypothesis of chordate origins, and genomic and embryonic studies in the past 20 years have supported this hypothesis. This hypothesis contends that the deuterostome ancestor was worm-like, with gill slits, very much like a chordate. In contrast, tunicates have a very derived adult body plan that evolved independently. Here, we review the current understanding of deuterostome phylogeny and supporting evidence for the relationships within each phylum. Then we discuss our hypothesis for chordate origins and evidence to support it. We explore some of the evolutionary changes that ascidians have made to their adult body plan and some of the key gene regulatory networks that have been elucidated in *Ciona*. Finally, we end with insights that we have gained from studying tailless ascidians for the past 30 years. We've found that differentiation genes, at the end of the gene regulatory networks, become pseudogenes and nonfunctional, even though they are still expressed in tailless ascidians. We expect that eventually these pseudogenes will not be expressed and the ascidian larval body plan is abandoned, leaving the embryo to develop directly into an adult.

Current Topics in Developmental Biology, Volume 141
ISSN 0070-2153
https://doi.org/10.1016/bs.ctdb.2020.10.001

1. Deuterostome phylogeny and chordate origins

The Deuterostomia cluster into two great clades—Chordata and Ambulacraria (Fig. 1) (Swalla, 2001; Swalla & Smith, 2008). Various animal groups have at one time or another been included in the deuterostomes due to developmental traits or phylogenomics, including chaetognaths, acoels and xenoturbella (Bourlat et al., 2006; Swalla & Smith, 2008). However, chaetognaths are now seen as an outgroup to the Lophotrochozoa (within the gnathiferan spiralian protostomes, Marlétaz, Peijnenburg, Goto, Satoh, & Rokhsar, 2019), and their deuterostome-like development has been attributed to a more ancestral-like pattern of development (Martín-Durán, Passamaneck, Martindale, & Hejnol, 2016). Likewise, including new taxa

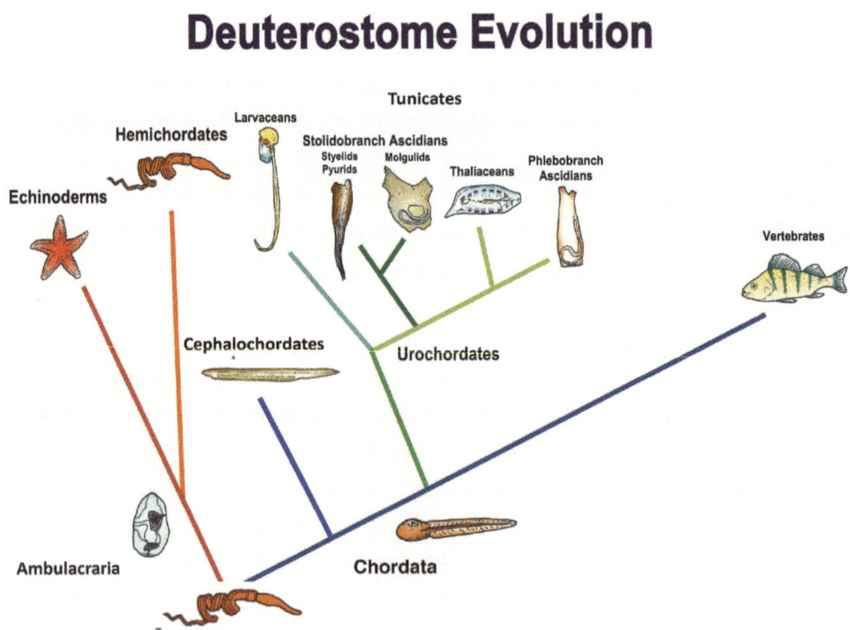

Fig. 1 Deuterostome Phylogeny, showing the current understanding of deuterostome relationships. Phylogenomic studies suggest that the tunicates are the sister group to the vertebrates and that lancelets (cephalochordates) are the basal group of chordates. The Ambulacraria consists of the echinoderms and hemichordates, which share a similar feeding larval type. *Modified from Swalla, B. J. (2001). Phylogeny of the urochordates: Implications for chordate evolution. In Hitoshi Sawada and Charles Lambert (Eds.),* Biology of ascidians (proceedings of the first international symposium on the biology of ascidians) *(pp. 219–224). Tokyo: Springer Verlag.*

and further phylogenomic analyses has suggested that the Acoela, Nemertodermatida, and *Xenoturbella*, all bilaterally symmetrical marine worms that lack several features common to most other bilaterians (for example, an anus, nephridia, and a circulatory system) are an outgroup to the bilateria (Cannon et al., 2016). Deuterostome development and evolution are important for understanding the origin of chordates, but the scanty fossil record of enteropneust hemichordates and tunicates makes it difficult to visualize the ancestor (Swalla & Smith, 2008).

Ambulacraria are made up of the Echinodermata and Hemichordata (Fig. 1). They have a bilateral larval body plan that corresponds with the animal-vegetal axis of the egg (Range, Angerer, & Angerer, 2013; Swalla, 2006). The enteropneust hemichordate larval and adult body plan has the same axes as the larvae (Brown, Prendergast, & Swalla, 2008; Röttinger & Lowe, 2012), while in echinoderms, the adult body plan is frequently pentaradial and has different axes than the larvae (Raff & Byrne, 2006; Swalla, 2006).

The other Deuterostome clade is the Chordata, distinguished by a notochord, dorsal neural tube, and a muscular, post-anal tail (Fig. 1). Phylogenomic analyses place the Tunicata as a sister group to the Vertebrata, with the Branchiostoma, or lancelets as the outgroup (Fig. 1; Delsuc, Philippe, Tsagkogeorga, et al., 2018). Clearly, the elongate Branchistoma adult body plan (Holland, 2015) and genome more closely resemble vertebrates, with extensive synteny found in their genomes, even though vertebrates have undergone two genome duplications (Putnam, Butts, Ferrier, et al., 2008).

2. Hemichordate development and evolution

Hemichordata are a sister group to Echinodermata within Ambulacraria (Fig. 1; Cannon et al., 2014). Hemichordates are currently organized into two different classes: the solitary, vermiform Enteropneusta and the colonial, branching Pterobranchia (Cannon et al., 2014; Tassia et al., 2016). Enteropneusts include two major groups, Ptychoderidae and Harrimaniidae (Cannon et al., 2014; Tassia et al., 2016). Hemichordates ("half chordates") have proved to be extremely useful for understanding chordate development and evolution (Brown et al., 2008; Cameron, Garey, & Swalla, 2000; Lowe et al., 2006; Chapter "Transitional chordates and vertebrate origins: Hemichordates" by Lowe, this volume). Enteropneust hemichordates, or acorn worms, likely resemble the deuterostome ancestor more closely than any other extant animal (Cameron et al., 2000; Tassia et al., 2016). Some direct-developing Harrimaniidae, like *Saccoglossus kowalevskii*, exhibit embryonic and larval spatial expression of central nervous system genes which code for structures in the same anterior-posterior axis order as that of vertebrates (Lowe et al., 2003; Röttinger & Lowe, 2012). Hemichordates exhibit bilateral symmetry throughout their life cycle (Röttinger & Lowe, 2012; Rychel & Swalla, 2009) and ptychoderid hemichordates undergo vertebrate-like neurulation (Luttrell, Konikoff, Byrne, Bengtsson, & Swalla, 2012).

Hemichordates are a sister taxon to Echinoderms, and both groups have species with remarkable regeneration (Rychel & Swalla, 2009). While some chordates are able to regenerate part of their body and nervous system, no chordate is able to regenerate all regions of its central nervous system after bisection (Luttrell, Su, & Swalla, 2018). *S. kowalevskii* lacks the regenerative abilities of indirect developing hemichordates, but *Ptychodera flava* exhibits an exceptional ability to regenerate its entire central nervous system and anterior structures in as little as 2 weeks in both planktonic larvae (Luttrell et al., 2018) and adults (Arimoto & Tagawa, 2015; Luttrell, Gotting, Ross, Alvarado, & Swalla, 2016; Rychel & Swalla, 2008). When an adult *P. flava* is cut in half, each half completely regenerates the missing parts, including the anterior proboscis, collar, and dorsal neural tube (Arimoto & Tagawa, 2015; Luttrell et al., 2016; Rychel & Swalla, 2008). *P. flava* shows complete CNS regeneration that may provide insights into achieving neural regeneration in other chordates (Luttrell et al., 2016). Cellular processes of anterior regeneration showed extensive programmed cell death, followed by cell proliferation, as new structures were generated, by using TUNEL and PCNA assays (Rychel & Swalla, 2008). *P. flava* regeneration is likely to be nerve-dependent, and gut endoderm remodeling plays an important role in regeneration (Rychel & Swalla, 2008). Transcriptome analyses of the genes that are upregulated and downregulated during the whole-body regeneration of *P. flava* have revealed significant differential gene expression between regenerating and control animals (Arimoto & Tagawa, 2015; Luttrell et al., 2016). Though adult *P. flava* have been a useful model for understanding adult regeneration, its natural yellow-brown color and thick tissue layers have made it difficult to study adult stem cell recruitment.

P. flava larvae are clear and also completely regenerate if part of the gut is left in the larval fragments (Luttrell et al., 2018). Robust regeneration in *P. flava* larvae has been described in all structures, including their nervous system, eyes, gut, mouth, and ciliary bands (Luttrell et al., 2018). Further studies on the regeneration of *P. flava* larvae may yield insight into the source of stem cells during regeneration.

In summary, hemichordates have contributed greatly to our understanding of deuterostome evolution, chordate origins and regeneration. There are a number of labs around the world continuing research on hemichordates and there are sure to be more important and interesting studies emerging (see Lowe, this volume). These studies will be facilitated by the recent whole genome sequencing of two model hemichordate species, *S. kowalevskii* and *P. flava* (Simakov, Kawashima, Marlétaz, et al., 2015).

3. Tunicate development and evolution

Tunicates are a fascinating, monophyletic group of animals that possess an outer tunic (Fig. 1; Delsuc et al., 2018; Shenkar & Swalla, 2011), named by Jean-Baptiste Lamarck in 1816 in the first invertebrate textbook "Histoire Naturelle des Animaux sans Vertébres." Lamarck correctly identified Thaliacea, Appendicularia and Ascidiacea as a monophyletic group (Lamark, 1816). Recent phylogenetic and phylogenomic studies have placed the pelagic and colonial Thaliacea nested within the Ascidiacea, in a monophyletic group that includes the colonial aplusobranch ascidians, the solitary and colonial phlebobranch ascidians (Fig. 1, Delsuc et al., 2018; Swalla, Cameron, Corley, & Garey, 2000). Assuming the ancestral ascidians were all sessile, then Thaliacea would have evolved a free-living, planktonic lifestyle from a colonial ascidian ancestor (Delsuc et al., 2018; Swalla et al., 2000). The solitary and pelagic Appendicularia have very small genomes and short lifetimes, resulting in long branches for most of their genome that make them difficult to place phylogenetically (Delsuc et al., 2018; Swalla et al., 2000).

There are nearly 3000 species of Ascidiacea (ascidians) described on the World Register of Marine Species (WoRMS) (Shenkar et al., 2020; Shenkar & Swalla, 2011) and it is estimated that there are probably another 1000 or so species that have not yet been described (WoRMS; Appeltans et al., 2012). Ascidians are chordates that usually have a tadpole stage with anterior sensory organs in the head, the ocellus and otolith, and a tail, which contains a notochord, dorsal neural tube and muscle cells (Swalla, 2004). Ascidian egg cleavage is determinant and invariant, and much is known about the gene networks for key larval ascidian tissues, especially the notochord cells that develop in the center of the tail (Di Gregorio, 2020), muscle cells (Razy-Krajka & Stolfi, 2019), and the nervous system connectome (Ryan, Lu, & Meinertzhagen, 2016). After metamorphosis, the tadpole larval tail is absorbed, undergoes programmed cell death and is used as an energy source for the developing adult (Karaiskou, Swalla, Sasakura, & Chambon, 2015).

After Lamarck correctly identified the tunicates (Lamark, 1816), they were subsequently considered mollusks until Kowalevsky (1866) described the ascidian embryos and the tadpole larvae and noted that they have a notochord, properly placing the tunicates within the Chordata. The notochord, one of the defining tissues of the chordates, has been extensively studied in ascidians (Di Gregorio, 2020; Jiang & Smith, 2007). The notochord cells in solitary ascidians are initially specified at the 32-cell stage by the tbox transcription factor, *Brachyury* (Yasuo & Satoh, 1993), and *FoxD* (Imai, Satoh, & Satou, 2002). Eight presumptive notochord cells are initially specified, just anterior to the endoderm, then two secondary cells are specified, making a total of ten notochord precursors on either side of the bilateral embryo (Fig. 2; Yasuo & Satoh, 1993). The ten notochord precursor cells undergo two rounds of division, making forty cells that converge and extend into a single row of cells, elongating the tail primordium that then is extended by the subsequent swelling of the cells (Jiang & Smith, 2007). Endoderm can be converted to notochord by overexpression of *brachyury*, but not other germ layers, such as mesodermal muscle or ectodermal epidermis (Takahashi et al., 1999). Ascidian embryos have allowed the characterization of the notochord gene network from initial *Brachyury* expression to the identification of the downstream genes that give the notochord its structural properties (Di Gregorio, 2020; Jiang & Smith, 2007). The notochord developmental gene network begins when maternal β-catenin is translocated from the cytoplasm to the nucleus of mesendodermal cells during the 16 cell stage, and directly activates *FoxD* through TCF/LEF binding sites (Imai et al., 2002). The activation and evolution of the notochord gene network in ascidians has been recently extensively reviewed (Di Gregorio, 2020) and can be followed from early development until the eventual caspase-dependent death of the notochord during ascidian metamorphosis (Fig. 3; Karaiskou et al., 2015).

The solitary ascidian nonfeeding tadpole larva lacks a heart, blood, and digestive system and therefore has been an excellent model system for understanding developmental gene networks, as these tissues are specified in the embryo, but do not form as organs until they metamorphose (Davidson, 2007; Franchi & Ballarin, 2017; Karaiskou et al., 2015). After hatching, ascidian larvae swim up toward light, using their light-sensing organ, the ocellus (Tsuda, Sakurai, & Goda, 2003). Once they are competent to undergo to metamorphose, the tadpole larvae then swim toward the sea

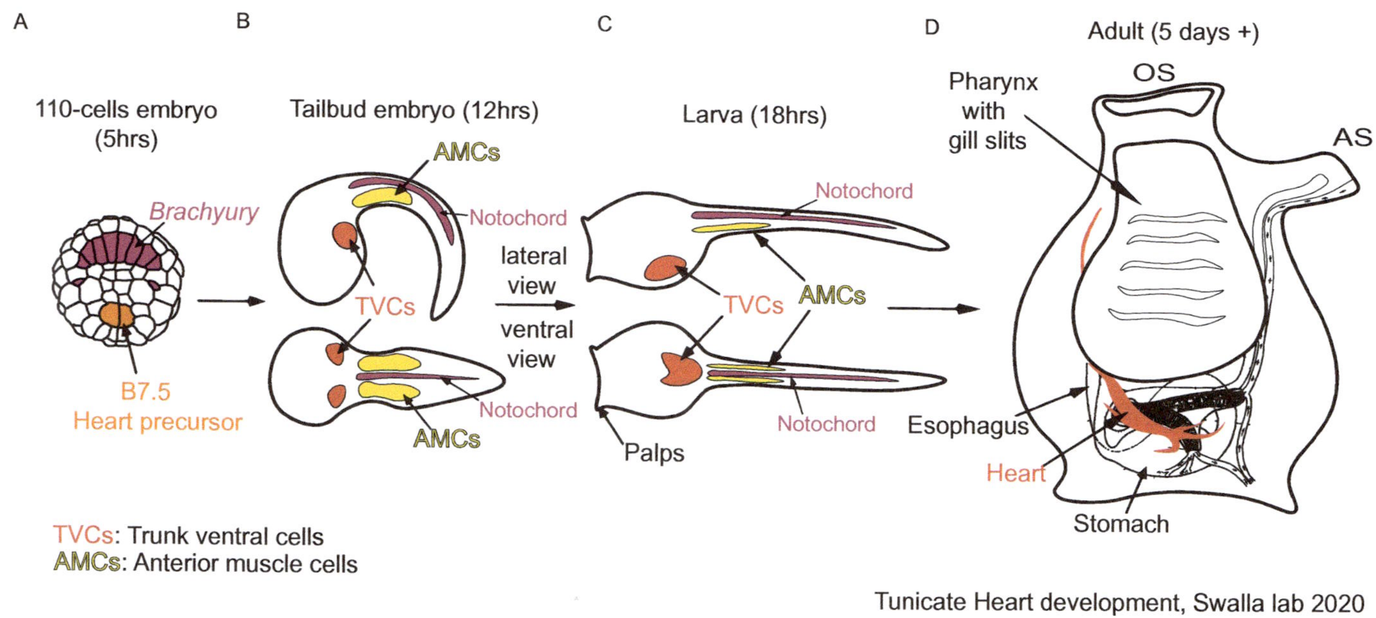

Fig. 2 See legend on next page.

floor, using its the otolith, a gravity sensing organ, to guide them (Tsuda et al., 2003). After metamorphosis, the adult ascidian will reorganize a U-shaped gut, with the incurrent and excurrent siphons pointed upwards, and remain sessile through the end of its life (Karaiskou et al., 2015). The mouth of the ascidian remains the mouth, or incurrent siphon, of the adult; but the excurrent siphon is derived from the otic placode that fuses in the center from either side of the larval head in *Ciona* larvae (Kourakis, Newman-Smith, & Smith, 2010). Thus, ascidians, loop their anus out their "ear" and remain sessile as adults, an odd choice for a chordate.

Phylogenomic evidence places tunicates as the closest invertebrate relative to vertebrates (Delsuc et al., 2018), yet adult ascidians clearly have modified adult body plans that look nothing like the larvae (Figs. 2 and 3). Most of the adult tissues are set aside as primordia during embryonic development and continue development through the larval stage to the final adult organ or tissue (Davidson, 2007; Karaiskou et al., 2015). We will review some of the analyses of gene networks that have given insight into the modifications of

Fig. 2 Ascidian heart and notochord development from embryo to larva to adult. Heart development through ascidian embryonic stages with color-coded cell types labeled by arrows. Legend for color labels shared in this diagram: trunk ventral cells (TVCs; red), anterior muscle cells (AMCs; yellow), notochord and *Brachyury* expression (Pink). (A) Diagram of the bilaterally symmetrical *Ciona* 110-cell embryo stage developed around 5h. A pair of heart lineage cells B7.5 cells emerge in the posterior, color-coded in orange. Total of 10 notochord precursor cells are anterior. (B) Diagram of the tailbud embryo developed around 12h, B7.5 lineage cells have divided into two different cell types: TVCs and AMCs, located on each lateral side of the embryo. Notochord precursor cells have divided twice and developed into a 40 cell notochord, located in the tail of the embryo. (C) Diagram of the swimming larva developed around 18h. At this stage, TVCs fuse ventrally together, while AMCs stays on the lateral side. Palps also form at this stage, necessary for larva settlement and metamorphosis. (D) Diagram of an adult *Ciona* ascidian developed above 5 days. The true heart starts forming after ascidian enter the juvenile stage (about 3 days of development, see Fig. 3). At this stage, half of the B7.5 lineage cells (TVCs) generate a two-compartment heart with two pacemaker cells on both sides, color-coded in red. The 40 cells notochord and the other half of the B7.5 lineage cells (AMCs) have gone through programmed cell death with the tail during metamorphosis. Oral siphon (OS) and atrial siphon (AS) are fully developed at this stage, located at the top of the adult ascidian. Details of the heart and digestive system of the adult ascidian are shown, arrows indicate fluid and blood flow in one direction, though ascidian hearts will periodically reverse their beat and pump blood in the opposite direction (Razy-Krajka & Stolfi, 2019). *Panel (A–C) redrawn from Satou, Y., Imai, K. S., & Satoh, N. (2004). The ascidian Mesp gene specifies heart precursor cells.* Development, 131*(11), 2533–2541; Razy-Krajka, F., & Stolfi, A. (2019) Regulation and evolution of muscle development in tunicates.* EvoDevo, 10*(1), 13*

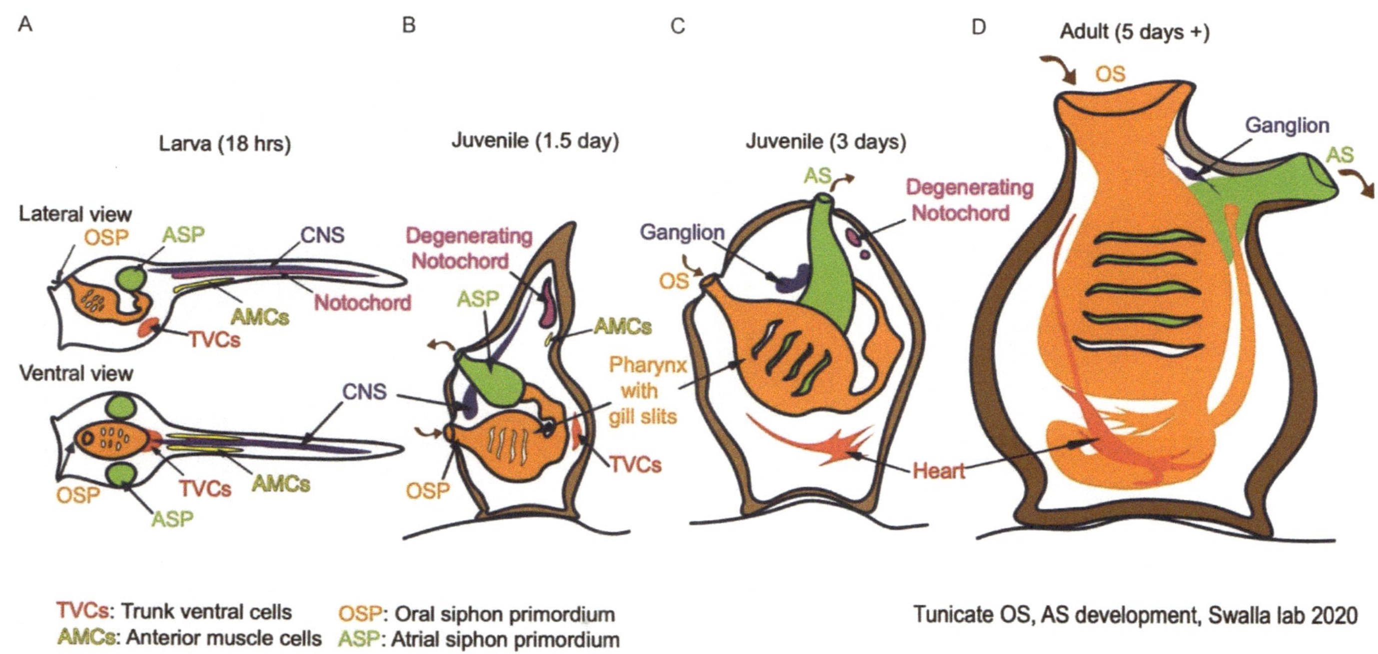

Fig. 3 See legend on next page.

morphology that have occurred during tunicate evolution for several different adult tissues, the innate immune system, the heart, branchial slits and the excurrent siphons.

Ascidian tadpole larvae lack blood and immune cells when they hatch from the chorion and then develop their innate immunity as they are swimming, even though some complement transcripts are made in the embryonic endoderm (Davidson & Swalla, 2002; Franchi & Ballarin, 2017). In fact, competency to settle in *Boltenia villosa* may be directly due to the development of the complement system, because inhibition of the complement system blocks metamorphosis (Roberts et al., 2007). In the adult ascidian, the immune system is active in the tunic, the hemolymph and the gut, where non-self molecules and cells may be found (Franchi & Ballarin, 2017). In fact, a recent link between cell surface integrins and the complement system may further our knowledge of

Fig. 3 Adult ascidian siphon development from embryo to larva to adult. This diagram indicates different ascidian developmental stages with color-coded cell types labeled by arrows. Color labels shared in this diagram: trunk ventral cells (TVCs), heart (red), anterior muscle cells (AMCs; yellow), oral siphon primordium (OSP) and oral siphon (orange), atrial siphon primordium (ASP) and atrial siphon (green), Central Nervous System (CNS) and Ganglion (dark blue), Notochord (pink), Brown arrows indicate the flow of water, through the OS, into the pharynx, out through the AS. (A) Diagram of the swimming larva developed around 18h. OSP locates on the ventral side of the larva, while ASPs and AMCs locate on each lateral side of the larva. CNS and notochord locate in the tail. (B) Diagram of the settled juvenile ascidian developed around one and a half days. With metamorphoses occurs at this stage, the first gill slits and pharynx had formed, TVCs cell proliferation in becoming a true heart. Ascidian larval tails go through programmed cell death, so degenerating notochord and AMS are shown in the diagram. Siphon formation also occurs at this stage, ASP on each side fuse ventrally together and undergo invagination to form a single opening atrial siphon rudiment, which connects to the pharynx through ciliated gill slits (Kourakis & Smith, 2007). (C) Diagram of the settled juvenile ascidian developed around one 3 days. OS and AS has formed completed, AMCs have completely degenerated, while a part of the notochord still remains. CNS has begun the transformation into the ganglion, the true heart has form at this stage. All body organs slowly rotate as development taking place. (D) Diagram of the fully developed adult ascidian after 5 days. OS and AS have taken place at the top of the ascidian, the true heart formation complete. Ganglion formation complete, locate between OS and AS, the remain notochord has also completely degenerated. *Panel (A) redrawn from Kourakis, M. J., & Smith, W. C. (2007). A conserved role for FGF signaling in chordate otic/atrial placode formation.* Developmental Biology, 312*(1), 245–257.Panel (D) redrawn from Davidson, B., & Swalla, B. J. (2002). A molecular analysis of ascidian metamorphosis reveals elements of an innate immune response.* Development, 129*(20), 4739–4751.*

how innate immunity may influence evolution and development in tunicates and other animals (Merle, Singh, Rahman, & Kemper, 2020).

Heart primordia are set aside in ascidian embryos as early as the gastrula stage, but heart development is delayed until after the tadpole larvae have hatched, and several muscle lineages are specified (Fig. 2; Davidson, 2007; Satou et al., 2004; Racioppi, Wiechecki, & Christiaen, 2019). The ascidian heart primordia are specified as two posterior cells during gastrulation by the expression of a transcription factor *mesp*. Later, during the larval stage, the *mesp* positive heart primordia divide and separate into two lineages—the heart cells in the head and the anterior tail muscle cells in the tail (Fig. 2; Davidson, 2007; Satou et al., 2004). After larval settlement, the heart primordia divide again and one set of cells becomes the heart, while the others become the muscle that contracts the excurrent siphon (Fig. 2; Racioppi et al., 2019; Razy-Krajka & Stolfi, 2019). This is similar to what has been found for the heart primordia in vertebrates, except that in vertebrates there is a split in the heart field and the sister cells become the pharyngeal muscle (Razy-Krajka & Stolfi, 2019). The split between cardiac and pharyngeal muscle precursors happens much earlier (late neurula stage) in *Ciona*, and is coordinated by the asymmetric cell division that places all of the cell's FGF receptors into the presumptive cardiac precursors (Cota & Davidson, 2015). Single cell and ATAC-sequencing have revealed an *FGF-Foxf* pathway works in progenitor cells to facilitate cardiopharyngeal-specific patterns of chromatin accessibility (Racioppi et al., 2019). Multiple cis-regulatory elements were required to activate two key transcription factor genes, *EBF* and *Tbx1/10* that together determine the cardiopharyngeal cell fate (Racioppi et al., 2019).

Ascidian larvae have two spherical shaped atrial siphon primordia located on each lateral side of the larva, inside the epidermis of the body wall (Katz, 1983). These were identified as placodes by gene expression and TEM analysis (Manni et al., 2004; Rigon et al., 2018). One oral siphon primordium is located next the "mouth" of the non-feeding larva and will become the incurrent siphon of the adult (Fig. 3A; Graham & Shimeld, 2013). During the continuing ascidian metamorphosis process, atrial primordia on both sides of the larval head undergo invagination and fuse together to form a single atrial opening siphon, leading to an atrial cavity which connects to the pharynx through ciliated gill slits (Fig. 3B and C; Graham & Shimeld, 2013; Kourakis et al., 2010; Kourakis & Smith, 2007). This fusion requires retinoic-driven *Hox 1* expression in the epidermis, as is also found for the otic placodes in vertebrates (Sasakura et al., 2012). When *Hox 1* is knocked out, the atrial siphon never develops and the juveniles do not survive long (Sasakura et al., 2012). The oral primordium develops into a single oral opening pharynx siphon with gill slits. As ascidians keep developing, both the atrial siphon and the oral siphon rotate from the lateral side of the body, toward the top of the body and eventually become a shared atrium that is observed in the adult (Fig. 3D; Graham & Shimeld, 2013; Kourakis et al., 2010; Sasakura et al., 2012).

3.1 Evolution of coloniality in tunicates

Ascidians are also a great model system to study the evolution of coloniality, as three of the major clades contain colonial species (Delsuc et al., 2018; Shenkar & Swalla, 2011). Solitary ascidians tend to be the dominate form near the poles, while colonial ascidians tend to be the major tunicates of complex coral reefs near the equator (Hiebert, Vieira, Dias, Tiozzo, & Brown, 2019). Colonial ascidians consist of individual zooids that share blood, nutrients and gametes (Brown & Swalla, 2012; Kassmer, Delany Rodriguez, & De Tomaso, 2016; Manni et al., 2019). Colonial ascidians can reproduce sexually, by outcrossing with another colony, or asexually, by various forms of budding (Brown & Swalla, 2012; Manni et al., 2019). Colonial ascidians have been used to study immunology, the evolution of self-nonself incompatibility systems, and whole-body regeneration (Kassmer et al., 2016).

3.2 Evolution of tailless ascidians

In one ascidian family, the Molgulidae, tailless ascidians have independently evolved several times (Huber, Burke da Silva, Bates, & Swalla, 2000; Maliska, Pennell, & Swalla, 2013). We are investigating two closely related species, *Molgula oculata* and *Molgula occulta* that hybridize easily in the lab (Fig. 4; Swalla & Jeffery, 1990, 1996; Swalla, Makabe, Satoh, & Jeffery, 1993). *M. oculata* is a tailed ascidian, similar to other solitary ascidian larvae, such as *Ciona*. In contrast, *M. occulta* has a tailless larva, lacking a tail, notochord, and muscle cells. When an egg of *M. occulta* is fertilized with the sperm of *M. oculata*, hybrid larvae develop that frequently contain a small tail and a sensory pigment spot, whereas the reverse hybrid fertilization gives normal tadpole larvae (Swalla & Jeffery, 1990; Swalla et al., 1993). These results suggest that tailless larvae have loss of function mutations in their genomes that cause taillessness, probably both maternal and zygotic. Both the egg cytoplasm and the zygotic genome contribute to the phenotype

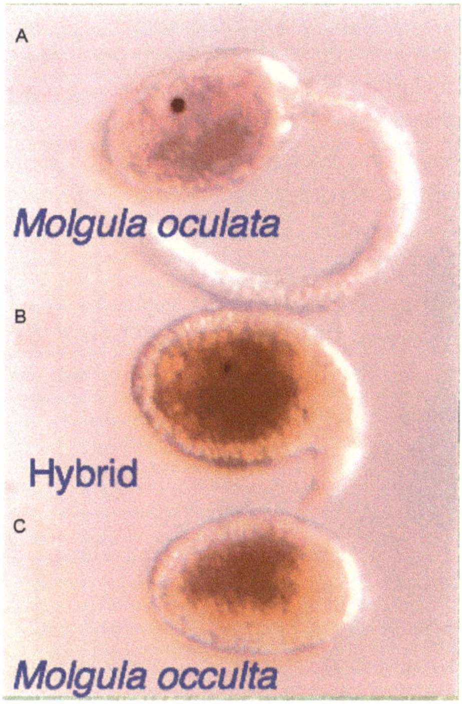

Fig. 4 Two species of ascidian larvae that are closely related and a hybrid between the two species. (A) *Molgula oculata* is a tailed species with an otolith (gravity sensing organ) in the head, a notochord with 40 cells found in the tail, while (C) *Molgula occulta* larvae hatch, but lack the otolith and the tail. (B) In contrast, a hybrid larva made by fertilizing a *M. occulta* egg with *M. oculata* sperm shows a pigmented spot and a short notochord of 20 cells. These two molgulid sister species are found off the coast of Roscoff, France and their genome and transcriptome sequences are available in ANISEED (Brozovic et al., 2016). *Republished with permission from Swalla B. J., & Jeffery W. R. (1996). Requirement of the Manx gene for expression of chordate features in a tailless ascidian larva.* Science, 274, *1205–1208.*

of the hybrids, as seen in the variability of the hybrid larvae (Swalla & Jeffery, 1990). Depending on the egg clutch, more or less hybrid features were seen, such as an otolith or short tail, or both (Swalla & Jeffery, 1990; Jeffery, 2004). We have subsequently searched for maternal and zygotic proteins that are necessary and sufficient to make tailed hybrids in the cross with *M. occulta* eggs.

A subtractive hybridization between RNA from the eggs of *M. occulta* and *M. oculata* allowed the characterization of *Manx*, which is expressed maternally and zygotically (Swalla et al., 1993). When *Manx* RNA was knocked out by antisense oligonucleotides, the hybrids reverted to the tailless phenotype (Swalla & Jeffery, 1996). *Manx* encodes a zinc-finger transcription factor which is vital for the development of chordate features in molgulid larva (Swalla & Jeffery, 1996). *Manx* is present in the *Ciona intestinalis* and other ascidian genomes, but it becomes increasingly difficult to establish homology outside of the tunicates because of the nature of the zinc finger conserving the spacing of cysteines, rather than specific amino acids. It was previously reported that in the molgulids, *Bobcat*, a DEAD-box RNA helicase gene now known to be the paralog of mammalian *DDX5*, is located within the first intron of *Manx*, with the two genes sharing two small noncoding exons (Fig. 5) (Swalla, Pederson, Just, & Jeffery, 1999). This gene complex was the first to be found in ascidians to show trans-splicing and also to be alternatively spliced to form multiple protein transcripts. Subsequently, trans-splicing has been reported in both *Ciona* (Sierro, Li, Suzuki, Yamashita, & Nakai, 2009) and the distantly related *Oikopleura* (Ganot, Kallesøe, Reinhardt, Chourrout, & Thompson, 2004) genome, so it is likely common in all tunicate genomes.

We have expanded the mapping of this gene complex to include a third gene, *SSNA1*, closely 5′ upstream of *Manx* and *Bobcat/DDX5*. *SSNA1* (Sjogren syndrome nuclear antigen 1) is a nuclear gene highly conserved within eukaryotes. This small coiled-coil protein lacks a nuclear localization signal, and it is thought that its small size of 14kD would allow it to diffuse through nuclear pores into the cytoplasm, but instead it remains in the nucleus (Ramos-Morales, Infante, Fedriani, Bornens, & Rios, 1998). *SSNA1*, an adaptor protein, is localized to the centrosome and sperm flagella (Pfannenschmid et al., 2003) and has been shown to be involved in targeting

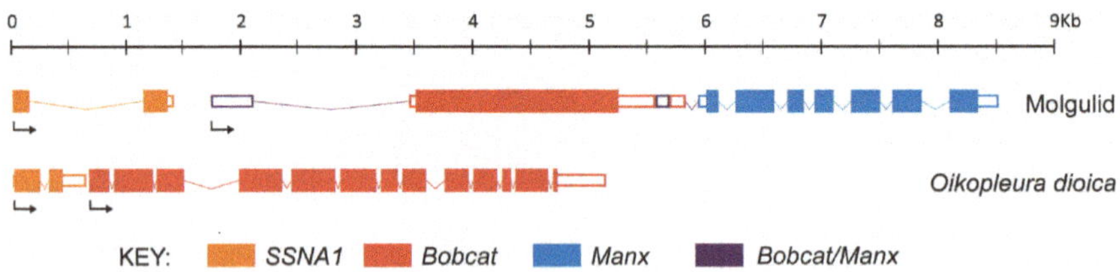

Fig. 5 *SSNA1/Bobcat/Manx* gene complex found in molgulid ascidians and *Oikopleura dioica*. *SSNA1*, shown in orange, is upstream and contains two exons. In molgulids, the first exon of both *Bobcat* and the long minor *Manx* transcript share a 5′ UTR, which is indicated as an open green box; minor *Manx* also shares its second exon in its 5′ UTR with the 3′ UTR of the *Bobcat* transcript. In *O. dioica*, *SSNA1* also contains two exons, and is just 35 base pairs upstream of *Bobcat* containing 13 exons. Noncoding UTRs are depicted as open boxes, coding exons as solid boxes. *SSNA1* is orange, *Manx* is red, *Bobcat* is blue, *Bobcat* and *Manx* shared exons are purple.

ATPase spastin to microtubules in the centrosome (Errico, Claudiani, D'Addio, & Rugarli, 2004).

The expression of *Bobcat/DDX5* in *M. oculata* and *M. occulta* was also found to be required for neural sensory organ, secondary muscle cell, and tail development (Swalla et al. (1999)). *DDX5* (formerly *p68*) and differentially spliced *DDX17* are paralogous DEAD-box RNA helicases found in vertebrates that are involved in the processing of mRNA and microRNA, both functions essential for development in mice (Fukuda et al., 2007). *DDX5* and *DDX17* have been found to co-activate transcription factors for muscle gene expression and differentiation (Caretti et al., 2006), as well as regulate alternative splicing. This implies that these genes may be involved in the development of muscle cells within tailed molgulid tadpole larvae. The gene sequence of *Bobcat* in *M. oculata* and *M. occulta* is interesting because it lacks introns in its coding region (Swalla et al. (1999)) that are found in non-molgulid homologs (Fig. 5), suggesting a previous retrotransposition event that inserted the processed RNA within the first intron of the *Manx* gene. We have learned that this unusual gene complex is found in all molgulids, but further functional testing will be necessary to discern the necessity of each of these genes for larval tissues during molgulid development.

We sequenced the transcriptomes of *M. occulta*, *M. oculata* and hybrid embryos during gastrulation, neurulation and tailbud formation to evaluate the expression of known developmental gene networks in the 40 notochord cells (Di Gregorio, 2020) and 38–42 muscle cells (Stolfi et al., 2014). The transcriptomes and genomes of *M. oculata*, *M. occulta* and *M. occidentalis* are available on ANISEED (Brozovic et al., 2016; Stolfi et al., 2014). We were surprised to find notochord-specific gene expression virtually unchanged in the anural tunicate *M. occulta*, because the larva never makes a notochord. However, some hybrid *M. occulta* X *M. oculata* larvae show a notochord perfectly converged and extended, albeit with only 20 cells, instead of the normal 40 notochord cells (Swalla & Jeffery, 1990; Takada et al., 2002).

Analysis of the genes underlying muscle development (Jeffery, Swalla, Ewing, & Kusakabe, 1999) and the otolith pigment cell formation (Racioppi et al., 2017) have shown that many of these genes in the tailless *M. occulta* have become pseudogenes, no longer encoding for muscle actin or tyrosinase. In these cases, it is now clear that the sperm of the tailed *M. oculata* carry one copy of an intact gene and this one copy can show normal expression in hybrid embryos (Jeffery et al., 1999; Racioppi et al., 2017). It is not yet known how many expressed pseudogenes that there are in the tailless

M. occulta genome, and it may be that when a tailless species evolves, many of the differentially expressed genes become inactive as selection is relaxed. In this case, it would seem that becoming tailless would almost certainly be disadvantageous and rapidly be selected against evolutionarily. However, if the process of becoming tailless increases speciation rates, then molgulid ascidians may be continually speciating and staying ahead of the eventual extinction rates (Maliska et al., 2013). This hypothesis can only be tested with an increased number of tailless ascidian species in the analyses to provide statistical power, but tailless species continue to be discovered in the molgulids (Maliska & Swalla, 2010), so it may eventually be possible.

The *C. intestinalis* neural connectome has been previously described by Ryan et al. (2016), through examination of thin serial sections from the larval brain to investigate left-right asymmetry as well as the number of the neurons synapsing inside the brain. The connectome revealed 177 neurons in the larval brain and 49 synapses with other cells (Ryan et al., 2016). In total, the tadpole larva of *C. intestinalis*'s tiny brain is approximately 330 cells (Nicol & Meinertzhagen, 1991). We compare the tailed *C. intestinalis* (redrawn from Ryan et al., 2016) larval neurons with the tailless *M. occulta* species in Fig. 6. In both species, sensory brain vesicle/brain (BV), connected to the motor ganglion (MG) within the middle of the larval trunk. The central nervous system (CNS) and notochord are shown in Fig. 5 next to the MG and BV in both species. In the tailed *C. intestinalis* larva, a caudal nerve cord (CNC, gray) in the tail is located directly above the notochord, colored red. The relay neurons of the brain vesicle, including PNS relay neurons (pnsRNs), photoreceptor relay neurons (PRRNs), antenna relay neurons (AntRNs) are shown inside *C. intestinalis*'s brain; specific cell types including ocellus and otolith pigment cells (Oc & Ot), antenna cells (Ant), coronet cells (Cor) and photoreceptor cells (PR) are also shown in *C. intestinalis*'s brain. Specific neural gene expression, including *Neurogenin*, *Onecut* and *Ebf* are shown in *M. occulta*'s brain. Motor neurons (MNs), interneurons (INs) and descending decussating neuron (ddN) are shown in both species' MGs in Fig. 6. *C. intestinalis* larvae have tails, which allow them to swim up toward the surface of the water toward light, using the ocellus (Tsuda et al., 2003). *M. occulta*, on the other hand, does not have a tail; therefore, the larvae cannot swim. Interestingly, all the neurons that were found in *C. intestinalis* seem to be present in *M. occulta*, even though the larvae never swim. Even though the connectome of *M. occulta* has not been delineated, the conserved number and location of specific neuronal cell types were expected if the CNS is still intact. Still, further work on the *M. occulta* connectome is needed

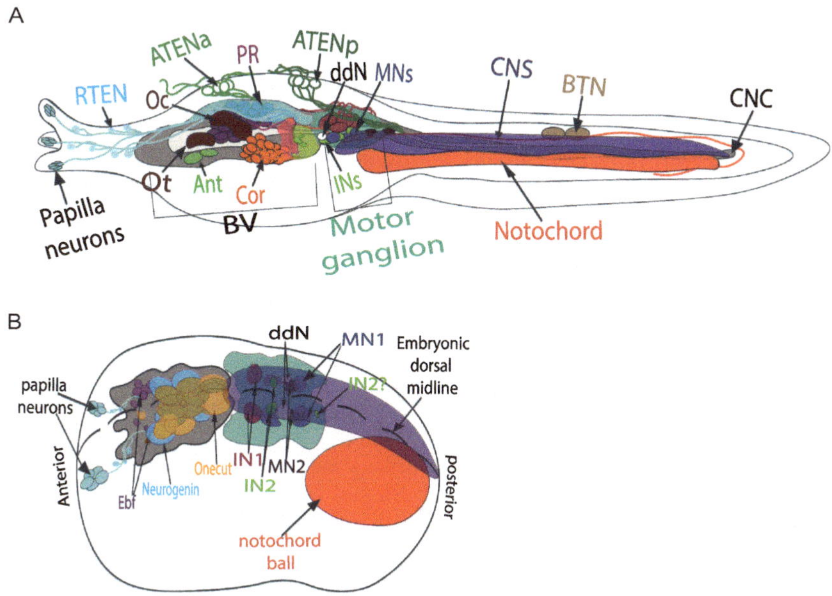

Fig. 6 The connectome of a larva of the swimming tailed ascidian species *Ciona intestinalis* (top; Ryan et al., 2016) and the larval neurons of a non-swimming tailless species *Molgula occulta* (bottom; Lowe et al., 2020). Both diagrams indicate the whole larva with color-coded cell types labeled by arrows. Legend for color labels in both diagrams: brain (gray), motor ganglion (teal), CNS (central nervous system; dark blue), and notochord (red). Tailless species' size is about one half to one-third of the tail species' size. (A) Diagram of the connectome of a *C. intestinalis* larva. Types of relay neurons (RNs) are shown as color-coded territories in the brain (gray): pnsRNs (light cyan), PRRNs (light Magenta), and AntRNs (light green). (B) Diagram of a whole tailless *M. occulta* larva. The notochord cells (red) fail to undergo convergent extension, and a notochord ball structure is seen. Embryonic dorsal midline indicated by black dashed line. Different developing nervous system cells types of *M. occulta* were color-coded. MN1 (blue), MN2 (dark brown), IN1 (plum), IN2 (green), and ddN (white) cells are located on either side of the motor ganglion (Lowe et al., 2020). *Panel (A) redrawn from Ryan, K., Lu, Z., & Meinertzhagen, I. A. (2016). The CNS connectome of a tadpole larva of Ciona intestinalis (L.) highlights sidedness in the brain of a chordate sibling.* eLife, *5, e16962.*

to understand if the tailless ascidian nervous system and their synaptic networks are present and functional. In summary, it appears that much of the neural circuitry and the notochord gene networks are intact in the tailless ascidians, in spite of the fact that the larvae never swim. The notochord cells form a "notoball" that expresses many of the intact notochord genes, including the upstream notochord initial transcription factor *brachyury*, (Takada et al., 2002) and many of the downstream genes as well. There are at least two possibilities for these results. One is that the relaxation of selection on the ascidian larval gene networks is stochastic, and it was random that the muscle gene networks (Jeffery et al., 1999) and pigment cell networks (Racioppi et al., 2017) are lost early in the evolutionary process in these particular tailless species. An alternative hypothesis is that the notochord gene network and the neural networks are used later during adult development and so the selection to keep the gene network intact is stronger. Distinguishing between these hypotheses will need further sequencing and analyses of larval and adult gene regulatory networks in other species of tailless ascidians.

Acknowledgments

Billie J. Swalla thanks Scott Gilbert for putting together this exciting volume on studies of Evolution and Development and his encouragement throughout the writing and publishing of this manuscript. We thank Megan M. McCarthy, an undergraduate researcher in the Swalla Lab, for reading drafts of the manuscript and providing feedback and suggestions. Much of the recent research described on tailless ascidians was supported by the National Science Foundation (NSF) BEACON NSF Science and Technology Center to BJS. This manuscript is dedicated to the memory of Alexander Fodor and the years that he spent as an integral member of the Swalla Lab in the Biology Department at the University of Washington and at Friday Harbor Laboratories.

Glossary

Ambulacraria A clade of deuterostome animals that includes the echinoderms and the hemichordates.

Ascidians A paraphyletic group of animals (having a single common ancestor) whose adult members are sessile and ensheathed in a tunic. Most ascidians have chordate tadpole larvae with a notochord and tail muscle cells, and a head with a light-sensing organ (ocellus) and balancing organ (otolith).

Chordates A group of animals that share the following traits at some stage of their life cycle: A hollow, dorsal nerve, notochord, muscular post-anal tail and gill slits. Lancelets, tunicates and vertebrates are all considered chordates.

Deuterostomes "Second mouth"—A group of animals that includes Chordata and Ambulacraria (echinoderms and hemichordates). The anus forms prior to the mouth during embryonic development in this group.

Echinoderms "Spiny skinned"—A group of animals that includes crinoids, sea stars, brittle stars, sea cucumbers and sea urchins.

Hemichordates "Half-chordate"—A group of animals, sister group to echinoderms, that possess an anterior proboscis, a neck region and a posterior abdomen, with gill slits. The solitary enteropneust worms and colonial pterobranchs make up this clade.

Tunicates A group of animals that share an outer "tunic" consisting of cellulose and protein. Ascidians, thaliaceans, and larvaceans are the three groups that make up tunicates, although phylogenetic analyses place the thaliaceans within the ascidians and the larvaceans are very difficult to place, given their substantial genomic changes.

References

Appeltans, W., et al. (2012). The magnitude of global marine species diversity. *Current Biology, 22*, 1–14.

Arimoto, A., & Tagawa, K. (2015). Hedgehog expression during development and regeneration in the hemichordate, *Ptychodera flava*. *Zoological Science, 32*(1), 33–37.

Bourlat, S. J., Juliusdottir, T., Lowe, C. J., Freeman, R., Aronowicz, J., Kirschner, M., et al. (2006). Deuterostome phylogeny reveals monophyletic chordates and the new phylum Xenoturbellida. *Nature, 444*(7115), 85–88.

Brown, F. D., Prendergast, A., & Swalla, B. J. (2008). Man is but a worm: Chordate origins. *Genesis, 46*, 605–613.

Brown, F. D., & Swalla, B. J. (2012). Evolution and development of budding by stem cells: Ascidian coloniality as a case study. *Developmental Biology, 369*(2), 151–162.

Brozovic, M., Martin, C., Dantec, C., Dauga, D., Mendez, M., Simion, P., et al. (2016). ANISEED 2015: A digital framework for the comparative developmental biology of ascidians. *Nucleic Acids Research, 44*, D808–D818. https://doi.org/10.1093/nar/gkv966.

Cameron, C. B., Garey, J. R., & Swalla, B. J. (2000). Evolution of the chordate body plan: New insights from phylogenetic analyses of deuterostome phyla. *Proceedings of the National Academy of Sciences of the United States of America, 97*, 4469–4474.

Cannon, J. T., Kocot, K. M., Waits, D. S., Weese, D. A., Swalla, B. J., Santos, S. R., et al. (2014). Phylogenomic resolution of the hemichordate and echinoderm clade. *Current Biology, 24*, 1–6.

Cannon, J. T., Vellutini, B. C., Smith, J., Ronquist, F., Jondelius, U., & Hejnol, A. (2016). Xenacoelomorpha is the sister group to Nephrozoa. *Nature, 530*, 89–93.

Caretti, G., Schiltz, R. L., Dilworth, F. J., Di Padova, M., Zhao, P., Ogryzko, V., et al. (2006). The RNA helicases p68/p72 and the noncoding RNA SRA are coregulators of MyoD and skeletal muscle differentiation. *Developmental Cell, 11*, 547–560.

Cota, C. D., & Davidson, B. (2015). Mitotic membrane turnover coordinates differential induction of the Heart progenitor lineage. *Developmental Cell, 34*(5), 505–519.

Davidson, B. (2007). *Ciona intestinalis* as a model for cardiac development. *Seminars in Cell & Developmental Biology, 18*(1), 16–26.

Davidson, B., & Swalla, B. J. (2002). A molecular analysis of ascidian metamorphosis reveals elements of an innate immune response. *Development, 129*(20), 4739–4751.

Delsuc, F., Philippe, H., Tsagkogeorga, G., et al. (2018). A phylogenomic framework and timescale for comparative studies of tunicates. *BMC Biology, 16*, 39.

Di Gregorio, A. (2020). Chapter eleven—The notochord gene regulatory network in chordate evolution: Conservation and divergence from *Ciona* to vertebrates. *Current Topics in Developmental Biology, 139*, 325–374.

Errico, A., Claudiani, P., D'Addio, M., & Rugarli, E. I. (2004). Spastin interacts with the centrosomal protein NA14, and is enriched in the spindle pole, the midbody and the distal axon. *Human Molecular Genetics, 13*, 2121–2132.

Franchi, N., & Ballarin, L. (2017). Immunity in protochordates: The tunicate perspective. *Frontiers in Immunology, 8*, 1664–3224.

Fukuda, T., Yamagata, K., Fujiyama, S., Matsumoto, T., Koshida, I., Yoshimura, K., et al. (2007). DEAD-box RNA helicase subunits of the Drosha complex are required for processing of rRNA and a subset of microRNAs. *Nature Cell Biology, 9*, 604–611.

Ganot, P., Kallesøe, T., Reinhardt, R., Chourrout, D., & Thompson, E. M. (2004). Spliced-leader RNA trans splicing in a chordate, *Oikopleura dioica*, with a compact genome. *Molecular and Cellular Biology, 24*, 7795–7805.

Graham, A., & Shimeld, S. M. (2013). The origin and evolution of the ectodermal placodes. *Journal of Anatomy, 222*(1), 32–40.

Hiebert, L. S., Vieira, E. A., Dias, G. M., Tiozzo, S., & Brown, F. D. (2019). Colonial ascidians strongly preyed upon, yet dominate the substrate in a subtropical fouling community. *Proceedings of the Royal Society B*, *286*, 20190396.
Holland, L. Z. (2015). Evolution of basal deuterostome nervous systems. *The Journal of Experimental Biology*, *218*, 637–645.
Huber, J. L., Burke da Silva, K., Bates, W. R., & Swalla, B. J. (2000). The evolution of anural larvae in molgulid ascidians. *Seminars in Developmental Biology*, *11*(6), 419–426.
Imai, K. S., Satoh, N., & Satou, Y. (2002). An essential role of a *FoxD* gene in notochord induction in *Ciona* embryos. *Development*, *129*(14), 3441–3453.
Jeffery, W. R., Swalla, B. J., Ewing, N., & Kusakabe, T. (1999). Evolution of the ascidian anural larva: Evidence from embryos and molecules. *Molecular Biology and Evolution*, *16*, 646–654.
Jeffery, W. R. (2004). Evolution and development of brain sensory organs in molgulid ascidians. *Evolution & Development*, *6*(3), 170–179.
Jiang, D., & Smith, W. C. (2007). Ascidian notochord morphogenesis. *Developmental Dynamics*, *236*(7), 1748–1757.
Karaiskou, A., Swalla, B. J., Sasakura, Y., & Chambon, J.-P. (2015). Metamorphosis in solitary ascidians. *Genesis*, *53*, 34–47.
Kassmer, S. H., Delany Rodriguez, D., & De Tomaso, A. W. (2016). Colonial ascidians as model organisms for the study of germ cells, fertility, whole body regeneration, vascular biology and aging. *Current Opinion in Genetics & Development*, *39*, 101–106.
Katz, M. J. (1983). Comparative anatomy of the tunicate tadpole, *Ciona intestinalis*. *The Biological Bulletin*, *164*(1), 1–27.
Kourakis, M. J., Newman-Smith, E., & Smith, W. C. (2010). Key steps in the morphogenesis of a cranial placode in an invertebrate chordate, the tunicate *Ciona savignyi*. *Developmental Biology*, *340*(1), 134–144.
Kourakis, M. J., & Smith, W. C. (2007). A conserved role for FGF signaling in chordate otic/atrial placode formation. *Developmental Biology*, *312*(1), 245–257.
Kowalevsky, A. (1866). Entwicklungsgeschichte der einfachen Ascidien. *Mémoires de l'Académie Impériale des Sciences de Saint Pétersbourg, Septième Série*, 7(10), 1–19.
Lamark, J. B. (1816). *Histoire naturelle des animaux sans vertébres. Tome III. Tuniciers* (pp. 80–130). Paris: Déterville.
Lowe, C. J., Terasaki, M., Wu, M., Freeman, R. M., Jr., Runft, L., Kwan, K., et al. (2006). Dorsoventral patterning in hemichordates: Insights into early chordate evolution. *PLoS Biol.*, *4*, e291.
Lowe, C. J., Wu, M., Salic, A., Evans, L., Lander, E., Stange-Thomann, N., et al. (2003). Anteroposterior patterning in hemichordates and the origins of the chordate nervous system. *Cell*, *113*, 853–865.
Luttrell, S. M., Gotting, K., Ross, E., Alvarado, A. S., & Swalla, B. J. (2016). Head regeneration in hemichordates is not a strict recapitulation of development. *Developmental Dynamics*, *245*(12), 1159–1175.
Lowe, E. K., Racioppi, C., Peyriéras, N., Ristoratore, F., Ruggiu, M., Christiaen, L., et al. (2020). A cis-regulatory change underlying the motor neuron-specific loss of terminal selector gene expression in immotile tunicate larvae. *Evolution and Development* (in revision).
Luttrell, S., Konikoff, C., Byrne, A., Bengtsson, B., & Swalla, B. J. (2012). Ptychoderid hemichordate neurulation without a notochord. *Integrative and Comparative Biology*, *52*, 829–834.
Luttrell, S. M., Su, Y. H., & Swalla, B. J. (2018). Getting a head with *Ptychodera flava* larval regeneration. *The Biological Bulletin*, *234*(3), 152–164.
Maliska, M. E., Pennell, M. W., & Swalla, B. J. (2013). Developmental mode influences diversification in ascidians. *Biology Letters*, *9*, 20130068.
Maliska, M. E., & Swalla, B. J. (2010). *Molgula pugetiensis* is a Pacific tailless ascidian within the Roscovita clade of molgulids. *Biological Bulletin*, *219*, 277–282.

Manni, L., Anselmi, C., Cima, C., Gasparini, F., Voskoboynik, A., Martini, M., et al. (2019). Sixty years of experimental studies on the blastogenesis of the colonial tunicate *Botryllus schlosseri*. *Developmental Biology*, *448*(2), 293–308.

Manni, L., Lane, N., Joly, J., Gasparini, F., Tiozzo, S., Caicci, F., et al. (2004). Neurogenic and non-neurogenic placodes in ascidians. *Journal of Experimental Zoology. Part B, Molecular and Developmental Evolution*, *302*, 483–504.

Marlétaz, F., Peijnenburg, K. T. C. A., Goto, T., Satoh, N., & Rokhsar, D. S. (2019). A new spiralian phylogeny places the enigmatic arrow worms among Gnathiferans. *Current Biology*, *29*(2), 312–318.

Martín-Durán, J. M., Passamaneck, Y. J., Martindale, M. Q., & Hejnol, A. (2016). The developmental basis for the recurrent evolution of deuterostomy and protostomy. *Nature Ecology & Evolution*, *1*, 0005.

Merle, N., Singh, P., Rahman, J., & Kemper, C. (2020). Integrins meet complement: The evolutionary tip of an iceberg orchestrating metabolism and immunity. *British Journal of Pharmacology*, 1–17.

Nicol, D., & Meinertzhagen, I. A. (1991). Cell counts and maps in the larval central nervous system of the ascidian *Ciona intestinalis* (L.). *The Journal of Comparative Neurology*, *309*, 415–429.

Pfannenschmid, F., Wimmer, V. C., Rios, R. M., Geimer, S., Kröckel, U., Leiherer, A., et al. (2003). Chlamydomonas DIP13 and human NA14: A new class of proteins associated with microtubule structures is involved in cell division. *Journal of Cell Science*, *116*, 1449–1462.

Putnam, N., Butts, T., Ferrier, D., et al. (2008). The amphioxus genome and the evolution of the chordate karyotype. *Nature*, *453*, 1064–1071.

Racioppi, C., Valoroso, M. C., Coppola, U., Lowe, E. K., Brown, C. T., Swalla, B. J., et al. (2017). Evolutionary loss of melanogenesis in the tunicate *Molgula occulta*. *EvoDevo*, *8*(1), 1–11.

Racioppi, C., Wiechecki, K. A., & Christiaen, L. (2019). Combinatorial chromatin dynamics foster accurate cardiopharyngeal fate choices. *eLife*, *8*, e49921.

Raff, R. A., & Byrne, M. (2006). The active evolutionary lives of echinoderm larvae. *Heredity*, *97*(3), 244–252.

Ramos-Morales, F., Infante, C., Fedriani, C., Bornens, M., & Rios, R. M. (1998). NA14 is a novel nuclear autoantigen with a coiled-coil domain. *Journal of Biological Chemistry*, *273*, 1634–1639.

Range, R. C., Angerer, R. C., & Angerer, L. M. (2013). Integration of canonical and noncanonical *Wnt* signaling pathways patterns the neuroectoderm along the anterior–posterior axis of sea urchin embryos. *PLoS Biology*, *11*(1), e1001467.

Razy-Krajka, F., & Stolfi, A. (2019). Regulation and evolution of muscle development in tunicates. *EvoDevo*, *10*(1), 13.

Rigon, F., Gasparini, F., Shimeld, S. M., Candiani, S., & Manni, L. (2018). Developmental signature, synaptic connectivity and neurotransmission are conserved between vertebrate hair cells and tunicate coronal cells. *The Journal of Comparative Neurology*, *526*(6), 957–971.

Roberts, B., Davidson, B., MacMaster, G., Ma, E., Smith Wallace, S., Lockhart, V., et al. (2007). A complement response may activate metamorphosis in the ascidian *Boltenia villosa*. *Development Genes and Evolution*, *217*(6), 449–458.

Röttinger, E., & Lowe, C. (2012). Evolutionary crossroads in developmental biology: Hemichordates. *Development*, *139*(14), 2463–2475.

Ryan, K., Lu, Z., & Meinertzhagen, I. A. (2016). The CNS connectome of a tadpole larva of Ciona intestinalis (L.) highlights sidedness in the brain of a chordate sibling. *eLife*, *5*, e16962.

Rychel, A. L., & Swalla, B. J. (2008). Anterior regeneration in the hemichordate *Ptychodera flava*. *Developmental Dynamics, 237*(11), 3222–3232.

Rychel, A. L., & Swalla, B. J. (2009). Regeneration in hemichordates and echinoderms. In V. Matranga, & B. Rinkevich (Eds.), *Stem cells in marine organisms* (pp. 245–265). Dordrecht: Springer.

Sasakura, Y., Kanda, M., Ikeda, T., Horie, T., Kawai, N., Ogura, Y., et al. (2012). Retinoic acid-driven *Hox1* is required in the epidermis for forming the otic/atrial placodes during ascidian metamorphosis. *Development, 139*, 2156–2160.

Satou, Y., Imai, K. S., & Satoh, N. (2004). The ascidian *Mesp* gene specifies heart precursor cells. *Development, 131*(11), 2533–2541.

Shenkar, N., Gittenberger, A., Lambert, G., Rius, M., Moreira da Rocha, R., Swalla, B. J., et al. (2020). Ascidiacea world database. In *Ascidiacea* World Register of Marine Species. http://www.marinespecies.org/aphia.php?p=taxdetails&id=1839. Accessed on 2020-04-16.

Shenkar, N., & Swalla, B. J. (2011). Global diversity of Ascidiacea. *PLoS One, 6*(6), e20657.

Sierro, N., Li, S., Suzuki, Y., Yamashita, R., & Nakai, K. (2009). Spatial and temporal preferences for trans-splicing in *Ciona intestinalis* revealed by EST-based gene expression analysis. *Gene, 430*, 44–49.

Simakov, O., Kawashima, T., Marlétaz, F., et al. (2015). Hemichordate genomes and deuterostome origins. *Nature, 527*, 459–465.

Stolfi, A., Lowe, E. K., Racioppi, C., Ristoratore, F., Brown, C. T., Swalla, B. J., et al. (2014). Divergent mechanisms regulate conserved cardiopharyngeal development and gene expression in distantly related ascidians. *eLife, 3*, e03728.

Swalla, B. J. (2001). Phylogeny of the urochordates: Implications for chordate evolution. In H. Sawada, & C. Lambert (Eds.), *Biology of ascidians (proceedings of the first international symposium on the biology of ascidians)* (pp. 219–224). Tokyo: Springer Verlag.

Swalla, B. J. (2004). Protochordate gastrulation: Lancelets and ascidians. In C. Stern (Ed.), *Gastrulation* (pp. 139–149). Cold Spring Harbor Press.

Swalla, B. J. (2006). Building divergent body plans with similar genetic pathways. *Heredity, 97*, 235–243.

Swalla, B. J., Cameron, C. B., Corley, L. S., & Garey, J. R. (2000). Urochordates are monophyletic within the deuterostomes. *Systematic Biology, 49*, 122–134.

Swalla, B. J., & Jeffery, W. R. (1990). Interspecific hybridization between an anural and urodele ascidian: Differential expression of urodele features suggests multiple mechanisms control anural development. *Developmental Biology, 142*, 319–334.

Swalla, B. J., & Jeffery, W. R. (1996). Requirement of the *Manx* gene for expression of chordate features in a tailless ascidian larva. *Science, 274*, 1205–1208.

Swalla, B. J., Makabe, K. W., Satoh, N., & Jeffery, W. R. (1993). Novel genes expressed differentially in ascidians with alternate modes of development. *Development, 119*, 307–318.

Swalla, B. J., Pederson, E. L., Just, M. L., & Jeffery, W. R. (1999). A multigene locus containing the manx and bobcat genes is required for development of chordate features in the ascidian tadpole larva. *Development, 126*, 1643–1653.

Swalla, B. J., & Smith, A. B. (2008). Deciphering deuterostome phylogeny: Molecular, morphological and palaeontological perspectives. *Philosophical Transactions of the Royal Society of London. Series B, Biological Sciences, 363*(1496), 1557–1568.

Takada, N., York, J., Davis, J. M., Schumpert, B., Yasuo, H., Satoh, N., et al. (2002). *Brachyury* expression in tailless molgulid ascidian embryos. *Evolution and Development, 4*(3), 205–211.

Takahashi, H., Hotta, K., Erives, A., Di Gregorio, A., Zeller, R. W., Levine, M., et al. (1999). *Brachyury* downstream notochord differentiation in the ascidian embryo. *Genes & Development, 13*(12), 1519–1523.

Tassia, M. G., Cannon, J. T., Konikoff, C. E., Shenkar, N., Halanych, K. M., & Swalla, B. J. (2016). The global diversity of Hemichordata. *PLoS One*, *11*(10), e0162564.
Tsuda, M., Sakurai, D., & Goda, M. (2003). Direct evidence for the role of pigment cells in the brain of ascidian larvae by laser ablation. *The Journal of Experimental Biology*, *206*, 1409–1417.
Yasuo, H., & Satoh, N. (1993). Function of the vertebrate *T* gene. *Nature*, *364*, 582–583.

CHAPTER SIX

Evolution of new cell types at the lateral neural border

Jan Stundl[a,†], Paola Y. Bertucci[b,†], Antonella Lauri[b,‡], Detlev Arendt[b,*], and Marianne E. Bronner[a,*]

[a]Division of Biology and Biological Engineering, California Institute of Technology, Pasadena, CA, United States
[b]European Molecular Biology Laboratory, Heidelberg, Germany
*Corresponding authors: e-mail address: arendt@embl.de; mbronner@caltech.edu

Contents

Abstract

During the course of evolution, animals have become increasingly complex by the addition of novel cell types and regulatory mechanisms. A prime example is represented by the lateral neural border, known as the neural plate border in vertebrates, a region of the developing ectoderm where presumptive neural and non-neural tissue meet. This region has been intensively studied as the source of two important embryonic cell types unique to vertebrates—the neural crest and the ectodermal placodes—which contribute to diverse differentiated cell types including the peripheral nervous system, pigment cells, bone, and cartilage. How did these multipotent progenitors originate in animal evolution? What triggered the elaboration of the border during the course of chordate

† These authors contributed equally.
‡ Present address: Genetics and Rare Diseases Research Division, Ospedale Pediatrico Bambino Gesù, IRCCS, Rome, Italy

Current Topics in Developmental Biology, Volume 141
ISSN 0070-2153
https://doi.org/10.1016/bs.ctdb.2020.11.005

evolution? How is the lateral neural border patterned in various bilaterians and what is its fate? Here, we review and compare the development and fate of the lateral neural border in vertebrates and invertebrates and we speculate about its evolutionary origin. Taken together, the data suggest that the lateral neural border existed in bilaterian ancestors prior to the origin of vertebrates and became a developmental source of exquisite evolutionary change that frequently enabled the acquisition of new cell types.

1. The lateral neural border in bilaterians

Bilaterian embryos undergo gastrulation, a fundamental process that transforms the simple multicellular single-layered embryo into a complex functional organism with three segregated germ layers: ectoderm, endoderm and mesoderm. Ectodermal precursors become localized to the outer surface of developing embryo, while endomesodermal cells are internalized. The ectoderm becomes further subdivided into two compartments: (i) the neuroectoderm, which mostly gives rise to the central nervous system and (ii) the non-neural ectoderm, which gives rise primarily to the epidermis and associated glands.

The region where neural and non-neural ectoderm are juxtaposed forms the "lateral neural border" (LNB), a transition zone that plays different roles in different organisms (Li et al., 2017). In chordates, the LNB takes the shape of a thickened plate and is referred to as the "neural plate border" (NPB). Most notably, the NPB is the site of origin of neural crest and ectodermal placode cells, both of which are unique to vertebrate embryos. These developmental cell types have been essential for providing portions of the cranial skeleton, advanced sensory structures, and body pigmentation, among many other features. Non-vertebrate chordates also have an LNB but lack *bona fide* neural crest and placodes, which raises intriguing questions regarding how the NPB has become elaborated during the course of vertebrate evolution. Finally, in recent years the lateral neural border of non-chordate bilaterians has become a new focus of attention because it is specified by a conserved set of transcription factors and, as in chordates, gives rise to peripheral mechanosensory neurons, some of which are migratory. Furthermore, more recent studies of LNB development and fate in annelids reveal that their LNB contains multipotent progenitors that give rise to diverse cell types, among them central and peripheral (including mechanosensory) neurons, peripheral glial cells, pigment cells, spinning glands, and also chaetoblasts, which produce the eponymous chaetae (bristles) of the "bristle worms." In terms of multipotency, this is reminiscent of the vertebrate situation and calls for a re-evaluation of classical concepts of neural plate border evolution.

Towards this aim, we will first review the knowledge on vertebrate NPB specification and neural crest and placode development in vertebrates, focusing on descendant cell types and important signaling events. We will then discuss the evolutionary changes that may have enabled the elaboration of the border and the advent of new cell types during the course of vertebrate evolution. Next, we describe the neural plate border of non-vertebrate chordates, focusing on conserved and disparate features in ascidians, the sister group of vertebrates, and in amphioxus, the most basal chordate. Finally, we review recent findings in non-chordate bilaterians that provide new insights concerning the evolutionary origins of the NPB and set the stage for the evolution of the vertebrate neural crest and placode.

2. The vertebrate neural plate border—Neural crest and cranial placodes

Neural crest and placodes are two defining features of vertebrates that are primarily associated with their highly specialized "new head," endowed with sense organs that do not exist in non-vertebrate chordates (Gans & Northcutt, 1983; Northcutt, 2005; Northcutt & Gans, 1983). Both prospective neural crest and cranial placode cells are specified within the neural plate border, then they become internalized by ingression or invagination and migrate over long distances in all vertebrates (Fig. 1A and B) (Ahrens & Schlosser, 2005; Green, Simoes-Costa, & Bronner, 2015; Pla & Monsoro-Burq, 2018; Schlosser, 2015; York & McCauley, 2020) prior to differentiation into defined structures (Baker & Schlosser, 2005; Coppola et al., 2009; Fleenor & Begbie, 2014; Grocott, Tambalo, & Streit, 2012; Steventon, Mayor, & Streit, 2014). Although both embryonic neural crest and placode cells share several common characteristics in terms of developmental origin and contribution to sensory structures, there also are several fundamental differences between them. Among these differences are (1) after initial induction, neural crest and placodes develop in quite different ways and often form distinct cell types; and (2) neural crest cells form along the entire antero-posterior axis of vertebrate embryos, whereas ectodermal placodes are confined to the vertebrate head.

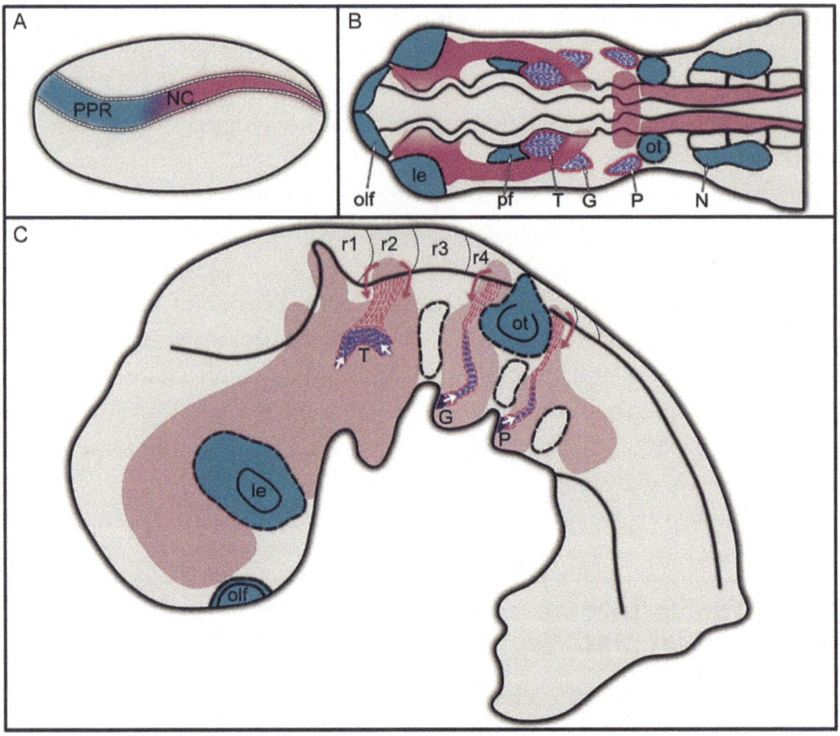

Fig. 1 NPB-derived cell populations and their interactions during the formation of cranial ganglia in vertebrates. (A) Neural crest (blue) and cranial placodes (purple) arise from the NPB (white dashed lines), but the boundary between both cell populations is not sharp. The preplacodal region (PPR) contains precursors for all sensory placodes. (B) A cartoon of cranial neural crest cells migration (purple) and position of cranial placodes (blue) in the chick embryo. Dorsal view. (C) Diagram showing interactions between neural crest cells and cranial placodes during the formation of trigeminal (T), geniculate (G), and petrosal cranial ganglia. Purple arrows show the direction of neural crest cells migration and white arrows the direction of migration from the placodes. Lateral view. Le, lens; N, nodose placode; olf, olfactory placode; ot, otic placode; pf, profundal placode; r1-4, rhombomere 1-4. *Modified from Fleenor, S. J., & Begbie, J. (2014). Neural crest cell and placode interactions in cranial PNS development.* Neural Crest Cells, *Elsevier Inc. https://doi.org/10.1016/B978-0-12-401730-6.00009-0.*

2.1 The role of signaling events in neural plate border specification

In vertebrates, the segregation of neural and non-neural ectodermal domains is regulated by bone morphogenetic protein (BMP)-signaling, with high levels of BMP in non-neural ectoderm, and low levels activating the neural program thus forming the neurectoderm (De Robertis & Sasai, 1996; Liem, Tremml, Roelink, & Jessell, 1995; Wilson & Hemmati-Brivanlou, 1995). In contrast, intermediate levels of BMP expression appear to define the NPB and are required for the neural crest induction (Faure, Barbara, Roberts, & Whitman, 2002; Pla & Monsoro-Burq, 2018; Wu, Ramel, Howell, & Hill, 2011). Modulating intracellular factors CKIP-1 and Smurf1 in turn alters levels of BMP-signaling and neural crest gene expression, illustrating that the maintenance of intermediate BMP-signaling levels is critical for proper neural crest induction (Piacentino & Bronner, 2018). Besides BMP, fibroblast growth factor (FGF) and WNT signals from surrounding tissues play an important role in NPB specification (Fig. 2A) (De Robertis & Kuroda, 2004; García-Castro, Marcelle, & Bronner-Fraser, 2002; Garnett, Square, & Medeiros, 2012; Groves & LaBonne, 2014; Knecht & Bronner-Fraser, 2002; Stuhlmiller & García-Castro, 2012). The interplay of all three pathways regulates expression of a large suite of transcription factors termed "NPB specifier genes," e.g., *Msx1/2*, *Pax3/7*, *Zic1*, *Gbx2*, *Id3*, *Dlx5*, *Tfap2*, *Dmbx1*, *Lhx5*, *Hairy2* and *Prdm1* (Fig. 2B) (Green et al., 2015; Maharana & Schlosser, 2018; Martik & Bronner, 2017; Moody & Lamantia, 2015; Pla & Monsoro-Burq, 2018). Studies in various vertebrates have shown that the expression of "NPB specifiers" is conserved across vertebrates (Basch, Bronner-Fraser, & García-Castro, 2006; Betters, Charney, & Garcia-Castro, 2018; Groves & LaBonne, 2014; Khudyakov & Bronner-Fraser, 2009; Meulemans & Bronner-Fraser, 2004; Nichane, Ren, Souopgui, & Bellefroid, 2008), which suggests that specification of the NPB domain likely evolved prior to the appearance of defining vertebrate features, such as the neural crest and cranial placodes (Yu, Meulemans, Mckeown, & Bronner-Fraser, 2008). After their initial expression, the "NPB specifiers" enhance mutual cross-regulatory interactions leading to a feed-forward NPB stabilization loop that defines the border domain (Milet & Monsoro-Burq, 2012; Pla & Monsoro-Burq, 2018; Simões-Costa & Bronner, 2015).

After initial establishment of the NPB and adjacent non-ectodermal and neural ectodermal domains, prospective populations of neural crest and cranial placodes begin to be determined in a combinatorial manner by a separate set of transcription factors ("neural crest specifiers" and "cranial placode specifiers," respectively) (Fig. 2B) (Bhat, Kwon, & Riley, 2013; Pla & Monsoro-Burq, 2018; Thiery, Buzzi, & Streit, 2020). Activation of the neural crest specification program requires not only intermediate BMP levels but also the cooperation between Wnt-signaling and the NPB specification

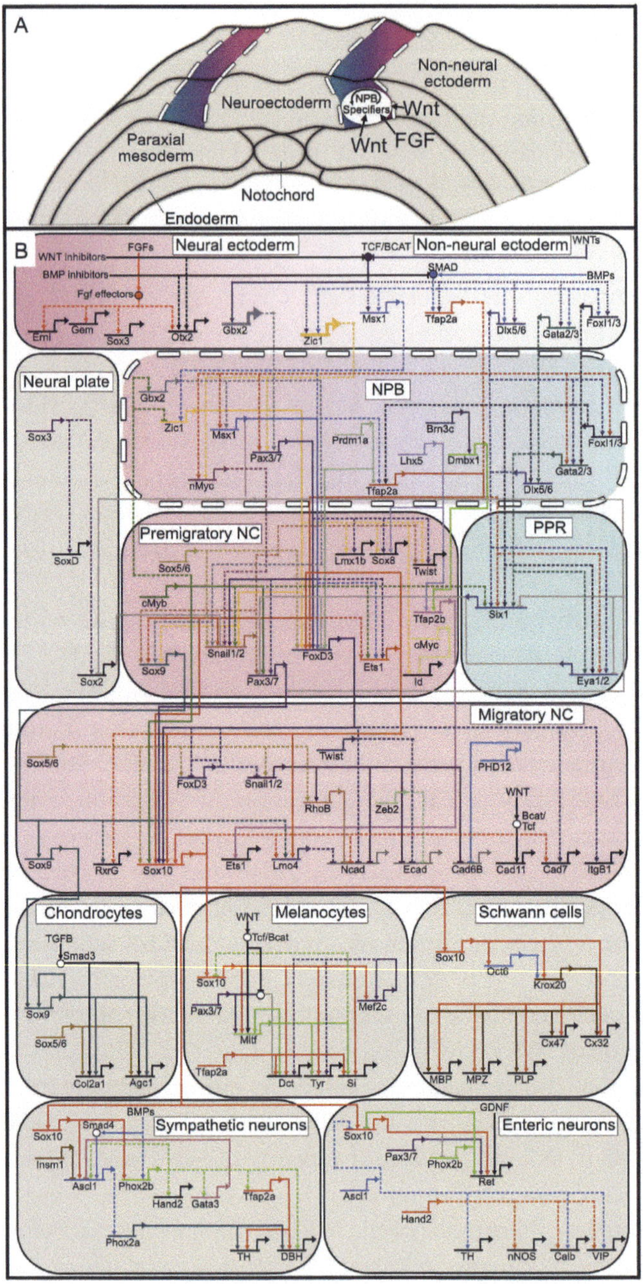

Fig. 2 Vertebrate neural plate border (NPB) specification and its underlying GRN modules. (A) The interplay of BMP, FGF, and WNT-signaling activates a core set of "NPB specifiers" that define NPB (blue-purple; marked by white dashed lines). (B) Gene regulatory network controls various steps of neural crest formation, including modules reflecting specification of the NPB, neural crest (NC; blue) and preplacodal region (PPR; purple) from which arise individual cranial placodes. Later migratory neural crest cells arise and differentiate into diverse cell types ranging from craniofacial chondrocytes to Schwann cells and different types of peripheral neurons.

module. Wnt-signaling activates the transcription factor and Wnt effector, Axud1, which in turn forms a complex with "NPB specifier" genes *Msx1* and *Pax3/7*. This complex activates the "neural crest specifier" *FoxD3* by binding to its enhancers. In addition, the experimental loss of *Axud1* causes the loss of several other "neural crest specifiers," such as *Sox9*, *Sox10*, and *Ets1* (Simões-Costa & Bronner, 2015). Thus, the Wnt-effector *Axud1* is a critical link between Wnt-signaling and its downstream targets to establish the neural crest stem cell niche in the dorsal neural tube. At this timepoint, there is transient expression of *Draxin*, which acts like a molecular rheostat that influences the timing of cranial neural crest cells epithelial mesenchymal transition (EMT) by transiently inhibiting canonical Wnt-signaling. Therefore, an appropriate level of Wnt-signaling is essential for the initiation of cranial neural crest EMT (Hutchins & Bronner, 2018). Interestingly, in contrast to the neural crest, the cranial placode program is activated in the absence or low level of Wnt-signaling (Brugmann, Pandur, Kenyon, Pignoni, & Moody, 2004; Hong & Saint-Jeannet, 2007; Patthey, Edlund, & Gunhaga, 2009; Patthey, Gunhaga, & Edlund, 2008). Thus, the Wnt-signaling levels are one important event determining whether neural crest or cranial placode fates arise within the NPB.

2.2 Segregation of neural crest and placodes

The developmental and evolutionary relationship of neural crest and placodal progenitors has been subject of much discussion (Baker & Bronner-Fraser, 1997; Schlosser, 2008; Schlosser, Patthey, & Shimeld, 2014). Two different models have been proposed to explain the developmental relationship of neural crest and placode cells: (i) The "binary competence model" suggests that the neural crest is induced from the lateral-most neuroectoderm while the preplacodal region (PPR) arises spatially distinct from non-neuroectoderm alongside the NPB (Ahrens & Schlosser, 2005; Maharana & Schlosser, 2018; Pieper, Ahrens, Rink, Peter, & Schlosser, 2012; Schlosser et al., 2014). (ii) The "neural plate border state model" assumes that both cell populations are first co-specified as NPB and then segregate into neural crest and PPR via the activity of lineage "specifiers" (Brugmann et al., 2004; Glavic et al., 2004; Litsiou, Hanson, & Streit, 2005; Maharana & Schlosser, 2018; Meulemans & Bronner-Fraser, 2004; Schlosser et al., 2014). Supporting the latter, no clear demarcations of segregated domains corresponding to prospective neural crest versus placode cells have been observed (Fig. 1A) (Roellig,

Tan-Cabugao, Esaian, & Bronner, 2017), and naïve NPB cells seem to have the developmental potential to contribute to both prospective neural crest or placode cells (Roellig et al., 2017), suggesting a common developmental and evolutionary origin of these embryonic populations. In line with this, Horie et al. recently have shown that the lateral neural border of *Ciona* exhibits a compartmentalization similar to the neural plate ectoderm in vertebrates; thus, it might serve as a common source for the evolution of both neural crest and placodes (Horie et al., 2018).

2.3 Cell types arising from the neural plate border

Neural crest precursors lose their neuroepithelial properties and undergo an EMT, which transforms them into individual migratory mesenchymal cells (Barriga, Maxwell, Reyes, & Mayor, 2013; Hutchins & Bronner, 2018; Nieto, 2011; Piacentino, Li, & Bronner, 2020; Theveneau & Mayor, 2012). They then migrate long distances in three migratory streams in the head region and in a segmental chain-like migration in the trunk to their final destinations. Upon differentiation, they generate multiple cell types including chondrocytes and osteoblasts that give rise to the bone and cartilage of the face, vascular smooth muscle cells, diverse autonomic and sensory neurons (including enteric, sympathetic, parasympathetic and different kinds of sensory cells), chromaffin cells, all of the peripheral glia, and skin melanocytes throughout the body (Green et al., 2015; Hall & Gillis, 2012), as well as adipocytes (Billon et al., 2007).

The horseshoe-shaped PPR in turn is located around the anterior neural plate. It gives rise to all cranial placodes, which are recognizable as transient ectodermal thickenings at stereotypical locations in the vertebrate head (Bhattacharyya et al., 2004; Grocott et al., 2012; Pieper et al., 2011; Schlosser, 2015; Streit, 2007; Thiery et al., 2020). The PPR progressively regionalizes into anterior (adenohypophyseal, olfactory and lens placodes), intermediate (trigeminal placode), and posterior (otic placode, lateral line, and epibranchial placodes) domains (Fig. 1A and B) (Moody & Lamantia, 2015; Schlosser, 2015). Similar to neural crest cells, most placodal cells move away from the remaining ectoderm via ingression or invagination to form the variety of sensory organ cell types in the head. These include the crystalline lens cells, the neuroendocrine cells of the anterior pituitary gland, the olfactory sensory neurons, inner ear structural and neural elements, water current- and electrical field- sensing neurons of the lateral line of fishes, and most of the neurons of the cranial sensory ganglia (Fig. 1C)

(Chen & Streit, 2012; Cvekl and Duncan, 2007; Lwigale, 2001; Ohyama et al., 2007; Piotrowski & Baker, 2014).

Thus, both neural crest and placodes cells internalize, migrate and differentiate into various cell types; moreover, their mutual interactions are essential for morphogenesis of the cranial peripheral nervous system (PNS) and head sense organs (Fig. 1C) (Coppola et al., 2009; D'Amico-Martel and Noden, 1983; Fleenor and Begbie, 2014; Freter et al., 2013; Steventon et al., 2014; York & McCauley, 2020).

2.4 Evolutionary origin of neural crest and placodes

The Cambrian Explosion is one of the most significant events in the history of life on earth. During this time, major metazoan lineages emerged, including our own vertebrate phylum (Marshall, 2006; Wray, Levinton, & Shapirot, 1996). Vertebrates have since attained an extraordinary degree of specialization and morphological diversity. The advent of vertebrates is often connected with acquisition of neural crest and cranial placodes (Gans and Northcutt, 1983; Northcutt, 2005; Northcutt & Gans, 1983) that shaped the body-plan of the vertebrate ancestor. Early vertebrates transitioned from passive filter-feeders to becoming active predators, with a well-developed cranial endoskeleton containing jaws and teeth, odontodes that formed a dermal armor, improved sensory organs, and many other features involving new cell types (Adameyko and Lallemend, 2010; Baker and Bronner-Fraser, 2001; Gans and Northcutt, 1983; Green et al., 2015; Schlosser, 2006; Simões-Costa & Bronner, 2015).

Bona fide neural crest and placode cells are first evident in jawless vertebrates (hagfishes and lampreys), which form a monophyletic group, Cyclostomata (Kuraku, Hoshiyama, Katoh, Suga, & Miyata, 1999; McCauley & Bronner-Fraser, 2003; Modrell et al., 2014; Ota et al., 2007; Stock & Whitt, 1992). Many fossils suggest that the body plan of the early vertebrate ancestor was similar to the body-plan of extant jawless fish (Janvier, 1996; Janvier & Janvier, 2008). While the fossil record of early vertebrates is quite limited, several spectacular soft-bodied fossils are well-preserved such as *Haikouella*, *Haikouichthys*, *Myllokunmingia* or *Metaspriggina* from the Chengjiang fauna and Burgess Shale, and they are generally assumed to be the closest relatives of all extant vertebrates. Despite the lack of fossilized mineralized tissue in these Cambrian chordates, they still share several characteristics with extant vertebrates such as cartilaginous olfactory capsules or skeleton associated with gill slits (Chen et al., 1999;

Morris and Caron, 2014; Shu et al., 1999, 2003). This may suggest that the neural crest and placodes already evolved at the base of the vertebrate radiation and the early vertebrate evolution is tightly connected with the evolution of neural crest and placodes (Green et al., 2015; Schlosser, 2015).

Some other evolutionary novelties derived from the neural crest, such as bone, calcified cartilage, and dentine, are absent in cyclostomes and unique to the second division of vertebrates-the jawed vertebrates (gnathostomes). They likely arose in the stem gnathostomes, represented by the armored ostracoderms (still jawless) and the jawed placoderms (Brazeau and Friedman, 2015). These already possessed dentine, one of the first hard substances to evolve, which is exclusively produced by neural crest-derived odontoblasts in extant vertebrates (Chai et al., 2000; Hall & Gillis, 2012). Dentine pulps formed part of the first dermal armor, composed of tooth-like odontodes ("dermal teeth"), dermal plates or both (Smith et al., 1996). Odontodes form from an odontogenic zone producing the dentine, often covered by a cap of enamel or enameloid; and an underlying osteogenic zone forming dermal bone (Donoghue, Sansom, & Downs, 2006; Huysseune & Sire, 1998; Keating et al., 2015; Ørvig, 1977; Sire et al., 2009; Sire & Huysseune, 2003). While the dermal armor covered the entire body of the fossil stem gnathostomes, it has regressed in extant gnathostomes, represented by the teeth in the cranial region, and scales, dermal denticles and scutes in the post-cranial region of several vertebrate species (Gillis et al., 2017; Sire & Huysseune, 1996, 2003). It is thus plausible that neural crest produced skeletal derivatives first along the entire body, and that this potential was later lost from the trunk, as appears to be the case in amniotes (Le Douarin, 1982). As case in point, there are trunk neural crest-derived dermal denticles in the little skate (Gillis, et al., 2017). Similarly, turtles have an extensive bony exoskeleton, the plastron (Cebra-Thomas et al., 2007, 2013; Clark et al., 2001; Gilbert, Bender, Betters, Yin, & Cebra-Thomas, 2007), thought to be the evolutionary the oldest part of the turtle shell (Schoch & Sues, 2015). Trunk neural crest cells that express neural crest markers such as *FoxD3*, PDGFRa, and HNK-1, arise from the neural tube and migrate ventrally to the prospective plastron region, indicating that they are likely to contribute to this structure (Cebra-Thomas et al., 2007, 2013). Thus, the diversity of odontodes-like structures in extant living vertebrates likely arose via combinations of conservation and loss of the odontogenetic and osteogenic compartments (Gillis, et al., 2017; Sire et al., 2009; Sire & Huysseune, 2003).

Taken together, the co-option of neural crest-derived odontoblasts into the skin of a common gnathostome ancestor might represent a key neural crest novelty associated with gnathostome evolution, giving rise to teeth and odontodes which comprise the dermal skeleton. Interestingly, it has been proposed that odontoblasts were originally sensory receptors transmitting external signals from the surrounding water—given that dentine is able to detect mechanical, temperature and osmotic changes (Allard, Magloire, Couble, Maurin, & Bleicher, 2006; Baker, 2008; Gans and Northcutt, 1983; Khatibi Shahidi et al., 2015; Lumsden, 1987; Magloire et al., 2004; Okumura et al., 2005). Thus, the dermal armor of the earliest vertebrates created a shield that protected sensory processes. In addition, the dermal armor contains pores and canals that strongly indicate the presence of electro-receptive or lateral line systems, highly reminiscent of those observed in the bones and scales of extant living bony fishes (Ørvig, 1972; Tarby and Webb, 2003; Webb et al., 2014).

3. Neural crest- and placode-like cells in non-vertebrate chordates

Vertebrates belong to the phylum Chordata, which is characterized by a dorsal hollow nerve cord, notochord, pharyngeal gill slits and post-anal tail. In addition to vertebrates, chordates comprise two other non-vertebrate subphyla, Cephalochordata (commonly known as amphioxus, or lancelet) and Tunicata (also known as Urochordata; including larvaceans, thaliaceans, and ascidians). Based on morphological similarities, amphioxus was classically considered to be the closest living relative of vertebrates. However, phylogenomic studies changed this concept, showing that the sister group of vertebrates are actually tunicates (Delsuc et al., 2018; Delsuc, Brinkmann, Chourrout, & Philippe, 2006; Delsuc, Tsagkogeorga, Lartillot, & Philippe, 2008; Putnam et al., 2008; Telford, Budd, & Philippe, 2015). This phylogenetic relationship was further supported by the presence of neural crest- and placode-like rudiments in the larval stages of tunicates (e.g., Abitua et al., 2012, 2015; Cao et al., 2019; Horie et al., 2018; Stolfi et al., 2015), which are absent in amphioxus. Therefore, for a closer understanding of the evolution of NPB-cell populations and their ancestry in vertebrates, it is essential to examine how these domains are established in non-vertebrate chordates.

Tunicates constitute a diverse group of marine filter feeders that have undergone remarkable evolutionary changes in body plan, genome organization, and life strategies. Their chordate traits are manifested by their tadpole-like larvae, which have a dorsal hollow neural tube, a notochord, striated heart muscles, and tail. In comparison to vertebrate embryos, they have an invariant and small number of cells, which makes it possible to follow the fates of single cells during development (Holland 2016; Lemaire, 2011; Satoh, 2014, 2016). Ascidian neural patterning does not involve a BMP morphogen gradient (Darras & Nishida, 2001; Lemaire et al., 2008) but rather the FGF/ERK, Nodal, and Delta-Notch signaling pathways (Hudson, 2016). Interestingly, the ascidian NPB is defined by a subset of the transcription factors involved in vertebrate NPB specification, such as *Msx*, *Zic*, and *Pax3/7* (Imai, Levine, Satoh, & Satou, 2006; Li et al., 2017; Stolfi et al., 2015; Zhao et al., 2019).

Putative tunicate homologs of neural crest cells derived from the NPB were first described by Abitua et al. (2012). The descendant cells of the a9.49 cell lineage give rise to pigment cells of the ocellus and otolith. These cells express NPB and neural crest "specifier" genes such as *Msx*, *Zic*, *Pax3/7*, *Snail*, *Ets*, *Id*, and *FoxD*, but they are not capable of long-range migration. In addition, misexpression of *Twist*, a transcription factor important in cranial neural crest formation in frog and fish, conveys migratory abilities to the a9.49 cell lineage (Abitua et al., 2012). Another piece of evidence that tunicates have neural crest-like cells is the bipolar tail neurons in the trunk of ascidian tadpole larvae. These ascidian neurons have a set of features similar to the vertebrate neural crest-derived mechanosensory neurons of the dorsal root ganglia, such as long-range migration to their final destinations, expression of vertebrate orthologues of *Neurog*, *Islet*, *Asic*, and also comparable mechanism of synapse formation (Stolfi et al., 2015).

In addition to there being putative neural crest cell homologies ascribed to tunicate cells, several studies have speculated that tunicates might also have rudiments of cranial placodes (Abitua et al., 2015; Cao et al., 2019; Horie et al., 2018; Ikeda et al., 2013; Mazet and Shimeld, 2005; Wagner and Levine, 2012). Molecular genetic evidence of neural plate formation revealed that ascidian embryos express vertebrate orthologues of *Six1/2*, *FoxC*, *Dmrt*, *Foxg* and *Eya* in the "preplacodal-like" domain (PPLD) (Abitua et al., 2015; Horie et al., 2018; Liu and Satou, 2019; Tresser et al., 2010). The first evidence that the ascidian PPLD gives rise to a placode-like structure, which produces ciliated primary sensory cells that might be evolutionary related to those derived from vertebrate olfactory placodes, was provided by Abitua et al. (2015). In addition, it was subsequently shown that ascidian neural crest-like cells, the bipolar tail neurons, could be transformed into the placode-like palp sensory cells, which arise from the "preplacodal" domain (Horie et al., 2018). Taken together, these data support the hypothesis that both neural crest and placodes have a common evolutionary origin from the NPB.

In contrast to tunicates, amphioxus lacks a cell population that is obviously homologous to neural crest or placodes. However, they do clearly display an NPB-like domain between the neural plate and non-neural ectoderm, which exhibits expression of "NPB specifiers" similar to those of vertebrates, including *Msx, Pax3/7, Zic* and *Dlx* (Gostling & Shimeld, 2003; Holland, Schubert, Kozmik, & Holland, 1999; Sharman, Shimeld, & Holland, 1999; Wada, Holland, Sato, Yamamoto, & Satoh, 1997; Yu et al., 2008). This suggests that the NPB specification module is at least partially conserved across chordates. In addition, specification of the amphioxus NPB-like region also requires BMP-signaling activity, however at higher levels (Kozmikova et al., 2013).

Interestingly, the amphioxus NPB-like progenitors transiently express the "neural crest specifier" *Snail*, even though no cells undergoing EMT or emigrating from the amphioxus neural tube have been detected (Langeland, Tomsa, Jackman, & Kimmel, 1998; Yu et al., 2008). Other NPB and neural crest specifiers (such as *Ap2* and *SoxE*) are expressed in the mesoderm or non-neural ectoderm, respectively, but are missing from the neural plate border (Meulemans and Bronner-Fraser, 2002; Yasui, Zhang, Uemura, Aizawa, & Ueki, 1998; Yu, 2010; Yu, Holland, & Holland, 2002; Yu, et al., 2008). The lack of the majority of "neural crest specifiers" in the NPB and the absence of migrating cells supports the hypothesis that the full module of the neural crest gene regulatory network was complemented from other cell types to the neural tube in a common ancestor of tunicates and vertebrates (Meulemans and Bronner-Fraser, 2005; Yu, 2010).

Another hypothesis based on amphioxus data assumes that the neural crest-descendant cell types evolved from the neuroectodermal pigmented photosensory cells. In this scenario, the amphioxus pigmented Hesse organs (dorsal ocelli) with their constituent photoreceptors and shading pigment cells represent a possible evolutionary homolog of vertebrate neural crest–derived sensory cells and pigment cells (Baker and Bronner-Fraser, 1997; Ivashkin & Adameyko, 2013). Appealing as it is, this hypothesis would necessitate that the Hesse ocelli emerge from the NBP, both in development and evolution—which remains to be shown.

In contrast to vertebrates and ascidians, amphioxus has no ectodermal domain with a specific gene expression profile homologous to the preplacodal region. Interestingly, the key placode "specifiers," *Six1/2* and *Eya* (Schlosser et al., 2014), are expressed in the epidermal sensory neurons situated along the entire length of the larval body (Kozmik et al., 2007). These sensory neurons have a "placodal" expression profile, undergo epithelial-to-mesenchymal transition and migrate like vertebrate placode-derived neurons but they do not arise from the NPB (Kaltenbach, Yu, & Holland, 2009; Lu, Luo, & Yu, 2012). This would imply that either these amphioxus cells are not homologous to the placode-derived sensory neurons, or that they are homologous to placodal derivatives but that a "developmental takeover" occurred during chordate evolution (Arendt, 2008).

There is a similar situation regarding amphioxus Hatschek's pit where *Six1/2* and *Eya* are also expressed (Kozmik et al., 2007). Traditionally, Hatschek's pit has been proposed as the amphioxus homologue of the vertebrate adenohypophyseal placode (anterior pituitary), a hypothesis supported by the similar gene expression patterns of *Pitx*, *Pax6*, *Islet*, *Lhx3* and *Trk* (Benito-Gutiérrez, Nake, Llovera, Comella, & Garcia-Fernàndez, 2005; Glardon et al., 1998; Hatschek, 1881; Jackman, Langeland, & Kimmel, 2000; Patthey et al., 2014; Schlosser, 2017; Yasui et al., 2000; Wang, Zhang, Yasui, & Saiga, 2002). However, the Hatschek's pit is formed by the left (Hatschek's) diverticulum derived from the endodermal pouch, which then fuses with the surface ectoderm (Hatschek and Tuckey, 1893). Thus, Hatschek's pit develops from mixed endodermal and ectodermal cells and therefore has a different embryonic origin from the ectodermal adenohypophysis of vertebrates. In consequence, Hatschek's pit is either not homologous to the adenohypophyseal placode, or it is homologous to adenohypophyseal placode but has changed its developmental origin during chordate evolution.

In summary, these results suggest an evolutionary scenario in which the NPB specification gene regulatory network (GRN) module was already established in the common chordate ancestor. The subsequent split of this module into neural crest- and placode-specific submodules within the NPB region may have occurred in the common ancestor of tunicates and vertebrates. Despite the conserved molecular identity of amphioxus peripheral sensory neurons, including *Six1/2* and *Islet*, these neurons do not appear to be developmentally derived from the NPB. However, given the prevalent origin of peripheral sensory neurons from the lateral neural border in non-chordate animals (see below), the sensory lineage in amphioxus may have been secondarily modified in evolution.

4. The lateral neural border domain in non-chordate deuterostomes

Previous studies have shown a strictly conserved molecular machinery specifying the LNB formation across chordates (e.g., Abitua et al., 2012; Groves & LaBonne, 2014; Martik et al., 2019; Sauka-Spengler, Meulemans, Jones, & Bronner-Fraser, 2007). However, whether the expression of the "NPB specifiers" is conserved also amongst non-chordate deuterostomes such as hemichordates and echinoderms, has been a matter of intense comparative research since it would shed light on the evolution of both the LNB and LNB-derived cell types.

Although a considerable number of conserved "NPB specifiers" are expressed in hemichordate and sea urchin embryos and developing larvae, their developmental role in the context of the ciliated larval body plan appears to be different from that of their chordate orthologs (Saudemont et al., 2010; Su et al., 2019). Nevertheless, differential expression of *nkx2*, *pax*, and *msx* genes along the dorsoventral axis is a shared feature of echinoderm and hemichordate indirect development (Su et al., 2019), opening up the possibility that at least parts of the NPB-GRN are conserved in non-chordate deuterostomes. (Fig. 3). Adding to this, BMP signaling activity has been demonstrated to be involved in setting up this network in non-chordate deuterostomes, as well. Here, BMP signaling is essential for the expression of "dorsal" marker genes including *msx* and *dlx* (Saudemont et al., 2010). Furthermore, transient overactivation of BMP-signaling in hemichordate and sea urchin embryos leads to a mediolateral re-organization of the neurogenic ectoderm similar to that of the chordate neural plate (Su et al., 2019). In addition, BMP appears to exert an anti-neurogenic effect similar to that in vertebrates (Su et al., 2019).

Not much is known regarding the cell types descending from the lateral NPB-like ectoderm in both groups. The expression of the sensory marker *pax2/5/8* in this region in sea urchin (Su et al., 2019) might suggest the presence of sensory cells, but further investigations will be needed to clarify this. Notably, the mediodorsal *msx*+ region present in both directly (Lowe et al. 2006) and indirectly (Su et al., 2019) developing enteropneust hemichordates spatially coincides with the later presence of the dorsal collar and cord in adults, which contains sensory neurons (Nomaksteinsky et al., 2009). This

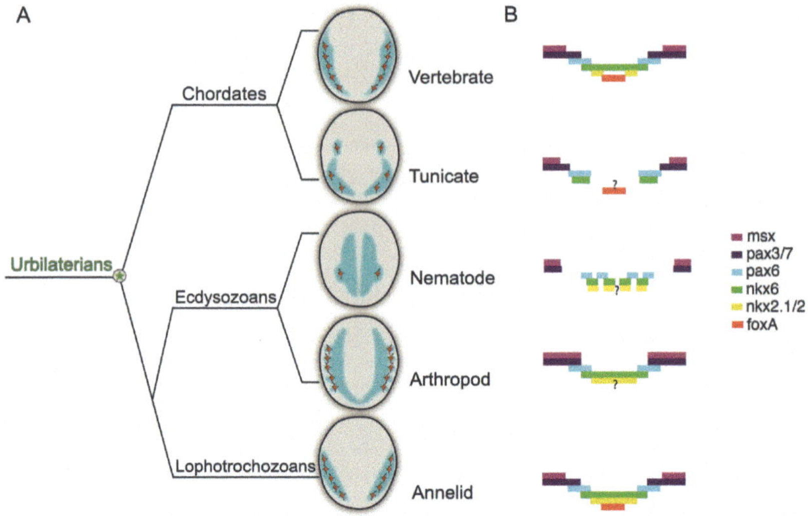

Fig. 3 Precursors of neural crest and cranial placodes might be already present in a common ancestor of bilaterians. (A) Simplified phylogenetic tree of Bilaterians showing a distribution of NPB-like domains (cyan) based on the expression of conserved "NPB specifiers," Msx/msh/vab-15. The orange cells in each NPB-like domain represent lateral neural border-derived neuroblasts that undergo EMT and migrate to differentiate into PNS neurons. The schematic drawings of neurulae represent the dorsal view of chordates and the ventral view of lophotrochozoans and ecdysozoans; based on Li et al. (2017). (B) Available gene expression data for mediolateral neuroectodermal patterning depicted for representative species. Question marks indicate missing data. *Modified from Arendt, D. (2018). Animal evolution: Convergent nerve cords? Current Biology, 28(5), R225–R227. https://doi.org/10.1016/j.cub.2018.01.056.*

is especially interesting given the location of sensory precursor cells in amphioxus in similar locations (Benito-Gutiérrez & Arendt, 2009).

5. The lateral neural border in protostomes

A possible conservation of ectodermal specification has also been addressed in the protostome branch of the Bilateria. Initial studies focused on dorsoventral patterning in the fruit fly *Drosophila melanogaster* and the nematode *Caenorhabditis elegans*, both belonging to the Ecdysozoan superphylum. Intriguingly, a spatial sequence of regionally restricted domains of ectodermal specification genes including *nkx2.1/2, nkx6, pax6, pax3/7, msx* and *dll* has been detected in both model organisms (fly and nematode in Fig. 3), indicating conservation of overall dorsoventral patterning across Bilateria. In particular, in these organisms a similar combination of transcription factors, e.g., *Zic*, *Pax3/7*, and *Msx*, appears to specify the LNB (Abitua et al., 2012; Arendt, 2018; Arendt, Tosches, & Marlow, 2016; D'Alessio & Frasch, 1996; Li et al., 2017; Pla & Monsoro-Burq, 2018; Sauka-Spengler et al., 2007; Yu, et al., 2008). In line with this, comparative expression studies revealed broad conservation of ectodermal specification also in the second large protostome superphylum, the Lophotrochozoa (annelid in Fig. 3), although in varying morphological contexts (Arendt, 2018; Denes et al., 2007; Martín-Durán et al., 2018).

Zooming in on the lateral *Zic*, *Pax3/7*, *Msx* domain, a more detailed molecular characterization of the LNB has recently been presented for the marine annelid *Platynereis dumerilii* (Fig. 4). This study reveals that the annelid LNB expresses orthologs of virtually all vertebrate "NPB specifiers"; besides *zic*, *pax3/7* and *msx* the LBN also expresses *gbx*, *prdm1*, *gfap2*, *dlx*, and *gata1/2/3*. This supports further the evolutionary conservation of an NPB-like domain at the lateral edge of the neural ectoderm across Bilateria. Importantly, the annelid LNB also expresses transcription factors specifying neural crest or the preplacodal region in the vertebrates (Fig. 4A). For the neural crest, this includes *SoxE* and *Snail*, whereas *FoxD*

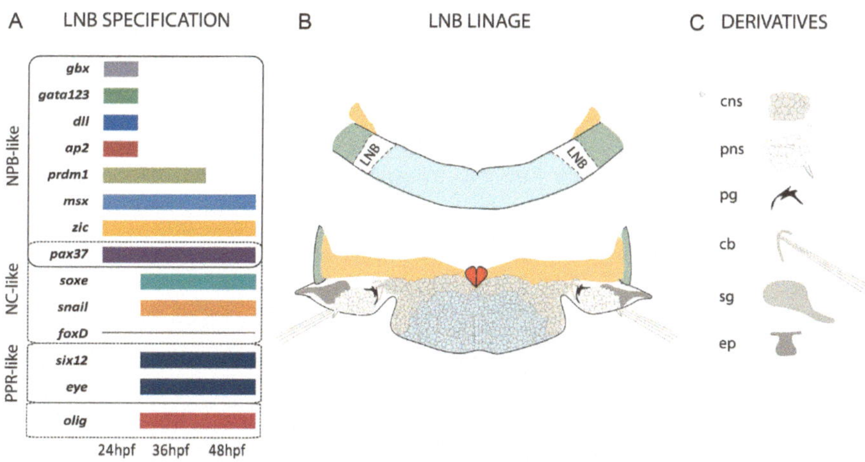

Fig. 4 Molecular identity, lineage, and cellular descendants of the LNB in the annelid *Platynereis dumerilii*. After Bertucci et al. (in prep.). (A) Expression of conserved transcription factors in the annelid LNB. (B) The developing ventral body side after 2 and 6 days of development. White, LNB; blue, neuroectoderm; green, epidermal ectoderm; yellow, mesoderm; red, axochord. (C) Cellular derivatives of the LNB in 6-days-old worms. cns, central nervous system; pns, peripheral nervous system; pg, pigment cells; cb, chaetoblasts; sg, spinning gland; ep, epidermal cell.

appears to be absent (Bertucci et al., in prep.). Interestingly, both placodal genes *Six1/2* and *Eya* are also expressed in the *Platynereis* LNB. This would indicate that (i) the annelid LNB presents a mixed NPB-neural crest-placode identity. This identity involves a close-to-complete complement of transcription factors, comparable to that found in the vertebrate or ascidian NPB. It also indicates that (ii) no further segregation into crest or placode-like cells occurs from within this region—in line with the absence of these developmental cell types from non-chordate deuterostomes and with the proposed common evolutionary origin of placodes and crest cells from the NPB (Horie et al., 2018).

5.1 A conserved role of BMP signaling in ectodermal patterning

Studies in protostome model systems have likewise revealed a role of BMP signaling in ectodermal patterning. In insects, for instance, the expression of neural specification genes is sensitive to threshold-dependent Bmp-mediated repression (Mizutani, Meyer, Roelink, & Bier, 2006). Just as in vertebrates, *Drosophila Dpp* acts in two phases (Bier, 1997; reviewed in Lowe et al., 2006). In the first phase, Dpp triggers the segregation of neural ectoderm from the epidermal ectoderm. In the second phase, Dpp contributes to mediolateral neural ectodermal patterning by differentially down- or upregulating neural specification genes (Mizutani et al., 2006; Rusten, Cantera, Kafatos, & Barrio, 2002). However, the specification of the neural ectoderm appears to be independent of BMP signaling in the nematode (Lu et al., 2012; Savage et al., 1996); and similarly, BMPs do not show a general antineurogenic effect in the marine annelid *Platynereis dumerilii*, (Denes et al., 2007). In this annelid, application of exogenous Bmp4 downregulates *Pdu-pax3/7*, whereas *ath*, a marker for peripheral sensory cells, is upregulated in a dose-dependent manner. Together, these findings suggest that the induction of neural ectoderm to LNB by the lowest levels of BMP signaling activity in the neural ectoderm may have been a patterning mechanism present in the bilaterian ancestor (Arendt and Nübler-Jung, 1994; Denes et al., 2007; Holley et al., 1995; Lowe et al., 2006; Martín-Durán et al., 2018; Su et al., 2019). Of note, the annelid LNB also gets enlarged by ectopic Wnt signaling (Lauri, 2013; Bertucci et al., in prep.), reminiscent of the role of Wnt signaling in vertebrate NPB patterning (Muroyama et al. 2002).

5.2 Cell types arising from the lateral neural border

If, indeed, an NPB-like, LNB region occurred in the lateral neural ectoderm of the developing bilaterian ancestor—what cell types did it give rise to? A first answer to this question comes from comparative studies showing that *Msx* orthologs are essential for the specification of sensory peripheral neurons in most bilaterian species that have been examined (Li et al., 2017; Zhao et al., 2019). This is reminiscent of vertebrates and tunicates, wherein peripheral sensory neurons arise from *msx*+ NPB-derived precursors. For example, in the fly, the *Msx* ortholog *Msh*, is strongly expressed in cells of the LNB encompassing CNS and PNS neuroblasts (Arendt & Nübler-Jung, 1999; D'Alessio & Frasch, 1996; Isshiki et al., 1997; Li et al., 2017). The lateral clusters of PNS neuroblasts thereby create the chordotonal organs, which develop from migratory cell precursors arising in that domain (D'Alessio & Frasch, 1996). In *msh*-mutant fly embryos, the migration of chordotonal precursors is significantly desynchronized, and the chordotonal organ has fewer neurons and glial cells. Thus, the fly ortholog of *Msx* has a significant influence on the development of PNS mechanosensory cells that arise from the LNB. The nematode ortholog of vertebrate *Msx*, *vab-15*, is also involved in the formation of the LNB domain and in the specification of the LNB-derived neuroblasts, when it is expressed in embryonic AB.p(lr)apapaa cell and in its descendent Q neuroblast. Loss of function mutations of *vab-15* result in a defect of the migration of Q neuroblast and failure of mechanosensory neuron formation (Du and Chalfie, 2001; Li et al., 2017). Interestingly, it has also been shown that some nematode neuroblasts express orthologs of several neural crest migratory genes, such as *ADAM13* gene, which is necessary for neural crest specification and migration (Kee et al., 2007; Meulemans Medeiros, 2013). As in the fly and nematode, mechanosensory neurons of the PNS also arise from the LNB in the annelid *Platynereis dumerilii* (Bertucci et al., in prep.), similar to NPB-descendant peripheral mechanosensory neurons of vertebrates, bipolar tail neurons of tunicates (Stolfi et al., 2015), or Q neuroblasts of nematodes (Sulston and Horvitz, 1977). However, even though the pericarya of these cells move inside to form part of the peripheral ganglia, they appear not to undergo full EMT (Bertucci et al., in prep.). Thus, the LNB domain in protostomes can produce both migratory and non-migratory descendants that appear to be specified by conserved terminal selector genes.

Yet, the complement of cell types arising from the LNB can be much broader than peripheral sensory neurons, as revealed by lineage tracing of

LNB-descendant cell types and subsequent mapping of cells into a multimodal, cellular-resolution atlas in *Platynereis* (Bertucci et al., in prep.; Vergara et al., 2020). This work reveals a population of multipotent progenitors residing in the annelid LNB that specifically expresses the bHLH transcription factor *olig* (Fig. 4A). Knockdown of *olig* abolishes a variety of peripheral cell types that descend from the LNB, including peripheral neurons, pigmented cells, cirri-forming chaetoblasts and spinning glands (Fig. 4C). These cell types subsequently express conserved terminal selectors shared with chordate NPB-derivatives (Bertucci et al., in prep.): For instance, sensory peripheral neurons express transcription factors involved in sensory specification including the POU and LIM homeodomain factors *pouIV* and *islet*, and peripheral neurons and ensheating glia-like cells express the paired-type homeodomain factor *Phox2* (Bertucci et al., in prep.), known to play a conserved role in peripheral nervous system circuit formation in the vertebrates (Bertucci & Arendt, 2013; Nomaksteinsky et al., 2013). LNB-derived annelid pigment cells in turn express the conserved master regulator of melanocyte specification gene *mitf* as has been shown in ascidian pigmented cells derived from the NPB-like region (Abitua et al., 2012). Interestingly, the annelid chaetoblasts are related in lineage to the pigment cells and also show a similar molecular identity, suggesting a sister cell type relationship. This goes in concert with overlapping metabolic pathways involved in melanin and chitin synthesis (Bertucci et al., in prep.). Thus, in addition to peripheral neurons, this work identifies glial and pigment cells as candidate ancestral cell types arising from the LNB in bilaterian ancestors. At the same time, it highlights both striking similarities and differences in the molecular profile of the multipotent progenitor cells that reside in the annelid LNB.

6. Cell type evolution in the lateral neural border region

As the boundaries between homology and convergent phenomena can be fuzzy, structures across phyla can be incorrectly described as homologous. However, the data from various bilaterians suggest that LNB domains share expression of conserved "NPB specifiers," and give rise to similar cell types. Taken together, the most parsimonious explanation is that cells of the LNB domains of non-chordate bilaterians reflect common traits with the vertebrate NPB, and that the LNB thus represents the evolutionary precursor of the region with neural crest and placode-forming abilities. This would imply that an LNB region already existed in the last common ancestor of deuterostomes and protostomes. Integrating available information, this region minimally expressed the transcription factors *Msx, Pax3/7,* and *Zic;* but may also have expressed *Prdm1, Tfap2, Gata1/2/3, Gbx*, and *Dll* as present in vertebrates and annelids. It may have given rise to mechanosensory neurons (Li et al., 2017; Zhao et al., 2019) and some kind of light-sensitive pigment cells (Baker and Bronner-Fraser, 1997; Ivashkin & Adameyko, 2013). Further comparative work also including bilaterian outgroups will help reconstructing the full GRN and identify all descendant cell types of the ancestral LNB.

Taking these conserved aspects as a starting point, what were the major evolutionary innovations that occurred in the LNB that ultimately led to the emergence of neural crest and placodes and their various descendant cell types in the vertebrate linage? We can envisage three possible paths of cell type evolution that are not mutually exclusive:

(1) Elaboration of transitory developmental cell types: The resident cell types of the LNB most likely emerged as persisting embryonic multipotent stem cells that subsequently diversified and specialized. For example, the ancestral bilaterian LNB changed into the chordate NBP, concomitant with the emergence of the neural plate that internalizes or cavitates to give rise to the neural tube; and their NBP subsequently split during the course of vertebrate evolution to generate the crest and head placodes. This involved manifold morphological change, manifest in the divergent cellular shapes and locations of neural crest and placodes, and changes and divergence of the corresponding GRNs as discussed above. Other pre-existing properties of the LNB cell types, such as EMT and migratory behavior, were also subject to evolutionary modification in the diverse bilaterian lineages.

(2) Elaboration of terminal fates: This involves the modification of pre-existing and the acquisition of new cell types that descend from the LNB linage. This could have occurred via cell type divergence or division of labor (Arendt, 2008), as suggested for diversification of peripheral neurons into dorsal root ganglia, sympathetic and parasympathetic neurons in the vertebrates (Arendt, 2008), for the origin of neural crest-derived adrenal chromaffin cells or the emergence of chaetoblasts most likely from pre-existing pigmented precursor cell types in annelids (Bertucci et al., in prep.). Beyond that, the LNB could have given rise to entirely new descendant cell types by neofunctionalization. For this, one or more regulatory factors may have been coopted into NPB cell types in the chordate ancestor (Abitua et al., 2012; Ivashkin & Adameyko, 2013; Stolfi, et al., 2015). Later in vertebrate evolution this may have enabled the evolution of new fates,

such as dentine-producing odontoblasts. This scenario is particularly exciting as it shows the stunning evolutionary potential of pre-existing multipotent developmental progenitor cell types.

(3) Developmental "takeover" of pre-existing lineages: in this scenario the highly plastic LNB descendant cell types take over the development of pre-existing cell types (that previously originated from other sources) (Arendt, 2008). This is evident for vertebrate cell types of mixed embryonic origin, including dermal bones, smooth muscles and gland cells that are only partially neural crest-derived, or neural crest-derived in changing proportions in different species. In some cases, the developmental "takeover" of cell types by the neural crest may be complete, and coupled to the acquisition of evolutionary novelty (Arendt, 2008). For instance, the deuterostome cartilaginous pharyngeal skeleton (Rychel & Swalla, 2007), which is neural crest-derived in vertebrates, likely predates the neural crest and its development, and thus may have been originally generated by mesodermal cells and then secondarily taken over by the neural crest. Another possible example for neural crest developmental takeover is the enteric nervous system, which shows molecular similarities between annelids and vertebrates (Brunet, Lauri, & Arendt, 2015), yet does not appear to form from the LNB in annelids (Lauri, 2013), indicating that its neural crest derivation in vertebrates is a secondary phenomenon. As case in point, the enteric nervous system of lamprey is not directly derived from vagal neural crest cell as in gnathostomes, but rather from neural crest-derived Schwann cell precursors that later migrate along nerves to populate the intestines (Green, Uy, & Bronner, 2017).

References

Abitua, P. B., Gainous, T. B., Kaczmarczyk, A. N., Winchell, C. J., Hudson, C., Kamata, K., et al. (2015). The pre-vertebrate origins of neurogenic placodes. *Nature*, *524*, 462–465. https://doi.org/10.1038/nature14657.

Abitua, P. B., Wagner, E., Navarrete, I. A., & Levine, M. (2012). Identification of a rudimentary neural crest in a non-vertebrate chordate. *Nature*, *492*, 104–107. https://doi.org/10.1038/nature11589.

Adameyko, I., & Lallemend, F. (2010). Glial versus melanocyte cell fate choice: Schwann cell precursors as a cellular origin of melanocytes. *Cellular and Molecular Life Sciences*, *67*, 3037–3055. https://doi.org/10.1007/s00018-010-0390-y.

Ahrens, K., & Schlosser, G. (2005). Tissues and signals involved in the induction of placodal Six1 expression in Xenopus laevis. *Developmental Biology*, *288*, 40–59. https://doi.org/10.1016/j.ydbio.2005.07.022.

Allard, B., Magloire, H., Couble, M. L., Maurin, J. C., & Bleicher, F. (2006). Voltage-gated sodium channels confer excitability to human odontoblasts: Possible role in tooth pain transmission. *The Journal of Biological Chemistry*, *281*, 29002–29010. https://doi.org/10.1074/jbc.M601020200.

Arendt, D. (2008). The evolution of cell types in animals: Emerging principles from molecular studies. *Nature Reviews Genetics*, *9*, 868–882. https://doi.org/10.1038/nrg2416.

Arendt, D. (2018). Animal evolution: Convergent nerve cords? *Current Biology*, *28*(5), R225–R227. https://doi.org/10.1016/j.cub.2018.01.056.
Arendt, D., & Nübler-Jung, K. (1994). Inversion of dorsoventral axis? *Nature*, *371*, 26.
Arendt, D., & Nübler-Jung, K. (1999). Comparison of early nerve cord development in insects and vertebrates. *Development*, *126*, 2309–2325.
Arendt, D., Tosches, M. A., & Marlow, H. (2016). From nerve net to nerve ring, nerve cord and brain—Evolution of the nervous system. *Nature Reviews Neuroscience*, *17*, 61–72. https://doi.org/10.1038/nrn.2015.15.
Baker, C. V. H. (2008). The evolution and elaboration of vertebrate neural crest cells. *Current Opinion in Genetics & Development*, *18*, 536–543. https://doi.org/10.1016/j.gde.2008.11.006.
Baker, C. V. H., & Bronner-Fraser, M. (1997). The origins of the neural crest. Part II: An evolutionary perspective. *Mechanisms of Development*, *69*, 13–29.
Baker, C. V. H., & Bronner-Fraser, M. (2001). Vertebrate cranial placodes I. Embryonic induction. *Developmental Biology*, *232*, 1–61.
Baker, C. V. H., & Schlosser, G. (2005). The evolutionary origin of neural crest and placodes. *Journal of Experimental Zoology. Part B: Molecular and Developmental Evolution*, *304B*, 269–273. https://doi.org/10.1002/jez.b.21060.270.
Barriga, E. H., Maxwell, P. H., Reyes, A. E., & Mayor, R. (2013). The hypoxia factor Hif-1α controls neural crest chemotaxis and epithelial to mesenchymal transition. *The Journal of Cell Biology*, *201*, 759–776. https://doi.org/10.1083/jcb.201212100.
Basch, M. L., Bronner-Fraser, M., & García-Castro, M. I. (2006). Specification of the neural crest occurs during gastrulation and requires Pax7. *Nature*, *441*, 1–5. https://doi.org/10.1038/nature04684.
Benito-Gutiérrez, E., & Arendt, D. (2009). CNS evolution: New insight from the mud. *Current Biology*, *19*(15), R640–R642. https://doi.org/10.1016/j.cub.2009.06.020.
Benito-Gutiérrez, È., Nake, C., Llovera, M., Comella, J. X., & Garcia-Fernàndez, J. (2005). The single AmphiTrk receptor highlights increased complexity of neurotrophin signalling in vertebrates and suggests an early role in developing sensory neuroepidermal cells. *Development*, *139*, 2191–2202. https://doi.org/10.1242/dev.01803.
Bertucci, P., & Arendt, D. (2013). Somatic and visceral nervous systems - An ancient duality. *BMC Biology*, *11*(54). https://doi.org/10.1186/1741-7007-11-54.
Betters, E., Charney, R. M., & Garcia-Castro, M. I. (2018). Early specification and development of rabbit neural crest cells. *Developmental Biology*, *444*, S181–S192. https://doi.org/10.1016/j.ydbio.2018.06.012.
Bhat, N., Kwon, H.-J., & Riley, B. B. (2013). A gene network that coordinates preplacodal competence and neural crest specification in zebrafish. *Developmental Biology*, *373*, 107–117. https://doi.org/10.1016/j.ydbio.2012.10.012.
Bhattacharyya, S., Bailey, A. P., Bronner-Fraser, M., & Streit, A. (2004). Segregation of lens and olfactory precursors from a common territory: cell sorting and reciprocity of Dlx5 and Pax6 expression. *Developmental Biology*, *271*, 403–414. https://doi.org/10.1016/j.ydbio.2004.04.010.
Bier, E. (1997). Anti-neural-inhibition: A conserved mechanism for neural induction. *Cell*, *89*(5), 681–684. https://doi.org/10.1016/s0092-8674(00)80250-0.
Billon, N., Iannarelli, P., Monteiro, M. C., Glavieux-Pardanaud, C., Richardson, W. D., Kessaris, N., & Dupin, E. (2007). The generation of adipocytes by the neural crest. *Development*, *134*(12), 2283–2292. https://doi.org/10.1242/dev.002642.
Brazeau, M. D., & Friedman, M. (2015). The origin and early phylogenetic history of jawed vertebrates. *Nature*, *520*, 490–497. https://doi.org/10.1038/nature14438.
Brugmann, S. A., Pandur, P. D., Kenyon, K. L., Pignoni, F., & Moody, S. A. (2004). Six1 promotes a placodal fate within the lateral neurogenic ectoderm by functioning as both a transcriptional activator and repressor. *Development*, *131*, 5871–5881. https://doi.org/10.1242/dev.01516.

Brunet, T., Lauri, A., & Arendt, D. (2015). Did the notochord evolve from an ancient axial muscle? The axochord hypothesis. *BioEssays*, *37*(8), 836–850. https://doi.org/10.1002/bies.201500027.

Cao, C., Lemaire, L. A., Wang, W., Yoon, P. H., Choi, Y. A., Parsons, L. R., et al. (2019). Comprehensive single-cell transcriptome lineages of a proto-vertebrate. *Nature*, *571*, 349–354.

Cebra-Thomas, J. A., Betters, E., Yin, M., Plafkin, C., McDow, K., & Gilbert, S. F. (2007). Evidence that a late-emerging population of trunk neural crest cells forms the plastron bones in the turtle *Trachemys scripta*. *Evolution and Development*, *9*(3), 267–277. https://doi.org/10.1111/j.1525-142X.2007.00159.x.

Cebra-Thomas, J. A., Terrell, A., Branyan, K., Shah, S., Rice, R., Gyi, L., … Gilbert, S. F. (2013). Late-emigrating trunk neural crest cells in turtle embryos generate an osteogenic ectomesenchyme in the plastron. *Developmental Dynamics*, *242*(11), 1223–1235. https://doi.org/10.1002/dvdy.24018.

Chai, Y., Jiang, X., Ito, Y., Bringas, P., Jr., Han, J., Rowitch, D. H., et al. (2000). Fate of the mammalian cranial neural crest during tooth and mandibular morphogenesis. *Development*, *127*, 1671–1679.

Chen, J.-Y., Huang, D.-H., & Li, C.-W. (1999). An early Cambrian craniate-like chordate. *Nature*, *402*, 518–522.

Chen, J., & Streit, A. (2012). Induction of the inner ear: Stepwise specification of otic fate from multipotent progenitors. *Hearing Research*. https://doi.org/10.1016/j.heares.2012.11.018.

Clark, K., Bender, G., Murray, B. P., Panfilio, K., Cook, S., Davis, R., … Gilbert, S. F. (2001). Evidence for the neural crest origin of turtle plastron bones. *Genesis*, *31*(3), 111–117. https://doi.org/10.1002/gene.10012.

Coppola, E., Rallu, M., Richard, J., Dufour, S., Riethmacher, D., Guillemot, F., et al. (2009). Epibranchial ganglia orchestrate the development of the cranial neurogenic crest. *Proceedings of the National Academy of Sciences of the United States of America*, *107*, 2066–2071. https://doi.org/10.1073/pnas.0910213107.

Cvekl, A., & Duncan, M. K. (2007). Genetic and epigenetic mechanisms of gene regulation during lens development. *Progress in Retinal and Eye Research*, *26*, 555–597. https://doi.org/10.1016/j.preteyeres.2007.07.002.

D'Alessio, M., & Frasch, M. (1996). msh may play a conserved role in dorsoventral patterning of the neuroectoderm and mesoderm. *Mechanisms of Development*, *58*, 217–231.

D'Amico-Martel, A., & Noden, D. M. (1983). Contributions of placodal and neural crest cells to avian cranial peripheral ganglia. *The American Journal of Anatomy*, *166*, 445–468.

Darras, S., & Nishida, H. (2001). The BMP signaling pathway is required together with the FGF pathway for notochord induction in the ascidian embryo. *Development*, *128*(14), 2629–2638.

De Robertis, E. M., & Kuroda, H. (2004). Dorsal-ventral patterning and neural induction in Xenopus embryos. *Annual Review of Cell and Developmental Biology*, *20*, 285–308. https://doi.org/10.1146/annurev.cellbio.20.011403.154124.

De Robertis, E. M., & Sasai, Y. (1996). A common plan for dorsoventral patterning in Bilateria. *Nature*, *380*, 37–40. https://doi.org/10.1038/380037a0.

Delsuc, F., Brinkmann, H., Chourrout, D., & Philippe, H. (2006). Tunicates and not cephalochordates are the closest living relatives of vertebrates. *Nature*, *439*, 965–968. https://doi.org/10.1038/nature04336.

Delsuc, F., Philippe, H., Tsagkogeorga, G., Simion, P., Tilak, M.-K., Turon, X., et al. (2018). A phylogenomic framework and timescale for comparative studies of tunicates. *BMC Biology*, *16*, 1–14. https://doi.org/10.1186/s12915-018-0499-2.

Delsuc, F., Tsagkogeorga, G., Lartillot, N., & Philippe, H. (2008). Additional molecular support for the new chordate phylogeny. *Genesis*, *46*(11), 592–604. https://doi.org/10.1002/dvg.20450.

Denes, A. S., Jekely, G., Steinmetz, P. R., Raible, F., Snyman, H., Prud'homme, B., ... Arendt, D. (2007). Molecular architecture of annelid nerve cord supports common origin of nervous system centralization in bilateria. *Cell*, *129*, 277–288. https://doi.org/10.1016/j.cell.2007.02.040.

Donoghue, P. C. J., Sansom, I. J., & Downs, J. P. (2006). Early evolution of vertebrate skeletal tissues and cellular interactions, and the canalization of skeletal development. *Journal of Experimental Zoology Part B Molecular and Developmental Evolution*, *306*(3), 278–294. https://doi.org/10.1002/jez.b.21090.

Du, H., & Chalfie, M. (2001). Genes regulating touch cell development in Caenorhabditis elegans. *Genetics*, *158*, 197–207.

Faure, S., Barbara, P. D. S., Roberts, D. J., & Whitman, M. (2002). Endogenous patterns of BMP signaling during early chick development. *Developmental Biology*, *244*, 44–65. https://doi.org/10.1006/dbio.2002.0579.

Fleenor, S. J., & Begbie, J. (2014). Neural crest cell and placode interactions in cranial PNS development. *Neural Crest Cells*. Elsevier Inc. https://doi.org/10.1016/B978-0-12-401730-6.00009-0.

Freter, S., Fleenor, S. J., Freter, R., Liu, K. J., & Begbie, J. (2013). Cranial neural crest cells form corridors prefiguring sensory neuroblast migration. *Development*, *140*, 3595–3600. https://doi.org/10.1242/dev.091033.

Gans, C., & Northcutt, R. G. (1983). Neural crest and the origin of vertebrates: A new head. *Science*, *220*, 268–274.

García-Castro, M. I., Marcelle, C., & Bronner-Fraser, M. (2002). Ectodermal Wnt function as a neural crest inducer. *Science*, *297*(5582), 848–851. https://doi.org/10.1126/science.1070824.

Garnett, A. T., Square, T. A., & Medeiros, D. M. (2012). BMP, Wnt and FGF signals are integrated through evolutionarily conserved enhancers to achieve robust expression of Pax3 and Zic genes at the zebrafish neural plate border. *Development*, *139*, 4220–4231. https://doi.org/10.1242/dev.081497.

Gilbert, S. F., Bender, G., Betters, E., Yin, M., & Cebra-Thomas, J. A. (2007). The contribution of neural crest cells to the nuchal bone and plastron of the turtle shell. *Integrative and Comparative Biology*, *47*(3), 401–408. https://doi.org/10.1093/icb/icm020.

Gillis, J. A., Alsema, E. C., & Criswell, K. E. (2017). Trunk neural crest origin of dermal denticles in a cartilaginous fish. *Proceedings of the National Academy of Sciences of the United States of America*, *114*, 13200–13205. https://doi.org/10.1073/pnas.1713827114.

Glardon, S., Holland, L. Z., Gehring, W. J., & Holland, N. D. (1998). Isolation and developmental expression of the amphioxus Pax-6 gene (AmphiPax-6): Insights into eye and photoreceptor evolution. *Development*, *125*, 2701–2710.

Glavic, A., Honoré, S. M., Feijóo, C. G., Bastidas, F., Allende, M. L., & Mayor, R. (2004). Role of BMP signaling and the homeoprotein iroquois in the specification of the cranial placodal field. *Developmental Biology*, *272*(1), 89–103. https://doi.org/10.1016/j.ydbio.2004.04.020.

Gostling, N. J., & Shimeld, S. M. (2003). Protochordate Zic genes define primitive somite compartments and highlight molecular changes underlying neural crest evolution. *Evolution & Development*, *5*, 136–144. https://doi.org/10.1046/j.1525-142x.2003.03020.x.

Green, S. A., Simoes-Costa, M., & Bronner, M. E. (2015). Evolution of vertebrates as viewed from the crest. *Nature*, *520*, 474–482. https://doi.org/10.1038/nature14436.

Green, S. A., Uy, B. R., & Bronner, M. E. (2017). Ancient evolutionary origin of vertebrate enteric neurons from trunk-derived neural crest. *Nature*, *544*, 88–91. https://doi.org/10.1038/nature21679.

Grocott, T., Tambalo, M., & Streit, A. (2012). The peripheral sensory nervous system in the vertebrate head: A gene regulatory perspective. *Developmental Biology*, *370*, 3–23. https://doi.org/10.1016/j.ydbio.2012.06.028.

Groves, A. K., & LaBonne, C. (2014). Setting appropriate boundaries: Fate, patterning and competence at the neural plate border. *Developmental Biology*, *389*, 2–12. https://doi.org/10.1016/j.ydbio.2013.11.027.

Hall, B. K., & Gillis, J. A. (2012). Incremental evolution of the neural crest, neural crest cells and neural crest-derived skeletal tissues. *Journal of Anatomy*, *222*, 1–13. https://doi.org/10.1111/j.1469-7580.2012.01495.x.

Hatschek, B. (1881). Studien über die Entwickelung des Amphioxus. *Arbeiten aus dem 2007 Institut zu Wien*, *4*, 1–88.

Hatschek, B., & Tuckey, J. (1893). *The amphioxus and its development*. Sonnenschein & Co: Swan.

Holland, L. Z. (2016). Tunicates. *Current Biology*, *26*(4), R146–R152. https://doi.org/10.1016/j.cub.2015.12.024.

Holland, L. Z., Schubert, M., Kozmik, Z., & Holland, N. D. (1999). AmphiPax3/7, an amphioxus paired box gene: Insights into chordate myogenesis, neurogenesis, and the possible evolutionary precursor of definitive vertebrate neural crest. *Evolution & Development*, *1*(3), 153–165. https://doi.org/10.1046/j.1525-142x.1999.99019.x.

Holley, S. A., Jackson, P. D., Sasal, Y., Lu, B., De Robertis, E. M., Hoffmann, F. M., et al. (1995). A conserved system for dorsal-ventral patterning in insects and vertebrates involving sog and chordin. *Nature*, *376*, 249–253. https://doi.org/10.1038/376249a0.

Hong, C.-S., & Saint-Jeannet, J.-P. (2007). The activity of Pax3 and Zic1 regulates three distinct cell fates at the neural plate border. *Molecular Biology of the Cell*, *18*, 2192–2202. https://doi.org/10.1091/mbc.E06.

Horie, R., Hazbun, A., Chen, K., Chen, C., Levine, M., & Horie, T. (2018). Shared evolutionary origin of vertebrate neural crest and cranial placodes. *Nature*, *560*, 228–232. https://doi.org/10.1038/s41586-018-0385-7.

Hudson, C. (2016). The central nervous system of ascidian larvae. *WIREs Developmental Biology*. https://doi.org/10.1002/wdev.239.

Hutchins, E. J., & Bronner, M. E. (2018). Draxin acts as a molecular rheostat of canonical Wnt signaling to control cranial neural crest EMT. *The Journal of Cell Biology*, *217*, 3683–3697. https://doi.org/10.1083/jcb.201709149.

Huysseune, A., & Sire, J. (1998). Evolution of patterns and processes in teeth and tooth-related tissues in non-mammalian vertebrates. *European Journal of Oral Sciences*, *106*, 437–481. https://doi.org/10.1111/j.1600-0722.1998.tb02211.x.

Ikeda, T., Matsuoka, T., & Satou, Y. (2013). A time delay gene circuit is required for palp formation in the ascidian embryo. *Development*, *140*, 4703–4708. https://doi.org/10.1242/dev.100339.

Imai, K. S., Levine, M., Satoh, N., & Satou, Y. (2006). Regulatory blueprint for a chordate embryo. *Science*, *312*, 1183–1188. https://doi.org/10.1126/science.1123404.

Isshiki, T., Takeichi, M., & Nose, A. (1997). The role of the msh homeobox gene during Drosophila neurogenesis: Implication for the dorsoventral specification of the neuroectoderm. *Development*, *124*, 3099–3109.

Ivashkin, E., & Adameyko, I. (2013). Progenitors of the protochordate ocellus as an evolutionary origin of the neural crest. *EvoDevo*, *4*, 1–15. https://doi.org/10.1186/2041-9139-4-12.

Jackman, W. R., Langeland, J. A., & Kimmel, C. B. (2000). Islet reveals segmentation in the amphioxus hindbrain homolog. *Developmental Biology*, *220*, 16–26. https://doi.org/10.1006/dbio.2000.9630.

Janvier, P. (1996). *Early vertebrates*. Oxford: Oxford University Press.

Janvier, A., & Janvier, P. (2008). Early jawless vertebrates and cyclostome origins early jawless vertebrates and cyclostome origins. *Zoological Science*, *25*, 1045–1056. https://doi.org/10.2108/zsj.25.1045.

Kaltenbach, S. L., Yu, J.-K., & Holland, N. D. (2009). The origin and migration of the earliest-developing sensory neurons in the peripheral nervous system of amphioxus. *Evolution & Development*, *11*, 142–151. https://doi.org/10.1111/j.1525-142X.2009.00315.x.

Keating, J. N., Marquart, C. L., & Donoghue, P. C. J. (2015). Histology of the heterostracan dermal skeleton: Insight into the origin of the vertebrate mineralised skeleton. *Journal of Morphology*, *276*, 657–680. https://doi.org/10.1002/jmor.20370.

Kee, Y., Hwang, B. J., Sternberg, P. W., & Bronner-Fraser, M. (2007). Evolutionary conservation of cell migration genes: From nematode neurons to vertebrate neural crest. *Genes & Development*, *21*, 391–396. https://doi.org/10.1101/gad.1509307. Genomic.

Khatibi Shahidi, M., Krivanek, J., Kaukua, N., Ernfors, P., Hladik, L., Kostal, V., & Fried, K. (2015). Three-dimensional imaging reveals new compartments and structural adaptations in odontoblasts. *Journal of Dental Research*, *94*(7), 945–954. https://doi.org/10.1177/0022034515580796.

Khudyakov, J., & Bronner-Fraser, M. (2009). Comprehensive spatiotemporal analysis of early chick neural crest network genes. *Developmental Dynamics*, *238*, 716–723. https://doi.org/10.1002/dvdy.21881.

Knecht, A. K., & Bronner-Fraser, M. (2002). Induction of the neural crest: A multigene process. *Nature Reviews Genetics*, *3*, 453–461. https://doi.org/10.1038/nrg819.

Kozmik, Z., Holland, N. D., Kreslova, J., Oliveri, D., Schubert, M., Jonasova, K., et al. (2007). Pax–Six–Eya–Dach network during amphioxus development: Conservation in vitro but context specificity in vivo. *Developmental Biology*, *306*, 143–159. https://doi.org/10.1016/j.ydbio.2007.03.009.

Kozmikova, I., Candiani, S., Fabian, P., Gurska, D., & Kozmik, Z. (2013). Essential role of Bmp signaling and its positive feedback loop in the early cell fate evolution of chordates. *Developmental Biology*, *382*(2), 538–554. https://pubmed.ncbi.nlm.nih.gov/23933491/.

Kuraku, S., Hoshiyama, D., Katoh, K., Suga, H., & Miyata, T. (1999). Monophyly of Lampreys and hagfishes supported by nuclear DNA—Coded genes. *Journal of Molecular Evolution*, *49*, 729–735. https://doi.org/10.1007/pl00006595.

Langeland, J. A., Tomsa, J. M., Jackman, W. R., Jr., & Kimmel, C. B. (1998). An amphioxus *snail* gene: expression in paraxial mesoderm and neural plate suggests a conserved role in patterning the chordate embryo. *Development Genes and Evolution*, *208*, 569–577. https://doi.org/10.1007/s004270050216.

Lauri, A. (2013). *The evolution of the neural crest from an annelid perspective: Conserved cell types and signaling pathways in Platynereis dumerili. Heidelberg.* Centre for Organismal Studies Heidelberg. http://www.ub.uni-heidelberg.de/archiv/15534.

Le Douarin, N. (1982). *The neural crest.* Cambridge University Press.

Lemaire, P. (2011). Evolutionary crossroads in developmental biology: The tunicates. *Development*, *138*, 2143–2152. https://doi.org/10.1242/dev.048975.

Lemaire, P., Smith, W. C., & Nishida, H. (2008). Ascidians and the plasticity of the chordate developmental program. *Current Biology*, *18*, 620–631. https://doi.org/10.1016/j.cub.2008.05.039.

Li, Y., Zhao, D., Horie, T., Chen, G., Bao, H., Chen, S., et al. (2017). Conserved gene regulatory module specifies lateral neural borders across bilaterians. *Proceedings of the National Academy of Sciences of the United States of America*, *114*, E6352–E6360. https://doi.org/10.1073/pnas.1704194114.

Liem, K. F., Jr., Tremml, G., Roelink, H., & Jessell, T. M. (1995). Dorsal differentiation of neural plate cells induced by BMP-mediated signals from epidermal ectoderm. *Cell*, *82*, 969–979. https://doi.org/10.1016/0092-8674(95)90276-7.

Litsiou, A., Hanson, S., & Streit, A. (2005). A balance of FGF, BMP and WNT signalling positions the future placode territory in the head. *Development*, *132*, 4051–4062. https://doi.org/10.1242/dev.01964.

Liu, B., & Satou, Y. (2019). Foxg specifies sensory neurons in the anterior neural plate border of the ascidian embryo. *Nature Communications*, *10*, 1–10. https://doi.org/10.1038/s41467-019-12839-6.

Lowe, C. J., Terasaki, M., Wu, M., Jr., Freeman, R. M., Runft, L., Kwan, K., et al. (2006). Dorsoventral patterning in hemichordates: Insights into early chordate evolution. *PLoS Biology*, *4*, e291. https://doi.org/10.1371/journal.pbio.0040291.

Lu, T.-M., Luo, Y.-J., & Yu, J-K. (2012). BMP and Delta/notch signaling control the development of amphioxus epidermal sensory neurons: Insights into the evolution of the peripheral sensory system. *Development*, *139*, 2020–2030. https://doi.org/10.1242/dev.073833.

Lumsden, A. G. S. (1987). The neural crest contribution to tooth development in the mammalian embryo. In P. F. Maderson (Ed.), *Developmental and evolutionary aspects of the neural crest* (pp. 261–300). New York: Wiley.

Lwigale, P. Y. (2001). Embryonic origin of avian corneal sensory nerves. *Developmental Biology*, *239*, 323–337. https://doi.org/10.1006/dbio.2001.0450.

Magloire, H., Couble, M., & Romeas, A. (2004). Odontoblast primary cilia: facts and hypotheses. *Cell Biology International*, *28*, 93–99. https://doi.org/10.1016/j.cellbi.2003.11.006.

Maharana, S. K., & Schlosser, G. (2018). A gene regulatory network underlying the formation of pre-placodal ectoderm in Xenopus laevis. *BMC Biology*, *16*, 1–18. https://doi.org/10.1186/s12915-018-0540-5.

Marshall, C. R. (2006). Explaining the cambrian "explosion" of animals. *Annual Review of Earth and Planetary Sciences*, *34*, 355–384. https://doi.org/10.1146/annurev.earth.33.031504.103001.

Martik, M. L., & Bronner, M. E. (2017). Regulatory logic underlying diversification of the neural crest. *Trends in Genetics*, *33*, 715–727. https://doi.org/10.1016/j.tig.2017.07.015.

Martik, M. L., Gandhi, S., Uy, B. R., Gillis, J. A., Green, S. A., Simoes-Costa, M., et al. (2019). Evolution of the new head by gradual acquisition of neural crest regulatory circuits. *Nature*, *574*, 675–678. https://doi.org/10.1038/s41586-019-1691-4.

Martín-Durán, J. M., Pang, K., Børve, A., Lê, H. S., Furu, A., & Cannon, J. T. (2018). Convergent evolution of bilaterian nerve cords. *Nature*, *553*, 45–50. https://doi.org/10.1038/nature25030.

Mazet, F., & Shimeld, S. M. (2005). Molecular evidence from ascidians for the evolutionary origin of vertebrate cranial sensory placodes. *Journal of Experimental Zoology. Part B. Molecular and Developmental Evolution*, *304B*, 340–346. https://doi.org/10.1002/jezmde.b.21054.evolutionary.

McCauley, D. W., & Bronner-Fraser, M. (2003). Neural crest contributions to the lamprey head. *Development*, *130*, 2317–2327. https://doi.org/10.1242/dev.00451.

Meulemans, D., & Bronner-Fraser, M. (2002). Amphioxus and lamprey AP-2 genes: Implications for neural crest evolution and migration patterns. *Development*, *129*, 4953–4962.

Meulemans, D., & Bronner-Fraser, M. (2004). Gene-regulatory interactions in neural crest evolution and development. *Developmental Cell*, 7, 291–299. https://doi.org/10.1016/j.devcel.2004.08.007.

Meulemans, D., & Bronner-Fraser, M. (2005). Central role of gene cooption in neural crest evolution. *Journal of Experimental Zoology. Part B Molecular and Developmental Evolution*, *304B*, 298–303. https://doi.org/10.1002/jez.b.21047.GENE.

Meulemans Medeiros, D. (2013). The evolution of the neural crest: New perspectives from lamprey and invertebrate neural crest-like cells. *WIREs Developmental Biology*, *2*, 1–15. https://doi.org/10.1002/wdev.85.

Milet, C., & Monsoro-Burq, A. H. (2012). Neural crest induction at the neural plate border in vertebrates. *Developmental Biology*, *366*, 22–33. https://doi.org/10.1016/j.ydbio.2012.01.013.

Mizutani, C. M., Meyer, N., Roelink, H., & Bier, E. (2006). Threshold-dependent BMP-mediated repression: A model for a conserved mechanism that patterns the neuroectoderm. *PLOS Biology, 4(10), e313.* https://doi.org/10.1371/journal.pbio.0040313.

Modrell, M. S., Hockman, D., Uy, B., Buckley, D., Sauka-Spengler, T., Bronner, M. E., et al. (2014). Evolution of developmental control mechanisms a fate-map for cranial sensory ganglia in the sea lamprey. *Developmental Biology*, *385*, 405–416. https://doi.org/10.1016/j.ydbio.2013.10.021.

Moody, S. A., & Lamantia, A. (2015). Transcriptional regulation of cranial sensory placode development, 1st ed, neural crest and placodes. *Current Topics in Developmental Biology*, *111*, 301–350. https://doi.org/10.1016/bs.ctdb.2014.11.009.

Morris, S. C., & Caron, J. (2014). A primitive fish from the Cambrian of North America. *Nature*, *512*, 419–422. https://doi.org/10.1038/nature13414.

Muroyama, Y., Fujihara, M., Ikeya, M., Kondoh, H., & Takada, S., (2002). Wnt Signaling Plays an Essential Role in Neuronal Specification of the Dorsal Spinal Cord. *Genes & Development*, 548–553. https://doi.org/10.1101/gad.937102.

Nichane, M., Ren, X., Souopgui, J., & Bellefroid, E. J. (2008). Hairy2 functions through both DNA-binding and non-DNA-binding mechanisms at the neural plate border in Xenopus. *Developmental Biology*, *322*, 368–380. https://doi.org/10.1016/j.ydbio.2008.07.026.

Northcutt, R. G. (2005). The new head hypothesis revisited. *Journal of Experimental Zoology (Mol Dev Evol)*, *304B*, 274–297. https://doi.org/10.1002/jez.b.21063.NEW.

Nieto, A. M. (2011). The ins and outs of the epithelial to mesenchymal transition in health and disease. *Annual Review of Cell and Developmental Biology*, *27*, 347–376.

Nomaksteinsky, M., Kassabov, S., Chettouh, Z., Stoeklé, H.-C., Bonnaud, L., Fortin, G., … Brunet, J.-F. (2013). Ancient origin of somatic and visceral neurons. *BMC Biology*, *11*(53). https://doi.org/10.1186/1741-7007-11-53.

Nomaksteinsky, M., Röttinger, E., Dufour, H. D., Chettouh, Z., Lowe, C. J., Martindale, M. Q., & Brunet, J.-F. (2009). Centralization of the deuterostome nervous system predates chordates. *Current Biology*, *19*(15), 1264–1269. https://doi.org/10.1016/j.cub.2009.05.063.

Northcutt, R. G., & Gans, C. (1983). The genesis of neural crest and epidermal placodes: A reinterpretation of vertebrate origins. *The Quarterly Review of Biology*, *58*, 1–28. https://doi.org/10.1086/413055.

Ohyama, T., Groves, A. K., & Martin, K. (2007). The first steps towards hearing: Mechanisms of otic placode induction. *The International Journal of Developmental Biology*, *51*, 463–472. https://doi.org/10.1387/ijdb.072320to.

Okumura, R., Shima, K., Muramatsu, T., Nakagawa, K.-I., Shimono, M., Suzuki, T., et al. (2005). The odontoblast as a sensory receptor cell? The expression of TRPV1 (VR-1) channels. *Archives of Histology and Cytology*, *68*, 251–257. https://doi.org/10.1679/aohc.68.251.

Ørvig, T. (1972). The Latero-sensory Component of the Dermal Skeleton in Lower Vertebrates and Its Phyletic Significance. *Zoologica Scripta*, *1*, 139–155. https://doi.org/10.1111/j.1463-6409.1972.tb00672.x.

Ørvig, T. (1977). A survey of odontodes ('dermal teeth') from developmental, structural, functional, and phyletic points of view. In S. M. Andrews, R. S. Miles, & A. D. Wallar (Eds.), *Linnean society symposium series: Vol. 4. Problems in vertebrate evolution* (pp. 53–75).

Ota, K. G., Kuraku, S., & Kuratani, S. (2007). Hagfish embryology with reference to the evolution of the neural crest. *Nature*, *446*, 672–675. https://doi.org/10.1038/nature05633.

Patthey, C., Edlund, T., & Gunhaga, L. (2009). Wnt-regulated temporal control of BMP exposure directs the choice between neural plate border and epidermal fate. *Development, 136*, 73–83. https://doi.org/10.1242/dev.025890.

Patthey, C., Gunhaga, L., & Edlund, T. (2008). Early development of the central and peripheral nervous systems is coordinated by Wnt and BMP signals. *PLoS One, 3*, 2–11. https://doi.org/10.1371/journal.pone.0001625.

Patthey, C., Schlosser, G., & Shimeld, S. M. (2014). The evolutionary history of vertebrate cranial placodes–I: Cell type evolution. *Developmental Biology, 389*, 82–97. https://doi.org/10.1016/j.ydbio.2014.01.017.

Piacentino, M. L., & Bronner, M. E. (2018). Intracellular attenuation of BMP signaling via CKIP-1/Smurf1 is essential during neural crest induction. *PLoS Biology, 1*, 1–25. https://doi.org/10.1371/journal.pbio.2004425.

Piacentino, M. L., Li, Y., & Bronner, M. E. (2020). Epithelial-to-mesenchymal transition and different migration strategies as viewed from the neural crest. *Current Opinion in Cell Biology, 66*, 43–50. https://doi.org/10.1016/j.ceb.2020.05.001.

Pieper, M., Ahrens, K., Rink, E., Peter, A., & Schlosser, G. (2012). Differential distribution of competence for panplacodal and neural crest induction to non-neural and neural ectoderm. *Development, 139*, 1175–1187. https://doi.org/10.1242/dev.074468.

Pieper, M., Eagleson, G. W., Wosniok, W., & Schlosser, G. (2011). Origin and segregation of cranial placodes in Xenopus laevis. *Developmental Biology, 360*, 257–275. https://doi.org/10.1016/j.ydbio.2011.09.024.

Piotrowski, T., & Baker, C. V. H. (2014). The development of lateral line placodes: Taking a broader view. *Developmental Biology, 389*(1), 68–81. https://doi.org/10.1016/j.ydbio.2014.02.016.

Pla, P., & Monsoro-Burq, A. H. (2018). The neural border: Induction, specification and maturation of the territory that generates neural crest cells. *Developmental Biology, 444*, S36–S46. https://doi.org/10.1016/j.ydbio.2018.05.018.

Putnam, N. H., Butts, T., Ferrier, D. E. K., Furlong, R. F., Hellsten, U., Kawashima, T., et al. (2008). The amphioxus genome and the evolution of the chordate karyotype. *Nature, 453*, 1–7. https://doi.org/10.1038/nature06967.

Roellig, D., Tan-Cabugao, J., Esaian, S., & Bronner, M. E. (2017). Dynamic transcriptional signature and cell fate analysis reveals plasticity of individual neural plate border cells. *eLife, e21620*, 1–24. https://doi.org/10.7554/eLife.21620.

Rusten, T. E., Cantera, R., Kafatos, F. C., & Barrio, R. (2002). The role of TGF beta signaling in the formation of the dorsal nervous system is conserved between Drosophila and chordates. *Development, 129*(15), 3575–3584.

Rychel, A. L., & Swalla, B. J. (2007). Development and evolution of chordate cartilage. *Journal of Experimental Zoology Part B: Molecular and Developmental Evolution, 308*(3), 325–335. https://doi.org/10.1002/jez.b.21157.

Satoh, N. (2014). *Developmental genomics of ascidians*. Wiley-Blackwell.

Satoh, N. (2016). Two decades of ascidian developmental biology: A personal research story. *Current Topics in Developmental Biology, 117*, 289–300. https://doi.org/10.1016/bs.ctdb.2015.11.016. https://doi.org/10.1016/bs.ctdb.2015.11.016.

Saudemont, A., Haillot, E., Mekpoh, F., Bessodes, N., Quirin, M., Lapraz, F., … Lepage, T. (2010). Ancestral regulatory circuits governing ectoderm patterning downstream of Nodal and BMP2/4 revealed by gene regulatory network analysis in an echinoderm. *PLoS Genetics, 6*(12). https://doi.org/10.1371/journal.pgen.1001259.

Sauka-Spengler, T., Meulemans, D., Jones, M., & Bronner-Fraser, M. (2007). Ancient evolutionary origin of the neural crest gene regulatory network. *Developmental Cell, 13*, 405–420. https://doi.org/10.1016/j.devcel.2007.08.005.

Savage, C., Das, P., Finelli, A. L., Townsend, S. R., Sunt, C. Y., Bairdt, S. E., et al. (1996). Caenorhabditis elegans genes sma-2, sma-3, and sma-4 define a conserved family of transforming growth factor 3 pathway components. *Proceedings of the National Academy of Sciences of the United States of America*, *93*, 790–794. https://doi.org/10.1073/pnas.93.2.790.

Schlosser, G. (2006). Induction and specification of cranial placodes. *Developmental Biology*, *294*, 303–351. https://doi.org/10.1016/j.ydbio.2006.03.009.

Schlosser, G. (2008). Do vertebrate neural crest and cranial placodes have a common evolutionary origin? *BioEssays*, *30*, 659–672. https://doi.org/10.1002/bies.20775.

Schlosser, G. (2015). Vertebrate cranial placodes as evolutionary innovations—The Ancestor's Tale. *Current Topics in Developmental Biology*, *111*, 235–300. https://doi.org/10.1016/bs.ctdb.2014.11.008.

Schlosser, G. (2017). From so simple a beginning—what amphioxus can teach us about placode evolution. *The International Journal of Developmental Biology*, *61*, 633–648. https://doi.org/10.1387/ijdb.170127gs.

Schlosser, G., Patthey, C., & Shimeld, S. M. (2014). The evolutionary history of vertebrate cranial placodes II. Evolution of ectodermal patterning. *Developmental Biology*, *389*, 98–119. https://doi.org/10.1016/j.ydbio.2014.01.019.

Shu, D.-G., Conway Morris, S., Han, J., Zhang, Z.-F., Yasui, K., Janvier, P., et al. (2003). Head and backbone of the Early Cambrian vertebrate Haikouichthys. *Nature*, *421*, 775–778. https://doi.org/10.1038/nature01354.1.

Schoch, R. R., & Sues, H.-D. (2015). A Middle Triassic stem-turtle and the evolution of the turtle body plan. *Nature*, *523*, 584–587. https://doi.org/10.1038/nature14472.

Sharman, A. C., Shimeld, S. M., & Holland, P. W. (1999). An amphioxus Msx gene expressed predominantly in the dorsal neural tube. *Development Genes and Evolution volume*, *209*(4), 260–263. https://doi.org/10.1007/s004270050251.

Shu, D., Luo, H., Morris, S. C., Zhang, X., Hu, S., Chen, L., et al. (1999). Lower Cambrian vertebrates from south China. *Nature*, *402*, 42–46.

Simões-Costa, M., & Bronner, M. E. (2015). Establishing neural crest identity: A gene regulatory recipe. *Development*, *142*, 242–257. https://doi.org/10.1242/dev.105445.

Sire, J., Donoghue, P. C. J., & Vickaryous, M. K. (2009). Origin and evolution of the integumentary skeleton in non-tetrapod vertebrates. *Journal of Anatomy*, *241*, 409–440. https://doi.org/10.1111/j.1469-7580.2009.01046.x.

Sire, J., & Huysseune, A. (2003). Formation of dermal skeletal and dental tissues in fish: A comparative and evolutionary approach. *Biological Reviews*, *78*, 219–249. https://doi.org/10.1017/s1464793102006073.

Sire, J., & Huysseune, A. (1996). Structure and development of the odontodes in an armoured catfish, Corydoras aeneus (Siluriformes, Callichthyidae). *Acta Zoologica*, *77*, 51–72. https://doi.org/10.1111/j.1463-6395.1996.tb01252.x.

Smith, M. P., Sansom, I. J., & Repetski, J. E. (1996). A histology of the first fish. *Nature*, *380*, 702–704.

Steventon, B., Mayor, R., & Streit, A. (2014). Neural crest and placode interaction during the development of the cranial sensory system. *Developmental Biology*, *389*, 28–38. https://doi.org/10.1016/j.ydbio.2014.01.021.

Stolfi, A., Ryan, K., Meinertzhagen, I. A., & Christiaen, L. (2015). Migratory neuronal progenitors arise from the neural plate borders in tunicates. *Nature*, *527*, 371–374. https://doi.org/10.1038/nature15758.

Stock, D. W., & Whitt, G. S. (1992). Evidence from 18S ribosomal RNA sequences that lampreys and hagfishes form a natural group. *Science*, *257*(5071), 787–789. https://doi.org/10.1126/science.1496398.

Streit, A. (2007). The preplacodal region: an ectodermal domain with multipotential progenitors that contribute to sense organs and cranial sensory ganglia. *The International Journal of Developmental Biology*, *51*, 447–461. https://doi.org/10.1387/ijdb.072327as.

Stuhlmiller, T. J., & García-Castro, M. I. (2012). FGF/MAPK signaling is required in the gastrula epiblast for avian neural crest induction. *Development, 139*, 289–300. https://doi.org/10.1242/dev.070276.

Su, Y., Chen, Y., Ting, H., Fan, T., Lin, C., Wang, K., et al. (2019). BMP controls dorsoventral and neural patterning in indirect-developing hemichordates providing insight into a possible origin of chordates. *Proceedings of the National Academy of Sciences of the United States of America, 116*, 12925–12932. https://doi.org/10.1073/pnas.1901919116.

Sulston, J. E., & Horvitz, H. R. (1977). Post-embryonic cell lineages of the nematode, Caenorhabditis elegans. *Developmental Biology, 56*, 110–156. https://doi.org/10.1016/0012-1606(77)90158-0.

Tarby, M. L., & Webb, J. F. (2003). Development of the supraorbital and mandibular lateral line canals in the Cichlid, Archocentrus nigrofasciatus. *Journal of Morphology, 255*, 44–57. https://doi.org/10.1002/jmor.10045.

Telford, M. J., Budd, G. E., & Philippe, H. (2015). Phylogenomic insights into animal evolution. *Current Biology, 25*, R876–R887. https://doi.org/10.1016/j.cub.2015.07.060.

Theveneau, E., & Mayor, R. (2012). Neural crest delamination and migration: From epithelium-to-mesenchyme transition to collective cell migration. *Developmental Biology, 366*, 34–54. https://doi.org/10.1016/j.ydbio.2011.12.041.

Thiery, A., Buzzi, A. L., & Streit, A. (2020). Cell fate decisions during the development of the peripheral nervous system in the vertebrate head, 1st ed. *Current Topics in Developmental Biology, 139*, 127–167. https://doi.org/10.1016/bs.ctdb.2020.04.002. Elsevier Inc. https://doi.org/10.1016/bs.ctdb.2020.04.002.

Tresser, J., Chiba, S., Veeman, M., El-Nachef, D., Newman-Smith, E., Horie, T., et al. (2010). doublesex/mab3 related-1 (dmrt1) is essential for development of anterior neural plate derivatives in Ciona. *Development, 137*, 2197–2203. https://doi.org/10.1242/dev.045302.

Vergara, H. M., Pape, C., Meechan, K. I., Zinchenko, V., Genoud, C., Wanner, A. A., … Arendt, D. (2020). Whole-body integration of gene expression and single-cell morphology. *bioRxiv*. https://doi.org/10.1101/2020.02.26.961037.

Wada, H., Holland, P. W., Sato, S., Yamamoto, H., & Satoh, N. (1997). Neural tube is partially dorsalized by overexpression of HrPax-37: The ascidian homologue of Pax-3 and Pax-7. *Developmental Biology, 187*, 240–252. https://doi.org/10.1006/dbio.1997.8626.

Wagner, E., & Levine, M. (2012). FGF signaling establishes the anterior border of the Ciona neural tube. *Development, 139*, 2351–2359. https://doi.org/10.1242/dev.078485.

Wang, Y., Zhang, P.-J., Yasui, K., & Saiga, H. (2002). Expression of Bblhx3, a LIM-homeobox gene, in the development of amphioxus Branchiostoma belcheri tsingtauense. *Mechanisms of Development, 117*(1–2), 315–319. https://doi.org/10.1016/s0925-4773(02)00197-1.

Webb, J. F., Bird, N. C., Carter, L., & Dickson, J. (2014). Comparative development and evolution of two lateral line phenotypes in Lake Malawi Cichlids. *Journal of Morphology, 275*, 678–692. https://doi.org/10.1002/jmor.20247.

Wilson, P. A., & Hemmati-Brivanlou, A. (1995). Induction of epidermis and inhibition of neural fate by Bmp-4. *Nature, 376*, 331–333. https://doi.org/10.1038/376331a0.

Wray, G. A., Levinton, J. S., & Shapirot, L. H. (1996). Molecular evidence for deep precambrian divergences among Metazoan phyla. *Science, 274*, 568–574. https://doi.org/10.1126/science.274.5287.568.

Wu, M. Y., Ramel, M., Howell, M., & Hill, C. S. (2011). SNW1 is a critical regulator of spatial BMP activity, neural plate border formation, and neural crest specification in vertebrate embryos. *PLoS Biology, 9*, e1000593. https://doi.org/10.1371/journal.pbio.1000593.

Yasui, K., Zhang, S. C., Uemura, M., Aizawa, S., & Ueki, T. (1998). Expression of a twist-related gene, Bbtwist, during the development of a lancelet species and its relation to cephalochordate anterior structures. *Developmental Biology, 195*, 49–59. https://doi.org/10.1006/dbio.1997.8834.

Yasui, K., Zhang, S., Uemura, M., & Saiga, H. (2000). Left-right asymmetric expression of BbPtx, a Ptx-related gene, in a lancelet species and the developmental left-sidedness in deuterostomes. *Development, 127*, 187–195.

York, J. R., & McCauley, D. W. (2020). The origin and evolution of vertebrate neural crest cells. *Open Biology, 10*, 190285. https://doi.org/10.1098/rsob.190285.

Yu, J.-K. S. (2010). The evolutionary origin of the vertebrate neural crest and its developmental gene regulatory network—Insights from amphioxus. *Zoology, 113*, 1–9. https://doi.org/10.1016/j.zool.2009.06.001.

Yu, J.-K., Holland, N. D., & Holland, L. Z. (2002). An amphioxus winged helix/forkhead gene, AmphiFoxD: Insights into vertebrate neural crest evolution. *Developmental Dynamics, 225*, 289–297. https://doi.org/10.1002/dvdy.10173.

Yu, J.-K., Meulemans, D., Mckeown, S. J., & Bronner-Fraser, M. (2008). Insights from the amphioxus genome on the origin of vertebrate neural crest. *Genome Research, 18*, 1127–1132. https://doi.org/10.1101/gr.076208.108.18.

Zhao, D., Chen, S., & Liu, X. (2019). Lateral neural borders as precursors of peripheral nervous systems: A comparative view across bilaterians. *Development, Growth & Differentiation, 61*, 58–72. https://doi.org/10.1111/dgd.12585.

CHAPTER SEVEN

Evo-devo studies of cyclostomes and the origin and evolution of jawed vertebrates

Shigeru Kuratani[a,b,*]

[a]Laboratory for Evolutionary Morphology, RIKEN Center for Biosystems Dynamics Research (BDR), Kobe, Hyogo, Japan
[b]Evolutionary Morphology Laboratory, RIKEN Cluster for Pioneering Research (CPR), Kobe, Hyogo, Japan
[*]Correspondence author: e-mail address: shigeru.kuratani@riken.jp

Contents

Abstract

Modern vertebrates consist of two sister groups: cyclostomes and gnathostomes. Cyclostomes are a monophyletic jawless group that can be further divided into hagfishes and lampreys, which show conspicuously different developmental and morphological patterns. However, during early pharyngula development, there appears to be a stage when the embryos of hagfishes and lampreys resemble each other by showing an "ancestral" craniofacial pattern; this pattern enables morphological comparison of hagfish and lamprey craniofacial development at late stages. This cyclostome developmental pattern, or more accurately, this developmental pattern of the jawless grade

Current Topics in Developmental Biology, Volume 141
ISSN 0070-2153
https://doi.org/10.1016/bs.ctdb.2020.11.011

of vertebrates in early pharyngula was very likely shared by the gnathostome stem before the division of the nasohypophyseal placode led to the jaw and paired nostrils. The craniofacial pattern of the modern jawed vertebrates seems to have begun in fossil ostracoderms (including galeaspids), and was completed by the early placoderm lineages. The transition from jawless to jawed vertebrates was thus driven by heterotopy of development, mainly caused by separation and shift of ectodermal placodes and resultant ectomesenchymal distribution, and shifts of the epithelial-mesenchymal interactions that underlie craniofacial differentiation. Thus, the evolution of the jaw was not a simple modification of the mandibular arch, but a heterotopic shift of the developmental interactions involving not only the mandibular arch, but also the premandibular region rostral to the mandibular arch.

1. Introduction—Cyclostomes, agnathans, and gnathostomes

Among vertebrates, which are often regarded as an independent phylum (Irie, Kuratani, & Satoh, 2018), cyclostomes used to be considered to represent one of the most primitive groups. Cope (1889) was the first to divide vertebrates into gnathostomes and agnathans, believing that they arose by dichotomy. However, hagfishes lack some traits used to define vertebrates, and were often placed more basally to the rest of vertebrates. In these old systems, lampreys and gnathostomes were classed as vertebrates, and vertebrates and hagfishes were grouped as "cranites." However, it has become clear by modern molecular analyses that modern cyclostomes consist of lampreys and hagfishes, together representing a monophyletic sister group of another monophyletic group, gnathostomes (Fig. 1; reviewed by Kuraku & Kuratani, 2006).

The name "cyclostomes" stems from the fact that both lampreys and hagfishes have a round mouth (see "Anatomy of cyclostomes" section; Fig. 1). Thus, "cyclostomes" is not synonymous with "agnathans" (or "jawless vertebrates"). The name "gnathostomes" primarily means "animals with jaws"; however, some of their stem group were jawless, namely ostracoderms, which represent agnathans obtained as fossils from the pre-Devonian era (see Janvier, 1996). This mismatch occurs because the divergence of a lineage is not always simultaneous with the acquisition of the apomorphic trait that defines its crown group(s) (in this case, jawed vertebrates). Before the dichotomy of cyclostomes and gnathostomes, there were also basal animals with vertebrate-like features, such as *Haikouichthys* and *Myllokunmingia* from the late Cambrian, that are now recognized as the vertebrate stem (Janvier, 2008).

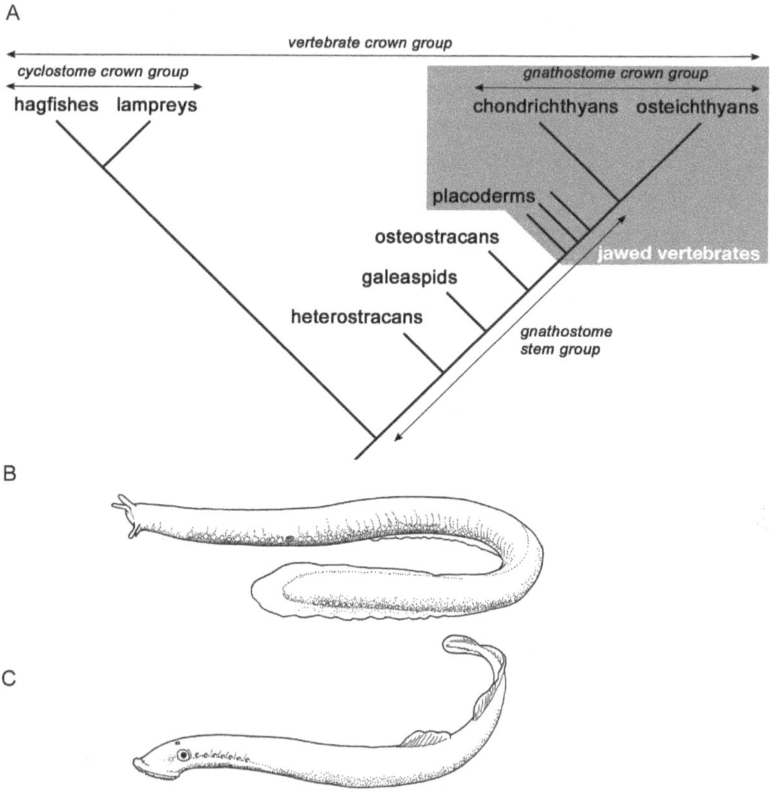

Fig. 1 (A) Phylogenetic tree showing some fossil groups and extant gnathostomes and cyclostomes. The stem group of gnathostomes contains only fossil taxa: heterostracans, galeaspids, osteostracans, and placoderms. Note that only a part of gnathostomes can be called "jawed vertebrates" (shaded area). (B) External morphology of a hagfish. (C) External morphology of a lamprey. *Panel (A) Modified from Kuratani, S., & Ahlberg, P. E. (2018). Evolution of the vertebrate neurocranium: Problems of the premandibular domain and trabecula.* Zoological Letters, 4, *1. Panel (C) Modified from Kuratani, S. (2017).* Evolutionary morphology: Bauplan and embryonic development of vertebrates. *New Edition. (in Japanese), Tokyo: Tokyo Univ. Press.*

The cyclostomes show notochord, overt cranium and head, median fins, and dorsally located central nervous system and so forth to define vertebrates, but also the following traits that are generally regarded as ancestral: (1) lack of jaw; (2) lack of occipital arch; (3) lack of trabecula; (4) undifferentiated epaxial and hypaxial muscles; (5) lack of paired fins and shoulder girdles; (6) lack of overt hypobranchial muscles; (7) monorhiny; (8) lack of sympathetic trunk, (9) lack of myelin sheaths in peripheral nerves; (10) lack of trapezius muscles and accessory nerves; (11) lack of clear stomach; (12) lack of thymus (but see Bajoghli et al., 2011); and (13) lack of cerebellum. Many of these traits are potentially shared by the gnathostome stem, and thus could define agnathans as a grade. Some of them are possibly coupled to each other developmentally (see below), implying that many gnathostome-defining traits were obtained almost simultaneously during the transition from ostracoderms to placoderms (the first jawed fish).

As seen in the scenario proposed for the evolution of gene regulatory mechanisms behind neural crest acquisition, comprehensive gene expression analyses and advances in genomics have brought new ideas about the establishment of vertebrate-specific traits (see Green & Bronner, 2014; Pascual-Anaya et al., 2018 and references therein). Also, the monophyly of cyclostomes has been confirmed by molecular phylogenetic analysis. In this review, I will summarize the evolution of early vertebrates centering on the head morphology from developmental and embryological perspectives.

2. Anatomy of cyclostomes

2.1 Lamprey

The mouth of lampreys is circular, and functions as a sucker (Fig. 2). This structure is called the "oral funnel". The mouth is ornamented by a number of horny teeth. The oral cavity leads to the esophagus on the dorsal side and the pharynx on the ventral side, i.e., the esophagus opens directly into the oral cavity rather than through the pharynx (Fig. 2B). This topography is brought about by a secondary modification that occurs during the metamorphosis from larvae to juveniles. Rostral to the pharynx, there is a pump-like structure called the velum. The external nostril opens on the dorsal aspect of the head, ending as a blind sac that does not open to the pharynx (Fig. 2B); the adenohypophysis adjoins this blind sac. At the base of the pharynx, a feeding device called the lingual apparatus is present.

Among cyclostomes, lampreys produce telolecithal eggs, which go through holoblastic cleavage. Lamprey larvae were classically regarded as representing an independent species and were given the genus name *Ammocoetes,* until their true identity was discovered (reviewed by Kearn, 2004). The larval stage lasts for several years in freshwater, and the larvae grow to a dozen centimeters in length. The larvae have a cartilaginous skeleton but no vertebral elements. The neurocranium is only weakly developed. In addition, there is a unique tissue of extracellular matrix called "mucocartilage," which becomes the supporting tissue in the first

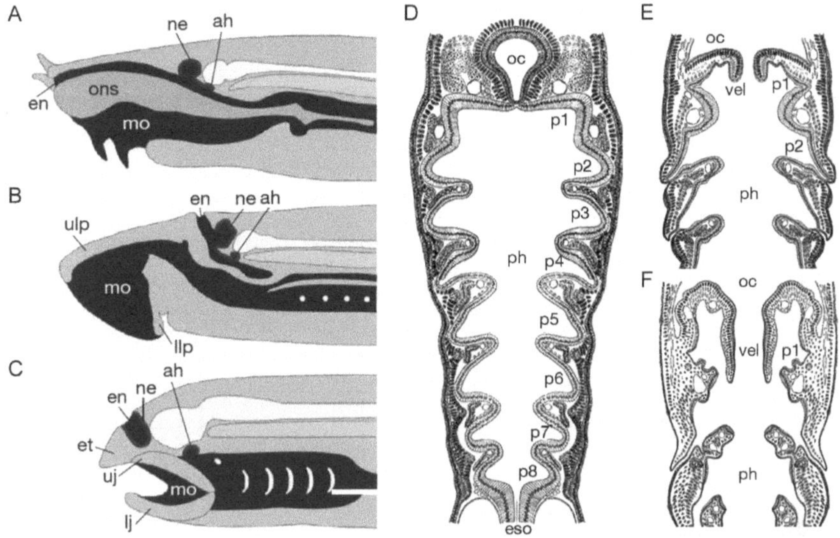

Fig. 2 Comparative anatomy of the vertebrate head. (A–C) Comparison of sagittal sections of a hagfish (A), adult lamprey (B), and a simplified jawed vertebrate (C) shown as highly schematized diagrams. In the hagfish, the nasal duct leads posteriorly to the pharynx and is separated from the mouth opening (mo) or oral cavity anteriorly by a horizontal septum called the oronasal septum (ons), which is the homolog of the lamprey upper lip (ulp; see the text). In the lamprey, the nasal duct ends blindly and does not lead to the pharynx. In jawed vertebrates, the nasal cavity is separated from the adenohypophysis that opens in the oral cavity. (D–F) Development of the lamprey velum (vel) at three different stages of developing lamprey embryos. Note that the velum arises in the oropharyngeal membrane, between the ectodermal lining of the oral cavity (oc) and endodermal epithelium of the pharynx (ph). ah, adenohypophysis; en, external nostril; et, ethmoidal region; llp, lower lip; ne, nasal epithelium; p1–3, pharyngeal pouches; ul, upper jaw (maxillary part). *Panel (D–F) Modified from Kuratani, S. (2017).* Evolutionary morphology: Bauplan and embryonic development of vertebrates. *New Edition. (in Japanese), Tokyo: Tokyo Univ. Press.*

and second pharyngeal arches, and is replaced by chondrocytes during metamorphosis. The mucocartilage and cellular cartilage are two cell types that share the same evolutionary origin and have similar mechanisms for differentiation through fibroblast growth factor (FGF) signaling (reviewed by De Beer, 1937; Jandzik et al., 2014 and references therein). The oral cavity of the ammocoete larva leads to the esophagus via the pharynx. Located at the bottom of the pharynx is an excretory organ, called an "endostyle," which is similar to that of invertebrate chordates; the endostyle secretes mucus to facilitate filter feeding.

2.2 Hagfish

Unlike lampreys, hagfishes go through direct development without larval stages (Dean, 1899). All hagfish species are deep-sea dwellers in temperate zones. Eggs are large and ellipsoid with bundles of filiform processes with hooks on both poles that enable the eggs to stick together to form clusters (Fig. 1). The adult hagfish has no eye lenses or lateral line system. However, a recent study of hagfish embryos has revealed the presence of a pan-placodal domain (Oisi, Ota, Fujimoto, & Kuratani, 2013a; Schlosser, 2005), delaminating neural crest cells, and vestigial vertebrate elements that express marker genes known in jawed vertebrate embryos (Ota, Fujimoto, Oisi, & Kuratani, 2011; Ota, Kuraku, & Kuratani, 2007). Thus, the apparent absence of the vertebral column in hagfishes is most likely a secondary state, and the absence of this feature alone cannot place the hagfishes as an out-group of vertebrates any longer. Beneath the skin, there is an extensive venous sinus, which used to be thought to represent an open vascular system. However, this sinus develops by a secondary fusion of embryonic blood vessels, and the blood is always in a lumen lined with endothelium. Thus, the closed vascular system, which is not found in invertebrate chordates (e.g., amphioxus and ascidians), is a trait unique to vertebrates.

The gill pores open on both sides at the center of the body (anatomically corresponding to the anterior part of the trunk) and vary in number among species (Eptatretidae have more than one, and in Myxinidae all the gill pores are connected to form one single duct that opens on either side of the body). The mouth opens in the front of the head, and there is no structure reminiscent of the oral funnel, but four pairs of barbels develop around the mouth (Fig. 2A).

A massive lingual apparatus like that of the lamprey develops in the pharyngeal floor; it can open and close a pair of sharp horny teeth on the dental plate associated with the apparatus. The external nostril opens anteriorly just above the mouth and leads to the pharynx via the olfactory epithelium and adenohypophysis. The nasal duct and oral cavity are separated by a horizontal septum called the oronasal septum, whose posterior end leads to the velum. A long esophagus-like canal follows the oral cavity rostral to the branchial arches, thus corresponding to the posterior part of the "pharynx". This part of the pharynx moves very posteriorly, creating a structure similar to the cervical region of amniotes (e.g., mammals, reptiles, and birds) (Oisi, Fujimoto, Ota, & Kuratani, 2015). Due to the posterior shift of gill pores, the anterior half of the animal can dive into carrion while breathing.

Although monophyly of cyclostomes is supported by molecular data, it cannot be denied that hagfishes give a more primitive impression than lampreys at the morphological level (reviewed by Janvier, 1996). However, through detailed analyses, many of the traits once thought to be absent in hagfishes are now known to be present (reviewed by Oisi, Ota, Fujimoto, & Kuratani, 2013b). In particular, the single semicircular canal, once thought to be a unique primitive character of hagfishes, has been shown to be shared by lampreys, and the condition of two semicircular canals was most likely present only in the gnathostome stem (Higuchi et al., 2019). Similarly, a close examination of embryonic morphology and the genome revealed that all vertebrates share a common set of segmental components in the brain primordium (Sugahara et al., 2016).

3. Mandibular arch

In this section, evolution of the jaw and some related structures will be considered. The mandibular arch, which is the major precursor of the jaw in gnathostomes, is the first pharyngeal arch in all vertebrate embryos and is located in the anterior most part of the embryonic pharynx. It develops lateral to the oropharyngeal membrane formed by the contact between the anterior end of the pharyngeal endoderm and the oral ectoderm. There are no pharyngeal pouches or slits in front of this arch (Fig. 2D–F). This arch has to be positionally specified for normal development to occur; Hox genes have pivotal roles in this process, as described in the next section.

3.1 Hox code

The Hox code is defined by the combination of a given set of expressed Hox genes at a particular level of the anterior–posterior axis of the embryo, and it is the direct result of spatial and temporal collinearity rules. The spatial rule is that the more 3′ a Hox gene lies in the Hox gene cluster, the more rostral the expression border of that gene (Fig. 3). This rule is largely conserved in Bilateria. The temporal rule is that the more 3′ a Hox gene lies in the cluster, the earlier the expression onset of that gene. This rule, called "whole-cluster temporal collinearity", applies particularly in vertebrates, and is conserved between cyclostomes and gnathostomes (Pascual-Anaya et al., 2018).

The whole genome duplication events that are thought to have occurred early in vertebrate evolution (Dehal & Boore, 2005)—the so-called 2R

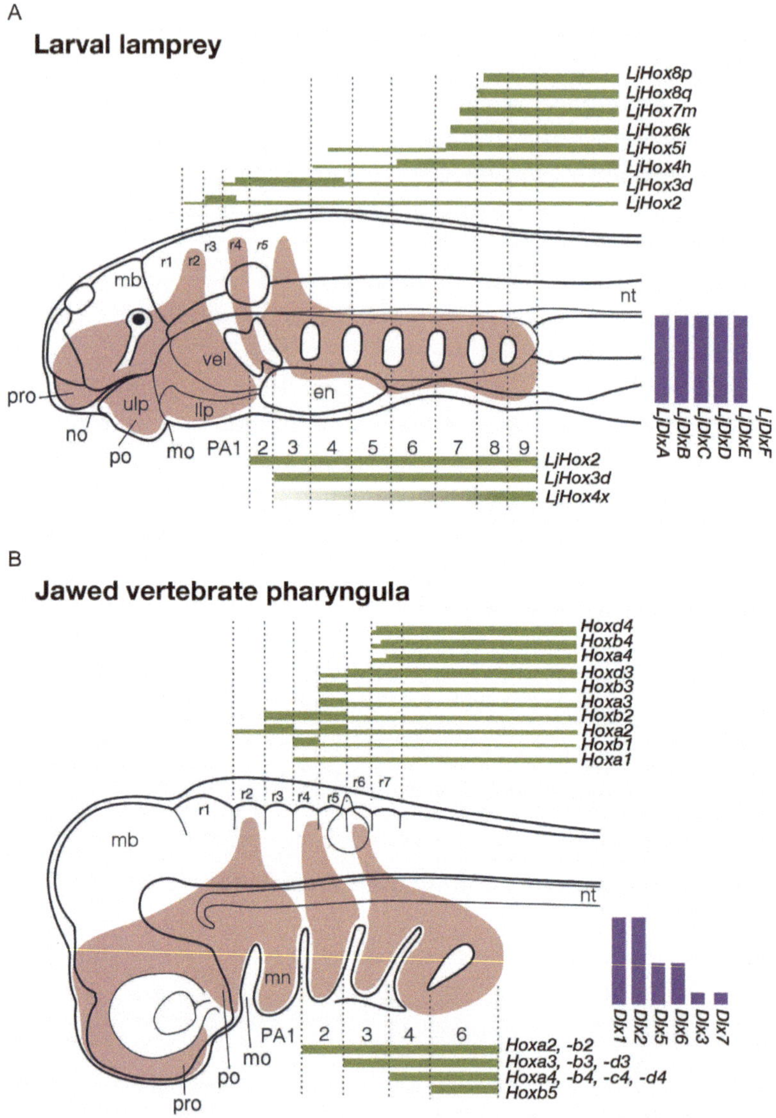

Fig. 3 Comparison of homeobox gene expression patterns in vertebrates. (A) lamprey pharyngula. (B) Pharyngula of jawed vertebrates. In both the lamprey and jawed vertebrate embryos, Hox gene transcripts show nested anteroposterior patterns of distribution (Hox code; light green) anteroposteriorly. The mandibular arch (PA1) expresses no Hox genes (Hox code default state). The Hox codes in the pharyngeal arch ectomesenchyme are similar in the rostral three PAs (PA1 to 3) in both lamprey and jawed vertebrate embryos. Mandibular arch neural crest cells (mn) form mesenchyme in the velum (vel) and lower lip (llp) in the lamprey, and in the upper and lower jaws in jawed vertebrates. Dlx genes are also expressed in CNC cells in the pharyngeal arches in both the animals (purple bars). en, endostyle; mb, midbrain; no, nostril; nt, notochord; ot, otocyst; p1 to p8, pharyngeal pouches; r1 to r7, rhombomeres; ulp, upper lip. *Panel (B) Modified from Kuratani, S. (2020). Neural crest and craniofacial evolution of early vertebrates. In B. F. Eams, D. M. Medeiros, I. Adameyko (Eds.),* Evolving neural crest cells *(pp. 219–242). Boca Raton, London, New York: CRC Press/Taylor & Francis Group.*

hypothesis—generated multiple Hox clusters, and therefore the vertebrate Hox code is more complex than that of its invertebrate counterparts. However, the recent genome sequencing of several cyclostome species has found than their genomes contain more Hox clusters than in the genomes of gnathostomes (up to six in both lampreys and hagfishes; Pascual-Anaya et al., 2018 and references therein), probably due to an extra genome duplication event. Hox code and Hox gene regulation, for instance in the hindbrain, have remained generally conserved between jawed and jawless vertebrates, despite their different genomic evolutions, suggesting the utmost importance of Hox genes in establishing vertebrate features.

In gnathostome embryos, the neural crest mesenchyme filling the mandibular arch does not express any Hox genes; this is termed the "Hox code default state". The mandibular arch of lamprey embryos also shows the default state. The patterns of the Hox paralogue group (PG) 2 and PG3 genes expressed in the second and third arches are also shared among vertebrates (see Takio et al., 2004 and references therein). Thus, the Hox code of the vertebrate pharyngeal arches was established before the divergence of cyclostomes and gnathostomes, i.e., by the latest common ancestor of vertebrates. There is no similar expression in invertebrate chordates, implying that the acquisition of the pharyngeal Hox code was close to the origin of the neural crest or the pharyngeal arch mesenchyme derived from the neural crest. In other words, it is very likely that the mandibular arch, hyoid arch, and the subsequent posterior ear pharyngeal arch had established morphological identities via their positional values long before the jaw was acquired. Thus, a scenario is possible in which the cyclostomes and gnathostomes each created a different oral apparatus on a common segmental template and pattern of Hox gene regulation (Parker et al., 2019).

3.2 Dlx code

In jawed vertebrate embryos, another set of homeobox genes, Dlx genes, establish a dorsoventrally nested expression pattern, or the Dlx code, in the pharyngeal arch mesenchyme (Fig. 3). This code specifies positional values within each arch along the dorsal–ventral (DV) axis. Manipulation of Dlx genes transforms the morphological pattern of the visceral skeleton. In particular, disruption of *Dlx5* and *-6*, which are normally expressed in the ventral half of the mandibular arch, leads to the loss of lower jaw identity, which is replaced by upper jaw identity (Depew, Lufkin, & Rubenstein, 2002). The expression of *Dlx1* and *-2* appear to function as the developmental default for the DV polarity of the arch, which, however, does not mean that it can be recognized as the evolutionary morphological default (see Köntges & Matsuoka, 2002).

Six Dlx genes (*DlxA-F*) are known in cyclostomes, but they do not show clear orthologies with any of the gnathostome Dlx genes (Fujimoto, Oisi, Kuraku, Ota, & Kuratani, 2013; Kuraku, Takio, Sugahara, Takechi, & Kuratani, 2010). According to the previous analysis, Dlx genes of the hagfish *Eptatretus burgeri* and those of lampreys appear to have experienced their own duplication events in each lineage. For example, the lamprey's *DlxA*, *-C*, *-D*, *-E* and the hagfish's *EbDlx2/3/5A* and *-B* and *EbDlx1/4/6A*, and *-B* belonging to the *Dlx2/3/5* clade have gone through independent duplication. Probably, the common ancestor of cyclostomes would have had a repertoire consisting of *Dlx1/4/6AB*, *Dlx2/3/5AB*, *Dlx1/4/6C*, and *Dlx2/3/5C* (reviewed by Kuraku et al., 2010).

Expression patterns of cyclostome Dlx genes have been observed in detail for lampreys (reviewed by Kuraku et al., 2010) and hagfishes (Fujimoto et al., 2013; Oisi et al., 2013b). The nested expression along the dorsoventral axis is clearly seen in all gnathostome embryos, whereas expressions of cyclostome Dlx genes widely overlap in the pharyngeal arch mesenchyme, with no clear DV-nested patterns, at least consistent with the dorsoventrally symmetrical pattern of the lamprey visceral skeleton (reviewed by Kuraku et al., 2010). There is also a report suggesting a slightly nested expression pattern centered around the lamprey gill pore: i.e., a cyclostome Dlx code (Cerny et al., 2010). Even if this code is demonstrated to exist, it may not explain the origin of the gnathostome Dlx code because the homologies of the Dlx genes between cyclostomes and gnathostomes are not clearly established. Moreover, the expression of *Hand* (a gene expressed in heart and neural crest derivatives) is restricted to the ventral pharynx, and endothelin signaling functions in morphogenesis of the visceral skeleton of lamprey embryos; inhibition of the endothelin signal results in selective deletion of cartilage in the lower part of the mandibular arch in lamprey larvae (Yao, Ohtani, Kuratani, & Wada, 2011). Therefore, although DV patterning exists in the lamprey pharynx, it seems that the gnathostome-type Dlx code was not established on it.

4. Evolution of the jaw

4.1 Theories of jaw evolution

Jaw evolution is typically understood as the homeotic differentiation of visceral arches (Fig. 4A; reviewed by Kuratani, 2012). Elementary comparative morphology has suggested that one of the anterior visceral arches has been

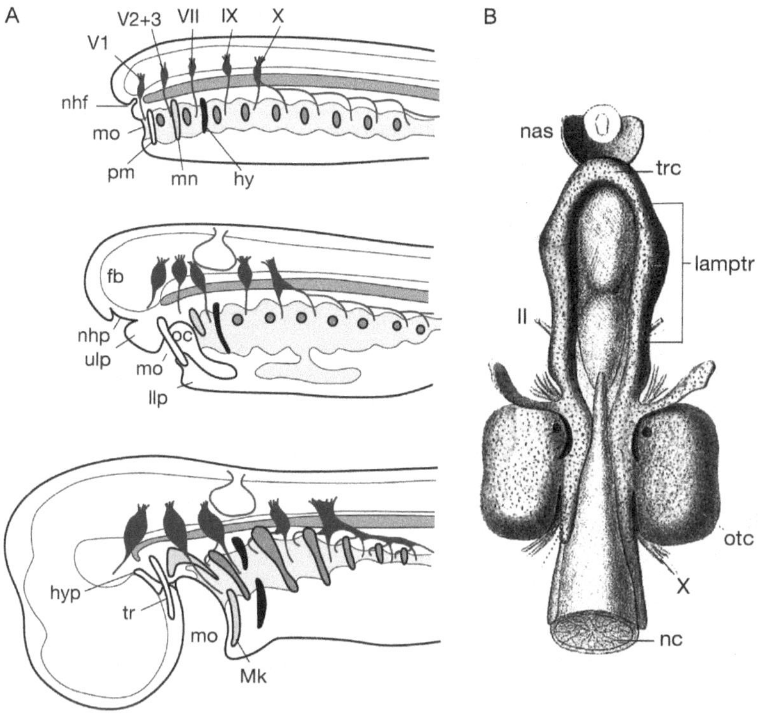

Fig. 4 Generalized scheme of the traditional understanding of jaw evolution. (A) Hypothetical evolutionary process of the vertebrate jaw. Top: Amphioxus-like hypothetical ancestor. Middle: Agnathan grade of ammocoete larva-like ancestor. Bottom: Late pharyngula stage of jawed vertebrate embryo. In this scenario, it was proposed that the jaw was obtained by modification of the second pharyngeal arch (mandibular arch), and the first pharyngeal arch (premandibular arch [pm]) was secondarily employed to form the rostral part of the neurocranium as trabecula (tr), to accommodate the enlarged forebrain (fb) in jawed vertebrates. The premandibular arch, however, has never been observed in the embryo nor in the fossil record. (B) So-called "trabecula" in the lamprey. hyp, adenohypophysis; II, optic nerve; llp, lower lip; Mk, Meckel's cartilage; mo, mouth; nas, nasal capsule; nc, notochord; nhf, nasohypophyseal foramen; ulp, upper lip; V-X, cranial nerves. *Panel (B) Modified from Kuratani, S., & Ahlberg, P. E. (2018). Evolution of the vertebrate neurocranium: Problems of the premandibular domain and trabecula.* Zoological Letters, *4, 1.*

evolutionarily transformed into a device for feeding, i.e., the jaw. This idea is part of the head segmentation theory (reviewed by Kuratani, 2012). Curiously, the expression "the jaw arises from a rostral arch" is truly appropriate for the description of the ontogenetic process of gnathostomes (particularly elasmobranchs), where the jaw differentiates directly from the mandibular arch. The evolutionary process, however, appears to have been more complicated than this.

The segmental theories were concerned with the question of whether premandibular arch(es) existed rostral to the mandibular arch, and if so, how many. Until the beginning of the 20th century, the ancestral existence of one premandibular arch was generally assumed, and was thought to have disappeared in modern animals. During the hypothetical evolutionary process from amphioxus-like state to shark-like state, the forebrain expanded rostrally and the premandibular arch differentiated into a skeletal element, called the "trabecula," to serve as a new cranial floor to support the forebrain, and then the second (or mandibular) arch became divided dorsoventrally into the upper and lower jaws (reviewed by Kuratani, 2012). In many of the classical theories, the premandibular region was assumed to be located anterior to the mouth, not the notochord. De Beer (1937) was the first to question how the premandibular arch shifted during evolution to form the trabecula (Fig. 4A).

In addition, it has been suggested that the vertebrate jaw is derived from the dorsoventrally divided velum of hypothetical agnathans (reviewed by Janvier, 1996). This scenario has some difficulties. First, it is not known if the velum was truly present in the ostracoderm ancestor. Second, the velum in cyclostomes arises in the oropharyngeal membrane, which disappears in the development of jawed vertebrate embryos. Third, as is the case with other derivatives of the pharyngeal arch, the velum arises as a part of a developmental module—in this case, the mandibular arch—and should come with other components such as the lingual apparatus, just like the lower jaw always comes with the upper jaw. The velum, together with the lingual apparatus, appears to represent a synapomorphy that can be used to define cyclostomes, but more paleontological evidence will be needed to substantiate this.

There is also a functional explanation for jaw evolution, i.e., the jaw occurred as an adaptation to increase respiratory efficiency rather than feeding ("ventilation theory" reviewed by Mallatt, 2008). In this model, it is considered that the jaw became larger by expansion of mandibular arch elements into the premandibular domain. Here, the ammocoete larva is employed as a model to represent an ancestral stage, where the traditional homologization remains, to identify counterparts of upper and lower jaws in the upper and lower lips in the ammocoete oral apparatus (see below). However, none of these theories can explain the origin of the trabecula and jaw at the same time.

4.2 Heterotopy theory

Evolutionary innovations often involve alterations of the ancestral developmental program, which in turn often disrupt developmental constraints (reviewed by Shigetani, Sugahara, & Kuratani, 2005). This is because changes in developmental patterns lead to a change in embryonic topography, which then causes changes in positional relationships of the primordia on which the criterion of homology relies. Therefore, a comparative morphological approach that reveals the homology of morphological elements cannot explain the acquisition of a true novelty (see Wagner and Kin in this issue).

The lamprey mandibular arch differentiates into a set of complex and multifunctional structures. These include the lingual apparatus, composed of a set of cartilage elements and muscles, at the bottom of the pharynx (Yalden, 1985). The velum, a pumping apparatus that sends water to the pharynx, is also derived from the mandibular arch within the oropharyngeal membrane (Fig. 2D–F, right; Kuratani, Nobusada, Horigome, & Shigetani, 2001; in gnathostomes, this membrane disappears during embryogenesis, and the velum will not be considered in the following discussion). This morphological complexity implies that the character state of the mandibular arch is highly derived, possibly representing a synapomorphy that could be used to define cyclostomes.

The outer part of the lamprey oral apparatus consists of upper and lower lips (Fig. 5), whose morphological identities are highly problematic. Focus of observation is the trigeminal crest cells (Fig. 6). These cells represent the first of three large cranial neural crest cell populations appearing in a late neurula. The trigeminal crest cells not only fill the mandibular arch, but also extend to populate the premandibular region and surround the forebrain. The trigeminal crest cells are divided into a total of three cell populations: the mandibular arch crest cells and the postoptic and preoptic subsets of the premandibular crest cells. The distribution and morphological pattern of the early trigeminal crest cells are common to all vertebrate embryos, and there is little difference between cyclostomes and gnathostomes in this regard.

In gnathostome embryos, preoptic crest cells are divided into medial and lateral cell populations by the presence of a pair of nasal placodes. The medial mesenchyme becomes the unpaired frontonasal process (or medial nasal prominence [MNP]) and the lateral mesenchyme becomes the lateral nasal prominence (LNP). In contrast, in lamprey embryos, the nasal placode remains unpaired in the central midline and is posteriorly continuous with the adenohypophyseal placode, which together form the so-called nasohypophyseal plate (Figs. 5 and 6). Thus, the lamprey preoptic crest cells remain as a single cell mass, resulting in a cyclostome-specific craniofacial pattern.

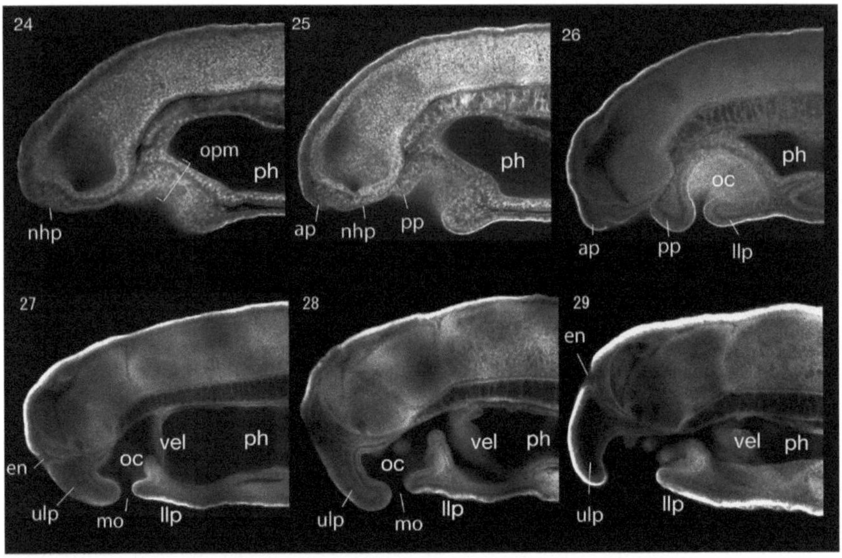

Fig. 5 Development of the lamprey head. Note that the upper lip (ulp) originates from the posthypophyseal process (pp) arising posterior to the nasohypophyseal plate (nhp). Velum (vel) develops in the oropharyngeal membrane (opm). Numbers represent Tahara's developmental stages (Tahara, 1988). ap, anterior nasal process; en, external nostril; llp, lower lip; mo, mouth. *From Kuratani, S. (2017).* Evolutionary morphology: Bauplan and embryonic development of vertebrates. *New Edition. (in Japanese), Tokyo: Tokyo Univ. Press.*

During embryogenesis of jawed vertebrates, the premandibular part of the trigeminal crest cells form the so-called prechordal cranium (Couly, Coltey, & Le Douarin, 1993), which is the rostral part of the neurocranium derived from the neural crest. The rostral neurocranium of modern gnathostomes consists of trabeculae that develop as a pair of rod-shaped cartilages and the lateral wall of the nasal capsule that lies lateral to the trabeculae (Fig. 7). Rostral to the trabeculae, a single median intertrabecula develops independently and fuses with the paired trabeculae to form the main part of the prechordal cranium (Wada, Nohno, & Kuratani, 2011; Fig. 7). The intertrabecula is specifically derived from the MNP, which comprises preoptic crest cells, whereas the trabeculae are derived from postoptic crest cells. In lamprey embryogenesis, on the other hand, postoptic crest cells form the upper lip (Fig. 6; Kuratani et al., 2001; Kuratani & Ahlberg, 2018). Although this anlage is often misidentified as a part of the mandibular arch, it belongs to the premandibular domain (Kuratani et al., 2001; Fig. 6). Thus, the mesenchyme in the lamprey upper lip is a tissue similar to the trabecula of

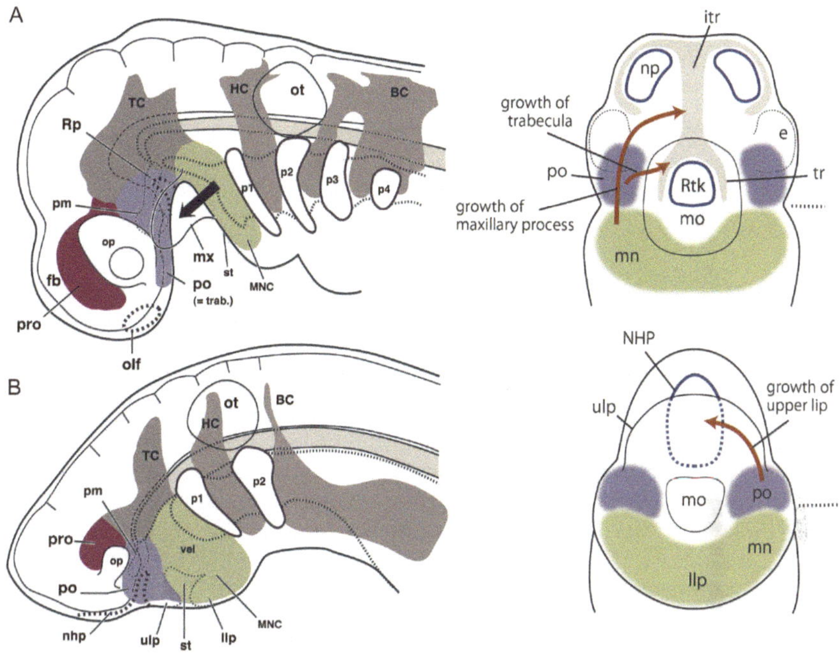

Fig. 6 Comparison of the distribution and development of cranial neural crest cells. (A) Jawed vertebrate pharyngula. Left: Trigeminal neural crest cells (TC) can be divided into three cell populations according to their distributions. The mandibular arch crest cells (MNC) are found in the mandibular arch; postoptic crest cells (po) are found posterior to the eye; and preoptic crest cells (pro) are found rostral to the eye. The latter two populations are located in the premandibular domain. Right: Growth of the crest-derived ectomesenchyme in craniofacial development in an older embryo. Arrows indicate directions of growth. Due to the splitting of the nasal placode (np), the intertrabecula (itr) can develop into internasal septum. (B) Lamprey embryo. Left: TC distribution similar to that in jawed vertebrates can be seen from the lateral view. Right: In an older embryo, however, postoptic crest cells form the upper lip (ulp), unlike in jawed vertebrates where the equivalent cells form trabecula (tr). BC, branchial arch crest cells; fb, forebrain; HC, hyoid crest cells; inf, infundibulum; llp, lower lip; mb, midbrain; mn & MNC, mandibular crest cells; MX, upper jaw; nhp & NHP, nasohypophyseal plate; op, eye; ot, otocyst; ph, pharynx; pm, premandibular mesoderm or prechordal plate; p1–4, pharyngeal slits 1–4; Rtk, Rp, Rathke's pouch; st, stomodaeum. *A (left) and B (left), modified from Kuratani, S. (2017).* Evolutionary morphology: Bauplan and embryonic development of vertebrates. *New Edition. (in Japanese), Tokyo: Tokyo Univ. Press. A (right) and B (right), modified from Kuratani, S. (2020). Neural crest and craniofacial evolution of early vertebrates. In B. F. Eams, D. M. Medeiros, I. Adameyko (Eds.),* Evolving neural crest cells *(pp. 219–242). Boca Raton, London, New York: CRC Press/Taylor & Francis Group.*

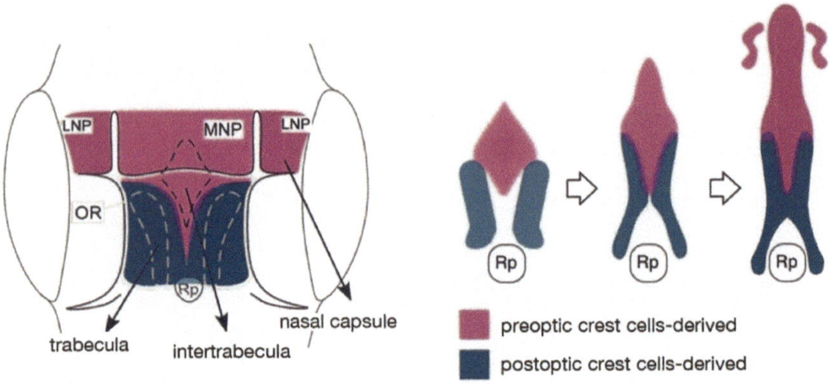

Fig. 7 Development of the trabecula. Left: Craniofacial primordia of an amniote post-pharyngula. The prechordal cranium of jawed vertebrates develop from preoptic crest cells in the lateral and medial nasal prominences (LNP and MNP, respectively) and the postoptic crest cells that will form the paired trabeculae. Right: Development of the prechordal cranium in the chick embryo. Rp, Rathke's pouch. *Modified from Wada, N., Nohno, T., & Kuratani, S. (2011). Dual origins of the prechordal cranium in the chicken embryo.* Developmental Biology, 356, *529–540.*

gnathostomes. This raises the question: what developmental module exactly does the structure called "trabecula" in lamprey correspond to?

De Beer (1937) once envisioned a homolog of the trabecula around the mouth of an intermediate animal that was like an ammocoete larvae (Fig. 4A). However, as mentioned above, lampreys have a skeletal element called a trabecula (Fig. 4B; reviewed by Kuratani & Ahlberg, 2018). It is an inverted U-shaped cartilage, like the frame of a tennis racket, that extends forward from both sides of the anterior end of the notochord. Both its apparent function (forming the cranial base at the level of the forebrain) and its shape (a pair of rod-like cartilages) are similar to trabeculae of gnathostomes, but the similarity in shape itself is not a requirement for morphological homology.

The first primordia of the lamprey trabeculae appear lateral to the notochord, dorsal to the mandibular arch, and secondarily extend rostrally to reach the forebrain, resembling the parachordal cartilage in gnathostomes more than the trabeculae of gnathostomes. This is consistent with the finding that lamprey trabecula primordia are labeled by microinjection of vital dyes into the mandibular mesoderm (reviewed by Kuratani & Ahlberg, 2018). Thus, the trabeculae in the lamprey belongs to the mesodermal chordal cranium. This means that the neurocranium of the lamprey is almost entirely derived from the head mesoderm.

There is a large difference between cyclostomes and gnathostomes, especially in the use of postoptic crest cells, which results in the loss of homology between the lamprey oral apparatus and the gnathostome jaw elements. The essential difference in head morphology between lamprey and gnathostome embryos can be ascribed to the pattern of placodes that influences the positional distribution of neural crest mesenchyme and the difference in the direction of ectomesenchymal growth in relation to them. The upper lip in the lamprey larvae therefore occupies a position very different from its cell lineage–level counterpart, the trabecula in gnathostomes, which are established concomitantly with diplorhiny; the formation of the trabecula in the prechordal cranium in the gnathostomes cannot be achieved without diplorhiny.

4.3 Genes that pattern the jaw and the heterotopy theory of jaw acquisition

The division of the mandibular arch into upper and lower regions is a prerequisite for the acquisition of the jaw. Developmentally, the molecular mechanism that provides positional values to each group of trigeminal crest cells is based on the epithelial-mesenchymal interaction. The genes involved are those encoding growth factors, such as BMP4 (bone morphogenic protein 4) and FGF8 (fibroblast growth factor 8), and homeobox genes *Dlx1*, *Msx1*, and *Barx1*, which are downstream of the growth factors and confer local mesenchymal identities. Thus, when the FGF8 distribution is expanded anteriorly, *Dlx1* and *Barx1* expression regions are also expanded anteriorly, resulting in a shift of trigeminal nerve innervation patterns and the division of premandibular and mandibular arch regions (reviewed by Noden & Francis-West, 2006). In the neurula, trigeminal crest cells are not yet positionally specified; they are specified through subsequent epithelial-mesenchymal interactions, depending on the distribution pattern of growth factors in the ectoderm (reviewed by Kuratani, 2020).

After the mandibular arch domain has been specified by the gene cascade described above, the same gene cascade functions as the proximodistal patterning mechanism within the mandibular arch. Levels of FGF8 specify the future jaw joint (reviewed by Shigetani et al., 2005). Then, the expression of *Dlx1/Barx1* is induced in the arch mesenchyme adjacent to it. *Bmp4* is expressed in the ectoderm corresponding to the tips of the upper and lower jaws, and *Msx1* is expressed in the mesenchyme immediately below these ectodermal regions. This positional distribution of homeobox gene expression in the mesenchyme of the mandibular arch specifies the proximodistal axis of the arch (reviewed by Noden & Francis-West, 2006).

The interesting thing about the homologues of *Bmp4*, *Fgf8*, *Dlx1*, and *Msx1* in lamprey embryos is that their expression patterns appear to suggest homology between the upper lip and upper jaw, and lower lip and lower jaw, in lampreys and gnathostomes, respectively (reviewed by Kuratani, 2012). However, as noted above, this homology does not hold when developmental patterns of lamprey and gnathostomes embryos are compared. This mean there is a non-homologous "shift" in morphological patterns and gene expressions. Cell–cell interactions based on homologous molecular cascades have generated functionally analogous structures rather than morphological homologs (reviewed by Shigetani et al., 2005); i.e., morphological homologies and gene homologies are decoupled from each other.

The heterotopy theory explains the evolution of the jaw as a novelty caused by the topographical shift of epithelial-mesenchymal interaction during the transition from jawless to jawed states (reviewed by Kuratani, 2012). In this transition, neural crest cells are eliminated at the site where the placode juxtaposes the central nervous system. According to this theory, (a) the change in the number and distribution of placodes was a major basis for evolutionary repatterning of the craniofacial ectomesenchyme and (b) character states such as jaw, adenohypophyseal position, and number of nostrils, are developmentally interconnected to each other in a single evolutionary change (heterotopy), or as a set of related changes starting from a single change.

4.4 Fossils

4.4.1 Shuyu—*Galeaspids*

Several prominent fossils appear to demonstrate the sequence of evolutionary events predicted by the heterotopy theory. Among the gnathostome stem group, is the group called galeaspids, which lived about 400 million years ago (mya) (Gai, Donoghue, Zhu, Janvier, & Stampanoni, 2011). Among them, the genus *Shuyu*, whose fossils have been found in China and Vietnam, had a single external nostril, but the nasal sac was divided into pairs intracranially, and the adenohypophysis gland opened into the oral cavity. This indicates that the nasal placode and adenohypophyseal placode were separated during development, and the distribution of the ectomesenchymal pattern would have been close to that of jawed vertebrates. Moreover, there was a true trabecula-like skeletal element around the adenohypophysis, but it had not yet formed the neurocranium at the level of the forebrain (Gai et al., 2011; Fig. 8A).

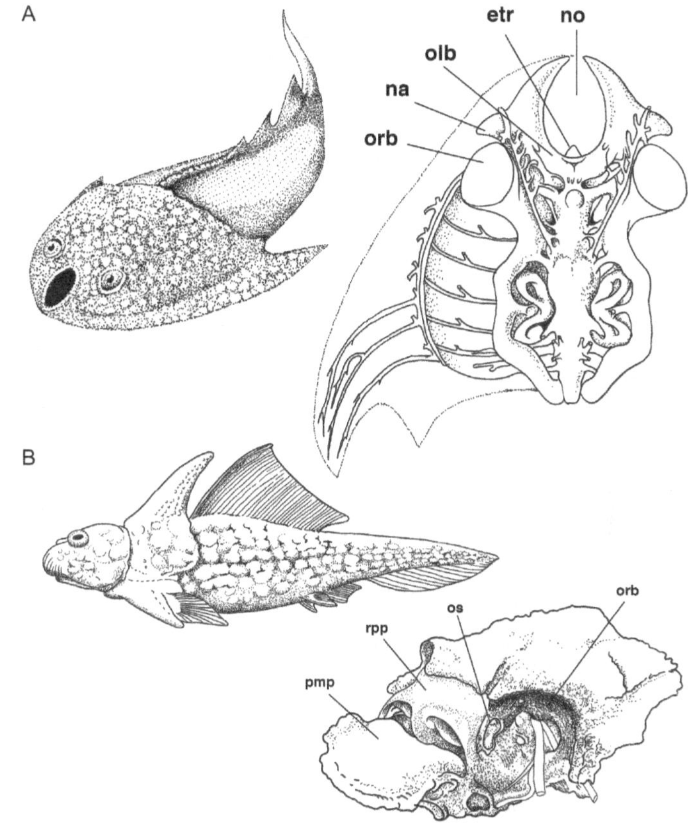

Fig. 8 Fossil records and evolution of the jawed vertebrate cranium. (A) *Shuyu*, an ostracoderm belonging to galeaspids. Left: external morphology. Right: Cranial morphology. There was only one nostril (no), but observation of the internal morphology revealed that the nasal sac (na) was divided into two parts, and the adenohypophysis opened to the oral cavity. A skeletal element resembling trabecular cartilage (etr) was also recognized, suggesting a transitional state toward the jawed vertebrate anatomy. (B) A basal placoderm, *Romundina*. Left: external morphology. Right: craniofacial anatomy. This animal had a jaw, but its neurocranium resembled that of an ostracoderm. The skeletal element (postpineal plate [rpp]) supporting the nasal capsule was apparently derived from the anterior nasal process, and was found rather posteriorly, similar to the nasal capsule of the lampreys and the pineal plate in *Cephalaspis* (a genus of osteostracans). The premedian plate (pmp) appears to have developed from the posthypophyseal process, which corresponds to the posterior part of trabecula of jawed vertebrates, while the structure looks similar to the upper lip of lamprey larvae. Olb, olfactory bulb; orb, orbit; os, stalk. *Panel (A) Modified from Gai, Z., Donoghue, P. C. J., Zhu, M., Janvier, P., & Stampanoni, M. (2011). Fossil jawless fish from China foreshadows early jawed vertebrate anatomy.* Nature, 476, *324–327.Panel (B) Modified from Dupret, V., Sanchez, S., Goujet, D., Tafforeau, P., & Ahlberg, P. E. (2014) A primitive placoderm sheds light on the origin of the jawed vertebrate face.* Nature, 507, *500–503.*

However, the above interpretation is inconsistent with the conventional theory that not the galeaspids, but osteostracans are placed basal to placoderms as a sister group of modern gnathostomes. In particular, there are synapomorphies such as perichondral bones, pectoral fins, and heterocercal tail that are shared by modern gnathostomes, but none of them are shared by galeaspids.

4.4.2 *Romundina*

One of the basal placodermi, *Romundina stellina*, lived in the early Devonian period in what is now known as Canada (Fig. 8B). It belongs to a group of primitive placoderms called Antiarchs. It had acquired a true jaw, but still had an agnathan-like craniofacial morphology—i.e., eyes located dorsally and approaching the midline, and a pair of nostrils opening immediately in front of the eyes—and classically it was often considered to belong to ostracoderms (reviewed by Dupret, Sanchez, Goujet, Tafforeau, & Ahlberg, 2014).

This fossil was re-examined by Dupret et al. (2014). The forebrain of *Romundina* was found to be compressed in the anterior–posterior direction, showing that the forebrain had not yet developed to the level of modern gnathostomes. In addition, the anterior cranium of this animal extended rostrally like a cyclostome upper lip, forming a protrusion called a "premedian plate" at the rostral tip, and the palatoquadrate articulated with the lateral aspect of the premedian plate. The position and shape of the anterior cranium of *Romundina* more closely resembled that of lampreys than of gnathostomes, and with the jaw removed, the entire skull would look like that of ostracoderms. There was also a skeletal module called a "rostropineal plate" behind the orbit and the external nostril (Fig. 8B below); this rostropineal plate is considered to be a derivative of the anterior nasal process.

If the above interpretation is correct, the composition of the heads of extant gnathostomes was already achieved in the head of *Romundina*, but the size and proportions of the skeletal modules remained very agnathan-like. The MNP was still poor, and premaxilla-like elements had not been established. That is, not only in modern gnathostomes, but also in primitive placoderms such as *Romundina*, the boundary between the MNP and trabecula is or was dorsally and posteriorly positioned, as in lampreys. If this boundary were to reach the ventral anterior tip around the snout like a belt conveyor, the cranium would become the typical pattern of jawed vertebrates. Dupret et al. (2014) state that changes in the proportions of various parts of the brain must have been involved in this change in boundary position. In other words, the change of neural crest mesenchyme and placode module organization that produced modern gnathostomes occurred first, allowing the evolution of the jaw, but a further shift of ectomesenchyme followed during adaptation to the subsequent enlargement of the brain, leading to the "face of the jawed fish" that we know *today*. Thus, the jaw would have been a key innovation for the formation of the face of the modern gnathostomes. Dupret et al. (2014) interpret their findings to mean that the *Romundina* head morphology represents a step following that of *Shuyu*.

5. Development of the hagfish

Before 2007, there had been only a few reports on hagfish embryos. A whole embryonic series of a hagfish species, *Bdellostoma (Eptatretus) stouti*, was obtained in 1899 by an American paleontologist, Bashford Dean who specialized in fossils of ostracoderms and placoderms. Unfortunately, since the eggshells were not removed but directly immersed in ethanol and fixed, the embryonic tissues were subjected to secondary deformation, making histological observation of these embryos extremely difficult.

This situation was overcome by the acquisition of hagfish embryos in a laboratory aquarium environment by K.G. Ota and co-workers; the embryos were obtained from a hagfish species, *Eptatretus burgeri*, that inhabits the relatively shallow seabed in the west part of Japan (reviewed by Kuratani & Ota, 2008). By 2013, nearly a whole series of embryos was obtained, and embryogenesis could be observed in detail at the histological level (Oisi et al., 2013a, 2013b).

5.1 Origin of the adenohypophysis

The most enigmatic discovery arising from the study of the Dean's embryonic collection had been the developmental origin of the hagfish adenohypophysis. The endodermal origin of the hagfish adenohypophysis was first put forth by Gorbman (1983), raising the possibility that hagfishes have a developmental plan quite different from that of other vertebrates, or have gone through a unique evolutionary change not shared by other vertebrates. Either hypothesis could have cast doubt on the monophyly of cyclostomes. However, the pattern and process of development of the *E. burgeri* embryos, described by Oisi et al. (2013a, 2013b) was very similar to those of other vertebrate embryos.

In the descriptions below, *E. burgeri* embryos are assumed to be representative of hagfish embryos in general. At the pre-pharyngular stage of *E. burgeri*, the epidermal ectoderm, which covers the forebrain, contacts the endoderm in front of the brain, forming the future oropharyngeal membrane. At this stage, the extraembryonic ectoderm wrinkles and lifts upward anterior to the head. As the forebrain expands, the oropharyngeal membrane folds into the ventral side of the forebrain to form the oral cavity (Fig. 9). In the early pharyngula, the oropharyngeal membrane disappears, and the extraembryonic ectodermal wrinkles fuse in the midline to form an ectodermal membrane that completely blocks the oral cavity from the exterior (Fig. 9). This ectodermal membrane, which is somewhat reminiscent of the amnion, is a structure independent of the oropharyngeal membrane, and develops in every *E. burgeri* embryo. On the other hand, the preoral gut adheres to the posterior part of the thalamus, which is located behind the oropharyngeal membrane and is derived from the endoderm. Thus, the morphology of *E. burgeri* pharyngula has the same pattern as the positional relationships shown by Rathke's pouch, preoral gut, anterior tip of the notochord, and hypothalamic primordium in other vertebrate pharyngulae.

The true adenohypophysis of the hagfish, like that of the lamprey, originates from the posterior half of the nasohypophyseal plate rostral to the oropharyngeal membrane, i.e., from the ectodermal region. This corresponds to the ventral midline of the pan-placodal domain, which is in contact with the ventral part of the forebrain or the hypothalamus that expresses *EbNkx2.1*. Thus, the positional relationship between adenohypophyseal primordium and other primordia is the same in *E. burgeri* embryos and lamprey embryos. The expression of regulatory genes also indicates that the true adenohypophysis is derived from the ectodermal lineage, whereas epithelium posterior to the real oropharyngeal membrane shows endoderm-specific gene expression (Oisi et al., 2013a). Therefore, the details of the development of hagfishes revealed by Oisi et al. (2013a, 2013b) support the notion that the two groups of cyclostomes are formed under a similar basic pattern and similar developmental programs.

5.2 The cyclostome pattern

A curious fact about hagfish development is that there is a stage where the shape of the embryonic head becomes very similar to that of a lamprey embryo (Fig. 10). This pattern appears at stage 45 (Dean, 1899) for hagfishes and stage 26 (Tahara, 1988) for lampreys (Fig. 10; Oisi et al., 2013a). At this

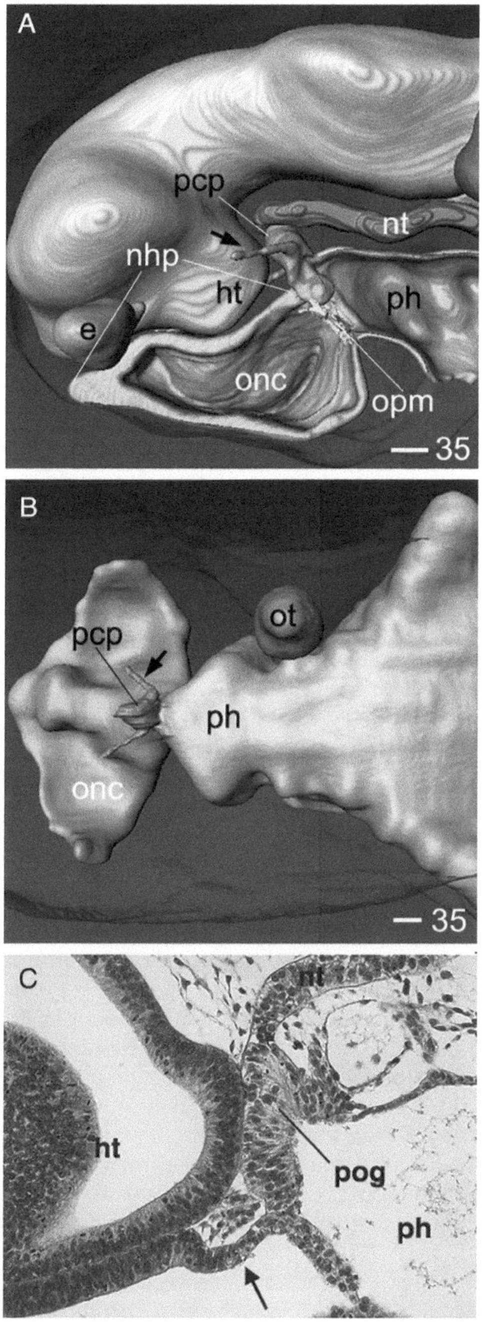

Fig. 9 See figure legend on next page.

time, the nasohypophyseal plate is found on the ventral aspect of the head, and cushion-like projections are formed anterior and posterior to it. These projections are called the anterior nasal process (ANP) and posthypophyseal process (PHP), respectively; the former contains preoptic crest cells and the latter contains postoptic crest cells and possibly ectomesenchyme of the mandibular arch. The PHP becomes the upper lip in lamprey embryos, whereas in hagfish embryos it differentiates into the oronasal septum. In both cases, the mouth opens ventral and caudal to this protrusion. This embryo pattern shared by the cyclostomes does not appear in the development of jawed vertebrates.

As has been shown many times, the PHP in lampreys expands to the anterior dorsal side as an upper lip, and simultaneously, the external nostril is pushed dorsally (Fig. 10). It is easy to derive the shape of the head of an ammocoete larva from the cyclostome pattern described above, and no major changes in morphogenesis in hagfishes occurs after this stage. However, since lamprey larvae undergo metamorphosis, adult lamprey organs such as the thyroid and lingual apparatus cannot be compared with the hagfish until the lamprey has undergone metamorphosis.

In later stage hagfish embryos, a complicated secondary deformation process takes place (Oisi et al., 2013a). Namely, the ANP extends rostrally together with the nasal duct, and the PHP extends anteroposteriorly (the oronasal septum is homologous to the lamprey upper lip), and the external nostril and the mouth are placed dorsoventrally at approximately the same level as each other (Figs. 2 and 11). In this way, hagfishes come to possess a very different craniofacial morphology from that of lampreys. The hagfish barbels (T1–4) form from the PHP, suggesting that these all correspond

Fig. 9 *Eptatretus burgeri* embryo at Dean's stage 35. (A) Lateral view of a sagittal section of a reconstructed embryo model. Premandibular mesoderm of this animal arises from the prechordal plate (pcp). (B) Dorsal view of the same embryo model. A pair of mesodermal chords are arising from the prechordal plate, similar to the premandibular cavity in the shark embryo. (C) The thickened prechordal plate is not Rathke's pouch found in jawed vertebrates, but it is an endodermal preoral gut (pog) to which the rostral tip of the notochord abuts. Below the gut, the oropharyngeal membrane is seen. An arrow indicates the posthypophyseal process. e, eye; ht., hypothalamus; nhp, nasohypophyseal plate; nt, notochord; onc, oronasal cavity; opm, oropharyngeal membrane; ot, otocyst; ph, pharynx. *Modified from Oisi, Y., Ota, K. G., Fujimoto, S., & Kuratani, S. (2013a). Craniofacial development of hagfishes and the evolution of vertebrates.* Nature, 493, *175–180, Oisi, Y., Ota, K. G., Fujimoto, S., & Kuratani, S. (2013b). Development of the chondrocranium in hagfishes, with special reference to the early evolution of vertebrates.* Zoological Science, 30, *944–961.*

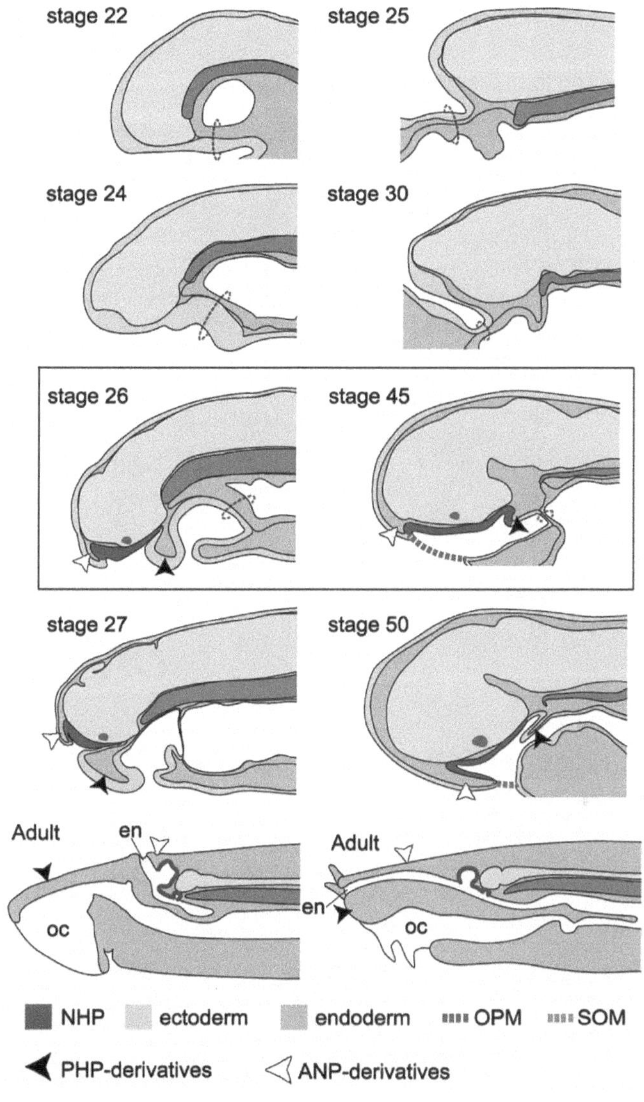

Fig. 10 Developmental sequences of lamprey and hagfish craniofacial primordia. At the early pharyngula stage (stage 26 [Tahara, 1988] for lampreys and stage 45 [Dean, 1899] for hagfish; outlined with a box), lamprey and hagfish craniofacial patterns become conspicuously similar to each other, showing the "cyclostome pattern" of craniofacial development, which is not shared by jawed vertebrates. After this stage, lamprey and hagfish will follow very different developmental pathways. *Modified from Oisi, Y., Ota, K. G., Fujimoto, S., & Kuratani, S. (2013a). Craniofacial development of hagfishes and the evolution of vertebrates.* Nature, 493, *175–180, Oisi, Y., Ota, K. G., Fujimoto, S., & Kuratani, S. (2013b). Development of the chondrocranium in hagfishes, with special reference to the early evolution of vertebrates.* Zoological Science, 30, *944–961.*

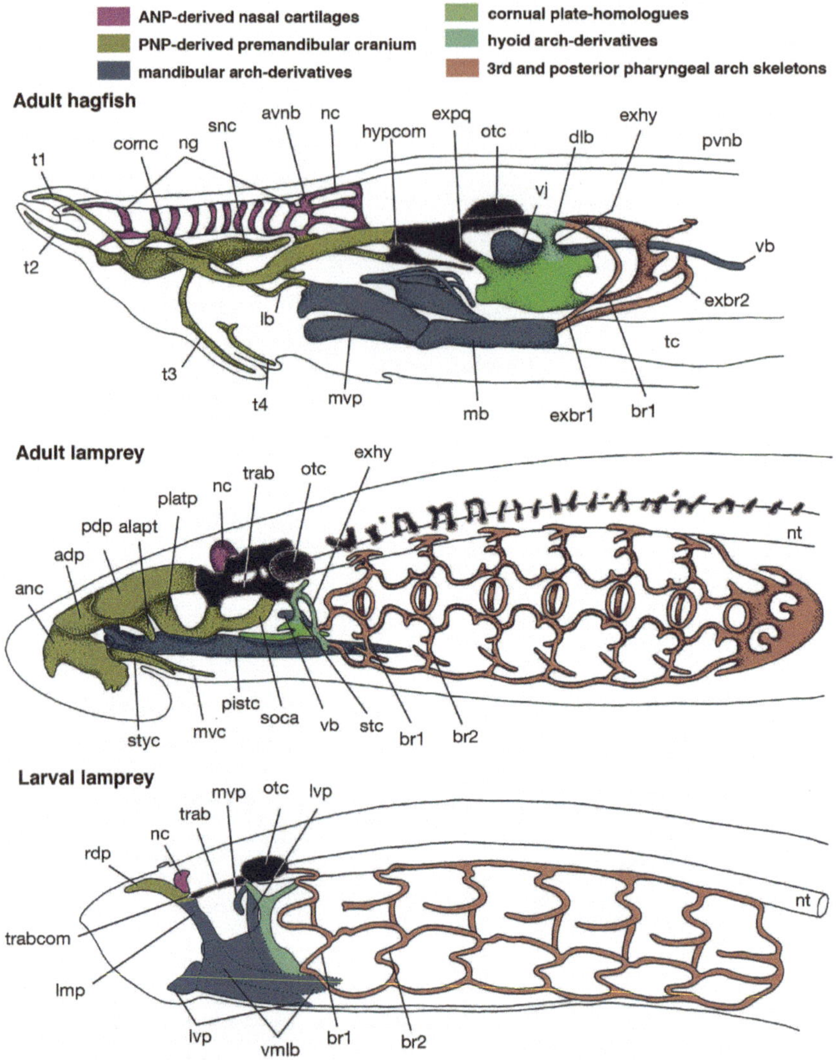

Fig. 11 Comparison of hagfish and lamprey craniums in terms of the embryonic derivation and cell lineage origins. adp, anterior dorsal plate; ah, adenohypophysis; alapt, anterior lateral apical cartilage; anc, annular cartilage; anp, anterior nasal process; avnb, anterior vertical nasal bar; br1–2, internal branchial arch 1–2 cartilage; con1, rostral commissure of dlb; con2, middle commissure of dlb; con3, posterior commissure of dlb; cornc, cornual cartilage; dp, dental plate primordium; dlb, dorsal longitudinal bar; e, eye; en, external nostril; exbr1, extrabranchiale 1; exbr2, extrabranchiale 2; exhy, extrahyal; expq, extrapalatoquadrate; hy, hyoid arch; hypcom, hypophyseal commissure; lb., labial cartilage; lp, lingual plate; lvp, latero-rostral part of basal plate; mb, medial part of basal plate; mo, mouth; mphp, PHP-derived mesoderm; mvc, medio-ventral cartilage; mvp, medio-rostral part of basal plate; nc, nasal capsule; ng, nasal duct cartilages; nhd, nasohypophyseal duct; nt, notochord; otc, otic capsule; palb, palatine bar; pch, parachordals; pdp, posterior dorsal plate; ph, pharynx; php, posthypophyseal process; pistc, piston cartilage; platp, posterior lateral plate; pom, periotic mesenchyme; ptr, posterior trabecula; snc, subnasal cartilage; soca, subocular arch; stc, styliform cartilage; styc, stylet cartilage; tc, tongue cartilage; t1–3, cartilaginous support for tentacles; trab, trabecula; vb, velar bar; vbrb, ventral branchial bar; vch, velum chamber; vj, joint caput for velum; vm, velum mesoderm and muscle. *Modified from Kuratani, S., Oisi, Y., & Ota, K. G. (2016). Evolution of the vertebrate cranium: Viewed from the hagfish developmental studies.* Zoological Science, 33, *229–238.*

to a part of the upper lip of lamprey. More specifically, T1, T3, and T4 correspond to the side wall of the upper lip, and T2 corresponds to the dorsomedial part of the upper lip. The homology of the head primordia presented above is consistent with the innervation patterns of trigeminal nerve branches (Oisi et al., 2013a); the first branch of the nerve innervates the medial part of PHP-derivatives and the ANP, and the second branch innervates the lateral part of the PHP. The latter contains motor fibers that control mandibular arch–derived muscles. The third branch innervates both the lower lip and the velum. Such patterns do not match those found in jawed vertebrates. Thus, we can see developmental modular structure in the distribution of mesenchyme and nerve branches, and homologization is possible based on this modularity during development.

As the hagfish development progresses further, the root of the PHP shifts posteriorly and finally disappears, making the nasohypophyseal duct posteriorly continuous with the pharynx, as in adult hagfishes. The conspicuous differences between adult hagfishes and lampreys are ascribed to these changes introduced rather late in hagfish ontogeny. The same is also true for the posterior shift of the pharynx in hagfishes. In this process, the glossopharyngeal and vagus nerves extend posteriorly for a long distance, sending out gill branches to innervate the branchial arches. Curiously, the shape of the contractile muscles of the arches and the cranial nerve branches to innervate them appear very similar to those in the lamprey, although their position is tremendously shifted compared with lamprey. The factors behind this shift in position, the accompanying changes in the body cavity of the trunk, and the changes that occur in the development of the hypobranchial musculature are not well understood.

The above findings suggest that most of the hagfish-specific traits are more derived than those in lampreys. In the past, lampreys and gnathostomes were thought to be sister groups to each other because they share more similar traits than hagfishes and lampreys. Based on the current understanding, however, the similarity between jawed vertebrates and lampreys can be regarded as plesiomorphy on the basis of the outgroup comparison with hagfishes showing the derived nature of hagfish-specific traits.

5.3 Evolution of the cranium

Is the cyclostome pattern specific to cyclostomes, or is the general craniofacial morphology of modern jawed vertebrates apomorphic? The key to answering this question is monorhiny. In addition to cyclostomes, the jawless gnathostome stem group also show monorhiny (reviewed by Janvier, 1996). These animals did not have developmental modules like MNP or LNP, nor did they develop a prechordal cranium, suggesting that their cranium probably developed from an embryonic morphology similar to that in cyclostomes; i.e., they possessed ANP-like, and PHP-like modules. If so, the cyclostome pattern can be understood to represent a primitive pattern that defines the agnathan grade.

The evolutionary novelty that separates agnathans from modern gnathostomes is related to the shift from monorhiny to diplorhiny, and the primary synapomorphy for jawed vertebrates is that nasal placodes are laterally separated into halves and rostrally dissociated from the adenohypophyseal placode. In fact, when we look at the gnathostome stem lineages, apparently monorhinous groups are found in ostracoderms, indicating that evolutionary polarity likely progressed from monorhiny to diplorhiny. *The change in the number of nostrils that occurred before jaw acquisition was the most fundamental evolutionary event in vertebrates, dividing vertebrates into grades roughly corresponding to agnathans and jawed vertebrates.*

A number of reports show comparisons of crania between lampreys and hagfishes, or between cyclostomes and modern gnathostomes; but every researcher found that such comparisons are not easy (reviewed by Kuratani, Oisi, & Ota, 2016). Because the cyclostome pattern is shared by lamprey and hagfish embryos, structural homology in adults can be attributed to this pattern. Therefore, below, I will first compare the chondrocranial elements among cyclostomes and infer the ancestral state or morphologically generalized skull morphotype of cyclostomes, and then I will compare this pattern with that of gnathostomes.

Comparison of cartilage cranial elements is made based on categories of nasal capsule (ANP-derivatives), visceral arch derivatives, posterior pituitary process derivatives, and mesodermal neurocranial derivatives. A comparison of such cartilage elements in the lingual apparatus has been reported by Yalden (1985). Based on this, a morphological comparison of the chondrocranium among cyclostomes has already been presented elsewhere (Kuratani et al., 2016). As expected, structures derived from modules not shared by modern gnathostomes are not found in cyclostomes, and even among cyclostomes are not always comparable especially in the fully formed state, either. Typically, structures derived from the PHP are comparable within cyclostomes as a single module (Fig. 11), and the equivalent module will be found

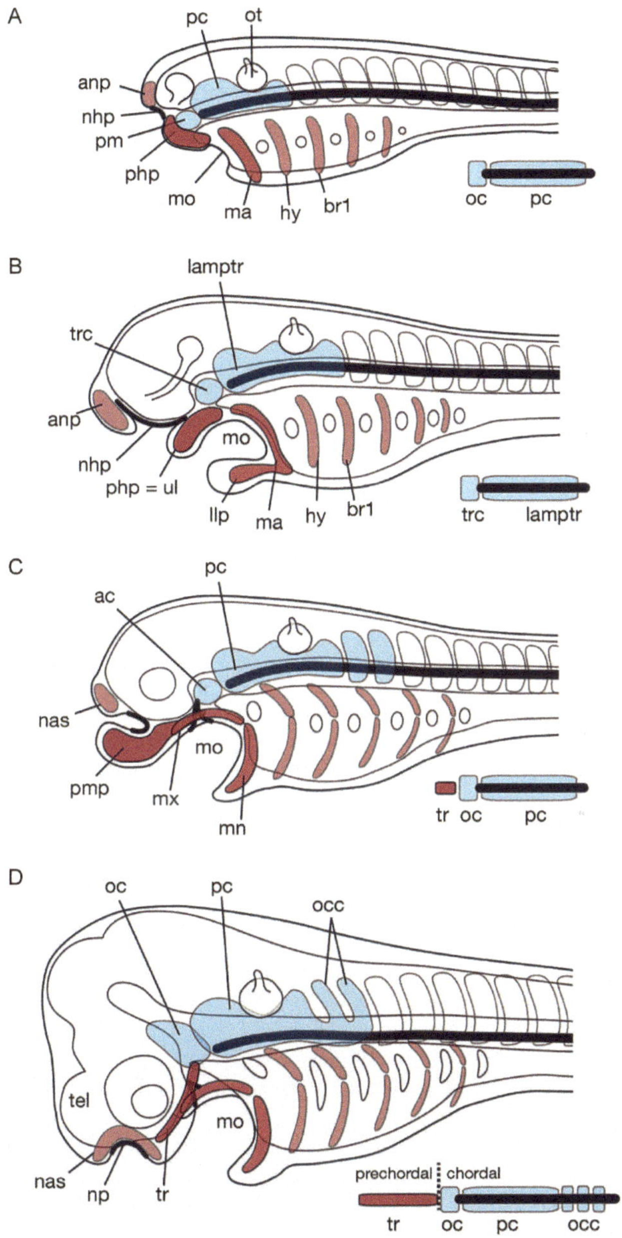

Fig. 12 See figure legend on next page.

in the trabecula derived from the postoptic crest cells. As mentioned in "The cyclostome pattern" section, however, the PHP in cyclostomes also incorporates tissue derived from the mandibular arch–derived muscles in the upper lip. There is no such tissue around the trabecula in modern gnathostomes.

Compared with cyclostome neurocrania (Figs. 11 and 12), the composition of the neurocranium of jawed vertebrates is more complicated due to the incorporation of postoptic crest cell–derivatives in the prechordal cranium. As shown in Fig. 12, it is very likely that the dual cell lineage composition of the cranium (mesoderm and neural crest cells) would have corresponded to the dual functional components of the ancestral cranium, namely, neuro- and viscerocranium. Enlargement of the forebrain and acquisition of the jaw required reorganization of these mesenchymal components, and now the interface of the cell lineages on the cranium no longer corresponds to the functional duality. In other words, the craniofacial developmental program in cyclostomes will tell us a lot about the ancestral developmental program of the cranium.

Fig. 12 Evolutionary sequence of the vertebrate neurocranium. Cephalic neural crest–derived skeletal elements are colored pink, and mesodermal elements are shown in light blue. Diagramatic composition of the neurocranium is presented on the right. (A) Hypothetical ancestor. Forebrain is weakly developed, and the paraxial mesoderm is restricted to the extension of the notochord. Neural crest–derived ectomesenchyme is seen around the nasal-hypophyseal placodes (nhp) as the anterior nasal process (anp) and posthypophyseal process (php). The neurocranium of this animal is formed only of mesoderm. (B) Idealized monorhinous jawless state, corresponding to the stem gnathostomes. The premandibular ectomesenchyme is used as a part of the oral apparatus, like the upper lip of the lamprey. (C) Early jawed stage like *Romundina*. The mandibular arch is dorsoventrally divided to differentiate into the upper jaw (mx) and the lower jaw (mn). (D) Idealized embryonic scheme of crown gnathostomes based on the elasmobranch morphology. The premandibular ectomesenchyme (anp and php) is forming the rostral neurocranium, and the entire neurocranium consists of a prechordal region derived from the neural crest (= trabecula, tr) and a chordal region derived from the mesoderm. The posterior part of the chordal cranium has incorporated occipital elements (occ), derived from rostral somites. ap, anterior process; br1, branchial arch 1; e, eye; hy, hyoid arch; lamptr, lamprey trabecula; llp, lower lip; ma, mandibular arch; mb, midbrain; mo, mouth; mx, maxillary process; nas, nasal capsule; nhp,; oc, orbital cartilage; ot, otic vesicle; pc, parachordal cartilage; pm, premandibular mesoderm; pmp, premedian plate in *Romundina*; tel., telencephalon; trc, transverse commissure in the lamprey (hypophyseal commissure in the hagfish); ul, upper lip. *Modified from Kuratani, S., & Ahlberg, P. E. (2018). Evolution of the vertebrate neurocranium: Problems of the premandibular domain and trabecula.* Zoological Letters, 4, *1.*

6. Perspectives

There still remain two curious questions about cyclostome development. The first question is how the endostyle and thyroid develop in cyclostomes. Although it has been suggested that an endostyle-like endodermal epithelial structure transiently appears in hagfish embryos, it is very likely that this is an artifact derived from inappropriate fixation. Among known vertebrate species, lampreys are the only group that possess an endostyle in their life cycle, and its homology with the endostyle in non-vertebrate chordates should be questioned. The second question is whether the indirect development of the lamprey (via ammocoete larva) or the direct development of the hagfish is the ancestral development pattern of vertebrates. This question could be rephrased as: what is the origin of vertebrate larva? It is unknown whether the ammocoete larva of the lamprey reflect the morphology of an unknown ancestral form, or whether the larval form reflects a secondarily inserted larval stage, newly invented in the lamprey lineage. Answering these questions will help us visualize the common ancestor of vertebrates and understand vertebrate evolutionary history. It is clear that the integration of paleontological and evolutionary developmental studies and comparative genomic studies will become more important than ever.

References

Bajoghli, B., Guo, P., Aghaallaei, N., Hirano, M., Stohmeier, C., McCurley, N., et al. (2011). A thymus candidate in lampreys. *Nature*, *470*, 90–94.

Cerny, R., Cattell, M., Sauka-Spengler, T., Bronner-Fraser, M., Yu, F., & Medeiros, D. M. (2010). Evidence for the prepattern/cooption model of vertebrate jaw evolution. *Proceedings of the National Academy of Sciences of the United States of America*, *107*, 17262–17267.

Cope, E. D. (1889). Synopsis of the families of Vertebrata. *The American Naturalist*, *1889*, 849–887.

Couly, G. F., Coltey, P. M., & Le Douarin, N. M. (1993). The triple origin of skull in higher vertebrates: A study in quail-chick chimeras. *Development*, *117*, 409–429.

De Beer, G. R. (1937). *The development of the vertebrate skull*. Press, New York: Oxford Univ.

Dean, B. (1899). On the embryology of Bdellostoma stouti. A genera account of myxinoid development from the egg and segmentation to hatching. In *Festschrift zum 70ten Geburtstag Carl von Kupffer* (pp. 220–276). Jena: Gustav Fischer.

Dehal, P., & Boore, J. L. (2005). Two rounds of whole genome duplication in the ancestral vertebrate. *PLoS Biology*, *3*, e314.

Depew, M. J., Lufkin, T., & Rubenstein, J. L. (2002). Specification of jaw subdivisions by *Dlx* genes. *Science*, *298*, 371–373.

Dupret, V., Sanchez, S., Goujet, D., Tafforeau, P., & Ahlberg, P. E. (2014). A primitive placoderm sheds light on the origin of the jawed vertebrate face. *Nature*, *507*, 500–503.

Fujimoto, S., Oisi, T., Kuraku, S., Ota, K. G., & Kuratani, S. (2013). Non-parsimonious evolution of the dlx genes in the hagfish. *BMC Evolutionary Biology, 13*, 15.
Gai, Z., Donoghue, P. C. J., Zhu, M., Janvier, P., & Stampanoni, M. (2011). Fossil jawless fish from China foreshadows early jawed vertebrate anatomy. *Nature, 476*, 324–327.
Gorbman, A. (1983). Early development of the hagfish pituitary gland: Evidence for the endodermal origin of the adenohypophysis. *American Zoologist, 23*, 639–654.
Green, S. A., & Bronner, M. E. (2014). The lamprey: A jawless vertebrate model system for examining origin of the neural crest and other vertebrate traits. *Differentiation, 87*, 44–51.
Higuchi, S., Sugahara, F., Oisi, Y., Pascual Anaya, J., Takagi, W., & Kuratani, S. (2019). Inner ear development in cyclostome and the evolution of vertebrate semicircular canals. *Nature, 565*, 347–350.
Irie, N., Kuratani, S., & Satoh, N. (2018). The phylum Vertebrata: A way of zoological recognition. *Zoological Letters, 4*, 32.
Jandzik, D., Hawkins, M. B., Cattell, M. V., Cerny, R., Square, T. A., & Medeiros, D. M. (2014). Roles for FGF in lamprey pharyngeal pouch formation and skeletogenesis highlight ancestral functions in the vertebrate head. *Development, 141*, 629–638.
Janvier, P. (1996). *Early vertebrates*. New York: Oxford: Clarendon Press.
Janvier, P. (2008). Early jawless vertebrates and cyclostome origins. *Zoological Science, 25*, 1045–1056.
Kearn, G. C. (2004). *Leeches, lice and lampreys* (pp. 318–338). Dordrecht: Springer.
Köntges, G., & Matsuoka, T. (2002). Jaws of the fates. *Science, 298*, 371–373.
Kuraku, S., & Kuratani, S. (2006). Timescale for cyclostome evolution inferred with a phylogenetic diagnosis of hagfish and lamprey cDNAs. *Zoological Science, 23*, 1053–1064.
Kuraku, S., Takio, Y., Sugahara, F., Takechi, M., & Kuratani, S. (2010). Evolution of oropharyngeal patterning mechanisms involving *Dlx* and *endothelins* in vertebrates. *Developmental Biology, 341*, 315–323.
Kuratani, S. (2012). Evolution of the vertebrate jaw from developmental perspectives. *Evolution & Development, 14*, 76–92.
Kuratani, S. (2020). Neural crest and craniofacial evolution of early vertebrates. In B. F. Eams, D. M. Medeiros, & I. Adameyko (Eds.), *Evolving Neural Crest Cells* (pp. 219–242). Boca Raton, London, New York: CRC Press/Taylor & Francis Group.
Kuratani, S., & Ahlberg, P. E. (2018). Evolution of the vertebrate neurocranium: Problems of the premandibular domain and trabecula. *Zoological Letters, 4*, 1.
Kuratani, S., Nobusada, Y., Horigome, N., & Shigetani, Y. (2001). Embryology of the lamprey and evolution of the vertebrate jaw: Insights from molecular and developmental perspectives. *Philosophical Transactions of the Royal Society, 356*, 15–32.
Kuratani, S., Oisi, Y., & Ota, K. G. (2016). Evolution of the vertebrate cranium: Viewed from the hagfish developmental studies. *Zoological Science, 33*, 229–238.
Kuratani, S., & Ota, G. K. (2008). The primitive versus derived traits in the developmental program of the vertebrate head: Views from cyclostome developmental studies. *Journal of Experimental Zoology. Part B, Molecular and Developmental Evolution, 310B*, 294–314.
Mallatt, J. (2008). Origin of the vertebrate jaw: Neoclassical ideas versus newer, development-based ideas. *Zoological Science, 25*, 990–998.
Noden, D. M., & Francis-West, P. (2006). The differentiation and morphogenesis of craniofacial muscles. *Developmental Dynamics, 235*, 1194–1218.
Oisi, Y., Fujimoto, S., Ota, K. G., & Kuratani, S. (2015). On the peculiar morphology and development of the hypoglossal, glossopharyngeal and vagus nerves and hypobranchial muscles in the hagfish. *Zoological Letters, 1*, 6.
Oisi, Y., Ota, K. G., Fujimoto, S., & Kuratani, S. (2013a). Craniofacial development of hagfishes and the evolution of vertebrates. *Nature, 493*, 175–180.

Oisi, Y., Ota, K. G., Fujimoto, S., & Kuratani, S. (2013b). Development of the chondrocranium in hagfishes, with special reference to the early evolution of vertebrates. *Zoological Science*, *30*, 944–961.

Ota, K. G., Fujimoto, S., Oisi, Y., & Kuratani, S. (2011). Identification of vertebra-like elements and their possible differentiation from sclerotomes in the hagfish. *Nature Communications*, *2*, 373.

Ota, K. G., Kuraku, S., & Kuratani, S. (2007). Hagfish embryology with reference to the evolution of the neural crest. *Nature*, *446*, 672–675.

Parker, H. J., De Kumar, B., Green, S. A., Prummel, K. D., Hess, C., Kaufman, C. K., et al. (2019). A Hox-TALE regulatory circuit for neural crest patterning is conserved across vertebrates. *Nature Communications*, *10*, 1–15.

Pascual-Anaya, J., Sato, I., Paps, J., Yandong, R., Sugahara, F., Higuchi, S., et al. (2018). Hagfish and lamprey Hox genes reveal conservation of temporal colinearity in vertebrates. *Nature Ecology & Evolution*, *2*, 859–866.

Schlosser, G. (2005). Evolutionary origins of vertebrate placodes: Insights from developmental studies and from comparisons with other deuterostomes. *Journal of Experimental Zoology. Part B, Molecular and Developmental Evolution*, *304*, 347–399.

Shigetani, Y., Sugahara, F., & Kuratani, S. (2005). A new evolutionary scenario of the vertebrate jaw. *BioEssays*, *27*, 331–333.

Sugahara, F., Pascual-Anaya, J., Oisi, Y., Kuraku, S., Aota, S., Adachi, N., et al. (2016). Evidence from cyclostomes for complex regionalization of the ancestral vertebrate brain. *Nature*, *531*, 97–100.

Tahara, Y. (1988). Normal stages of development in the lamprey, *Lampetra reissneri* (Dybowski). *Zoological Science*, *5*, 109–118.

Takio, Y., Pasqualetti, M., Kuraku, S., Hirano, S., Rijli, F. M., & Kuratani, S. (2004). Lamprey *Hox* genes and the evolution of jaws. *Nature*, *429*, 622.

Wada, N., Nohno, T., & Kuratani, S. (2011). Dual origins of the prechordal cranium in the chicken embryo. *Developmental Biology*, *356*, 529–540.

Yalden, D. W. (1985). Feeding mechanisms as evidence for cyclostome monophyly. *Zoological Journal of the Linnean Society*, *84*, 291–300.

Yao, T., Ohtani, K., Kuratani, S., & Wada, H. (2011). Development of lamprey mucocartilage and its dorsal–ventral patterning by endothelin signaling, with insight into vertebrate jaw evolution. *Journal of Experimental Zoology. Part B, Molecular and Developmental Evolution*, *316B*, 339–346.

CHAPTER EIGHT

Heading for higher ground: Developmental origins and evolutionary diversification of the amniote face

Zachary S. Morris[a,†] and Arhat Abzhanov[b,c,*]
[a]Museum of Comparative Zoology and Department of Organismic and Evolutionary Biology, Harvard University, Cambridge, MA, United States
[b]Department of Life Sciences, Imperial College London, Berkshire, United Kingdom
[c]Natural History Museum, London, United Kingdom
*Corresponding author: e-mail address: a.abzhanov@imperial.ac.uk

Contents

Abstract

Amniotes, a clade of terrestrial vertebrates, which includes all of the descendants of the last common ancestor of the reptiles (including dinosaurs and birds) and mammals, is one of the most successful group of animals on our planet. In addition to having an egg equipped with an amnion, an adaptation to lay eggs on land, amniotes possess a number of other major morphological characteristics. Chief among them is the amniote skull, which can be classified into several major types distinguished by the presence and number of temporal fenestrae (windows) in the posterior part. Amniotes evolved from ancestors who possessed a skull composed of a complex mosaic of small bones separated by sutures. Changes in skull composition underlie much of the large-scale evolution of amniotes with many lineages showing a trend in reduction of cranial

† Current address: Department of Earth & Planetary Sciences, Yale University, New Haven, CT, United States.

Current Topics in Developmental Biology, Volume 141
ISSN 0070-2153
https://doi.org/10.1016/bs.ctdb.2020.12.003

elements known as the "Williston's Law." The skull of amniotes is also arranged into a set of modules of closely co-evolving bones as revealed by modularity and integration tests. One of the most consistently recovered and at the same time most versatile modules is the "face," anatomically defined as the anterior portion of the head. The faces of amniotes display extraordinary amount of variation, with many adaptive radiations showing parallel tendencies in facial scaling, e.g., changes in length or width. This review explores the natural history of the amniote face and discusses how a better understanding of its anatomy and developmental biology helps to explain the outstanding scale of adaptive facial diversity. We propose a model for facial evolution in the amniotes, based on the differential rate of cranial neural crest cell proliferation and the timing of their skeletal differentiation.

1. Introduction

The invasion of the terrestrial realm by vertebrates occurred sometime in the Carboniferous (Benton, 2015; Clack, 2012) and involved the evolution of many critical anatomical and physiological features (Martin & Nagy, 1997; Martin & Sumida, 1997). However, this transition was arguably incomplete until the origin of fully terrestrial reproduction and the eponymous amniotic egg in Amniota. This clade, which includes all of the descendants of the common ancestor of mammals, lepidosaurs, turtles, crocodylians, and birds, has evolved an impressive diversity of lifestyles (Myhrvold et al., 2015), diets (Reilly, McBrayer, & White, 2001), social systems (Doody, Burghardt, & Dinets, 2012), reproductive modes (Blackburn, 2000; Pokorná & Kratochvíl, 2016), and anatomical features (Brugmann, Kim, & Helms, 2006; Young et al., 2014). Although there are many synapomorphies which distinguish extant amniote subclades, the skull is arguably one of the most divergent aspects of their anatomy, as exemplified by comparison of the beaked avian face, the flattened crocodylian snout, and the highly kinetic snake skull. This disparity is even more impressive when you consider the non-avian dinosaurs, various marine reptile lineages, and remarkably diverse stem lineages for all extant clades in the amniote fossil record. However, despite these major differences in both general form and skeletal conformation, there are also parallel trends in craniofacial evolution across amniote lineages—including similarities in the evolution of facial elongation (Fig. 1)—which suggest potential common evolutionary and/or underlying developmental mechanisms may exist. For this reason, an exploration of both divergent and parallel trends in the evolution and development of the amniote face and skull may hold important insight into the mechanisms underlying macroevolutionary processes.

Fig. 1 Similar patterns of variation in facial morphology are observed within different amniote clades despite the fundamental differences in the skull across Amniota. Image credits: Eric Kilby—Creative Commons Attribution-Share Alike 2.0 Generic (*Myrmecophaga tridactyla*); NPS Photo/Ken Conger—Creative Commons Attribution 2.0 Generic (*Canis lupus*) Alex Borisenko–Creative Commons Attribution-Share Alike 3.0 Unported (*Artibeus jamaicensis*); Vaughn Bodden (*Anolis maynardi*); Bernard Dupont—Creative Commons Attribution-Share Alike 2.0 Generic (*Tupinambis teguixin*); Diego Delso—Creative Commons Attribution-Share Alike 4.0 International (*Amblyrhynchus cristatus*); Drferry—Creative Commons Attribution-Share Alike 3.0 Unported (*Ensifera ensifera*); Laitche—Creative Commons Attribution-Share Alike 4.0 International (*Corvus macrorhynchos*); Gregory "Slobirdr" Smith—Creative Commons Attribution-Share Alike 2.0 (*Clangula hyemalis*); Przemek Pietrak—Creative Commons Attribution-Share Alike 3.0 Unported (*Gavialis gangeticus*); Gareth Rasberry—Creative Commons Attribution-Share Alike 4.0 International (*Alligator mississippiensis*); Morten Ross (*Caiman latirostris*).

2. Major trends in amniote evolution

Since Haeckel first conceptualized this group in 1866, there has been little disagreement about the composition of Amniota, with only a few extinct groups along the stem lineage (i.e., Seymouriamorpha, Diadectomorpha) or individual taxa known from incomplete specimens being in dispute (Klembara et al., 2020; Laurin & Reisz, 2020; Vallin & Laurin, 2004). There has, however, been considerable disagreement and multiple phylogenetic hypotheses for the relationships among amniote subclades (Gardiner, 1982; Gauthier, Kluge, & Rowe, 1988; Laurin & Reisz, 2020), particularly with regards to the position of turtles (e.g., Field et al., 2014; Li, Wu, Rieppel, & Wang, 2008; Lu, Yang, Dai, & Fu, 2013; Lyson et al., 2012) and several completely extinct lineages (Evans & Jones, 2010; Neenan, Klein, & Scheyer, 2013; Rieppel, 1993a). The earliest fossils clearly diagnosable to Amniota are from the Lower Pennsylvanian Joggins Formation of Nova Scotia (Reisz & Modesto, 1996). These deposits, which were previously considered to be as young as 316–313 Ma (Reisz & Müller, 2004) or 310.7 Ma $\pm$8.5 Myr (Van Tuinen & Hadly, 2004), are now understood to fall between 319–318 Ma (Benton et al., 2015). Importantly, these extinct species represent both stem mammals (Synapsida) and stem reptiles (Sauropsida), clearly demonstrating that amniotes had diverged by the Pennsylvanian. Recent phylogenetic analyses suggest that the split could have happened as early as 331 Ma (Ford & Benson, 2020). Tetrapods likely first invaded terrestrial environments in the Early to Middle Mississippian (360–340 Ma; Clack, 2012), meaning that the ancestral amniote appeared only 20–40 million years after the first vertebrates appeared on land. Extant amniotes possess many adaptations for terrestriality, including integumentary (Frolich, 1997), physiological (Martin & Nagy, 1997), and reproductive features (Packard & Seymour, 1997), but it is difficult to say precisely when these features first evolved or whether stem amniotes also possessed any of these modifications due to their lack of preservation (Martin & Sumida, 1997). The modifications to the skeleton, which facilitated improved terrestrial locomotion and feeding are, however, more well documented by stem and crown amniote fossils.

Ancestrally, tetrapods possessed dorsoventrally flat skulls composed of many small bones, with muscles which attached from the back of the skull onto the jaws for the limited force necessary to draw prey into the mouth (Clack, 2012; Lauder & Gillis, 1997). However, the transition to land meant that prehension of prey by the jaws (i.e., grabbing and biting) was much more important for feeding and resulted in many changes in musculoskeletal anatomy. One of the major anatomical innovations found in early amniotes were temporal fenestrae ("windows") in the posterior skull roof that accommodate larger and more strongly anchored (i.e., attached at lower angles) muscles for jaw closing (Holliday & Witmer, 2007; Tarsitano, Oelofsen,

Frey, & Riess, 2001). Historically, differences in the position and conformation (i.e., which bones bound an opening) of these temporal fenestra were used to categorize different forms into anapsid (no openings), synapsid (single upper opening), euryapsid (single lower opening), and diapsid (one upper and one lower opening) groups (Fig. 2). While the advent and application of phylogenetic methods to understanding evolutionary relationships supported some of these groupings, even early analyses (Benton, 1985; Gauthier et al., 1988) demonstrated that the anapsid skull condition was ancestral for amniotes and tetrapods (i.e., plesiomorphic) and that the euryapsid skull was modified from an ancestrally diapsid skull condition by multiple lineages independently (i.e., convergently). Additionally, more recent phylogenetic analyses of early amniotes have revealed that the base of the amniote tree shows considerable variation in the conformation of these openings (Ford & Benson, 2020). The complexity of cranial evolution notwithstanding, it has long been recognized that extant mammals cluster with most synapsid fossils, forming a clade known as Synapsida, and that lepidosaurs, crocodylians, and birds are descended from a common diapsid ancestor, hence belong to Diapsida (Fig. 2). Turtles (small red triangles in Fig. 2) have been traditionally aligned with a group of anapsid reptiles known as parareptiles, due to the lack of temporal fenestrae in modern species (Gauthier et al., 1988). However, more recent phylogenetic analyses utilizing different lines of genetic evidence (e.g., whole genome, miRNA) suggest that turtles are nested within Diapsida and are likely more closely related to Archosauria (crocodylians & birds) than lepidosaurs (Chiari, Cahais, Galtier, & Delsuc, 2012; Field et al., 2014). Re-evaluation of the anatomy of *Eunotosaurus africanus*, an extinct species often recovered as a stem turtle, suggests temporal fenestrae form early in ontogeny but are subsequently covered over (Bever, Lyson, Field, & Bhullar, 2015), potentially providing consilience between conflicting evidence from morphology and genetics. Importantly, temporal fenestrae evolved convergently across early amniote lineages and the organization of the temporal region was highly variable, potentially reflecting important ecological diversification among early amniotes (Hotton, Olson, & Beerbower, 1997) given the functional importance of these structures for the attachment of jaw musculature (Holliday & Witmer, 2007; Tarsitano et al., 2001) and coincident increase in dental diversity (Reisz, 2006). A proposed genetic mechanism suggests that expansion or reduction of skeletogenic mesenchyme in the temporal region during embryonic development is responsible for variation in the presence, size, and shape of temporal fenestrae (Tokita, Chaeychomsri, & Siruntawineti, 2013). While recent conflict among phylogenetic analyses (Ford & Benson, 2020; Klembara et al., 2020) suggests

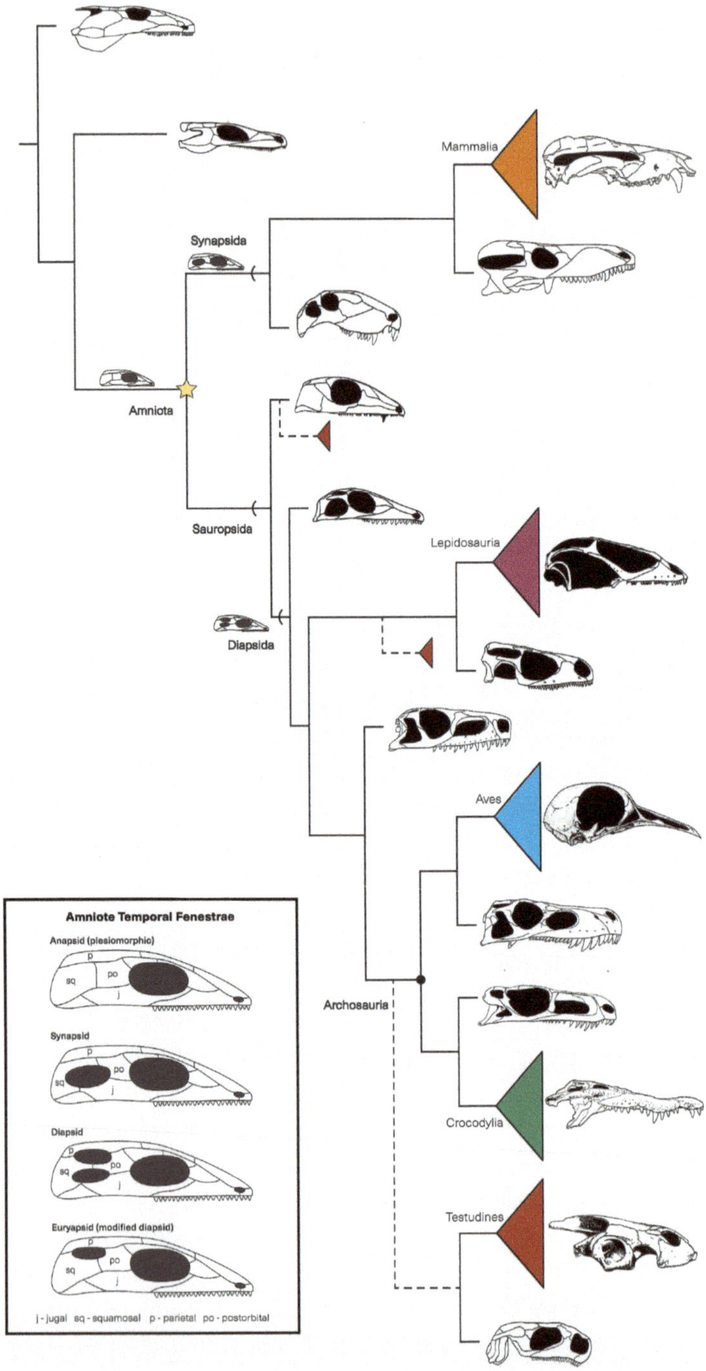

Fig. 2 The major relationships of Amniota and trends in the evolution of skull form, with alternative phylogenetic positions of turtles depicted with small red triangles. Inset shows

that more focused study of the early radiation of amniotes is needed, the relationships among extant clades are now fairly stable and suggest that these lineages have been independently evolving for over 250 million years (Nesbitt, 2011).

The divergence and subsequent diversifications of amniotes has given rise to an impressive variety of dietary ecologies and feeding modalities, but also rather unique forms, which often make direct anatomical and functional comparisons less clear. As mentioned previously, the ancestral tetrapod skull was built from many, small, irregular bones (Benton, 2015), with some forms possessing upwards of 52 distinct elements (not including the lower jaw). However, in most extant amniotes larger, fewer and, arguably, individually more morphologically and functionally complex bones comprise the skull (Gregory, 1935), with a domestic cat possessing only 29 distinct cranial bones. This trend toward fewer skull bones, known as "Willinston's law," is well documented in the evolution of mammals (Sidor, 2001) and birds (Sereno, 1997), but is detected to some degree in all amniote groups (Carroll, 1988; Estes, 1961; Esteve-Altava, Marugán-Lobón, Botella, & Rasskin-Gutman, 2013; Gaffney, 1979; Goodrich, 1958; Rieppel, 1993b)—as even crocodylians fuse the ancestrally paired frontals and parietals during their evolution. Similar trends have been described in ray-finned fishes (Ascarrunz, Sánchez-Villagra, Betancur-R, & Laurin, 2019), suggesting that this may reflect a broader trend across vertebrate evolution—although no studies (to our knowledge) have revealed such trends in sarcoptorygian or early tetrapod cranial evolution. Regardless of whether this reduction is indeed a common theme of vertebrate evolution more generally, within amniotes this pattern is quite obvious. Even a simple comparison of early synapsids and extant mammals, as exemplified by *Dimetrodon*, *Didelphis*, and *Canis* (Fig. 3), demonstrates that many bones around the eye and posterior jaw have been lost and the anterior regions of the head have been reshaped as well. The reduction in skull elements is most clear in birds and mammals, who also show parallel patterns of increased encephalization across their fossil records (Bhullar et al., 2012; Northcutt, 2011; Rowe, Macrini, & Luo, 2011), while other examples can be found

the differences in the conformation of the posterior skull among the major amniote temporal fenestrae groups. The anapsid condition, without openings, is found among early tetrapods and stem amniotes in addition to groups of crown amniotes such as parareptiles and turtles. The synapsid condition is found among mostly stem mammals, while all extant sauropsids (birds and reptiles) likely descend from a diapsid ancestor. The euryapsid condition is generally considered to have been derived from an ancestrally diapsid skull and is found among many convergently marine lineages of extinct reptiles.

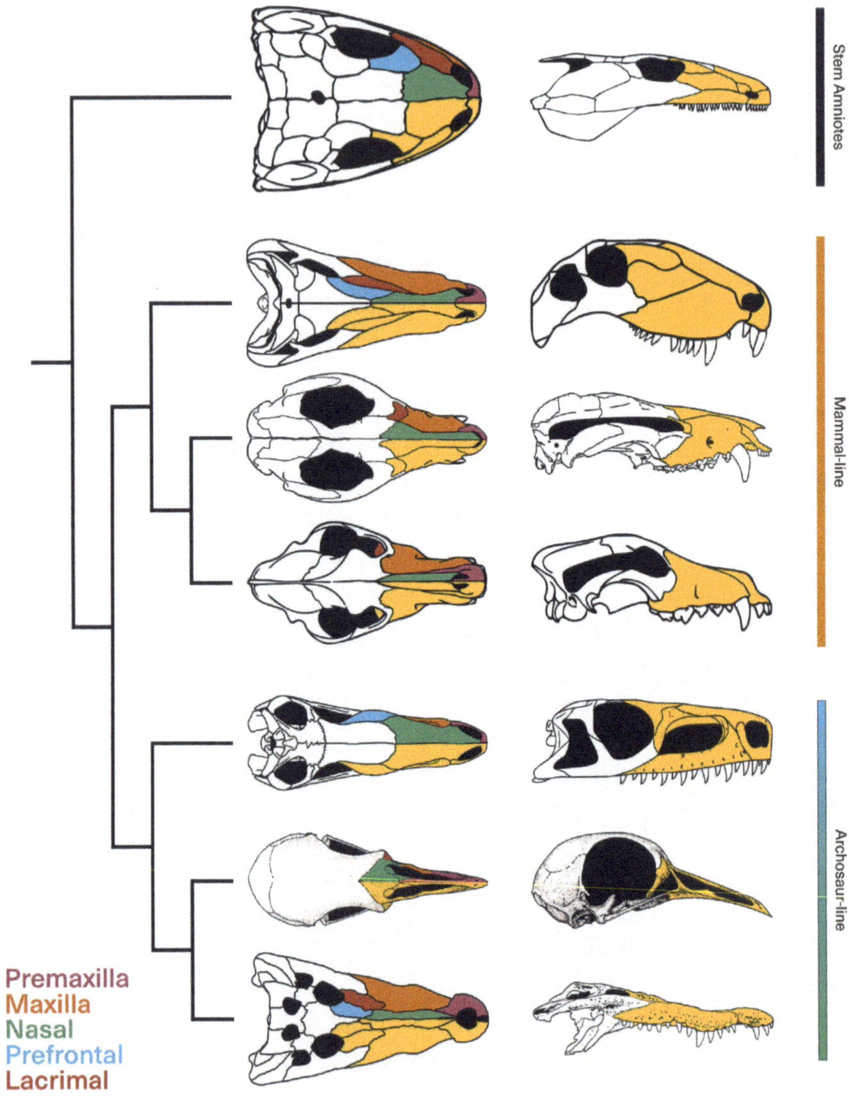

Fig. 3 Evolution of the amniote face and "Williston's Law." Dorsal and lateral reconstructions of the skull demonstrate that stem amniotes possessed many small and irregular skull bones, while multiple amniote lineages show progressive loss and fusion of cranial elements, particularly in the posterior skull (uncolored). By comparing the colored elements it is clear that both mammals and birds have reduced or fused elements together while crocodylians retain all elements of the face (yellow region). Note that in the bird beak all elements fuse and lack sutures, marking an extreme condition. Skull reconstructions: *Seymouria baylorensis* (Klembara et al., 2005), *Dimetrodon limbatus* (Romer & Price, 1940), *Didelphis virginiana* (Jollie, 1962), *Canis lupus* (original), *Euparkeria capensis* (Sookias et al., 2020), *Columba livia* (Procter & Lynch, 1993), and *Crocodylus porosus* (original).

in lineages which have evolved a burrowing lifestyle or undergone other major ecological shifts (Da Silva et al., 2018; Stayton, 2005). This trend in the "simplification" of overall cranial architecture has been shown to be associated with an increase in cranial asymmetry and individual bone complexity (Esteve-Altava et al., 2013; Lee, Esteve-Altava, & Abzhanov, 2020), potentially because it is more advantageous to combine several roles in fewer and larger multifunctional bones. In several cases the "loss" of certain cranial elements (e.g., the tabulars in mammals or pre-frontal in birds) have been demonstrated to occur via fusion with other elements during embryonic development (Koyabu, Maier, & Sánchez-Villagra, 2012; Smith-Paredes et al., 2018). It has been suggested that the fusion of smaller elements into larger, more architecturally complex bones may help with the distribution of greater mechanical forces, thereby increasing the structural integrity of the skull (Sidor, 2001). In birds and mammals the rapid growth of the brain also affects the rate and pattern of skull development, potentially forcing greater contact and spatial overlap of ossification centers resulting in greater fusion (Fabbri et al., 2017). In all cases, these fusions appear to reflect a tighter physical interaction and integration among previously more independent centers of ossification (Lee et al., 2020). This strongly implies that modifications in amniote skull composition, such as the reduction in number of skeletal elements and fusion of individual skull bones observed across amniote lineages are linked to changes in the modularity of the skull.

3. The face as a module of the amniote head

The connections, both developmental and physical, among cranial bones mean that the shape of one part of the skull is never truly independent of the rest of the skull. However, the distinct functions of different cranial regions such as the braincase and the jaws often result in divergent selective pressures. While the methods applied do vary, the preponderance of studies of cranial integration strongly suggest that several semi-independent regions each including multiple bones, known as modules, comprise the amniote skull and that changes in the degree of integration among these modules have been critical in the evolution of amniote lineages (e.g., Esteve-Altava et al., 2013; Felice et al., 2020; Lee et al., 2020). The integration, or strength of covariation, among elements has been hypothesized to have the potential to bias the evolution of skull form (Felice, Randau, & Goswami, 2018; Renaud, Pantalacci, & Auffray, 2011; Schluter, 1996), making understanding the macroevolutionary patterns of integration crucial (Felice, Watanabe, et al., 2019). Recent studies using high-density three-dimensional landmarking strategies (e.g., Felice & Goswami, 2018; Watanabe et al., 2019) have revealed that patterns of integration can remain remarkably static across evolutionary time and major changes in skull form. Surprisingly, "lizards" (non-serpentine squamates) and snakes share a conserved pattern of cranial integration, suggesting that despite the major re-structuring of the snake skull, conserved developmental and structural units persist across all squamates (Watanabe et al., 2019). Similar modules and patterns of integration can also be observed among archosaurs, although lineage specific trends can be detected (Felice, Watanabe, et al., 2019; Lee et al., 2020). One of the most consistently recovered modules, is the "face" or rostrum (Esteve-Altava et al., 2013; Felice & Goswami, 2018; Goswami, Milne, & Wroe, 2011; Watanabe et al., 2019). While the common usage of face somewhat muddies its precise usage, anatomically the face is defined as the anterior portion of the head, which contains the skeletal and soft tissues of the eyes, nose, and upper mouth. The facial skeleton attaches to the more posterior braincase and includes bones (both membraneous and endochondral) and internal cartilaginous structures (e.g., nasal septum) that provide structural support for the rest of this anatomical subdivision. Although the face is a developmentally composite structure (Tapadia, Cordero, & Helms, 2005), as shown via elegant cell lineage tracing experiments from the 20th century (Le Douarin, 1982), the facial skeleton itself is derived exclusively from a migratory population of neuroectodermal cells, called cranial neural crest (Langille & Hall, 1993; Le Douarin, 1982). The consistent recognition of the facial module across amniotes is quite remarkable when considering the key differences in the conformation and shape of the face across amniote lineages, including whether elements within the facial module itself fuse (Lee et al., 2020).

There have been many studies of cranial disparity within amniote subclades, including lepidosaurs (Da Silva et al., 2018; Watanabe et al., 2019), turtles (Claude, Pritchard, Tong, Paradis, & Auffray, 2004; Ferreira, Rincón, Solórzano, & Langer, 2015; Foth, Rabi, & Joyce, 2016), crocodylians (Morris, Vliet, Abzhanov, & Pierce, 2019; Pierce, Angielczyk, & Rayfield, 2008; Piras et al., 2010; Sadleir, 2009), birds (Bright, Marugán-Lobón, Cobb, & Rayfield, 2016; Felice & Goswami, 2018; Felice, Tobias, et al., 2019), mammals (Camacho et al., 2019; Goswami et al., 2011; Lu et al., 2013; Wroe & Milne, 2007), and even those sampling extinct lineages (e.g., Felice et al., 2020; Godoy et al., 2018; Pierce, Angielczyk, & Rayfield, 2009). Often these studies have focused on the relationship between changes in skull shape and ecological diversification (e.g., Felice, Tobias, et al., 2019; Foth et al., 2016), particularly in well-studied adaptive radiations (Campàs, Mallarino, Herrel, Abzhanov, & Brenner, 2010; Hedrick et al., 2020; Sanger et al., 2013; Tokita, Yano, James, & Abzhanov, 2016). Although there is considerable variation in the disparity of cranial regions across the diversity of amniotes, the facial skeleton is reliably documented as being distinctly variable. For instance, variation in beak shape has long been recognized to relate to ecological and dietary differentiation in birds (Lack, 1947), including strong associations with known recent adaptive radiations, such as Darwin's finches (Schluter & Grant, 1984) and Hawaiian honeycreepers (Tokita et al., 2016), and more weakly across larger clades (Bright et al., 2016; Felice, Tobias, et al., 2019). Quantification of whole skull shape across the diversity of extant birds shows disparity is achieved by correlated differences in facial proportions (length, width, and depth) and the facial module has evolved much faster than the rest of the avian skull (Felice, Tobias, et al., 2019). Interestingly, among dinosaurs this pattern appears to be a uniquely avian feature, as non-avian dinosaurs show much higher rates of cranial evolution across the skull and in the posterior skull in particular (Felice et al., 2020). A pattern similar to that detected in birds has long been recognized in crocodylians, with the vast majority of their cranial variation attributable to differences in the length and breadth of the snout (Brochu, 2001; McHenry, Clausen, Daniel, Meers, & Pendharkar, 2006; Pierce et al., 2008), which, in turn, may be related to differences in diet and/or hydrodynamics. Curiously, this variation is not observed in embryonic crocodylians, demonstrating that their face changes across development more dramatically than any other region of the skull (Morris et al., 2019). Several studies of squamate cranial anatomy have recovered a major dichotomy in form and structure between "lizards" and snakes, but differences in snout proportions are an important common aspect of adaptive disparity within both groups (Da Silva et al., 2018; Stayton, 2005; Watanabe et al., 2019). Importantly, similar face shapes are observed in fossorial snakes and "lizards" (e.g., amphisbaenians, annielids, and dibamids), demonstrating that these axes of facial proportions share important ecological correlations across squamates (Da Silva et al., 2018; Watanabe et al., 2019). Habitat and, to a lesser extent, diet have a significant effect on skull shape in turtles (Claude et al., 2004; Foth et al., 2016), but this is predominantly related to modifications to the palate (Ferreira et al., 2015) and post-temporal emargination for jaw muscles (Foth et al., 2016), with much less variation in the distinct facial skeleton. Clearly, the most variable aspects of cranial form for the majority of sauropsid lineages, with the notable exception of turtles, are the length and breadth of the face. Remarkably, across mammalian carnivores nearly identical axes of skull shape variation have been recovered (Goswami et al., 2011; Wroe & Milne, 2007) and the degree of facial elongation is the most variable major trait in several subgroups including bats (Camacho et al., 2019; Hedrick et al., 2020), cetaceans (Churchill, Geisler, Beatty, & Goswami, 2018; McCurry et al., 2017), and primates (Marroig & Cheverud, 2004). The patterns within mammalian subgroups do show some exceptions or deviations from the general trend, as clades with specialized feeding behaviors and mechanics, like rodents and felids, are much less variable in proportionate facial length than other mammalian clades (Álvarez, Perez, & Verzi, 2015; Goswami et al., 2011; Lu et al., 2013). However, there is also evidence that adult skull size is strongly correlated with facial length in mammals (Cardini & Polly, 2013), strongly suggesting that allometric or developmental factors may play a major role in the evolution of facial disparity. It is remarkable that differences between long/slender and short/wide faces are the most important components of cranial variation in nearly all extant amniote lineages (Fig. 1), despite the fact, for example, that the avian beak is primarily composed of fused premaxillae, which are only a small pair of elements in the tip of the face in most non-avian amniotes.

These studies taken together demonstrate that the facial module is a common feature of the adult amniote skull and that within amniote subclades this module has evolved along remarkably parallel trends in response to selective pressures on diet, habitat, and ecology, despite key differences in skull composition and other lineage-specific features. Indeed, correlated ecological and anatomical convergence in facial form appears to be one of the most common pattern observed within (Felice, Tobias, et al., 2019; Goswami et al., 2011; Pierce et al., 2008; Watanabe et al., 2019) and across amniote lineages, as exemplified by the evolution of long, slender snouts in whales, crocodylians and marine reptiles, such as ichthyosaurs (McCurry et al., 2017). The observation of these parallel trends within groups that possess dramatically distinct lifestyles and ecologies suggests that more fundamental principles are driving these trends rather than simply similar selective pressures. The morphological and functional qualities of adult amniote faces are established during ontogeny (embryonic development and later growth), suggesting that developmental mechanisms warrant exploration. To understand what developmental mechanisms may have been most important in creating observed evolutionary patterns, we need to perform detailed quantification of the embryonic origins and post-embryonic growth of the amniote face and skull. Recent comparative studies on the ontogeny of the skull in groups as divergent as dinosaurs, phyllostomid bats, and crocodylians have revealed that species with different facial proportions as adults generally share a similar face shape at the earliest stages of development (Bhullar et al., 2012; Camacho et al., 2019; Morris et al., 2019). Early birds, such as *Archaeopteryx*, and non-avian theropod dinosaurs have distinct faces as adults yet share the same part of the skull shape morphospace as hatchling birds (Bhullar et al., 2012). Adults of frugivorous, nectarivorous, carnivorous, and sanguivorous (blood-feeding) phyllostomid bats (Phyllostomidae) possess incredibly different face shapes, while their youngest embryos all cluster together near the reconstructed ancestral skull shape (Camacho et al., 2019). In crocodylians, all but the most extremely slender snouted species share a conserved embryonic skull shape (Morris et al., 2019). Differences in the species-specific rate of elongation and widening of the face generates the adult crocodylian diversity as they diverge from the early shared morphology during development. Reconstruction of the evolution of these developmental patterns in phyllostomid bats and crocodylians demonstrated that modifications to the timing and rate of craniofacial development are a key evolutionary mechanism for generating adult disparity in facial form. This phenomenon, known as *heterochrony*, allows relatively subtle alterations to the exact timing or rate of cranial *morphogenesis* to give rise to significant differences in adults (Alberch, Gould, Oster, & Wake, 1979; Gould, 1977; Reilly, Wiley, & Meinhardt, 1997). The impressive sexual dimorphism in facial length observed within and among some species of *Anolis* "lizards" have also evolved via sex-specific heterochronic changes in facial elongation (Sanger et al., 2013). The evolution of the avian skull from theropod ancestors mentioned earlier has been described as an example of *paedomorphosis*, a type of heterochrony in which juvenile features of the ancestors are retained in the adults of the descendants (Bhullar et al., 2012). Importantly for this discussion, formation of the unique avian beak and its subsequent scaling modifications have been ascribed to further heterochronic modifications within the facial module itself (Abzhanov, Protas, Grant, Grant, & Tabin, 2004; Bhullar et al., 2015; Campàs et al., 2010; Fritz et al., 2014; Tokita et al., 2016). These morphologically detectable heterochronic events are themselves expected to result from regulatory changes in developmental genes expression, as has been reported for changes in estrogen signaling associated with Anole facial length dimorphism (Sanger et al., 2014) or the novel mid-face expression domain of *Fgf8* that results in the fusion of the premaxillae to form the avian beak (Bhullar et al., 2015). While it is obvious that selective pressures due to ecology and biomechanical function served as an important backdrop for evolution of facial disparity, these studies emphasize the significance of heterochrony and other developmental mechanisms in explaining parallel trends of facial diversification across amniote lineages. However, such variation does not break class-level boundaries as a long-faced mammal, like an anteater, is clearly still a mammal and the long-snouted Indian Gharial remains a member of Crocodylia (Fig. 1). Thus, to understand the genesis of the key features underlying the macro-, meso-, and microevolutionary trends, a deeper understanding the developmental origins of the face is required.

4. Developmental origins of the face

The face, as mentioned previously, is a developmentally composite structure, which requires merging of different progenitor cell populations prior to differentiation into the adult skeletal form. The adult amniote face is composed of numerous structures and organs, produced, in turn, by multiple tissues (Le Douarin & Dupin, 2012). These include skin and epidermal appendages (e.g., scales, hair, feathers, keratin beaks), muscles, special sense organs, and the major skeletal elements (bone, teeth, and cartilage). The facial skeleton is the structural support for the face (Fig. 4A) and, elaborate keratinous horns and fleshy noses notwithstanding, is primarily responsible for the ultimate shape of the face. The first developmental hints of the face can be seen during the formation of the primitive streak during neurulation (Elinson & Kao, 1993). The progenitors that give rise to the entire face, the descriptively named cranial neural crest (CNC) cells, first appear as a distinct population of cells forming at the crest of tissue between the folding neural plate and the adjacent ectoderm during the involution of the incipient brain and spinal cord (Fig. 4B). Ultimately, the CNC contributes to all tissues of the face except for the epidermis and epithelia of the special sense organs (derived from facial ectoderm and cranial placodes) and the striated muscles of the face & tongue (which originate from cranial and somitic mesoderm, respectively; Le Douarin & Dupin, 2012).

During pre-skeletal cranial morphogenesis (Fig. 4B), the CNC cells undergo an epithelial-to-mesenchymal transition (EMT), detach from the basal lamina of the neural tube, and migrate in streams to the head and pharyngeal arches. The CNC cells that give rise to the facial skeleton are those from the premandibular and mandibular streams originating primarily from

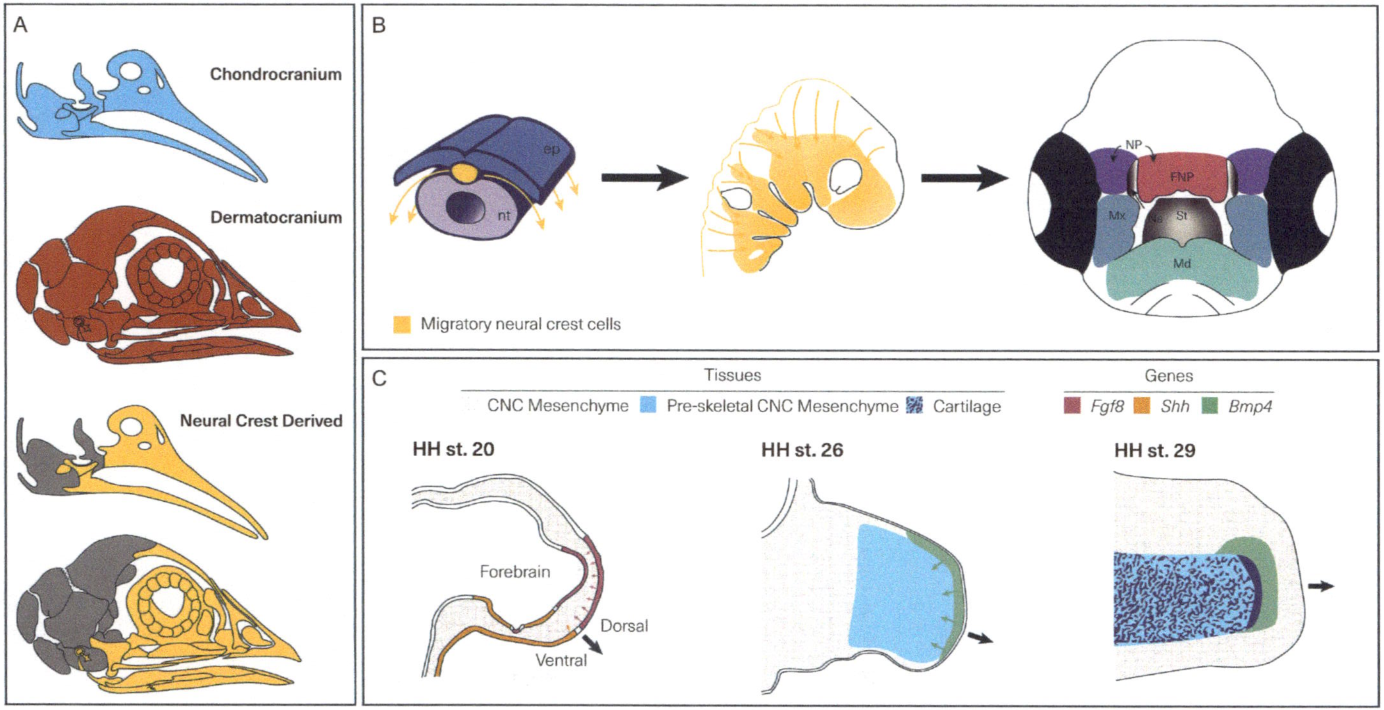

Fig. 4 Developmental origins of the facial skeleton. The anterior portions of both cartilaginous (blue) and ossified (red) portions of the skull are derived from cranial neural crest (CNC, yellow) cells (A), which migrate from near the neural tube during early development into the face and merge with cranial mesoderm in the early facial primordia (B). The early outgrowth of the face has been linked to signaling to undifferentiated mesenchyme in the embryonic mid-face (C), including apposing *Fgf8* and *Shh* signals in the FEZ and later *Bmp4* signaling. Abbreviations: Frontonasal prominence (FNP), mandibular arch (Ma), maxillary prominence (Mx), nares (Na), nasal prominence (NP), stomodeum or incipient mouth (St).

the developing mid-brain (mesencephalon), whose lack of Hox expression is critical for their ability to form cranial skeletal elements (Abzhanov, Tzahor, & Lassar, 2003; Creuzet, Couly, Vincent, & Douarin, 2002). After migration, these cell populations merge with cranial mesoderm and give rise to the first recognizable anatomical structures of the face (Fig. 4B)—the paired nasal prominences, maxillary processes, and mandibular arches. The paired nasal prominences fuse anterolaterally to the adjacent maxillary processes and to each other along the midline, forming the frontonasal prominence (FNP) and incipient nasal openings (Le Douarin & Dupin, 2012; Tapadia et al., 2005). Pre-skeletal cranial morphogenesis culminates at the "phylotypic stage" of development, which is noted for the remarkably similar anatomy of disparate amniote lineages (Abzhanov, 2013; Sander, 1983). Although it has been proposed that changes in facial proportions may be linked to differences in the formation of the facial primordia (Usui & Tokita, 2018), subsequent patterning, differentiation, and growth of the face has been shown to be the primary source of diversification in facial anatomy (Young et al., 2014).

The first skeletal structures of the face are cartilaginous, appearing initially as separate condensations in the ventral aspect of the head (Bellairs & Kamal, 1981). This includes the posterior parachordal base, which enwraps the anterior end of the notochord and the anterior paired trabecular cartilages (*trabeculae cranii*), which eventually connect and collectively form the cranial base (*basi cranii*; Langille & Hall, 1993). Nasal, optic, and otic capsule cartilages subsequently condense and merge with the cranial base as further shaping of the chondrocranium occurs. In the facial region, the trabecular cartilages merge anteriorly to form the nasal septum along the midline, separating the nasal capsules, and posteriorly merge with portions of the orbital cartilages to form the interorbital septum (Bellairs & Kamal, 1981). In many amniote lineages important components of the anterior chondocranium, such as the nasal septum, are maintained as cartilage (Bellairs & Kamal, 1981; Rieppel, 1993b). The anterior chondrocranium gives rise to the initial shape of the face, but at later developmental stages membraneous bones ossify in the dermis surrounding the cartilage and will provide the scaffolding for the adult face. The major bones of the face include elements external to the nasal capsule (e.g., maxilla, nasals) and overlying the orbit (e.g., frontals). However, some amniote lineages do form thin, sheet-like bones (e.g., septomaxilla) via endochondral ossification of midline facial cartilage (Bellairs & Kamal, 1981; Rieppel, 1993b). There is also considerable variation in the shape, size, and composition of ossifications in the orbital region, as birds and mammals tend to ossify the ethmoid region of the chondrocranium and the medial surface of the orbit while most reptiles retain a fully cartilaginous interorbital septum (Novacek, 1993; Rieppel, 1993b; Zusi, 1993). These differences may have important implications for mechanisms of growth at the boundary between the face and the braincase. But how are the complex suite of required cellular movements, differentiation, and growth coordinated and directed during development to give rise to facial diversity?

5. Molecular patterning of cranial morphogenesis

As briefly mentioned above, Hox patterning plays an important role in the ultimate fate of CNC descendants. In the pre-migratory neural fold, *Hox2a* is only expressed in posterior CNC populations while those of the prosencephalon, mesencephalon, and rhombomeres 1 and 2 lack *Hox2a* (Richman & Lee, 2003). This pattern of expression is maintained when CNC populations migrate, such that CNC in the face and first pharyngeal arch do not express *Hox2a* while more posterior pharyngeal arches do. In addition to neatly delimiting CNC populations that give rise to facial and jaw elements from those that do not, experimental gain-of-function *Hox2a* mis-expression in anterior CNC results in embryos lacking the facial skeleton (Creuzet et al., 2002) and loss of *Hox2a* expression in the CNC of second pharyngeal arch results in a duplication of upper and lower jaw bones (Tapadia et al., 2005). These results demonstrate that Hox patterning affects the ability of CNC populations to take on certain identities, but additional evidence suggests that local cues from endoderm and ectoderm encountered during and after migration are also critical for the final fate of CNC cells.

The fate of neural crest cells is, in part, defined by their migratory behaviors. Therefore, the molecular signals that influence the precise movements of CNC during migration have been of intense interest. While trunk neural crest cells migrate progressively as individuals, the CNC populations undergo EMT and migrate as multilayered groups of cells in close association with the overlying surface ectoderm (Theveneau & Mayor, 2011). The migration of CNC streams to specific regions in the head and face is governed by several processes, including EMT, cell-cell contact-mediated inhibition of protrusions and movement, chemotaxis, and restrictive signaling cues from adjacent cells/tissues (Tapadia et al., 2005; Theveneau & Mayor, 2011). The various signaling pathways, which transduce these cues into CNC populations (e.g., cadherins, non-canonical Wnt/PCP pathway, class 3-semaphorins, Sdf1, VEGF, FGF, metalloprotinases) demonstrate that the control of migration is incredibly complex but also that common cellular effectors (e.g., Neuropilin-1, RhoA, Rac1) integrate local cues to modulate dispersion

(Richman & Lee, 2003; Theveneau & Mayor, 2011). It has also been long recognized that CNC from rhombomeres 3 and 5 show limited migration potential and instead undergo apoptosis (Graham, Heyman, & Lumsden, 1993), likely playing an important role in segmenting and distinguishing the posterior pharyngeal arches. This pattern is determined by BMP signaling from the adjacent rhombomeres, with *Bmp4* and *Msx2* both becoming unregulated in rhombomeres 3 and 5 resulting in apoptosis of the neural crest cells (Graham, Koentges, & Lumsden, 1996). This, however, appears to be limited to the hindbrain as no evidence of substantial apoptotic differentiation of premandibular and mandibular CNC streams has been found.

Once the CNC cells have arrived in the face, signaling from both the endoderm and ectoderm are critical for correct development of the facial primordia and the facial skeleton. One of the most critical signaling centers is the frontonasal ectodermal zone (Fig. 4C; FEZ), which is a region in the facial ectoderm in which *Fgf8* and *Shh* expression domains are apposed dorsally and ventrally (Hu, Marcucio, & Helms, 2003). This signaling center itself appears to be induced by *Shh* signaling in the neurectoderm of the forebrain (Tapadia et al., 2005). Ectopic and inverted transplantation of the FEZ demonstrates that the apposed *Fgf8* and *Shh* signals cause dorsal and ventral differentiation respectively (Hu et al., 2003). Additionally, Shh is critical for maintenance of cellular proliferation (Jeong, Mao, Tenzen, Kottmann, & McMahon, 2004) and cartilage condensation and differentiation (Abzhanov & Tabin, 2004; Wada et al., 2005) during the outgrowth of the face. The pharyngeal endoderm is also a critical patterning center, as experimental rotation causes a shift in the direction of nasal capsule and facial bone formation (Couly, Creuzet, Bennaceur, Vincent, & Douarin, 2002). However, transplantation of the anterior neural plate (containing the CNC anterior to rhombomere 3) between quail and duck embryos, demonstrates that the shape of the facial skeleton, including positional and species-specific identity of its components, is at least partially programmed into the anterior CNC and not determined solely by later signals (Schneider & Helms, 2003). These experiments suggest that the migrating CNC itself may induce the specific organization of the FEZ, which subsequently coordinates growth (Richman & Lee, 2003).

6. Developmental origins of facial diversity

These studies of the early ontogenetic origins of the face have revealed many conserved molecular mechanisms directing the movement and later differentiation of the CNC-derived mesenchyme of the FNP. However, these developmental events are similar across amniotes, result in similarly shaped facial primordia (Abzhanov, 2013; Sander, 1983; Young et al., 2014), and do not explain the differences among and within amniote clades. However, the organization of the FEZ and subsequent signaling have been shown to differ between mammals and birds (Hu & Marcucio, 2009) and differences in FGF8 expression in the FEZ were critical for the assembly of the unique avian beak (Bhullar et al., 2015). Additionally, later ectodermal signaling (Fig. 4C) of *Bmp4* and *Calmodulin* to the derivatives of the CNC mesenchyme have been shown to direct facial length and depth in the faces of diverse clades of birds and fishes (Ahi, 2016; Mallarino et al., 2011). Although the cellular mechanisms involved in translating these differences in genetic patterning into facial disparity among and within amniote clades are poorly understood, recent studies suggest the organization and maintenance of cellular proliferation in the CNC mesenchyme of the FNP prior to differentiation may be important.

In passerine birds, it has been shown that a population of proliferating mesenchymal cells comprise an apical growth zone in the FNP which drives outgrowth of the embryonic face (Hamburger-Hamilton Stages 28-37; Fig. 5), with the prenasal cartilage differentiating and growing into the space posterior to this zone (Fritz et al., 2014). In zebra finch, this population of proliferative cells initially fills the mesenchyme of the FNP (HH28) and subsequently decreases in size and changes shape across later developmental stages (HH30-37). The progressive reduction of this growth zone creates the curvature of the beak (Fig. 6) by essentially setting the major beak shape parameters embryonically (Fritz et al., 2014). Once the apical growth zone disappears during zebra finch development, interstitial proliferation within the prenasal cartilage and ossification of the fused premaxilla appear to drive embryonic growth (Fig. 5; data from Fritz et al., 2014). Similar processes are presumed to maintain and facilitate post-hatching growth of the beak, although the genetic mechanisms maintaining the interstitial proliferation and ossification at the posterior end of the beak are poorly understood. The adaptive radiation of beak shape in Darwin's finches appears to be the main component of their evolutionary and ecological success as revealed with phylogenetic and morphometric studies (Reaney, Bouchenak-Khelladi, Tobias, & Abzhanov, 2020). Importantly, beak shapes of all species of Darwin's finches share common scaling parameters for the length and depth of the beak (Campàs et al., 2010). These scaling patterns appear to be determined by embryonic expression of key developmental genes, such

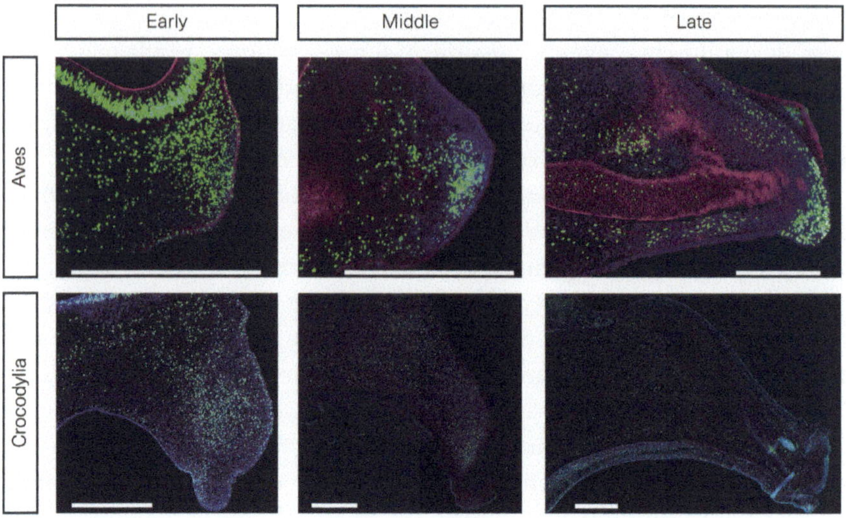

Fig. 5 Comparison of patterns of cell proliferation in representatives of Aves and Crocodylia. Mid-sagittal sections of the embryonic face of zebra finch (Aves) and American alligator (Crocodylia) across morphogenesis. The location and amount of cellular proliferation (green labeled cells) were assayed using EdU during incubation. The early morphogenesis phase (HH28 & FS14) involves an anterior growth zone with high cell proliferation in the pre-skeletal CNC mesenchyme. By middle stages (HH30 & FS17) cartilage differentiation begins to occur (shown in magenta in zebra finch), but an anterior growth zone is still present. By the later stages (HH35 & FS19) interstitial cell proliferation within the facial cartilage begins to take precedence, particularly in zebra finch where the apical growth zone soon disappears. Scale bar is 500 μm. Images for zebra finch are from Fritz, J. A., Brancale, J., Tokita, M., Burns, K. J., Hawkins, M. B., & Abzhanov, A. (2014). Shared developmental programme strongly constrains beak shape diversity in songbirds. *Nature Communications 5*, 3700.

as *Bmp4* and *Calmodulin*, in the anterior mesenchyme and subsequent prenasal cartilage of the beak (Abzhanov et al., 2004; Campàs et al., 2010; Mallarino et al., 2011). Although the diversification of Darwin's finches represents a combination of specific selective pressures and a unique genetic toolkit (Tokita et al., 2016), the genetic maintenance of this apical growth zone in the undifferentiated CNC mesenchyme of the face and alterations to the exact parameters of cell proliferation in the developing beak, are likely key developmental mechanisms for generating diversity in beak shape across all birds.

A similar growth zone in the anterior facial mesenchyme has been identified in the leaf-nosed bats (Phyllostomidae) during embryonic morphogenesis, although multiple growth zones can be identified and the pattern

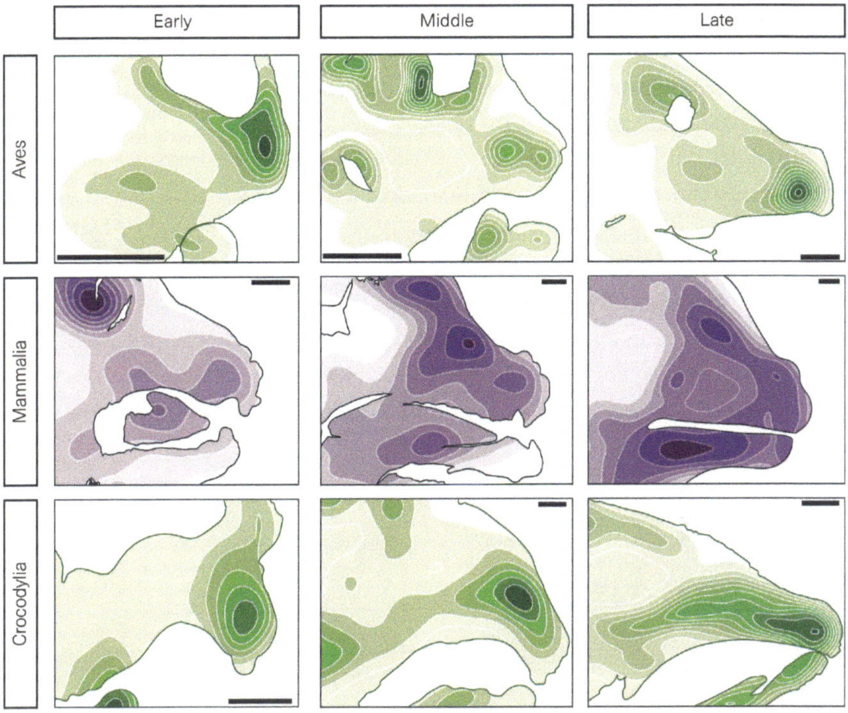

Fig. 6 The dynamics of the anterior facial growth zone. Cellular proliferation data for zebra finch (Aves), *Carollia perspicillata* (Mammalia), and American alligator (Crocodylia) are summarized in these heat maps. An anterior concentration of proliferation in the CNC mesenchyme is observed in all species during early morphogenesis. Although it decays in birds, this anterior growth zone persists in the nasal septum cartilage in mammals and crocodylians. Green colorscales based on EdU labeling and purple colorscale based on anti-PH3 antibody staining. Scale bar is 500 μm.

of local proliferation rate across the head is much more dynamic (Figs. 5 and 6; data from Camacho et al., 2020). The more complex pattern of proliferation may relate to the greater number of skeletal elements integrated in the bat face compared to the bird beak and the more dramatic changes in facial form observed during later skeletal growth in phyllostomid bats (Camacho et al., 2019). The evolution of skull shape in phyllostomids has been previously linked to heterochronic shifts during cranial development, specifically acceleration and prolonging of facial development (Camacho et al., 2019). Remarkably, the changes to the rate of cellular proliferation in this early growth zone and the timing of both mitotic activity and the onset of skeletal differentiation of the CNC-derived mesenchyme were identified as essential heterochronic mechanisms underlying the disparity of phyllostomid facial proportions (Camacho et al., 2020). A similar facial growth zone has been described in the mouse (Kaucka, Zikmund, Tesarova, & Gyllborg, 2017), suggesting that this zone of proliferative CNC mesenchyme may be important across all mammals. The initial perinatal growth of the snout in the mouse (Wealthall & Herring, 2006) also shows interstitial proliferation is distributed throughout the nasal septum (Fig. 7A) and is likely responsible for the majority of snout elongation during postnatal development. Intercalary proliferation within the cartilage and, potentially, aggregation and differentiation of nearby mesenchyme into existing cranial cartilages appear to be critical processes in the embryonic growth of the mouse face (Kaucka et al., 2017). Although more research is needed to validate these mechanisms and their connection to facial diversity, the

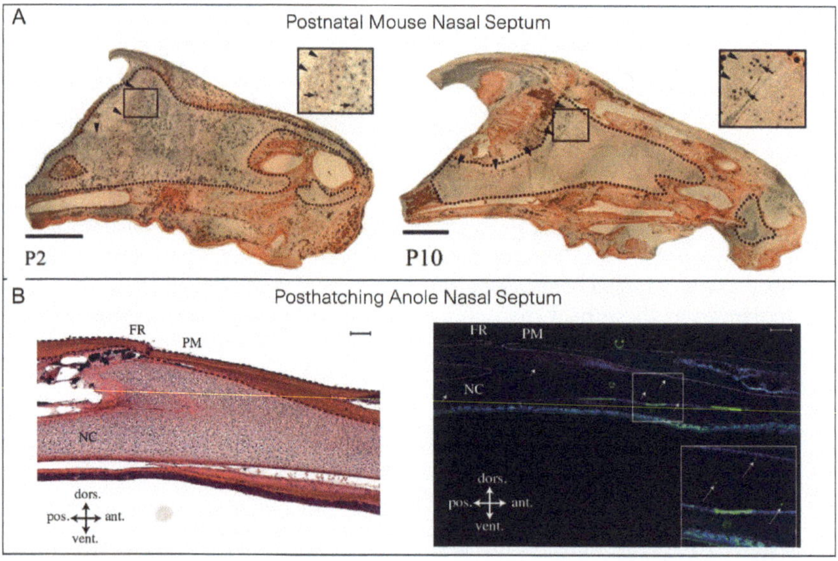

Fig. 7 Proliferation within the cartilaginous nasal septum is also critical to facial elongation post-embryonically in mouse (A) and the "lizard" *Anolis carolinensis* (B). At post-natal day two (P2) in mouse, interstitial cellular proliferation is ubiquitous (blue cells; Brdu labeling) and drives facial elongation until a later growth plate forms (P10). In juvenile *A. carolinensis* no growth plate exists in the face (histology on left), but interstitial cell proliferation (arrows to green EdU labeled cells on right) gives rise to differences in facial length among sexes. FR—frontal bone, PM—premaxillary bone, NC—nasal cartilage. Arrows indicate proliferating cells. *Modified from Wealthall, R. J., & Herring, S. W. (2006). Endochondral ossification of the mouse nasal septum.* The Anatomical Record: Advances in Integrative Anatomy and Evolutionary Biology 288A, *1163–1172 and Sanger, T. J., Seav, S. M., Tokita, M., Langerhans, R. B., Ross, L. M., & Losos, J. B. (2014). The oestrogen pathway underlies the evolution of exaggerated male cranial shapes in Anolis lizards.* Proceedings of the Royal Society B: Biological Sciences, 281, *20140329.*

combined evidence strongly suggests that the maintenance of cellular proliferation and changes in the timing of skeletal differentiation are important mechanism for the generation of the facial disparity across placental mammals.

Unfortunately, very little is known about growth zones or the cellular dynamics of CNC-derived mesenchyme during facial morphogenesis in reptiles. Preliminary data in *Alligator mississippiensis* suggest that a distinct proliferative growth zone exists during embryonic facial morphogenesis. More specifically, after facial prominence fusion (Ferguson stage 14; Ferguson, 1985), there is a recognizable population of mitotically active cells in the FNP mesenchyme in *Alligator* embryos which is very similar to the anterior facial growth zones (Fig. 5) described in zebra finch (Fritz et al., 2014) and phyllostomid bats (Camacho et al., 2020). At this stage of development, there are also no clear differences in relative facial length in different species of crocodylians, despite the fact that extreme slender snouted forms (e.g., *Tomistoma schlegelii*) feature distinctly longer and narrower snouts prior to hatching (Morris et al., 2019). During later embryonic morphogenesis and early skeletal differentiation, cellular proliferation is not quite so concentrated but it remains greatest in the anterior mid-face than other regions of the cranial base (Fig. 6). This is evident even with the transition of this growth zone from within undifferentiated mesenchyme to within a differentiated cartilage, the nasal septum, which would be expected to dramatically change the potential signaling and cell-cell communication mechanisms. Although the role of cellular proliferation has not been yet explored during post-hatching crocodylian development, we can draw from observations that low level interstitial proliferation within the nasal cartilage and ossification at sutures in the dermal skeleton jointly drive increases in snout length in the "lizard" *Anolis carolinensis* (Fig. 7B; Sanger et al., 2014). Although comparable periods of embryonic and post-hatching development in crocodylians and lizards have not been studied yet, these data match the trends seen in both birds and mammals and lead to a putative model for the developmental mechanisms which allow adaptive facial disparity to arise in the proper ecological settings.

7. Model for amniote facial developmental evolution

It is important to recognize that the data collected thus far sample very few species from distantly related groups at partially overlapping periods of facial development. However, when considered together they suggest a potential model for the generation of amniote facial disparity (Fig. 8).

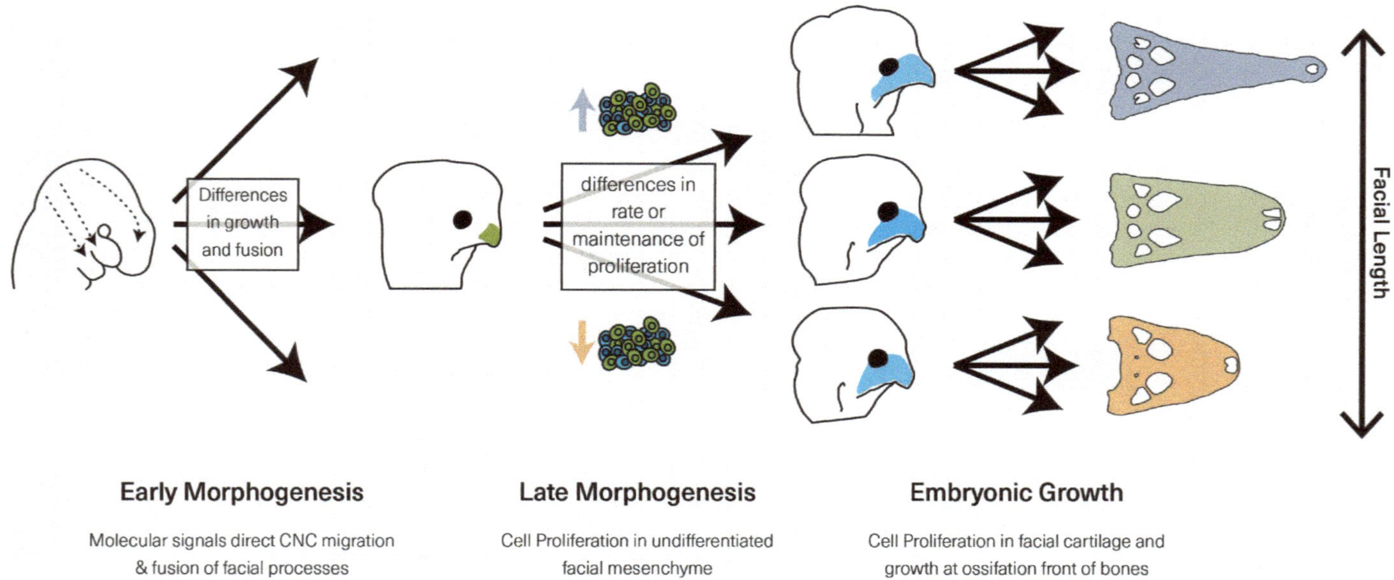

Fig. 8 Model of developmental mechanisms implicated in the origin of amniote facial disparity. Changes in early craniofacial morphogenesis, including patterning and direction of cranial neural crest (CNC) migration, and the growth and fusion of facial processes, are likely critical for explaining variation across Amniota. However, current data suggest that th changes in the timing and rate of cellular proliferation in the undifferentiated facial mesenchyme (of CNC origin) are important in the origin of different facial proportions within individual clades. How embryonic growth is directed is not well understood, but it is known to be critical to the formation of adult diversity in crocodylians and phyllostomid bats.

First, the early phase of craniofacial morphogenesis in which patterning and coordination of CNC migration occurs is likely critical for the differentiation among amniote lineages. The gene regulatory networks involved in patterning and directing CNC migration to the cranium clearly influence later events that result in the formation of developmentally distinct parts of the head, which can coalesce into distinct cranial modules like the face. The face as a distinct developmental module, identified by the migration of mostly midbrain-derived Hox-negative CNC cells, may be more broadly shared among tetrapods as it has been reported for amphibians (Pasqualetti, Ori, Nardi, & Rijli, 2000). These cells populate the FNP and other facial primordia providing a foundation for the evolution of this part of the head along a distinct evolutionary trajectory. By the time facial primordia have fused in turtles, crocodylians, and lizards, the genetic patterning of temporal fenestra has already happened (Tokita et al., 2013), osteogenic cell fates are clear, and the number and position of ossification centers is likely already established. Thus, developmental events which give rise to the major differences in skull composition among amniote lineages and are relevant to amniote macroevolutionary trends appear to occur during the earliest period of cranial morphogenesis. These patterning events coincide with formation of multiple independent signaling centers, such as the FEZ (Hu et al., 2003; Richman & Lee, 2003), which control the later stages of facial morphogenesis.

Second, an anterior growth zone is established in the undifferentiated CNC-derived mesenchyme of the FNP that directs initial outgrowth of the face. This defined growth zone is likely a conserved feature of amniote facial development and not an apomorphy of beak development in birds. The anterior growth zones are initially remarkably similar (Figs. 5 and 6) among reptiles (*Alligator*), birds (zebra finch), and mammals (phyllostomid bats). In birds this population of proliferative cells undergoes controlled and carefully timed reduction, generating the characteristic overall conical-shaped beak and controlling the scaling factors of the beak shape (Campàs et al., 2010; Fritz et al., 2014). In phyllostomid bats, increased rate of proliferation and earlier skeletal differentiation have been implicated in facial shortening while sustained maintenance of proliferation in pre-cartilage growth zones is hypothesized to result in longer faces (Camacho et al., 2020). Preliminary data suggests that crocodylians with different adult snout proportions potentially differ in the rate of proliferation in this growth zone prior to apparent differences in facial proportions, although more data are needed to understand the cellular dynamics across development. Taken as a whole, these observations suggest the anterior concentration of proliferation plays an important role in shaping the proportions of the face in all three amniote lineages, even if beak shape in passerine birds is more directly controlled by the dynamics of this single growth zone (Campàs et al., 2010; Fritz et al., 2014). Considering that heterochrony has played an important role in facial evolution in crocodylians and phyllostomids, we may expect that changes in the timing or rate of cell proliferation during beak development to contribute to the evolution of beaks with different scaling parameters. Indeed, it was already reported that the deep and broad bills of the ground finches within the Darwin's finch radiation can be explained by the earlier and stronger expression of *Bmp4* (Abzhanov et al., 2004) and *Shh* is known to have a key role in maintaining cellular proliferation and cartilage differentiation in the face (Abzhanov & Tabin, 2004; Jeong et al., 2004; Wada et al., 2005). From the available data it is clear that this anterior growth zone in the CNC derived mesenchyme is a common feature of amniote facial development. It appears that changes in the regulation of this proliferative growth zone and the skeletal differentiation of the CNC mesenchyme may be fundamental cellular processes for the evolution of facial disparity at the meso-evolutionary level. That common developmental mechanisms are responsible for the derivation of facial disparity within distantly related amniote lineages may explain why parallel trends are observed despite major differences in life history and ecology (Blackburn, 2000; Doody et al., 2012; Myhrvold et al., 2015; Pokorná & Kratochvíl, 2016; Reilly et al., 2001).

Third, after condensation and differentiation of cartilage occurs, proliferation is maintained throughout the mid-face cartilage but in a more distributed fashion (Fig. 6). Although the pre-chondrogenic growth zone is clearly important for deriving the initial differences in facial proportions (Fig. 8), modifications to this growth zone alone are not sufficient to explain changes in the shape of the face that emerge during later embryonic and post-natal allometric facial growth (Camacho et al., 2019; Morris et al., 2019; Sanger et al., 2013). This is probably due to the fundamental differences in the genetic control and cellular dynamics within differentiated skeletal tissues at these later stages. Once the mesenchymal tissues of the face have differentiated, subsequent changes in the shape, size, and conformation of skull elements occur via their independent growth. Cellular differentiation (both location and timing) continues to play an important role in the growth of skeletal tissues (Kaucka et al., 2017), but during this phase actively proliferating progenitor cell populations are traditionally thought to be more spatially and temporally restricted than in prior developmental stages (Cendekiawan, Wong, & Rabie, 2010). However, proliferation within the nasal and orbital septa is more spatially distributed and dynamic during the later stages of embryonic development in crocodylians and phyllostomid bats (Fig. 5). Distributed, interstitial proliferation within the mid-facial cartilage is similarly observed (Fig. 7) during later embryonic growth of zebra finch (Fritz et al., 2014) and mouse (Kaucka et al., 2017), but also during perinatal and postnatal growth in the mouse (Cendekiawan et al., 2010) and *Anolis* "lizards" (Sanger et al., 2014). Recent analysis in mouse suggests that intercalary proliferation or mesenchymal aggregation may be key mechanisms in sustained growth and reshaping of the nasal capsule during late embryonic development (Kaucka et al., 2017), rather than concentrated growth zones or plates. This suggests that the regulation of proliferation throughout the mid-face cartilage is a shared mechanism for fine-tuning the shape of the face during late embryonic and juvenile growth (Fig. 8). However, much of the details are currently obscured by our lack of comparative data on cellular processes at equivalent developmental time points in different species. Broader and deeper sampling among and within diverse amniote clades is fundamental for further characterization and identification of potential mechanisms driving evolution of facial disparity among large taxonomic groups, such as classes, and explaining ecomorphological diversification within such clades.

8. Conclusions

Since their emergence from their modest-sized lizard-like ancestors who lived among the Carboniferous forests and swamps, amniotes have undergone the most extraordinary evolutionary transformations and generated an astonishing array of diverse and disparate forms (Benton, 2015; Clack, 2012; Martin & Sumida, 1997). Arguably, their most important early invention was the larger eggs covered with fibrous and calcified shell and the extra-embryonic membranes surrounding the embryo, which allowed laying eggs in terrestrial environments (Packard & Seymour, 1997). This facilitated the invasion and occupation of environments that were not available to their amphibious ancestors. Representatives of this vast and highly successful group of vertebrate animals are found practically in all ecosystems and climatic zones across the globe from the deepest oceans to the high mountains, from the hot dry deserts to the cold wastes of Antarctica.

Several groups evolved powered flight and many returned to the aquatic environments, both marine and freshwater. Modern amniotes range in size from the tiny gecko lizard *Sphaerodactylus parthenopion* (Reptilia, Gekkonidae) weighing only around 0.117 g to the blue whale *Balaenoptera musculus* (Mammalia, Balaenopteridae), which can reach up to 173 tonnes (Smith & Lyons, 2013). However, the greatest hallmarks of the evolutionary success of this clade are the numerous adaptations for inhabiting the air, land, and water. Therefore, it is very important to understand the processes, both extrinsic and intrinsic, that promoted morphological diversification of such an evolutionarily significant group of animals at both large- and small-scale taxonomic levels.

Amniotes have a spectacular record of evolution, which spans over 300 million years and continues today by many living and thriving clades. As such, there is a tremendous disparity among and within these clades, reflecting both shared and unique evolutionary events, selective pressures, and ecological opportunities. Across the evolution of the entire clade several major macroevolutionary trends are observed, from the restructuring the muscles and posterior skull for terrestrial feeding (Holliday & Witmer, 2007; Lauder & Gillis, 1997) and the parallel trends in reduction of cranial elements known as "Williston's Law" (Esteve-Altava et al., 2013; Gregory, 1935). These macroevolutionary trends demonstrate the selective pressures and functional constraints operating on the amniote skull in deep time. However, the more recent radiations of extant amniote clades show clear connections between the anterior of the skull, the face, and ecological diversification (e.g., Da Silva et al., 2018; Felice, Tobias, et al., 2019; Ferreira et al., 2015; Goswami et al., 2011; Pierce et al., 2008; Watanabe et al., 2019). The anatomical, developmental, and evolutionary coherence and distinctiveness of the facial region of the skull have been recovered time and time again (Esteve-Altava et al., 2013; Felice, Tobias, et al., 2019; Lee et al., 2020; Watanabe et al., 2019). This module is reliably recovered using different modularity and integration tests and its appearance likely goes back to the dawn of the amniotes. Much of the adaptive success that amniotes enjoy, especially in the clades that underwent evolutionary radiations, can be to a large extent explained by the continued elaboration of the face at the species level. For example, phylogenetic analyses of Darwin's finches suggest that their beak shape changes most quickly and dramatically of any other morphological traits studied in this group (Reaney et al., 2020) and may be true of birds more broadly (Felice et al., 2020). Remarkably, these lineages also have revealed that heterochrony and change to developmental patterns were key mechanisms in the diversification of amniote facial form, in both classic adaptive radiations and larger clades (Camacho et al., 2019; Morris et al., 2019). The genetic mechanisms underlying the origins of these differences, at both the macro- and meso-evolutionary levels, have been the focus of much research and revealed many specific molecular pathways critical to the formation of variation in temporal fenestra and facial form alike.

When combined together the observations emerging from rather disparate groups of amniotes strongly suggest that the face, as a module, is indeed an ancient attribute of the amniote head, which displays similar evolutionary qualities and is built with a similar developmental genetic blueprint. Even with the limited and sporadic data collected thus far it is possible to layout a preliminary model for the developmental origins of facial diversity among and within amniote clades. Clearly, the earliest stages of facial morphogenesis lay down a genetic pattern, which goes on to direct later morphogenesis and growth via key cellular processes such as migration, differentiation, and proliferation. This phase (early morphogenesis in Fig. 8) is when differences in the patterning of temporal fenestra (Tokita et al., 2013) and the organization of the FEZ (Hu & Marcucio, 2009) become apparent. More study of this period of development is necessary to understand the precise mechanisms which generate the major differences among extant amniote clades and macroevolutionary trends in amniote cranial evolution. However, these events also result in the formation of conserved embryonic growth zones in the face and shared processes during later morphogenesis and growth, giving rise to parallel trends in the evolution of facial proportions within clades.

During late embryonic and post-embryonic stages (Fig. 8) the facial module is already well established and regulated semi-autonomously, relying on the set of internal morphogenetic factors, such as signaling molecules and downstream transcription factors. The nature of the molecular and cellular interactions within this module during morphogenesis imposes generative constraints but also provides certain possibilities. The data from birds, bats, and crocodylians all suggest that the control of the rate of proliferation and the timing of skeletal differentiation are critical mechanisms in directing facial development, likely giving rise the heterochronic evolution of cranial disparity in these groups (Camacho et al., 2019; Felice & Goswami, 2018; Morris et al., 2019). Maintenance of this growth zone appears to create the subsequent growth trajectories, except in the case of the apomorphic avian beak in which it takes on special importance (Campàs et al., 2010; Fritz et al., 2014) and unique patterning (Bhullar et al., 2015). This is a more extreme situation but all developing faces studied so far rely on a similar mechanism. It appears that the most accessible way to alter the shape of the face in response to the changing environmental challenges is by modifying the behaviors of the available cell populations, primarily by tweaking their rate and timing. Such limited developmental changes nonetheless produced rather impressive results as attested by the range of scaling variation of the facial structures in birds, crocodylians and phyllostomid bats. Thus, the general mode of morphological evolution of the amniote face may be via the heterochrony-driven scaling allometry, where the absolute and relative length, width and depth of the face can be altered rather quickly and dramatically, often providing access to the previously inaccessible resources. Moving forward, describing the precise differences in the cellular processes of facial morphogenesis in phylogenetically diverse amniotes and uncovering their genetic controls will ultimately allow a greater understanding of why and how amniotes have been able to become and remain so evolutionary successful.

References

Abzhanov, A. (2013). von Baer's law for the ages: lost and found principles of developmental evolution. *Trends in Genetics*, *29*(12), 712–722.

Abzhanov, A., & Tabin, C. J. (2004). *Shh* and *Fgf8* act synergistically to drive cartilage outgrowth during cranial development. *Developmental Biology*, *273*(1), 134–148.

Abzhanov, A., Protas, M., Grant, B. R., Grant, P. R., & Tabin, C. J. (2004). Bmp4 and morphological variation of beaks in Darwin's finches. *Science*, *305*, 1462–1465.

Abzhanov, A., Tzahor, E., & Lassar, A. B. (2003). Dissimilar regulation of cell differentiation in mesencephalic (cranial) and sacral (trunk) neural crest cells in vitro. *Development*, *130*, 4567–4579.

Ahi, E. P. (2016). Signalling pathways in trophic skeletal development and morphogenesis: Insights from studies on teleost fish. *Developmental Biology*, *420*(1), 11–31.

Alberch, P., Gould, S. J., Oster, G. F., & Wake, D. B. (1979). Size and shape in ontogeny and phylogeny. *Paleobiology*, *5*(3), 296–317.

Álvarez, A., Perez, S. I., & Verzi, D. H. (2015). The role of evolutionary integration in the morphological evolution of the skull of caviomorph rodents (Rodentia: Hystricomorpha). *Evolutionary Biology*, *42*, 312–327.

Ascarrunz, E., Sánchez-Villagra, M. R., Betancur-R, R., & Laurin, M. (2019). On trends and patterns in macroevolution: Williston's law and the branchiostegal series of extant and extinct osteichthyans. *BMC Evolutionary Biology*, *19*, 117.

Bellairs, A.d., & Kamal, A. E.-D. M. (1981). The chondrocranium and the development of the skull in recent reptiles. In C. Gans, & T. S. Parsons (Eds.), *vol. 11. Biology of the reptilia* (pp. 1–263). New York: Academic Press. (Morphology F), ch. 1.

Benton, M. J. (1985). Classification and phylogeny of the diapsid reptiles. *Zoological Journal of the Linnean Society*, *84*, 97–164.

Benton, M. J. (2015). *Vertebrate Palaeontology* (4th edn). Wiley Blackwell.

Benton, M., Donoghue, P., Vinther, J., Asher, R., Friedman, M., & Near, T. (2015). Constraints on the timescale of animal evolutionary history. *Palaeontologia Electronica*, *18*(1.1FC), 1–106.

Bever, G. S., Lyson, T. R., Field, D. J., & Bhullar, B.-A. S. (2015). Evolutionary origin of the turtle skull. *Nature, 525*, 239–242.

Bhullar, B.-A. S., Marugán-Lobón, J., Racimo, F., Bever, G. S., Rowe, T. B., Norell, M. A., et al. (2012). Birds have paedomorphic dinosaur skulls. *Nature, 487*, 223–226.

Bhullar, B.-A. S., Morris, Z. S., Sefton, E. M., Tok, A., Tokita, M., Namkoong, B., et al. (2015). A molecular mechanism for the origin of a key evolutionary innovation, the bird beak and palate, revealed by an integrative approach to major transitions in vertebrate history. *Evolution, 69*, 1665–1677.

Blackburn, D. G. (2000). Classification of the reproductive patterns of amniotes. *Herpetological Monographs, 14*, 371–377.

Bright, J. A., Marugán-Lobón, J., Cobb, S. N., & Rayfield, E. J. (2016). The shapes of bird beaks are highly controlled by nondietary factors. *Proceedings of the National Academy of Sciences, 113*, 5352–5357.

Brochu, C. A. (2001). Crocodylian snouts in space and time: Phylogenetic approaches toward adaptive radiation. *American Zoologist, 41*, 564–585.

Brugmann, S. A., Kim, J., & Helms, J. A. (2006). Looking different: Understanding diversity in facial form. *American Journal of Medical Genetics Part A, 140A*, 2521–2529.

Camacho, J., Heyde, A., Bhullar, B.-A. S., Haelewaters, D., Simmons, N. B., & Abzhanov, A. (2019). Peramorphosis, an evolutionary developmental mechanism in neotropical bat skull diversity. *Developmental Dynamics, 1*, 1–15.

Camacho, J., Moon, R., Smith, S. K., Lin, J. D., Randolph, C., Rasweiler, J. J., et al. (2020). Differential cellular proliferation underlies heterochronic generation of cranial diversity in phyllostomid bats. *EvoDevo, 11*, 1–17.

Campàs, O., Mallarino, R., Herrel, A., Abzhanov, A., & Brenner, M. P. (2010). Scaling and shear transformations capture beak shape variation in Darwin's finches. *Proceedings of the National Academy of Sciences, 107*, 3356–3360.

Cardini, A., & Polly, P. D. (2013). Larger mammals have longer faces because of size-related constraints on skull form. *Nature Communications, 4*, 2458.

Carroll, R. L. (1988). *Vertebrate paleontology and evolution*. New York: W. H. Freeman and Company.

Cendekiawan, T., Wong, R. W. K., & Rabie, A. B. M. (2010). Relationships between cranial base synchondroses and craniofacial development: A review. *The Open Anatomy Journal, 2*, 67–75.

Chiari, Y., Cahais, V., Galtier, N., & Delsuc, F. (2012). Phylogenomic analyses support the position of turtles as the sister group of birds and crocodiles (Archosauria). *BMC Biology, 10*, 65.

Churchill, M., Geisler, J. H., Beatty, B. L., & Goswami, A. (2018). Evolution of cranial telescoping in echolocating whales (Cetacea: Odontoceti). *Evolution, 72*, 1092–1108.

Clack, J. A. (2012). *Gaining ground: The origin and evolution of tetrapods* (2nd edn). Bloomington: Indiana University Press.

Claude, J., Pritchard, P. C. H., Tong, H., Paradis, E., & Auffray, J.-C. (2004). Ecological correlates and evolutionary divergence in the skull of turtles: A geometric morphometric assessment. *Systematic Biology, 53*, 933–948.

Couly, G., Creuzet, S., Bennaceur, S., Vincent, C., & Douarin, N. M. L. (2002). Interactions between Hox-negative cephalic neural crest cells and the foregut endoderm in patterning the facial skeleton in the vertebrate head. *Development, 129*, 1061–1073.

Creuzet, S., Couly, G., Vincent, C., & Douarin, N. M. L. (2002). Negative effect of Hox gene expression on the development of the neural crest-derived facial skeleton. *Development, 129*, 4301–4313.

Da Silva, F. O., Fabre, A.-C., Savriama, Y., Ollonen, J., Mahlow, K., Herrel, A., et al. (2018). The ecological origins of snakes as revealed by skull evolution. *Nature Communications, 9*, 1–11.

Doody, J. S., Burghardt, G. M., & Dinets, V. (2012). Breaking the social-non-social dichotomy: A role for reptiles in vertebrate social behavior research? *Ethology*, *119*, 95–103.

Elinson, R. P., & Kao, K. R. (1993). Axis Specification and Head Induction in Vertebrate Embryos. In J. Hanken, & B. K. Hall (Eds.), *vol. 1. The skull* (pp. 1–41). Chicago: University of Chicago Press. ch. 1.

Estes, R. (1961). Cranial anatomy of the cynodont reptile *Thrinaxodon liorhinus*. *Bulletin of the Museum of Comparative Zoology*, *125*, 165–180.

Esteve-Altava, B., Marugán-Lobón, J., Botella, H., & Rasskin-Gutman, D. (2013). Structural constraints in the evolution of the tetrapod skull complexity: Williston's law revisited using network models. *Evolutionary Biology*, *40*, 209–219.

Evans, S. E., & Jones, M. E. H. (2010). The origin, early history and diversification of lepidosauromorph reptiles. In S. Bandyopadhyay (Ed.), *New aspects of mesozoic biodiversity* (pp. 27–44). Heidelberg: Springer.

Fabbri, M., Koch, N. M., Pritchard, A. C., Hanson, M., Hoffman, E., Bever, G. S., et al. (2017). The skull roof tracks the brain during the evolution and development of reptiles including birds. *Nature Ecology & Evolution*, *1*, 1543–1550.

Felice, R. N., & Goswami, A. (2018). Developmental origins of mosaic evolution in the avian cranium. *Proceedings of the National Academy of Sciences*, *115*, 555–560.

Felice, R. N., Randau, M., & Goswami, A. (2018). A fly in a tube: Macroevolutionary expectations for integrated phenotypes. *Evolution*, *72*, 2580–2594.

Felice, R. N., Tobias, J. A., Pigot, A. L., & Goswami, A. (2019). Dietary niche and the evolution of cranial morphology in birds. *Proceedings of the Royal Society B*, *286*, 20182677.

Felice, R. N., Watanabe, A., Cuff, A. R., Noirault, E., Pol, D., Witmer, L. M., et al. (2019). Evolutionary integration and modularity in the Archosaur cranium. *Integrative and Comparative Biology*, *59*, 371–382.

Felice, R. N., Watanabe, A., Cuff, A. R., Hanson, M., Bhullar, B.-A. S., Rayfield, E. R., et al. (2020). Decelerated dinosaur skull evolution with the origin of birds. *PLoS Biology*, *18*, e3000801.

Ferguson, M. W. J. (1985). Reproductive biology and embryology of the crocodilians. In C. Gans, F. Billet, & P. F. A. Maderson (Eds.), *vol. 14. Biology of the reptilia* (pp. 329–493). New York: John Wiley & Sons. (Development A), ch. 5.

Ferreira, G. S., Rincón, A. D., Solórzano, A., & Langer, M. C. (2015). The last marine pelomedusoids (Testudines: Pleurodira): a new species of Bairdemys and the paleoecology of Stereogenyina. *PeerJ*, *3*, e1063.

Field, D. J., Gauthier, J. A., King, B. L., Pisani, D., Lyson, T. R., & Peterson, K. J. (2014). Toward consilience in reptile phylogeny: miRNAs support an archosaur, not lepidosaur, affinity for turtles. *Evolution & Development*, *16*, 189–196.

Ford, D. P., & Benson, R. B. J. (2020). The phylogeny of early amniotes and the affinities of Parareptilia and Varanopidae. *Nature Ecology & Evolution*, *4*, 57–65.

Foth, C., Rabi, M., & Joyce, W. G. (2016). Skull shape variation in extant and extinct Testudinata and its relation to habitat and feeding ecology. *Acta Zoologica*, *98*, 310–325.

Fritz, J. A., Brancale, J., Tokita, M., Burns, K. J., Hawkins, M. B., Abzhanov, A., et al. (2014). Shared developmental programme strongly constrains beak shape diversity in songbirds. *Nature Communications*, *5*, 3700.

Frolich, L. M. (1997). The role of the skin in the origin of amniotes: Permeability barrier, protective covering and mechanical support. In S. S. Sumida, & K. L. M. Martin (Eds.), *Amniote origins. Completing the transition to land* (pp. 327–352). San Diego: Academic Press.

Gaffney, E. S. (1979). Comparative cranial morphology of recent and fossil turtles. *Bulletin of the American Museum of Natural History*, *164*, 65–375.

Gardiner, B. G. (1982). Tetrapod classification. *Zoological Journal of the Linnean Society*, *74*, 207–232.

Gauthier, J., Kluge, A., & Rowe, T. (1988). Amniote phylogeny and the importance of fossils. *Cladistics*, *4*, 105–209.
Godoy, P. L., Ferreira, G. S., Montefeltro, F. C., Nova, B. C. V., Butler, R. J., & Langer, M. C. (2018). Evidence for heterochrony in the cranial evolution of fossil crocodyliforms. *Palaeontology*, *61*, 543–558.
Goodrich, E. S. (1958). *Studies on the structure and development of vertebrates*. New York: Dover Publications.
Goswami, A., Milne, N., & Wroe, S. (2011). Biting through constraints: cranial morphology, disparity and convergence across living and fossil carnivorous mammals. *Proceedings of the Royal Society B: Biological Sciences*, *278*, 1831–1839.
Gould, S. J. (1977). *Ontogeny and phylogeny*. Cambridge, MA: Harvard University Press.
Graham, A., Heyman, I., & Lumsden, A. (1993). Even-numbered rhombomeres control the apoptotic elimination of neural crest cells from odd-numbered rhombomeres in the chick hindbrain. *Development*, *119*, 233–245.
Graham, A., Koentges, G., & Lumsden, A. (1996). Neural crest apoptosis and the establishment of craniofacial pattern: An honorable death. *Molecular and Cellular Neuroscience*, *8*, 76–83.
Gregory, W. K. (1935). Willison's law' relating to the evolution of skulls bones in the vertebrates. *American Journal of Physical Anthropology*, *20*(2), 123–152.
Haeckel, E. (1866). *Generelle Morphologie der Organismen*. Berlin: Reimer.
Hedrick, B. P., Mutumi, G. L., Munteanu, V. D., Sadier, A., Davies, K. T. J., Rossiter, S. J., et al. (2020). Morphological diversification under high integration in a hyper diverse mammal clade. *Journal of Mammalian Evolution*, *27*, 563–575.
Holliday, C. M., & Witmer, L. M. (2007). Archosaur adductor chamber evolution: Integration of musculoskeletal and topological criteria in jaw muscle homology. *Journal of Morphology*, *268*, 457–484.
Hotton, N., III, Olson, E. C., & Beerbower, R. (1997). Amniote origins and the discover of herbivory. In S. S. Sumida, & K. L. M. Martin (Eds.), *Amniote origins. Completing the Transition to Land* (pp. 207–264). San Diego: Academic Press.
Hu, D., & Marcucio, R. S. (2009). A SHH-responsive signaling center in the forebrain regulates craniofacial morphogenesis via the facial ectoderm. *Development*, *136*, 107–116.
Hu, D., Marcucio, R. S., & Helms, J. A. (2003). A zone of frontonasal ectoderm regulates patterning and growth in the face. *Development*, *130*, 1749–1758.
Jeong, J., Mao, J., Tenzen, T., Kottmann, A. H., & McMahon, A. P. (2004). Hedgehog signaling in the neural crest cells regulates the patterning and growth of facial primordia. *Genes & Development*, *18*, 937–951.
Kaucka, M., Zikmund, T., Tesarova, M., & Gyllborg, D. (2017). Oriented clonal cell dynamics enables accurate growth and shaping of vertebrate cartilage. *eLife*, *6*, e25902.
Klembara, J., Hain, M., Ruta, M., Berman, D. S., Pierce, S. E., & Henrici, A. C. (2020). Inner ear morphology of diadectomorphs and seymouriamorphs (Tetrapoda) uncovered by high-resolution x-ray microcomputed tomography, and the origin of the amniote crown group. *Palaeontology*, *63*, 131–154.
Koyabu, D., Maier, W., & Sánchez-Villagra, M. R. (2012). Paleontological and developmental evidence resolve the homology and dual embryonic origin of a mammalian skull bone, the interparietal. *Proceedings of the National Academy of Sciences*, *109*, 14075–14080.
Lack, D. (1947). *Darwin's finches*. Cambridge, UK: Cambridge University Press.
Langille, R. M., & Hall, B. K. (1993). Pattern formation and the neural crest. In J. Hanken, & B. K. Hall (Eds.), *The skull* (pp. 77–111). Chicago: University of Chicago Press. vol. 1, ch. 3.
Lauder, G. V., & Gillis, G. B. (1997). Origin of the amniote feeding mechanism: experimental analyses of outgroup clades. In S. S. Sumida, & K. L. M. Martin (Eds.), *Amniote origins. Completing the Transition to Land* (pp. 169–206). San Diego: Academic Press.

Laurin, M., & Reisz, R. R. (2020). Amniota E. Haeckel 1866 [M. Laurin and R. R. Reisz], converted clade name. In K. de Queiroz, P. D. Cantino, & J. A. Gauthier (Eds.), *Phylonyms* (pp. 793–797). Boca Raton: CRC Press.

Le Douarin, N. M. (1982). *The neural crest*. Cambridge: Cambridge University Press.

Le Douarin, Nicole M., & Dupin, Elisabeth. (2012). The neural crest in vertebrate evolution. *Current Opinion in Genetics & Development, 22*, 381–389. https://doi.org/10.1016/j.gde.2012.06.001.

Lee, H. W., Esteve-Altava, B., & Abzhanov, A. (2020). Evolutionary and ontogenetic changes of the anatomical organization and modularity in the skull of archosaurs. *Scientific Reports, 10*, 16138.

Li, C., Wu, X.-C., Rieppel, O., & Wang, L.-T. (2008). An ancestral turtle from the Late Triassic of southwestern China. *Nature, 456*, 497–501.

Lu, B., Yang, W., Dai, Q., & Fu, J. (2013). Using genes as characters and a parsimony analysis to explore the phylogenetic position of turtles. *PLoS One, 8*, e79348.

Lyson, T. R., Sperling, E. A., Heimberg, A. M., Gauthier, J. A., King, B. L., & Peterson, K. J. (2012). MicroRNAs support a turtle + lizard clade. *Biology Letters, 8*, 104–107.

Mallarino, R., Grant, P. R., Grant, B. R., Herrel, A., Kuo, W. P., & Abzhanov, A. (2011). Two developmental modules establish 3D beak-shape variation in Darwin's finches. *Proceedings of the National Academy of Sciences, 108*, 4057–4062.

Marroig, G., & Cheverud, J. M. (2004). Did natural selection or genetic drift produce the cranial diversification of neotropical monkeys? *The American Naturalist, 163*, 417–428.

Martin, K. L. M., & Nagy, K. A. (1997). Water balance and the physiology of the amphibian to amniote transition. In S. S. Sumida, & K. L. M. Martin (Eds.), *Amniote Origins. Completing the transition to land* (pp. 399–423). San Diego: Academic Press.

Martin, K. L. M., & Sumida, S. S. (1997). An integrated approach to the origin of Amniotes: completing the transition to land. In S. S. Sumida, & K. L. M. Martin (Eds.), *Amniote origins. Completing the transition to land* (pp. 1–8). San Diego: Academic Press.

McCurry, M. R., Evans, A. R., Fitzgerald, E. M. G., Adams, J. W., Clausen, P. D., & McHenry, C. R. (2017). The remarkable convergence of skull shape in crocodilians and toothed whales. *Proceedings of the Royal Society B: Biological Sciences, 284*, 20162348–20162349.

McHenry, C. R., Clausen, P. D., Daniel, W. J. T., Meers, M. B., & Pendharkar, A. (2006). Biomechanics of the rostrum in crocodilians: A comparative analysis using finite-element modeling. *The Anatomical Record: Advances in Integrative Anatomy and Evolutionary Biology, 288A*, 827–849.

Morris, Z. S., Vliet, K. A., Abzhanov, A., & Pierce, S. E. (2019). Heterochronic shifts and conserved embryonic shape underlie crocodylian craniofacial disparity and convergence. *Proceedings of the Royal Society B: Biological Sciences, 286*, 20182389.

Myhrvold, N. P., Baldridge, E., Chan, B., Sivam, D., Freeman, D. L., & Ernest, S. K. M. (2015). An amniote life-history database to perform comparative analyses with birds, mammals, and reptiles. *Ecology, 96*, 3109.

Neenan, J. M., Klein, N., & Scheyer, T. M. (2013). European origin of placodont marine reptiles and the evolution of crushing dentition in Placodontia. *Nature Communications, 4*, 1621.

Nesbitt, S. J. (2011). The early evolution of Archosaurs: Relationships and the origin of major clades. *Bulletin of the American Museum of Natural History, 352*, 1–292.

Northcutt, R. G. (2011). Evolving large and complex brains. *Science, 332*, 926–927.

Novacek, M. J. 1993. Patterns of Diversity in the Mammalian Skull. in Hanken, J. and Hall, B. K. (eds.), *The skull*. University of Chicago Press, Chicago, vol. 2, ch. 7, pp. 438–545.

Packard, M. J., & Seymour, R. S. (1997). Evolution of the amniote egg. In S. S. Sumida, & K. L. M. Martin (Eds.), *Amniote origins. Completing the transition to land* (pp. 265–290). San Diego: Academic Press.

Pasqualetti, M., Ori, M., Nardi, I., & Rijli, F. M. (2000). Ectopic Hoxa2 induction after neural crest migration results in homeosis of jaw elements in Xenopus. *Development, 127*, 5367–5378.

Pierce, S. E., Angielczyk, K. D., & Rayfield, E. J. (2008). Patterns of morphospace occupation and mechanical performance in extant crocodilian skulls: A combined geometric morphometric and finite element modeling approach. *Journal of Morphology, 269*, 840–864.

Pierce, S. E., Angielczyk, K. D., & Rayfield, E. J. (2009). Morphospace occupation in thalattosuchian crocodylomorphs: skull shape variation, species delineation and temporal patterns. *Palaeontology, 52*, 1057–1097.

Piras, P., Colangelo, P., Adams, D. C., Buscalioni, A., Cubo, J., Kotsakis, T., et al. (2010). The Gavialis-Tomistoma debate: the contribution of skull ontogenetic allometry and growth trajectories to the study of crocodylian relationships. *Evolution & Development, 12*, 568–579.

Pokorná, M. J., & Kratochvíl, L. (2016). What was the ancestral sex-determining mechanism in amniote vertebrates? *Biological Reviews, 91*, 1–12.

Reaney, A. M., Bouchenak-Khelladi, Y., Tobias, J. A., & Abzhanov, A. (2020). *Ecological and morphological determinants of evolutionary diversification in Darwin's finches and their relatives.* Ecology and Evolution.

Reilly, S. M., Wiley, E. O., & Meinhardt, D. J. (1997). An integrative approach to heterochrony: The distinction between interspecific and intraspecific phenomena. *Biological Journal of the Linnean Society, 60*, 119–143.

Reilly, S. M., McBrayer, L. D., & White, T. D. (2001). Prey processing in amniotes: biomechanical and behavioral patterns of food reduction. *Comparative Biochemistry and Physiology. Part A, Molecular & Integrative Physiology, 128*, 397–415.

Reisz, R. R. (2006). Origin of dental occlusion in tetrapods: signal for terrestrial vertebrate evolution? *Journal of Experimental Zoology Part B: Molecular and Developmental Evolution, 306B*, 261–277.

Reisz, R. R., & Modesto, S. P. (1996). Archerpeton anthracos from the Joggins Formation of Nova Scotia: a microsaur, not a reptile. *Canadian Journal of Earth Sciences, 33*, 703–709.

Reisz, R. R., & Müller, J. (2004). Molecular timescales and the fossil record: a paleontological perspective. *Trends in Genetics, 20*, 237–241.

Renaud, S., Pantalacci, S., & Auffray, J.-C. (2011). Differential evolvability along lines of least resistance of upper and lower molars in island house mice. *PLoS One, 6*, e18951.

Richman, J. M., & Lee, S. (2003). About face: Signals and genes controlling jaw patterning and identity in vertebrates. *BioEssays, 25*, 554–568.

Rieppel, O. (1993a). Euryapsid relationships: a preliminary analysis. *Neus jahrbuch für Geologie und Paläontologie, Abhandlungen, 188*, 241–264.

Rieppel, O. 1993b. Patterns of diversity in the reptilian skull. in Hanken, J. and Hall, B. K. (eds.), The skull. University of Chicago Press, Chicago, vol. 2, ch. 7, pp. 344–390.

Rowe, T. B., Macrini, T. E., & Luo, Z.-X. (2011). Fossil evidence on origin of the mammalian brain. *Science, 332*, 955–957.

Sadleir, R. (2009). *A morphometric study of crocodylian ecomorphology through ontogeny and phylogeny.* PhD thesis University of Chicago.

Sander, K. 1983. The evolution of patterning mechanisms: gleanings from insect embryogenesis and spermatogenesis. in Goodwin, B. C., N. Holder, and Wylie CC. (eds.), Development and evolution. Cambridge University Press., Cambridge, pp. 137–159.

Sanger, T. J., Sherratt, E., McGlothlin, J. W., Brodie, E. D., Losos, J. B., & Abzhanov, A. (2013). Convergent evolution of sexual dimorphism in skull shape using distinct developmental strategies. *Evolution, 67*, 2180–2193.

Sanger, T. J., Seav, S. M., Tokita, M., Langerhans, R. B., Ross, L. M., Losos, J. B., et al. (2014). The oestrogen pathway underlies the evolution of exaggerated male cranial shapes in Anolis lizards. *Proceedings of the Royal Society B: Biological Sciences, 281*, 20140329.

Schluter, D. (1996). Adaptive radiation along genetic lines of least resistance. *Evolution, 50*, 1766–1774.

Schluter, D., & Grant, P. R. (1984). Ecological correlates of morphological evolution in a Darwin's finch, *Geospiza difficilis*. *Evolution, 38*, 856–869.

Schneider, R. A., & Helms, J. A. (2003). The cellular and molecular origins of beak morphology. *Science, 299*, 565–568.

Sereno, P. C. (1997). The origin and evolution of dinosaurs. *Annual Reviews of Earth and Planetary Sciences, 25*, 435–489.

Sidor, C. A. (2001). Simplification as a trend in synapsid cranial evolution. *Evolution, 55*, 1419–1442.

Smith, F. A., & Lyons, S. K. (2013). On being the right size: the importance of size in life history, ecology and evolution. In F. A. Smith, & S. K. Lyons (Eds.), *Animal body size* (pp. 1–10). Chicago: The University of Chicago Press.

Smith-Paredes, D., Núñez-León, D., Soto-Acuña, S., O'Connor, J., Botelho, J. F., & Vargas, A. O. (2018). Dinosaur ossification centres in embryonic birds uncover developmental evolution of the skull. *Nature Ecology & Evolution, 2*, 1966–1973.

Stayton, C. T. (2005). Morphological evolution of the lizard skull: A geometric morphometrics survey. *Journal of Morphology, 263*, 47–59.

Tapadia, M. D., Cordero, D. R., & Helms, J. A. (2005). It's all in your head: new insights into craniofacial development and deformation. *Journal of Anatomy, 207*, 461–477.

Tarsitano, S. F., Oelofsen, B., Frey, E., & Riess, J. (2001). The origin of temporal fenestrae. *South African Journal of Science, 97*, 334–336.

Theveneau, E., & Mayor, R. (2011). Collective cell migration of the cephalic neural crest: The art of integrating information. *Genesis, 49*, 164–176.

Tokita, M., Chaeychomsri, W., & Siruntawineti, J. (2013). Skeletal gene expression in the temporal region of the reptilian embryos: implications for the evolution of reptilian skull morphology. *SpringerPlus, 2*, 336.

Tokita, M., Yano, W., James, H. F., & Abzhanov, A. (2016). Cranial shape evolution in adaptive radiations of birds: comparative morphometrics of Darwin's finches and Hawaiian honeycreepers. *Philosophical Transactions of the Royal Society B: Biological Sciences, 372*, 20150481.

Usui, K., & Tokita, M. (2018). Creating diversity in mammalian facial morphology: a review of potential developmental mechanisms. *EvoDevo, 9*, 1–17.

Vallin, G., & Laurin, M. (2004). Cranial morphology and affinities of *Microbrachis*, and a reappraisal of the phylogeny and lifestyle of the first amphibians. *Journal of Vertebrate Paleontology, 24*, 56–72.

Van Tuinen, M., & Hadly, E. A. (2004). Error in estimation of rate and time inferred from the early amniote fossil record and avian molecular clocks. *Journal of Molecular Evolution, 59*, 267–276.

Young, N. M., Hu, D., Lainoff, A. J., Smith, F. J., Diaz, R., Tucker, A. S., et al. (2014). Embryonic bauplans and the developmental origins of facial diversity and constraint. *Nature, 141*, 1059–1063.

Wada, N., Javidan, Y., Nelson, S., Carney, T. J., Kelsh, R. N., & Schilling, T. F. (2005). Hedgehog signaling is required for cranial neural crest morphogenesis and chondrogenesis at the midline in the zebrafish skull. *Development, 132*, 3977–3988.

Watanabe, A., Fabre, A.-C., Felice, R. N., Maisano, J. A., Müller, J., Herrel, A., et al. (2019). Ecomorphological diversification in squamates from conserved pattern of cranial integration. *Proceedings of the National Academy of Sciences, 116*, 14688–14697.

Wealthall, R. J., & Herring, S. W. (2006). Endochondral ossification of the mouse nasal septum. *The Anatomical Record: Advances in Integrative Anatomy and Evolutionary Biology, 288A*, 1163–1172.

Wroe, S., & Milne, N. (2007). Convergence and remarkably consistent constraint in the evolution of the carnivore skull shape. *Evolution, 61*, 1251–1260.

Zusi, R. L. (1993). Patterns of diversity in the avian skull. In J. Hanken, & B. K. Hall (Eds.), *vol. 2. The skull* (pp. 391–437). Chicago: University of Chicago Press. ch. 8.

CHAPTER NINE

The origin of wing polyphenism in ants: An eco-evo-devo perspective

Lisa Hanna and Ehab Abouheif*
Department of Biology, McGill University, Montreal, QC, Canada
*Corresponding author: e-mail address: ehab.abouheif@mcgill.ca

Contents

Abstract

The evolution of eusociality, where solitary individuals integrate into a single colony, is a major transition in individuality. In ants, the origin of eusociality coincided with the origin of a wing polyphenism approximately 160 million years ago, giving rise to colonies with winged queens and wingless workers. As a consequence, both eusociality and wing polyphenism are nearly universal features of all ants. Here, we synthesize fossil, ecological, developmental, and evolutionary data in an attempt to understand the factors that contributed to the origin of wing polyphenism in ants. We propose multiple models and hypotheses to explain how wing polyphenism is orchestrated at

Current Topics in Developmental Biology, Volume 141
ISSN 0070-2153
https://doi.org/10.1016/bs.ctdb.2020.12.004

multiple levels, from environmental cues to gene networks. Furthermore, we argue that the origin of wing polyphenism enabled the subsequent evolution of morphological diversity across the ants. We finally conclude by outlining several outstanding questions for future work.

1. Introduction

Dramatic increases in biological complexity have occurred repeatedly during the history of life. Maynard-Smith and Szathmary (1997; Szathmary & Smith, 1995) proposed that increases in biological complexity originate and evolve through "major evolutionary transitions in individuality," which occur when individuals integrate to form a single, higher-level, individual that replicates independently and has a division of labor between its constituent parts (Buss, 2014; Taylor, Reuter, & Sumner, 2019; West, Fisher, Gardner, & Kiers, 2015). Prime examples of major transitions in individuality include unicellular organisms integrating to form a multicellular organism, distantly related organisms integrating to form an obligate endosymbiosis, and solitary individuals integrating to form a eusocial colony. Although much recent progress has been made, both theoretically and empirically, in understanding the evolutionary pathways toward major evolutionary transitions in individuality (Boomsma & Gawne, 2018; Maliet, Shelton, & Michod, 2015; Miller & Reeve, 2019; Moran, 2007; Rafiqi, Rajakumar, & Abouheif, 2020; Ruiz-Trillo & Nedelcu, 2015), much work remains in understanding the evolutionary and developmental steps leading to transitions in different taxonomic groups, which have occurred independently across the tree of life.

Major transitions in individuality are often associated with the evolution of a polyphenism, which is defined as the ability of a single genome to produce discrete phenotypes in response to environmental cues (Nijhout, 1999; Stearns, 1989). For example, the transition from unicellular to obligate (complex) multicellularity in animals was accompanied by a cellular polyphenism, in which differentiated cell types (e.g., over 200 in humans) are produced from the same genome in response to cues from the internal environment, like morphogen gradients (Marquez-Zacarias et al., 2020). Recent studies suggest that the origins of this cellular polyphenism may have even preceded the major transition to multicellularity in animals because the choanoflagellates, the closest living unicellular relatives of the animals, differentiate into five distinct cell types in response to environmental cues (Dayel et al., 2011; Mikhailov et al., 2009). Furthermore, the transition to eusociality, which is defined by individuals that obligately live in colonies with overlapping generations, cooperative brood care, and a reproductive division of labor between the reproductive queen and male caste and non-reproductive worker caste, is also associated with the evolution of a polyphenism (Wilson, 1971; Wilson & Holldobler, 2005a). Both ants (order Hymenoptera) and termites (order Isoptera), for example, have convergently evolved eusociality and a polyphenism in wing development producing winged queens and wingless workers in response to environmental cues (Wilson, 1971). As a consequence, all species of ants and termites are wing polyphenic and eusocial. Therefore, wing polyphenism has been proposed to be associated with, if not the basis for, the convergent origin of eusocial life in these two groups (Abouheif & Wray, 2002; Holldobler & Wilson, 1990; Korb, 2008; Nowak, Tarnita, & Wilson, 2010). Finally, in other insect groups where eusociality has evolved, including wasps and bees (order Hymenoptera), thrips (order Thysanoptera), aphids (order Hemiptera), and ambrosia beetles (order Coleoptera), the females are also polyphenic and develop into discrete reproductive or non-reproductive castes based on environment cues (Crespi, 1992; Foster, 1990; Kent & Simpson, 1992). Therefore, studying the evolution of polyphenisms in taxonomic groups before and after they have undergone major transitions in individuality is key to understanding the factors and steps leading to the origin and elaboration of these transitions.

Here we focus on the origin of wing polyphenism in ants. Ants belong to the order Hymenoptera, along with sawflies, bees, and wasps, and are one of the most ecologically dominant and evolutionarily successful groups of insects on the planet. The transition to eusociality occurred several independent times within the Hymenoptera. Bees and wasps display a wide range of social organization, from communal groups to a complex grade of eusocial organization (Michener, 1969; West-Eberhard, 1978; Wheeler, 1928b). In contrast, all species of ants are eusocial and wing polyphenic, where the females develop either into winged queens or wingless workers in response to environmental cues such as temperature and nutrition (Fig. 1) (Holldobler & Wilson, 1990; Wheeler, 1928b).

In the majority of ant species, the establishment of a new colony typically begins with mating flights where unmated winged males and queens emerge from their parental nest, fly, and then mate in the air. After mating, the queens search for a proper nest site to lay their eggs. Once a suitable site is found, the queen sheds her wings and begins to lay her first eggs, which typically develop into female workers to kick start her new colony.

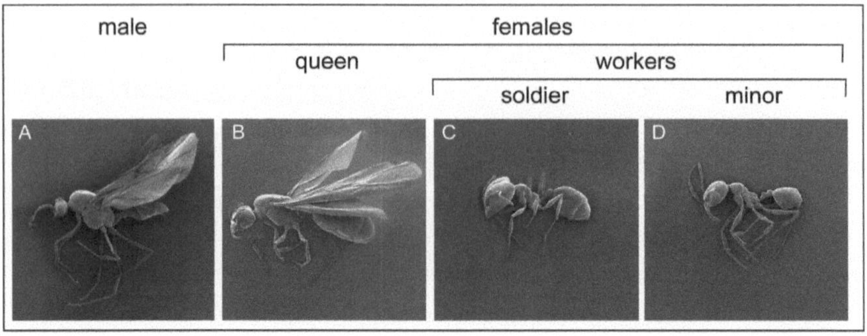

Fig. 1 Wing polyphenism in the female caste of ants. Scanning electron micrographs of adult castes of *Pheidole morrisi*. Unfertilized eggs laid by the queen develop into winged males (A). Fertilized eggs laid by the queen develop into females (B–D). Female embryos develop either into winged queens (B) or wingless workers (C and D) in response to temperature and photoperiod. In a subset of taxa, worker larvae develop into either wingless soldiers (C) or wingless minor workers (D) in response to nutritional cues.

Sex-determination in the Hymenoptera occurs through "haplodiploidy"—if the queen fertilizes her eggs, they are diploid and develop into females; if they are not fertilized, they are haploid and develop into males (Grimaldi & Engel, 2005; Holldobler & Wilson, 1990). The primary role of males is procreation, which is why male offspring are typically produced and present in colonies for only a short period of time once a year just before the mating flight of a colony takes place. In contrast, females are long-lived and their development depends on environmental cues like temperature and nutrition. In response to these environmental cues, female eggs will develop either into winged queens that disperse and establish their own nests, or into wingless workers that remain in the nest and carry out tasks such as foraging for food and caring for the brood.

In early ant evolution, queens and workers were similar in morphology and reproductive potential. The origin of wing polyphenism in ants, therefore, is thought to have prevented the wingless workers from participating in mating flights, while simultaneously allowing winged queens to do so. This may have been one of several factors that either facilitated or maintained the appearance of reproductive division of labor, which is a defining feature of eusociality. After the origin of wing polyphenism and eusociality in ants, a diverse array of mating strategies evolved, including the evolution of wingless queens and when the queen caste is permanently lost, workers that mate with males on the ground (Keller, Peeters, & Beldade, 2014; Peeters, 1991).

Studies in evolutionary developmental biology (evo-devo) have shown that morphological diversity in animals arose through changes in a relatively small number of highly conserved developmental regulatory genes, collectively known as the 'genetic toolkit' (Carroll, 2008; Gilbert, Opitz, & Raff, 1996; Shubin, Tabin, & Carroll, 2009). However, the development of polyphenic traits is largely influenced by environmental and social factors. Therefore, to understand the origins of wing polyphenism in ants it's important to assess the interactions between the environment, genes, development, and evolution. The emerging field of ecological evolutionary development (eco-evo-devo) integrates ecology with evolutionary development in order to explore how the environment interacts with genetic toolkits during development to generate phenotypic diversity over evolutionary time (Abouheif et al., 2014; Gilbert & Epel, 2015; Sultan, 2015; Toth & Rehan, 2017; West-Eberhard, 2003). Here, we take an eco-evo-devo approach to provide a holistic synthesis of the factors that may have led to the origin and subsequent evolution of wing polyphenism in ants. We hope that the knowledge gained by reconstructing the events leading to wing polyphenism in ants will potentially teach us about other major evolutionary transitions in individuality.

2. Hymenopteran clues for the origin of wing polyphenism in the Formicidae

Within the Hymenoptera, the ants, vespid wasps, crabronid wasps, and halictid bees all evolved eusociality independently (see stars in Fig. 2). Among these eusocial groups, ants are the only family (Formicidae) in which all species are both eusocial and wing polyphenic with winged queen and wingless worker castes within a colony. In this section, we explore whether wing polyphenism is a unique feature that originated in ants, or if it also evolved in other Hymenopteran groups. Furthermore, we informally explore the possible phylogenetic correlations between the origin of wing polyphenism and sexual wing dimorphism, a feature that is widespread among the Hymenoptera. We raise the possibility that the developmental capacity to generate sexual wing dimorphism was co-opted to generate a wing polyphenism among the ants. Finally, we explore environmental and ecological factors, as well as possible selection pressures, that may have facilitated the origin of wing polyphenism.

By mapping the occurrence of wing polyphenism on a phylogeny of the Hymenoptera, we discovered that wing polyphenism is not unique to ants,

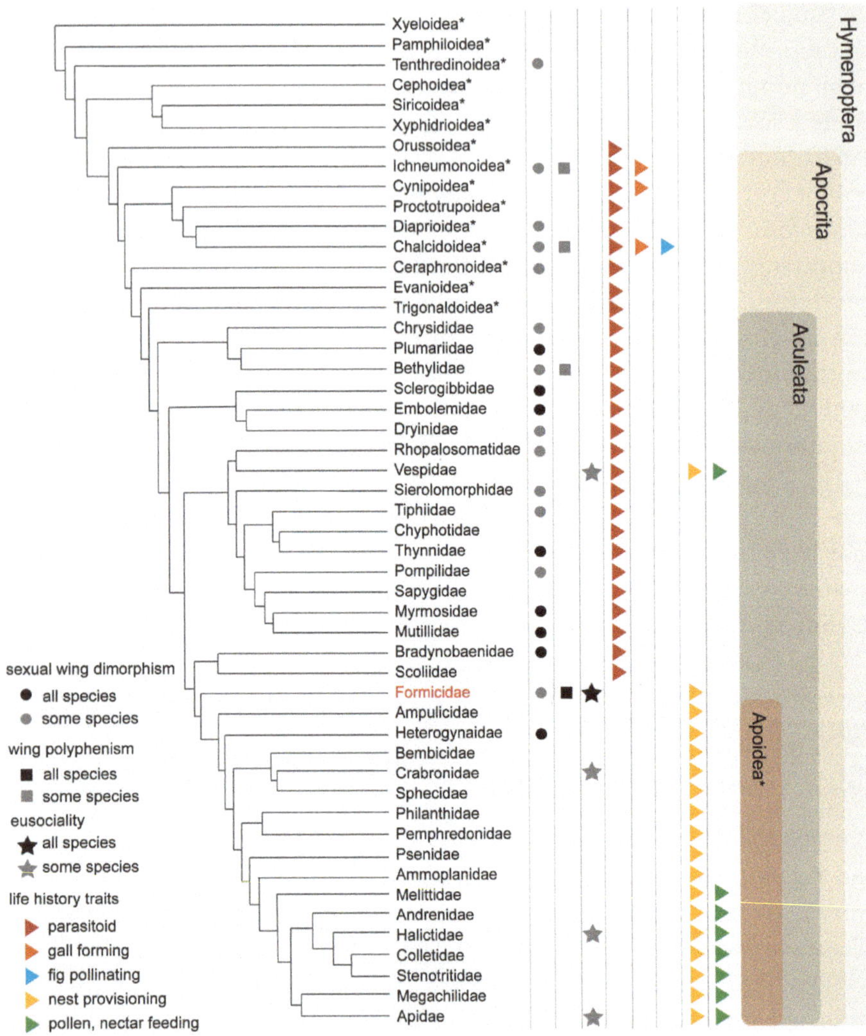

Fig. 2 Major life history and morphological traits informally mapped on a phylogeny of the main hymenopteran families and superfamilies. The characteristics include the evolution of sexual wing dimorphism (black and gray circles), wing polyphenism (black and gray squares), eusociality (black and gr a y stars), and various life history traits (colored triangles; see key for details). Phylogenetic relationships based on Branstetter, Danforth, et al. (2017) and Sann et al. (2018). Character states of wing dimorphism and polyphenism for each family (or superfamily) is based on Goulet and Huber (1993) and Smith (1993), and character states for life history traits are based on Whitfield (1998). Branch lengths do not represent evolutionary time. Hymenoptera (order); Apocrita (suborder); Aculeata (subclade). Asterisks indicate superfamily level.

but instead, evolved several times independently in a small number of species belonging to four different wasp families (see squares in Fig. 2). In the family Bethylidae (bethylid wasps), wing polyphenism in the females (winged and wingless morphs) evolved in two species, *Sclerodermus puparia* and *Sclerodermus guani* (Wei, Gao, Tang, Wang, & Yang, 2016; Yang, Wang, Yao, Gould, & Cao, 2012). In the superfamily Ichneumonoidea (ichneumon wasps), wing polyphenism evolved in males of the species *Gelis corruptor* where they can develop either into a fully winged morph or a morph with minute rudimentary wings (Salt, 1952). In the superfamily Chalcidoidea (chalcid wasps), wing polyphenism is found in all species of the genus *Melittobia* (family Eulophidae) and in various species of fig wasps (family Agaonidae). In *Melittobia*, females have a wing polyphenism such that they develop into a short-winged morph or a fully winged morph (Matthews & Gonzalez, 2008; Matthews, Gonzalez, Matthews, & Deyrup, 2009; Schmieder, 1933). The Agaonidae (fig wasps) have commonly evolved a wing polyphenism in males, with an estimate of five independent origins across the family (Cook, Compton, Allen Herre, & West, 1997). Furthermore, some species of fig wasps produce only wingless males (Hamilton, 1979). Overall, our informal character mapping of wing polyphenism shows that has it has evolved several times independently across the Hymenoptera in taxa that are distantly related to ants (Fig. 2). Despite this repeated evolution of wing polyphenism across the Hymenoptera, ants remain the only group in which it is a universal trait.

Sexual wing dimorphism refers to genetically induced variation in wing development between the sexes (Mayr, 1963). Because sexual wing dimorphism has frequently evolved within the Hymenoptera, we also informally mapped its occurrence to explore the phylogenetic correlation between the independent origins of sexual wing dimorphism and wing polyphenism (see black and gr ay circles and squares in Fig. 2). Sexual wing dimorphism has evolved in at least 19 wasp families, and in at least 8 of these families, it is present in all species (Fig. 2; Goulet & Huber, 1993; Reid, 1941). Among the sawflies, sexual wing dimorphism is extremely rare and found in about four species belonging to the Tenthredinoidea superfamily (Smith, 1993). In almost all cases where a sexual wing dimorphism has evolved, the females are the sex that is wingless or short-winged, whereas the males are winged. Majority of the taxa that develop a sexual wing dimorphism are part of the Apocrita suborder, which includes wasps, bees, and ants (Fig. 2). Within the Apocrita, the subclade Aculeata, with the exception of the superfamily Apoidea, contains the largest concentration of families where a sexual wing dimorphism has evolved (Fig. 2). Our informal mapping suggests that the evolution of a sexual wing dimorphism coincides with the evolution of a wing polyphenism in 2 wasp taxa: the genus *Melittobia* (superfamily Chalcidoidea) and the species *G. corruptor* (superfamily Ichneumonoidea). In *Melittobia*, the males are always short-winged, whereas the females can be short-winged or fully-winged (Matthews et al., 2009; Matthews & Gonzalez, 2008; Schmieder, 1933). In *G. corruptor*, the females are always wingless, whereas the males can be short-winged or fully-winged (Salt, 1952). In ants, wingless males have evolved infrequently, notably in the species *Hypoponera bondroiti* and in the genus *Cardiocondyla* (Heinze, Trindl, Seifert, & Yamauchi, 2005; Oettler, Suefuji, & Heinze, 2010; Yamauchi, Kimura, Corbara, Kinomura, & Tsuji, 1996). These species represent instances where the evolution of sexual wing dimorphism (winged female queens and wingless males) and wing polyphenism (winged queen and wingless workers) also overlap in ants. Moreover, these observations reveal that, for the majority of cases, the females are the wingless morph in both the evolution of sexual wing dimorphism across the Hymenoptera and wing polyphenism in ants. Furthermore, two out of four wasp taxa that have wing polyphenism within one sex also develop sexual wing dimorphism between the sexes.

Ants are a sister-taxon to the Apoidea superfamily, which constitutes sphecid wasps and bees (Fig. 2; Branstetter et al., 2017; Johnson et al., 2013; Peters et al., 2017). Since sexual wing dimorphism is rare in Formicidae and Apoidea (Fig. 2), we can infer that sexual wing dimorphism was absent in the ancestral ant lineages and their immediate wasp relatives. However, there may have existed an ancestral developmental potential, which is a latent genetic program for a specific trait that evolved in the ancestors of a group of organisms and is retained in the descendants. Such potentials can be subsequently realized resulting in the independent or parallel evolution of a trait in response to certain mutations or environmental perturbations (Fave et al., 2015; Rajakumar et al., 2012; West-Eberhard, 2003). The frequent and repeated evolution of sexual wing dimorphism in the Apocrita (see circles in Fig. 2) raises the possibility that the ancestral lineage leading the origin of ants may have possessed a latent developmental potential to generate a sexual wing dimorphism that was retained from the ancestor of the Apocrita. If the developmental potential to produce sexual wing dimorphism was present in early ants, or even in the common ancestor of Formicidae and Apoidea, then we propose the bold hypothesis that this potential may have been co-opted to give rise to wing polyphenism within the female caste of ants. A transition from genetically determined sexual dimorphism to an environmentally-induced wing polyphenism is possible if the developmental mechanism underlying the regulation of sexual dimorphism becomes sensitive to environmental stimuli (Schwander & Leimar, 2011). In the section titled "Evolution of wing polyphenism via genetic accommodation," we propose a mechanism through which such a transition could have evolved as well as a mechanism for the *de novo* origin of wing polyphenism in the absence of developmental potentials in the ancestral lineages that led to the evolution of ants.

Additionally, we explored if there are similar environmental factors and selective pressures that led to the parallel or convergent evolution of wing polyphenism across the five Hymenopteran groups (including ants). Temperature, photoperiod, nutrition, and brood size appear to be involved in regulating wing polyphenism across these five groups (Consoli & Vinson, 2004; Hamilton, 1979; Heinze, 2017; Innocent, Abe, West, & Reece, 2010; Wang, Wei, Yang, Jennings, & Duan, 2016; Wheeler, 1986). This would not be surprising given that these environmental factors are also found to trigger the development of a wing polyphenism in species outside of the Hymenoptera, such as aphids, rice planthoppers, and crickets (reviewed in Zhang, Brisson, & Xu, 2019). Furthermore, in many cases, the selection pressure behind the origin of a wing polyphenism is a trade-off between remaining in the nest or dispersing. In *Mellitobia* wasps, fig wasps, and *Cardiocondyla* ants, wing polyphenism is associated with a trade-off between different mating strategies—mating with conspecifics in the confines of the parental nest or dispersing to mate outside of the parental nest. Specifically, in the wasp genus *Mellitobia*, winged or wingless morph development depends on nutritional levels where decline in nutrition (due to an increase in brood size) leads to the production of winged females that disperse and mate outside of the parental nest (Consoli & Vinson, 2004; Innocent et al., 2010; Schmieder, 1933). Fig-pollinating wasps nest in the ovaries of fig flowers where the males have evolved a wing polyphenism linked to different mating strategies. The male fig wasps develop either into winged morphs that disperse or wingless morphs that mate inside the confines of the nest. The development of wingless males is found to be correlated with a large brood size (Hamilton, 1979). Finally, in the ant genus *Cardiocondyla,* the production of winged or wingless males is also due to selection between mating in the natal nest and dispersal (reviewed in Heinze, 2017). This is similar to the female wing polyphenism in ants where winged queens disperse and wingless workers remain in the parental nest.

The ecology and life-history traits of the Apocrita clade may have facilitated the evolution of sexual wing dimorphism. The Apocrita are characterized by having a restriction in their abdomen (known as "wasp-waist") and a parasitoid lifestyle (see red triangles in Fig 2). The females use their wasp-waist to bend their abdomen and lay their eggs on or in the larvae of other insects (Whitfield, 2003). The parasitoid female wasps commonly locate host larvae that are well concealed in ground nests, tree bark, or plant galls—a behavior suggested having evolved as selection against parasitizing exposed larvae that would be vulnerable to predators (Whitfield, 1998). Although this idea has not been formally tested, the evolution of a sexual wing dimorphism is correlated with a parasitoid lifestyle where the females burrow through soil or small spaces in search of host larvae (Deyrup, 1988; Kimsey, 1991; Quicke, 2015). Furthermore, it has been previously suggested that wingless females evolved as an adaptation to burrowing through cryptic spaces as (1) wings hamper movement and become abraded; and (2) energy used for development of flight muscles can be instead allocated toward muscles used for digging (Deyrup, 1988; Evans, 1969; Quicke, 2015). Therefore, a parasitoid lifestyle which involves navigating through concealed environments in search of hosts may have been a strong selective pressure for the repeated evolution of sexual wing dimorphism in wasps. This selection pressure is different from that which induces wing polyphenism in wasps (trade-off between mating strategies, see above). In ants, however, wing polyphenism and wing shedding behavior in queens after mating, is in part associated with their ability to inhabit ground and subterranean strata. Therefore, the selective pressure for generating sexual wing dimorphism in wasps (wingless female wasps) and wing polyphenism in ants (wingless worker caste) may have evolved in response to similar environmental factors and selective pressures.

In the Apoidea superfamily, which includes sphecid wasps and bees, the evolution of sexual wing dimorphism is rare (Fig. 2). Ecological or behavioral changes that occurred in Apoidea may account for the decreased selection for evolving wingless females in this group. The majority of sphecid wasps construct nests where they provision their young with hosts to consume as they develop (see yellow triangles in Fig. 2). Since the host or prey is translocated to the nest, the sphecid wasps no longer need to burrow through cryptic habitats to locate hosts for their young. In fact, many sphecid wasps parasitize adult insects as opposed to their well-concealed larvae (Goulet & Huber, 1993). Perhaps it was this ecological innovation (i.e., nest building and provisioning) that contributed to lowering the selection pressure to evolve a sexual wing polyphenism. In bees, in addition to nest building, the parasitoid lifestyle was completely substituted with provisioning the young with pollen and nectar (see green triangles in Fig. 2; Michener, 2000). Consequently, this suggests that there was no selection for developing a sexual wing dimorphism since flight is essential in this group for provisioning the larvae with food. Nest-building and pollen-collecting have also evolved in some species of vespid wasps (family Vespidae), which also coincides with the absence of sexual wing dimorphism in this group (Fig. 2; Goulet & Huber, 1993). Finally, the transition from a parasitoid lifestyle to that of gall-forming (where females lay their eggs on a plant and the developing larvae induce the formation of a protective gall where they feed on the plant tissue) and fig-pollinating (where females lay their eggs in the ovaries of fig flowers while simultaneously pollinating them) occurs in lineages where the evolution of a sexual wing dimorphism is sporadic (see orange and blue triangles in Fig. 2; Whitfield, 2003). Taken together, compared with a parasitoid lifestyle, the evolution of nest-building and provisioning, pollen-collecting, gall-forming, and fig-pollinating may have decreased the selective pressure to evolve a sexual wing dimorphism in Apoidea and other hymenopteran families.

3. Ant fossils provide evidence that wing polyphenism and eusociality in ants arose together during the Early Cretaceous

The oldest ant fossils found to date are from the Early Cretaceous dating approximately 100 million years ago (mya) (Barden, 2016; Perrichot, Lacau, Neraudeau, & Nel, 2008). The majority of ant fossils found in the Early Cretaceous are members of extinct and early-branching subfamilies. In addition, several of the Early Cretaceous fossils discovered belong to the crown group of ants, which are extant and derived ant lineages (Grimaldi & Agosti, 2000; Perrichot, 2019). These fossil findings, in addition to molecular analyses, place the origin of the crown group of ants between 111 and 169 mya (Borowiec, 2019; Brady, Schultz, Fisher, & Ward, 2006; Moreau & Bell, 2013; Moreau, Bell, Vila, Archibald, & Pierce, 2006). Therefore, the extinct and early-branching subfamilies of ants (stem group) must have originated earlier than 111–169 mya. These Early Cretaceous fossils, although rare, have been instrumental in revealing the origin and ancestral characteristics of ants, particularly the origins of wing polyphenism and

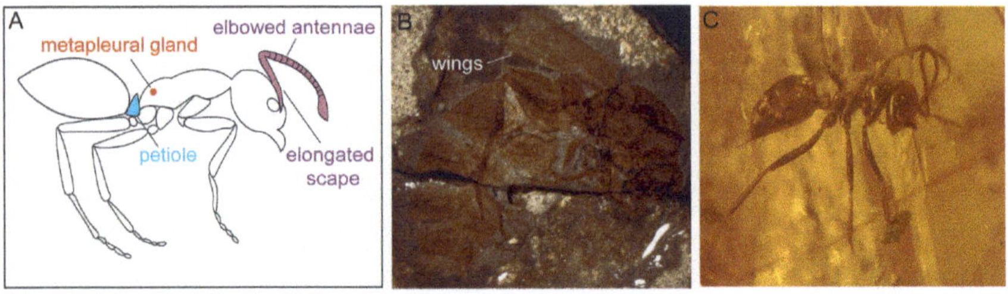

Fig. 3 Ant synapomorphies and Early Cretaceous ant fossils. (A) an illustration of the four morphological synapomorphies used to distinguish ants: metapleural gland (orange circle), petiole (blue triangle), elbowed antennae and elongated scape (purple). (B) rock imprint fossil of Armaniidae (*Armania robusta*). (C) *Sphecomyrma freyi* (image by Frank M. Carpenter, courtesy of Edward O. Wilson). *Panel B: reprinted from LaPolla, J.S., Dlussky, G.M., & Perrichot, V. (2013). Ants and the fossil record.* Annual Review of Entomology, 58, *609–630. https://doi.org/10.1146/annurev-ento-120710-100600 with permission.*

eusociality (Perrichot, Nel, Neraudeau, Lacau, & Guyot, 2008; Wilson, 1987; Wilson, Carpenter, & Brown Jr., 1967a).

The main morphological synapomorphies used to define ants (with the exception of secondary losses in certain taxa) include a metapleural gland, a petiole (constriction of the second abdominal segment), elongated scape (the first antennal segment), and elbowed antennae (Fig. 3A; Barden, 2016; Engel & Grimaldi, 2005; Wilson et al., 1967a). Additionally, some authors use non-morphological synapomorphies such as eusocial behavior and the presence of a wingless worker caste (Bolton, 2003; LaPolla et al., 2013). The metapleural gland, which is a uniquely evolved feature of ants, serves numerous functions from communication to antisepsis and hygiene (Yek & Mueller, 2011). It is regarded by many paleomyrmecologists as the primary synapomorphy that defines true ants (Barden, 2016; Bolton, 2003; Engel & Grimaldi, 2005; Ward, 2007; Wilson et al., 1967a)

A recently discovered fossil from the mid-Cretaceous, known as the Camelosphecia, represents the extinct sister clade to the ants (Boudinot, Perrichot, & Chaul, 2020). This clade, along with the ants (family Formicidae), compose s the superfamily Formicoidea. In addition to the lack of a metapleural gland, the Camelosphecia are only known from winged males and females, suggesting that wingless female workers are absent in this group (Boudinot et al., 2020) . The first appearance of what might be considered the earliest ants are the Armaniid ae (Fig. 3B). This group is known exclusively from poorly preserved rock impressions of winged individuals from the Albian (~100–110 mya), part of the Early Cretaceous

(Dlussky, 1983, 1999). These specimens have ant-like characteristics such as a poorly developed petiole and a short scape, but the poor preservation of impression fossils has made the presence of a metapleural gland in Armaniid ae extremely difficult to ascertain. Similar to Camelosphecia, no wingless specimens of Armaniid ae have been found to date that would indicate the presence of a worker caste (Dlussky, 1983, 1999). The lack of a wingless worker caste from the Armaniidae can be explained by one of the following: (1) these early ants were solitary; (2) they lived in groups but did not have a social structure; (3) there was a social structure but the females were not morphologically differentiated; and (4) they were morphologically differentiated (i.e., presence of wingless worker caste) but conditions did not allow workers to be fossilized. These types of fossils that are formed in lake sediments are biased to include aquatic insects or insects that fell into the lake while flying above (LaPolla et al., 2013). Therefore, if a wingless worker caste did exist in the Armaniid ae then it is improbable that it would have been fossilized in lake sediment.

One of the oldest unequivocal ant fossils is *Sphecomyrma freyi* (Fig. 3C) in New Jersey amber (dating ~92 mya) (Wilson et al., 1967a). This was the first discovery of a Cretaceous fossil that undoubtedly had a metapleural gland (Agosti, Grimaldi, & Carpenter, 1998). The primitive characteristics of *S. freyi,* in addition to its remarkable combination of ant-like and wasp-like features, led to the establishment of the Sphecomyrminae subfamily (Wilson et al., 1967a; Wilson, Carpenter, & Brown, 1967b), considered by some myrmecologists to be the most basal subfamily in Formicidae and the closest relative to all extant ants (Wilson et al., 1967a). The fact that *S. freyi* is known from two wingless workers preserved in the same resin strongly indicates that individuals in the colonies of this species were wing polyphenic and eusocial because it is improbable for two ants to be trapped in the same resin unless they were foraging together (LaPolla et al., 2013; Perrichot, Lacau, et al., 2008). This fossil discovery, in addition to the fact that wing polyphenism is present in all extant ant lineages, suggests that wing polyphenism evolved either just prior to or shortly after the origin of ants.

Following the discovery of the Cretaceous ant fossil *S. freyi*, further support for the early origin of wing polymorphism in ants came from the discovery of wingless females in numerous other Cretaceous fossils (Agosti et al., 1998; Barden & Grimaldi, 2014; Barden & Grimaldi, 2016; Engel & Grimaldi, 2005; Wilson et al., 1967a). Wilson's (1987) analysis of various wingless ant fossils showed that they morphologically group with modern ants and hence are part of a worker caste as opposed to their grouping with solitary wingless female wasps (see Section 2). Perhaps the strongest support for the existence of winged and wingless castes is the discovery of winged queens or males, and most importantly, the discovery of both winged queens or males and wingless workers from the same species (Barden & Grimaldi, 2012; Perrichot, Nel, et al., 2008). Numerous fully-winged males and females, as well as dealate females (queens that have shed their wings), have been discovered and classified into at least six distinct genera (Barden & Grimaldi, 2012; Barden & Grimaldi, 2016; Grimaldi, Agosti, & Carpenter, 1997; Perrichot, Nel, et al., 2008). Among these, a winged queen was discovered preserved in French amber (dating ~100 mya) alongside two wingless workers belonging to the Sphecomyrminae species *Haidomyrmodes mammuthus* (Perrichot, Nel, et al., 2008). This finding removes any uncertainty about the presence of wing polyphenic female caste in early ants with a eusocial lifestyle.

The fossils of *H. mammuthus* also suggest that the ancestral condition of ants are colonies with queens and workers with little size dimorphism and the principal morphological difference between them is the wing polyphenism (Barden & Grimaldi, 2016; Perrichot, Nel, et al., 2008). This is also supported by less derived ant subfamilies, such as the Ponerinae and Leptanillinae, which are generally characterized by colonies in which the queen and worker caste also show little size dimorphism or other morphological differences between them (Brady et al., 2006; Masuko, 1990; Peeters, 1997; Wilson & Holldobler, 2005b). This contrasts with the derived subfamilies, where the queen can be up to ten times larger than the workers (Peeters, 1997). The few species for which data is available show there is a correlation between the timing of caste determination and morphological divergence between castes; the earlier caste determination occurs, the more divergent queen and worker morphologies become (Fjerdingstad & Crozier, 2006; Wilson, 1954). Consequently, if the ancestral condition of ant colonies is the absence of pronounced queen-worker size dimorphism, then we can infer that caste determination in early ants may have occurred during a late stage of larval development. This makes less derived ant subfamilies with low queen-worker dimorphism a good model to infer the developmental steps to the origin of wing polyphenism ants. This topic will be further discussed in Section 5.

Finally, early Cretaceous fossils also provide evidence on the nature of eusociality in these early ant lineages. First, studies in crown-group ants (derived lineages) show that the primary role for the metapleural gland is in colony hygiene and chemical defense (Yek & Mueller, 2011), and therefore, its presence in early ants suggests that its evolution allowed them to live and defend their colonies in the ground or in rotting wood above the ground. Second, Barden and Grimaldi (2016) reported two remarkable samples of amber each containing an aggregation of 6 and 11 *Gerontoformica spiralis* workers. This finding, along with the presence of a prey item in one of the amber samples, led the authors to suggest that these ants were social and performed group search and recruitment of prey. Finally, a notable piece of amber also reported by Barden and Grimaldi (2016) contains two workers from different species engaged in a fight. This suggests the presence of warfare between different species and aggression towards invaders, which is a common feature of a social or cooperative lifestyle. Collectively, in all fossils where a social organization is apparent, the presence of wingless workers is also found, indicating that wing polyphenism and eusociality originated within a very short period of evolutionary time of one another.

4. The implications of originating first: Wing polyphenism versus eusociality

Although there is little doubt that the early ant lineages were both eusocial and wing polyphenic (Barden & Grimaldi, 2016; Perrichot, Nel, et al., 2008), it remains unclear which of these key traits originated before the other. If wing polyphenism followed the evolution of eusociality then it may have been instrumental to maintaining or reinforcing eusociality. On the other hand, if wing polyphenism originated first, then it may have facilitated the origin of eusociality. And of course, it remains possible that, although these two events evolved very close to one another in evolutionary time, they may have evolved independently and without mutually influencing each other. Here we explore the implications of three possible scenarios: (1) eusociality evolves before wing polyphenism; (2) eusociality evolves after wing polyphenism; or (3) eusociality evolves at the same time as wing polyphenism.

In all three scenarios (Fig. 4A–C1), we can assume that the ancestral wasp lineage leading to the ants was solitary, preyed on arthropods, and constructed/inhabited nests similar to the sphecid wasps, which are the closest relatives to ants (see Section 2; Branstetter, Danforth, et al., 2017; Johnson et al., 2013; Peters et al., 2017). In the first scenario or model, which we term the "eusociality first model," the formation of groups composed of parents and their daughter offspring may have evolved immediately for the purpose of sharing and defending a nest (Fig. 4A2; Nowak et al., 2010).

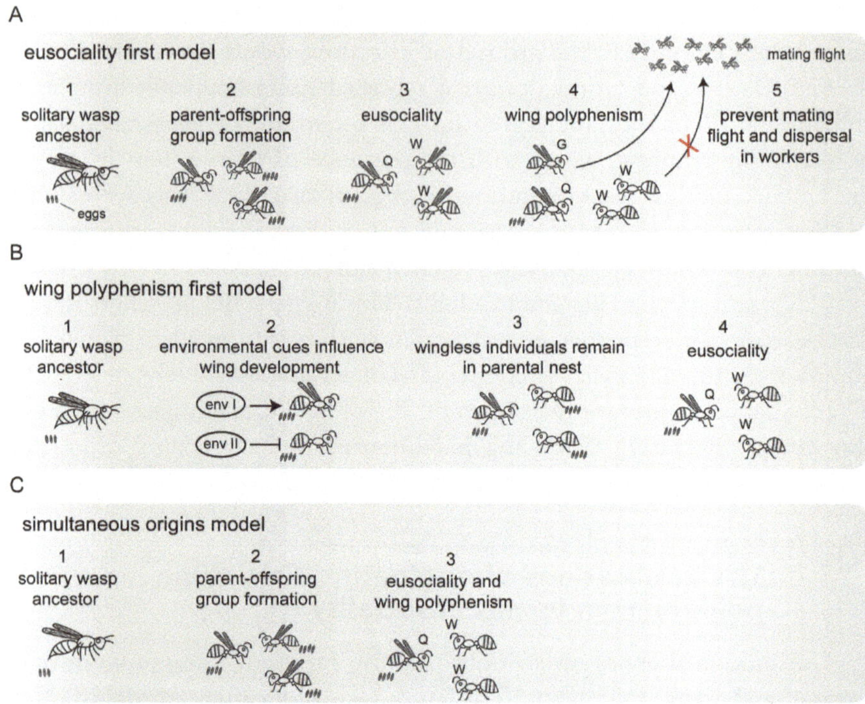

Fig. 4 Hypothetical scenarios for the origin of wing polyphenism and eusociality in the ancestral ants. We propose three models: (A) eusociality first model, (B) wing polyphenism first model, and (C) simultaneous origins model. In all the models, ants evolved from ancestral wasps that constructed nests where they laid their eggs (step 1 in A–C). In the eusociality first model (A): (step 2) formation of parent-offspring groups that shared a nest site; (step 3) division of labor, brood care, and overlapping generations amongst group members led to the evolution of eusociality with reproductive queen (Q) and nonreproductive offspring (worker, W); (step 4) eusociality, among other factors, contributed to the evolution of wing polyphenism with wingless workers and winged gynes (G) that will become future queens; (step 5) wing polyphenism prevents wingless workers from participating in mating flights, while allowing winged gynes the same capability. In the wing polyphenism first model (B): (step 2) wing polyphenism evolved in response to different environmental conditions (env I and env II); (step 3) the wingless morphs remain in the parental nest (group formation); (step 4) group living advances to eusociality. In the simultaneous origins model (C): (step 2) formation of parent-offspring groups that shared a nest site; (step 3) various ecological and/or social factors led to the simultaneous origin of wing polyphenism and eusociality.

Subsequent to the formation of such groups, cooperation and reproductive division of labor appeared leading to the origin of eusociality (Fig. 4A3). Although at this stage workers may have had the capacity to reproduce, social regulation by the queen suppressed worker reproduction. Next, the evolution of eusociality, where daughters remained in the nest and foraged on the ground, may have contributed to the origin of wing polyphenism in ants. The reproductive females (gynes) developed wings whereas the non-reproductive females (workers) did not (Fig. 4A4). The rise of a wingless worker caste may have enabled ants to inhabit new niches particularly underground strata. Finally, we propose that the origin of wing polyphenism further reinforced the reproductive division of labor within the colony by preventing workers from participating in mating flights and dispersing far from the nest (Fig. 4A5). Therefore, in this scenario, the origin of wing polyphenism is thought to have reinforced eusociality.

In the second scenario, which we term the "wing polyphenism first model," wing polyphenism evolved in response to different environmental conditions, such as nutritional levels, rearing density, and temperature. This would have led offspring from the same nest to develop either into winged or wingless morphs depending on environmental cues (Fig. 4A2). The development of daughter offspring without wings would have prevented them from dispersing, and as a consequence, they would have remained in the parental nest and mated with nestmates (Fig. 4B3). The sharing of a nest by parent-offspring groups eventually led to cooperation and reproductive division of labor, giving rise to the origin of eusociality (Fig. 4B4). Therefore, in this scenario, the origin of wing polyphenism first is thought to have directly influenced the formation of parent-offspring groups, ultimately leading to the origin of eusociality.

Finally, in the third scenario, which we term "simultaneous origins model," wing polyphenism and eusociality evolved simultaneously. First, the formation of parent-offspring groups (Fig. 4C2) may have evolved for the purpose of sharing and defending a nest (Fig. 4A2; Nowak et al., 2010). Ecological and/or social factors may have then led to the simultaneous evolution of wing polyphenism and eusociality (Fig. 4C3). Wing polyphenism may have either evolved because of ecological factors or through the manipulation of larval nutrition by the mother. At the same time, a reproductive division of labor may have evolved either through the aggression of dominant individuals toward "subordinate" individuals or through aggression of the mother towards her daughters. In this scenario, it remains unclear whether wing polyphenism and eusociality evolved independently and without influencing each other or whether they had mutual influence on each other and were causally linked.

The ancestral habitat of the early ants, which may have contributed to both the origin of wing polyphenism and eusociality, regardless of which of the three scenarios may be true, remains largely unknown. One hypothesis put forth by Wilson and Holldobler (2005b), called the "Dynastic Succession" hypothesis, proposes that ants evolved in leaf-litter and soil surface habitats and then subsequently spread to other strata (underground and canopy) during the rise of the angiosperms. This is supported by the presence of large eyes in the extinct ant genus *Sphecomyrma* indicating that early ants foraged above ground (Wilson et al., 1967a). In contrast, Lucky, Trautwein, Guenard, Weiser, and Dunn (2013) put forth an alternative hypothesis called the "Out of the Ground" hypothesis. Lucky et al. (2013) performed an ancestral state reconstruction of habitat transitions in ants and concluded that the ancestral ants were subterranean (nested and foraged below the surface). Therefore, colonization of leaf-litter and other strata would have evolved only subsequently.

The metapleural gland, a defining feature of ants that is present in all early ant fossils (Agosti et al., 1998; Barden, 2016), may provide insight into the ancestral habitat as well as the relationship between wing polyphenism evolution and eusociality. In extant ants, the gland has been shown to secrete antibiotic substances essential to overcoming the pressures from microbes and parasite transmission associated with living in colonies as well as nesting under the soil or in decomposing leaf-litter or rotting wood (Boomsma, Schmid-Hempel, & Hughes, 2005; Holldobler & Wilson, 1990; Yek & Mueller, 2011). Although there have been no studies on the function of the metapleural gland in early-branching ant lineages, it is reasonable to infer that its original function was for colony hygiene and immunity against soil microbes. The metapleural gland is located on the metapleuron plate of the thorax (Holldobler & Wilson, 1990), where its secretions can be easily accessed and spread to the body through coordinated leg movements (Fernandez-Marin, Zimmerman, Rehner, & Wcislo, 2006). We propose that the presence of wings on the thorax of workers would have blocked access to the metapleural gland, and as a consequence, would have hindered its evolution. If our hypothesis is correct, then the suppression of wings from the thorax of workers (origin of wing polyphenism) may have preceded, and was a necessary condition for, the evolution of the metapleural gland. This would also apply to queens because they shed their wings after mating. It therefore follows that if the metapleural gland is a necessary condition for the origin of eusociality because its antimicrobial and immune functions were necessary for group/colony living, then this would support the "wing polyphenism first model," where wing polyphenism preceded the origin of eusociality. Furthermore, if nesting in subterranean environments (under the soil surface) also depends on the presence of a metapleural gland,

and if we assume that the development of the metapleural gland was made possible by evolving a wing polyphenism, then wing polyphenism must have preceded underground nesting. This raises the possibility that the ancestral ants arose aboveground and moved to soil secondarily (as proposed by the Dynastic Succession hypothesis). These hypotheses regarding the metapleural gland can be tested by investigating whether or not there are biomechanical constraints placed on the metapleural gland by the presence of wings.

5. A model for the molecular and developmental mechanisms regulating wing polyphenism in basal ant lineages

Polyphenism typically begins with sensing an environmental cue, which is then converted into internal signals that alter the developmental trajectory of an organism (Nijhout, 1999). Various types of environmental cues have been shown to regulate wing polyphenism in insects, such as social interactions, pheromones, photoperiod, temperature, nutrition quality, and rearing density (Rajakumar et al., 2018; Zhang et al., 2019). In ants, nutrition, temperature, and social interactions are key environmental cues known to induce worker-queen determination and wing polyphenism (Penick & Liebig, 2012; Wheeler, 1986; Wheeler & Nijhout, 1984). After sensing the ecological or social cue s, internal signals such as insulin signaling and juvenile hormone are released during a sensitive period in development. This sensitive period can be as early as embryogenesis, such as in the genus *Pheidole* (Passera & Suzzoni, 1979), or as late as the last larval instar, such as in the ponerine species *Harpegnathos saltator* (Penick, Prager, & Liebig, 2012).

In ants, wings as well as other adult structures such as legs and antennae, develop from clusters of cells in the larvae called "imaginal discs." These imaginal discs proliferate during larval development, particularly the last larval instar, and terminally differentiate and evaginate during pupal development to become the adult structures (Held, 2002). Queen and male ants develop fully functional adult wings from wing imaginal discs (hereafter wing discs). However, adult workers, although completely wingless, develop vestigial or rudimentary wing discs during larval development (Fig. 5A; Dewitz, 1878; Wheeler & Nijhout, 1981a). The size and shape of the worker rudimentary wing discs varies between species, varies within subcaste of the same species, and correlates with the timing of queen-worker determination (Abouheif & Wray, 2002; Shbailat, Khila, & Abouheif, 2010; Wheeler & Nijhout, 1981b).

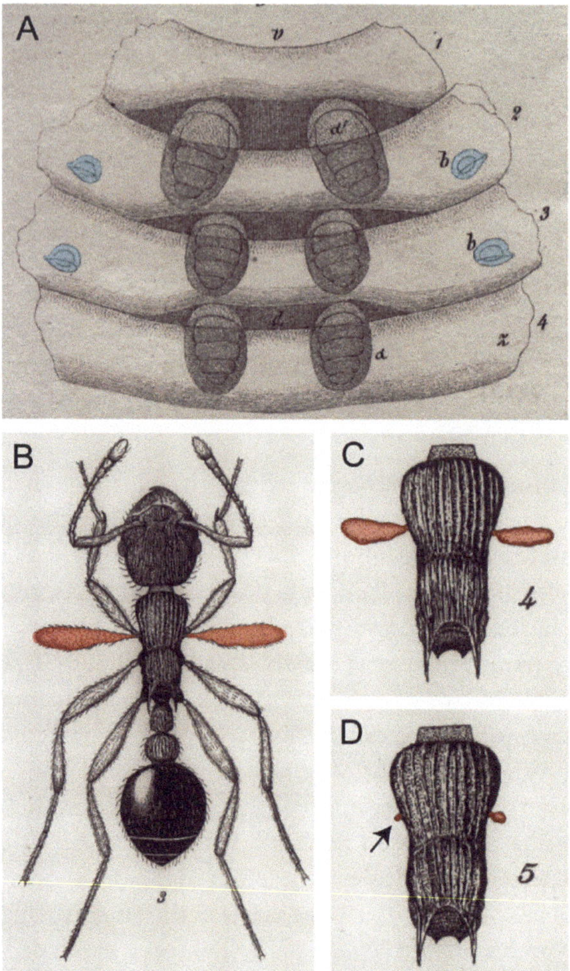

Fig. 5 Rudimentary wing discs and pterergates. (A) ventral view of worker-destined larva of the species *Formica rufa* (subfamily: Formicinae) showing rudimentary fore- and hindwing discs (pseudo-colored in blue). Drawing from Dewitz (1887). (B–C) adult workers with vestigial forewings (pseudo-colored in red) known as "pterergates" from the species *Myrmica rubra scabrinodis* (subfamily: Myrmicinae). Size as well as morphological features of the vestigial wings vary between workers. The left forewing in (D) is minute and nodule like (arrow). Drawings from Wheeler (1905). Small letters and numbers are from original images. See Dewitz (1887) and Wheeler (1905) for full descriptions.

Molecular studies investigating the genetic and developmental mechanisms underlying ant wing polyphenism have been predominantly conducted in the two most species-rich and highly-derived subfamilies: the Formicinae and the Myrmicinae (Fig. 6). These subfamilies are part of the Formicoid

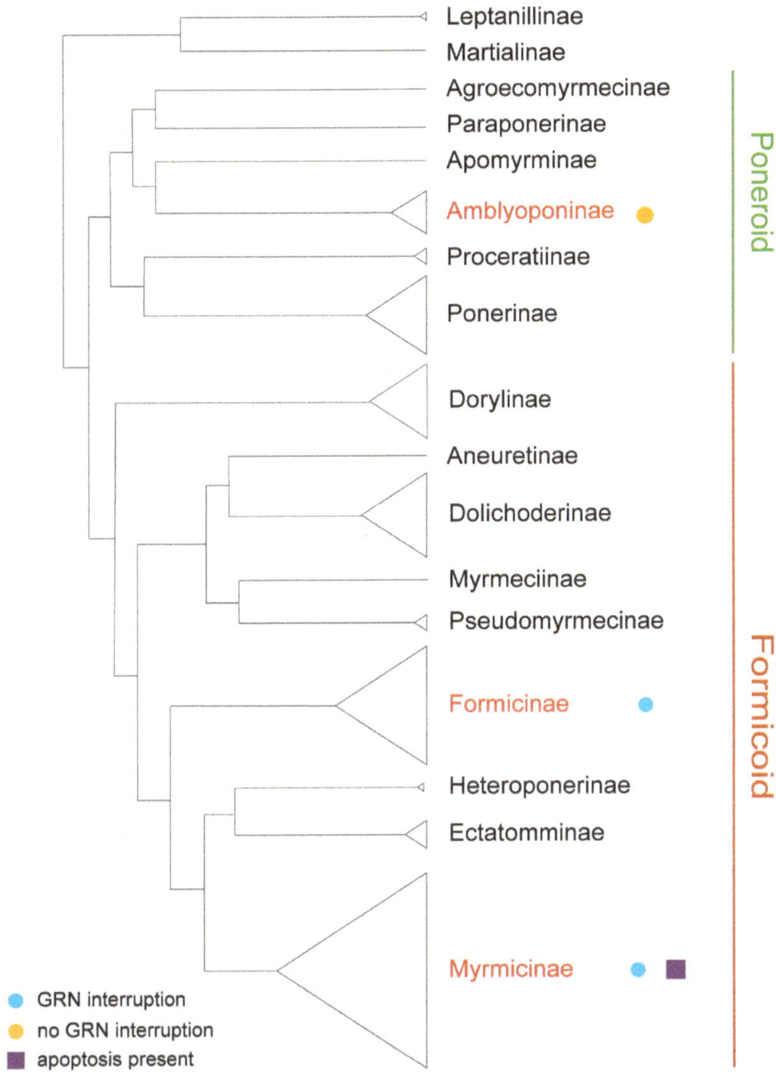

Fig. 6 A phylogeny of the major extant ant subfamilies showing the relationship between the Poneroid and Formicoid clades. Subfamilies colored in red are those where developmental studies on ant wing polyphenism have been conducted. Relative size of triangles indicates size of subfamilies, after Moreau and Bell (2013). Phylogenetic relationships among subfamilies follow (Borowiec, 2019; Branstetter, Longino, Ward, & Faircloth, 2017). Blue circles indicate presence of GRN interruption; yellow circle indicates absence of GRN interruption; purple square indicates presence of apoptosis. Apoptosis has only been tested in species belonging to Myrmicinae subfamily.

clade, which is composed of taxa that have derived developmental and social characteristics (Fig. 6; Borowiec, 2019; Brady et al., 2006). The Poneroid clade, on the other hand, is composed of subfamilies (particularly the Ponerinae) that exhibit less divergent characteristics such as predatory behavior, low queen-worker dimorphism, and small colony size (Peeters, 1997; Wilson & Holldobler, 2005b). These two clades, the Poneroid and Formicoid, have had the same amount of time to diverge from the ancestral phenotype, yet they drastically differ in their evolutionary rates and developmental and social characteristics. The less divergent characteristics of the Poneroid clade, especially the subfamily Ponerinae, reflect the presumed ancestral condition because they are similar to those found in the most basally branching clade of ants: the subfamilies Leptanillinae and Martialinae, and the extinct subfamily Sphecomyrminae (Masuko, 1990; Perrichot, Nel, et al., 2008; Perrichot, Wang, & Engel, 2016; Rabeling, Brown, & Verhaagh, 2008). Unfortunately, these early-branching subfamilies are rarely found in nature and have limited life history information and no developmental studies. Here, we use all available information from ants to propose a model of developmental mechanisms that may have facilitated the origin of wing polyphenism. Due to the limited number of studies on wing polyphenism in ants, we will also base our model on the mechanisms of wing development known from *Drosophila* and Lepidoptera, as well as from other insects that evolved a wing dimorphism.

5.1 Interruption in the wing gene regulatory network (GRN)

In insects, wing development is regulated by a series of transcription factors and signaling molecules (morphogens) that together comprise the wing gene regulatory network (GRN; Fig. 7A; Held, 2002). This GRN has been conserved in insects for at least 400 million years (Abouheif & Wray, 2002; Brisson, Ishikawa, & Miura, 2010; McCulloch et al., 2019; Tomoyasu, Arakane, Kramer, & Denell, 2009). Regulation of the wing GRN is well understood in the fruit fly *Drosophila melanogaster*. In *D. melanogaster*, the wing GRN in the wing discs begins with the activation of the transcription factors *apterous* (*ap*) and *engrailed* (*en*), which specify the formation of the dorsal-ventral (D-V) and the anterior-posterior (A-P) compartments, respectively (Cohen, 1993; Diaz-Benjumea & Cohen, 1993). In the D-V compartment, *ap* initiates the activation of *serrate* (*ser*) which then activates *wingless* (*wg*), while in the A-P compartment, *en* activates *hedgehog (hh)* which in turn activates *decapentaplegic* (*dpp*) (Diaz-Benjumea & Cohen, 1995;

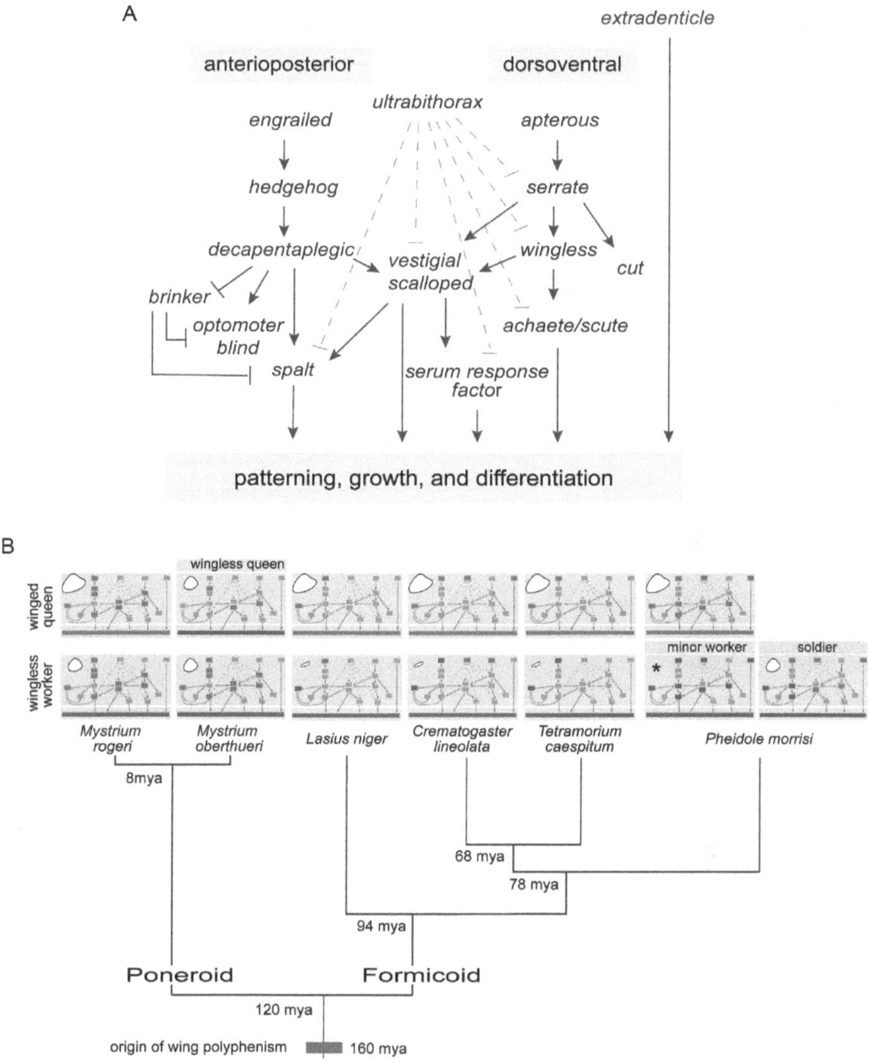

Fig. 7 Wing gene regulatory network in *Drosophila* and network interruption in ants. (A) the wing patterning network in *Drosophila melanogaster* during late larval development. Transcription factors and signaling molecules control the formation of the anterioposterior and dorsoventral compartments. Elements from both compartments regulate *vestigial* and *scalloped* to control the growth and differentiation of the wing discs. Arrows represent activation and perpendicular lines represent inhibition. Gray, dashed perpendicular lines are only active if *ultrabithorax* is expressed. Redrawn from Shbailat and Abouheif (2013) with permission. (B) a summary diagram re-drawn with some modifications from Behague et al. (2018) with permission, showing interruption points in the wing GRN of queens and workers belonging to Formicoid and Poneroid species. Workers belonging to Formicoid species exhibit interruptions in several wing genes (indicated by small, red squares), while workers from Poneroid species do not exhibit interruptions in any of the tested genes (indicated by small, green squares). Asterisk indicates absence of imaginal wing discs.

Kim, Irvine, & Carroll, 1995; Zecca, Basler, & Struhl, 1995). *Dpp* activates *spalt* (*sal*), which controls notum and hinge development, as well as vein positioning in the *D. melanogaster* wing discs (de Celis & Barrio, 2000; Grieder, Morata, Affolter, & Gehring, 2009). Additionally, *dpp* suppresses the expression of *brinker* (*brk*), which when upregulated, suppresses the activity of *Dpp* targets, such as *sal*, to prevent wing disc development (Campbell & Tomlinson, 1999; Martin, Moreno, & Morata, 2004; Minami, Kinoshita, Kamoshida, Tanimoto, & Tabata, 1999; Winter & Campbell, 2004). Ultimately, the D-V and A-P compartments together control the growth of the wing discs through the activation of the wing-specific selector gene complex *vestigial* (*vg*) and *scalloped* (*sd*) (Cohen, 1996; Couso, Knust, & Martinez Arias, 1995; Halder et al., 1998; Kim, Magee, & Carroll, 1997; Simmonds et al., 1998; Williams, Bell, & Carroll, 1991).

Abouheif and Wray (2002) conducted the first study investigating the differences in the wing GRN between winged queens and wingless workers in ants. In this study, the authors initially predicted that a common mechanism for interrupting wing development was present in all workers since wing polyphenism originated once in ants. An interruption is defined as any change in activation, suppression, or expression pattern (spatial/temporal) of any gene in the network when compared to its expression in winged queens or males of that species. Expression analysis of six genes in the wing GRN across four species belonging to the subfamilies Formicinae and Myrmicinae (Fig. 6) showed that different genes were interrupted in the wingless worker caste of different species, between the wingless subcastes (minors and majors) within a single species, and between the forewings and hindwings within a single caste (Fig. 6 blue circles; Fig. 7B). For example, *en* expression was absent in the species *Formica nitidiventris*, *Crematogaster lineolata*, and minor workers of *Pheidole morrisi*, but was expressed in the soldier caste of *P. morrisi* (Abouheif & Wray, 2002). This finding established that, despite the fact that wing polyphenism evolved once at the origin of ants, the genetic mechanism underlying its regulation is labile, even across relatively short time scales.

Since only a proportion of all the genes comprising the wing GRN were sampled, the presence of conserved non-labile genes that mediate wing suppression remains a possibility. *brk* was predicted to be such a gene since it prevents wing disc growth when upregulated (Martin et al., 2004) and is a repressor of *sal* (Winter & Campbell, 2004), which is the only interrupted gene in the soldier subcaste of *P. morrisi* (Abouheif & Wray, 2002). Nahmad, Glass, and Abouheif (2008) constructed a mathematical model to simulate the effect of upregulation or downregulation of *brk*, in addition to other genes in the network, on the growth of the rudimentary wing discs in wingless worker ants. The simulations showed that a 5- or 10-fold increase in *brk* expression led to a significant reduction in *sal* expression as well as a reduction in the size of the rudimentary wing discs. Based on this model, it was predicted that expression of *brk* would be upregulated in the rudimentary wing discs in worker larvae, and thus repressing genes such as *sal* to prevent wing disc growth. However, when formally tested, instead of upregulation, *brk* expression was found to be absent in the rudimentary wing discs of *P. morrisi* soldiers (Fig. 7B; Shbailat et al., 2010). Analysis of *brk* expression in the worker caste of three additional species, *Crematogaster lineolata*, *Tetramorium caespitum*, and *Lasius niger*, consistently showed its absence in the rudimentary wing discs of the worker caste (Fig. 7B; Shbailat & Abouheif, 2013). However, a study in the genus *Mystrium* (Subfamily Amblyoponinae, Poneriod clade; Fig. 6) showed the presence of *brk* expression (Fig. 7B; Behague et al., 2018), suggesting that interruption in *brk* expression (absence) is a common feature that evolved in derived lineages.

Analysis of *brk* and eight additional genes in the Poneroid ant genus *Mystrium* showed a similar expression pattern between the wing discs in queens and workers, suggesting a lack of interruption in the core part of the wing GRN of the worker caste of less-derived lineages (Fig. 6 yellow circle; Fig. 7B; Behague et al., 2018). However, the possibility remains that an interruption in the Poneroid clade may occur downstream of the network. If future studies in the Poneroid clade show that interruption indeed occurs downstream of the network, then it would reveal a compelling evolutionary pattern: late interruption in less-derived lineages and early interruption in more-derived lineages. Furthermore, if interruption does occur downstream of the network in the Poneroid clade, then it might account for the large size of the larval rudimentary wing discs in this group since the wing discs would substantially grow and differentiate before an interruption occurs that halts further development. In contrast, in the Formicoid subfamilies, which generally develop small rudimentary wing discs compared to Poneriod subfamilies, early interruption in the GRN does not allow the rudimentary wing discs enough time to proliferate considerably. However, an evolutionary reversion to this pattern is observed in the rudimentary wing discs in the genus *Pheidole* (subfamily Myrmicinae) (Rajakumar et al., 2012; Rajakumar et al., 2018). Another intriguing possibility is that in the Poneroid subfamilies there is no interruption in the wing GRN and an alternate mechanism, such as apoptosis, may be responsible for the elimination of the rudimentary wing discs. The two possibilities are not mutually exclusive and further studies in this clade are needed to reveal the precise developmental regulation of wing polyphenism in less-derived ant lineages that have ancestral-like characteristics.

5.2 Apoptosis

Apoptosis, also known as programmed cell death, is an important developmental mechanism underlying morphological evolution, and particularly is involved in the production of various polyphenic and dimorphic traits. For instance, apoptosis is involved in the degeneration of pupal wings in female tussock moths (Lobbia, Niitsu, & Fujiwara, 2003), resorption of pupal horns in female dung beetles (Kijimoto, Andrews, & Moczek, 2010), and elimination of male-specific gonads during embryonic development of female fruit flies (DeFalco et al., 2003); in all of these cases, apoptosis facilitates the production of sexual dimorphism. Is apoptosis also involved in generating a wing polyphenism in ants? And if so, is apoptosis induced by the interruption in the wing GRN? A study by Sameshima, Miura, and Matsumoto (2004) showed that apoptosis is present in the rudimentary wing discs of the soldier subcaste of *Pheidole megacephala*. Moreover, in the species *P. morrisi*, Shbailat et al. (2010) showed that the induction of apoptosis is spatially and temporally correlated with a change in expression patterns of two genes in the wing GRN: *sal* and *dpp*. Over five decades, genetic manipulations in *D. melanogaster* have shown that appendage patterning genes can induce apoptosis and that mutations in genes belonging to the wing GRN (such as *ap*, *wg*, *vg*) can induce cell death in the tissue of the wing discs (Fristrom, 1969; Giraldez & Cohen, 2003; James & Bryant, 1981; Sedlak, Manzo, & Stevens, 1984). Therefore, it is possible that interruption in the wing GRN may induce apoptosis to eliminate rudimentary wing discs. Furthermore, if multiple genes in the wing GRN are able to induce apoptosis then this may explain why interruption points are evolutionarily labile. Thus far, only two species across the ants (belonging to Myrmicinae subfamily) have been tested for the presence of apoptosis in the rudimentary wing discs of the worker caste (Fig. 6 purple square; Sameshima et al., 2004; Shbailat et al., 2010). Investigating the prevalence of apoptosis across the Formicidae is critical to determine if it is a conserved mechanism across the phylogeny, and if so, it may have important implications in elucidating the mechanism responsible for generating a wing polyphenism during the early origin of ants.

5.3 Differential gene expression outside the wing GRN

Many other genes outside of the wing GRN may be differentially expressed between queens and worker ants in order to initiate and regulate caste-specific development. Here we will examine some of these genes and infer what role they may have on the specific regulation and evolution of wing polyphenism.

A study by Klein et al. (2016) analyzed the expression of the sex-determining gene, *doublesex* (*dsx*), and showed that it was alternatively spliced between males and females and differentially expressed between the female caste (i.e., queens and workers) and between the male morphs (winged and wingless) during all life stages of the ant *Cardiocondyla obscurior* (subfamily Myrmicinae). This led to the hypothesis that *dsx* is co-opted from its function in the development of sex-specific traits to a novel function in modulating polyphenic traits within one sex. A similar function of *dsx* was also reported in the dung beetle *Onthophagus taurus*, where *dsx* promotes differential horn development between sexes (horned males and hornless females) and within the males (horned large males and hornless small males) (Kijimoto, Moczek, & Andrews, 2012). *Dsx* has also been shown to control sex-specific traits by altering sensitivity to juvenile hormone (JH) in the stag beetle *Cyclommatus metallifer* (Gotoh et al., 2014), suggesting that its role in ant wing polyphenism may also be through modulating JH sensitivity (Klein et al., 2016). Loehlin et al. (2010) showed that wing size reduction in the wasp *Nasonia* is correlated with an increase in *dsx* expression. Future studies should investigate the regulatory role of *dsx* in wasp wing dimorphism and polyphenism, as well as ant wing polyphenism. These studies raise the possibility that there is a shared proximate mechanism underlying sexual wing dimorphism in wasps and wing polyphenism in ants.

A transcriptomics analysis in *C. obscurior* also identified a series of genes involved in sphingolipid metabolism to be differentially regulated between sexes and within castes (Schrader, Simola, Heinze, & Oettler, 2015). Sphingolipids, found on cell membranes, function in various metabolic processes such as fat storage and nutrient utilization to control growth and cell death (reviewed in Kraut, 2011). Specifically, the genes *lace* and *schlank,* responsible for sphingolipid synthesis, were revealed to be weakly expressed in workers and wingless males in *C. obscurior* (Schrader et al., 2015). Interestingly, studies in *Drosophila* have shown that *lace* and *schlank* are required for the development of wing discs, where *lace* mutants exhibit incision in the wing margin and a high number of dead cells

(Adachi-Yamada et al., 1999), and downregulation of *schlank* affects *wg* activity, increases levels of apoptosis, and reduces cell size of the wing discs (Pepperl et al., 2013). Therefore, downregulation of these genes in worker ants and wingless males may be affecting wing development that leads to wingless morphs. Transcriptomics studies are likely to provide many more candidate genes that should be functionally tested for their role in regulating wing polyphenism in ants.

5.4 Juvenile hormone, ecdysone, and insulin signaling

JH and ecdysone regulate molting and metamorphosis in insects (Nijhout, 1994; Truman & Riddiford, 2002). Ecdysone is a steroid hormone that stimulates molting events, which represent the transition from one larval stage to the next, as well as metamorphosis, which represents larval to pupal transition. The level of JH during an ecdysone pulse determines whether this pulse will cause molting or metamorphosis: high JH results in molting, while low or absent JH results in metamorphosis (Emlen & Allen, 2004; Nijhout, 1994; Nijhout & Wheeler, 1982; Truman & Riddiford, 2002). JH and ecdysone, in addition to insulin signaling, also influence the body and imaginal disc growth in insects (reviewed in Edgar, 2006; Martin & Shearn, 1980; Mirth, Truman, & Riddiford, 2009; Shingleton, Das, Vinicius, & Stern, 2005; Tobler & Nijhout, 2010; Truman et al., 2006), and are involved in the regulation of various polyphenisms including caste polyphenism in social Hymenoptera and termites, male horn dimorphism in beetles, winged/wingless and sexual/asexual phases in aphids, solitary and gregarious phases in locusts, and color polymorphism in butterflies (reviewed in Hartfelder & Emlen, 2012; Nijhout & McKenna, 2018; Nijhout & Wheeler, 1982). Direct studies on the influence of JH, ecdysone, and insulin signaling on wing polyphenism in the worker caste of ants are limited relative to information about their role in other wing dimorphic insects (this topic was recently reviewed by Zhang et al., 2019). We will therefore synthesize what is known about the interactions between the environment, endocrine hormones, GRN interruption, and apoptosis in ants and other insects to propose a model for the developmental and genetic mechanisms that may have been present during the origin of wing polyphenism in ants.

In ants, as well as other social Hymenoptera, there is ample evidence that high levels of JH during a sensitive period can induce queen development (reviewed in Nijhout & Wheeler, 1982; Wheeler, 1986). Induction of queen development by JH would suggest that JH is positively regulating the development of wings among other queen-specific traits. In support of this, Rajakumar et al. (2012) showed that the application of the JH analog, methoprene, stimulates the growth of the rudimentary wing discs in the soldier subcaste of *P. morrisi*. In contrast, studies in Lepidoptera and *Drosophila* have demonstrated that JH inhibits the growth, differentiation, and evagination of the wing discs (Chihara, Petri, Fristrom, & King, 1972; Miner, Rosenberg, & Nijhout, 2000; Tobler & Nijhout, 2010; Truman et al., 2006). Furthermore, studies in insects with a wing dimorphism, such as the cricket *Gryllus rubens*, the brown treehopper *Nilaparvata lugens*, the stonefly *Zelandoperla fenestrate*, and the aphid *Megoura crassicauda*, collectively have shown that presumptive short-winged (in crickets and brown treehoppers) or wingless (in aphids and stoneflies) morphs have elevated levels of JH titers and that topical application of JH, or its analog, inhibits wing development and results in the production of short-winged or wingless morphs (Ayoade, Morooka, & Tojo, 1999; Bertuso, Morooka, & Tojo, 2002; Ishikawa, Gotoh, Abe, & Miura, 2013; Iwanaga & Tojo, 1986; McCulloch et al., 2019; Zera, 2004, 2006). The reversed role of JH that evolved in ants as compared with other insects from numerous orders suggests that when JH was co-opted to control queen-specific development in social Hymenoptera, it was also co-opted to positively regulate wing development in queens during the origins of ants and the evolution of a wing polyphenism.

Unlike JH, the molting hormone ecdysone has received less attention in caste-determination studies. The limited number of studies investigating its role have shown that in ants, worker-biased eggs and worker larvae have higher ecdysone levels than queen-biased eggs and queen larvae (Schwander et al., 2008; Suzzoni, Passera, & Strambi, 1980, 1983). Despite this correlation, the role of ecdysone in caste determination and ant wing polyphenism has yet to be investigated. Its role on wing development has been well elucidated from studies in *Drosophila* and the moth *Manduca sexta*, where it is found to induce growth and evagination of the wing discs (Chihara et al., 1972; Nijhout et al., 2007). Furthermore, ecdysone is known to upregulate numerous genes in the wing GRN during *Drosophila* metamorphosis (Li & White, 2003). However, this positive regulation of wing development is contrary to its role in wing morph differentiation in the pea aphid *Acyrthosiphon pisum*. A study by Vellichirammal, Madayiputhiya, and Brisson (2016) conducted functional testing of ecdysone, as well as its antagonists, and consistently showed that higher maternal ecdysone levels resulted in an increased percentage of wingless aphid progeny. This suggests that high ecdysone levels contribute to wing suppression. In several species of sexually wing dimorphic moths (winged males and wingless females) the injection of ecdysone was found to induce apoptosis in the female wing discs leading to their degeneration (Lobbia et al., 2003; Niitsu et al., 2014; Niitsu, Lobbia, Izumi, & Fujiwara, 2008). Similarly, in the beetle genus *Onthophagus*, which exhibits a horn dimorphism, ecdysone-induced apoptosis is implicated in regulating horn development in species and sex-specific manner (Kijimoto et al., 2010). Therefore, the role of ecdysone in ants may be similar to its role in regulating dimorphic traits in these insects. In taxa where GRN interruption is absent (Fig. 6 yellow circle; Behague et al., 2018), we propose that high levels of ecdysone in worker-destined larvae may be inducing apoptosis to eliminate or halt the growth of the wing discs.

The insulin/insulin-like signaling pathway (IIS) is a highly conserved nutrient-sensing pathway that serves various functions in metabolism, reproduction, longevity, cell size and, growth (reviewed in Edgar, 2006; Wu & Brown, 2006). IIS and its downstream target the TOR (target of rapamycin) pathway are known to be regulators of polyphenisms that are triggered by differential nutrient levels and involve differences in body and organ size (reviewed in Nijhout & McKenna, 2018). Specifically, insulin signaling regulates caste determination in the honey bee *Apis mellifera,* where high expression of IIS and TOR are found in queen-destined larvae, and their knockdown results in worker-like morphologies (de Azevedo & Hartfelder, 2008; Mutti et al., 2011; Patel et al., 2007; Wheeler, Buck, & Evans, 2006; Wheeler, Buck, & Evans, 2014; Wolschin, Mutti, & Amdam, 2011). Furthermore, knockdown of insulin receptors and TOR leads to reduced levels of JH in queen-destined larvae and a conversion to worker-like morphologies, revealing that JH is downstream of the nutrient-sensing pathways in establishing caste-determination (Mutti et al., 2011). In ants, studies investigating reproductive division of labor have revealed that insulin signaling is upregulated in the ovaries of adult reproductives in the genus *Diacamma* (Okada et al., 2010), and in the brains of adult reproductives from seven species sampled from the Poneroid and Formicoid clades (Chandra et al., 2018). These studies suggest that reproductive adults with high IIS emerge from larva and pupa that received high-quality nutrition and thus had high IIS levels. However, a study in the fire ant *Solonopsis invicta* has shown that IIS receptor expression levels may fluctuate between the larval and the adult stage (Lu & Pietrantonio, 2011). These disparate findings indicate that the role of IIS in ants is complex and requires further and more rigorous exploration.

5.5 A model for the developmental processes underlying the origin of wing polyphenism in early ants

Here, we synthesize known interactions between the environment, endocrine hormones, the wing GRN, *dsx*, and apoptosis in a model for the regulation of wing polyphenism at the origin of ants (Fig. 8). Although here we focus on the major endocrine hormones that regulate trait development, we acknowledge that other numerous signaling pathways are present and should be integrated into future models. The first component of the model proposed here is the sensing of environmental and social factors such as temperature, nutrition, pheromones, and worker behavior (Penick & Liebig, 2012; Wheeler, 1986; Wheeler & Nijhout, 1984). The nutritional regulation of

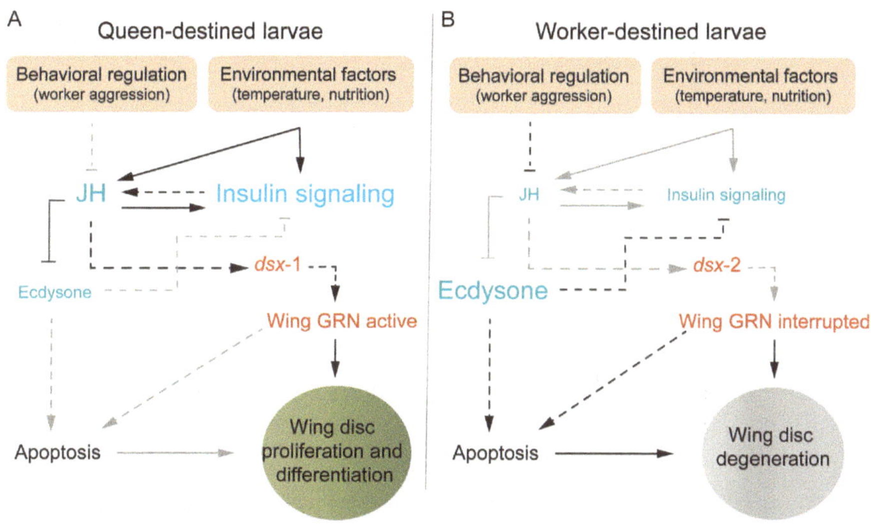

Fig. 8 A proposed model for the interactions between the environment, endocrine hormones, wing GRN, *dsx*, and apoptosis in the development of wing polyphenism in ants. Pathways for the development of a winged queen (A) and a wingless worker (B). Environmental and social cues regulate JH and insulin signaling levels. In turn, JH regulates ecdysone levels and wing GRN activation. Ecdysone and the wing GRN regulate apoptosis. Endocrine hormones are labeled in blue and genes/GRN are labeled in orange. Font size of the endocrine hormones indicates their relative levels. Pathways highlighted in black are activated and those highlighted in gray are inactivated. Arrows represent activation and perpendicular lines represent inhibition. Solid arrows and perpendicular lines indicate activation/inhibition relationships that have been directly tested in ants. Dashed arrows and perpendicular lines indicate activation/inhibition relationships that have been tested in other insects and await future testing in ants.

castes is predominantly used by more derived species where adult workers regurgitate food to immobile larvae (known as trophallaxis) and thus control larval nutritional levels (Cassill & Tschinkel, 1995; Wilson, 1971; Wilson & Holldobler, 2005b). In contrast, larvae from species belonging to the Poneroid clade are mobile and trophallaxis occurs much less frequently leaving adult workers with limited control over larval nutrition (Wilson & Holldobler, 2005b). In the Poneroid ant species *H. saltator* where trophallaxis is absent, Penick and Liebig (2012) demonstrated that behavioral regulation in the form of worker aggression is involved in caste determination. Specifically, JH-treated larvae that become queen-destined are aggressed (through biting) by adult workers which leads to a reversion to worker development, possibly through inhibition of JH (Penick & Liebig, 2012). Therefore, we have included worker aggression in our model as one of the factors that may have regulated caste determination and wing polyphenism during the early evolution of ants.

These environmental and social cues are sensed by hormones such as IIS and JH. The absence of worker aggression may lead to an increase in JH synthesis (Fig. 8A), while presence of worker aggression might lead to JH reduction (Fig. 8B). High nutritional levels may also be directly sensed by JH and cause its increased synthesis (Fig. 8A). A study in the ant *Pogonomyrmex rugosus* has shown that IIS may be downstream of JH (Libbrecht et al., 2013), therefore, an increase in JH production may lead to an increase in IIS. Nutritional levels may also be directly sensed by IIS, which can in turn positively regulate JH as shown in *Drosophila* and honeybees (Mutti et al., 2011; Tatar et al., 2001; Tu, Yin, & Tatar, 2005). If future studies provide evidence for the latter, then a positive feedback mechanism between JH and IIS may exist in ants. Therefore, high nutritional levels, as well as the absence of worker aggression, lead to an increase in JH levels and consequently queen development (Fig. 8A). On the other hand, low nutrition and/or the presence of worker aggression result in a decrease in JH synthesis and the activation of worker development (Fig. 8B).

How might differential levels of JH between queen- and worker-destined larvae control wing polyphenism? We present two non-mutually exclusive possibilities. First, high levels of JH in queen-destined larvae induce wing growth through transcriptional regulation of the wing GRN (Fig. 8A). Conversely, in worker-destined larvae, low JH levels are unable to activate the wing GRN leading to interruption points and the degeneration of the rudimentary wing discs (Fig. 8B). This hormonal regulation of wing GRN may be mediated by different isoforms of *dsx*, as suggested by

Klein et al. (2016). Second, JH may be influencing wing development through the modulation of other factors and hormones such as ecdysone. Ecdysone levels, contrary to JH, are high in worker-destined larvae compared with queen-destined larvae (Schwander et al., 2008; Suzzoni et al., 1980, 1983). It has been shown in the ant *P. rugosus*, as well as in *Drosophila*, that JH inhibits ecdysone production (Libbrecht et al., 2013; Mirth et al., 2014; Richard & Gilbert, 1991). Therefore, we propose that high JH in queen-destined larvae reduces ecdysone production (Fig. 8A), whereas in worker-destined larvae, low JH levels are unable to suppress ecdysone and hence its increased levels (Fig. 8B). Ecdysone, in turn, may be regulating insulin signaling. Numerous studies in *Drosophila* have shown that ecdysone antagonizes insulin, whereas insulin can increase ecdysone production (Colombani et al., 2005; Orme & Leevers, 2005). This regulatory interaction between ecdysone and IIS remains to be tested in ants. However, if future studies show that ecdysone antagonizes IIS in ants, then it might reveal a positive feedback loop where ecdysone contributes to its own synthesis by inhibiting IIS and preventing JH increase (Fig. 8B). More importantly, ecdysone regulates dimorphic traits in insects through the induction of apoptosis (Kijimoto et al., 2010; Lobbia et al., 2003). Therefore, we propose that high levels of ecdysone in worker-destined larvae may be activating apoptosis that leads to the degeneration of the rudimentary wing discs (Sameshima et al., 2004; Shbailat et al., 2010). Alternatively, apoptosis may also be induced through an interruption in the wing GRN (Fig. 8B), as suggested by the spatiotemporal correlation between GRN interruption and apoptosis pattern in the ant *P. morrisi* (Shbailat et al., 2010).

6. Origin of wing polyphenism via genetic accommodation

Since the 1800s, and perhaps earlier, entomologists have tried to understand the evolutionary origin of wing polyphenism in ants. Early work, notably by Dewitz (1878) and Wheeler (1905, 1916), relied on morphological and developmental observations of the female caste to elucidate the origin of wing polyphenism in ants. Wheeler (1905) discovered rarely appearing anomalous workers with small vestigial wings (he termed "pterergates") in a colony from the derived subfamily Myrmicinae (Fig. 5B–D). In addition to pterergates, the wide range of wing morphologies that have evolved in the queen caste across species (i.e., winged, short-winged, wingless) led Wheeler to dismiss the idea that wing polyphenism evolved in a single saltational step. Instead, Wheeler (1905, 1916) proposed that the ancestral ant lineage would have exhibited higher variation in the degree of wing development in adult workers, of which only the extremes (winged and wingless) survived in many taxa. Both Wheeler (1916) and Dewitz (1878) recognized the relative ease with which wing development can be altered in ants, leading Wheeler (1916) to propose that the winged queen and wingless worker caste evolved through the gradual suppression of intermediate forms. His theory (Wheeler, 1916) was largely based on those of Darwin (1859), Emery (1894), and later Wilson (1953) for how discrete worker subcastes, such as minor workers and soldiers (Fig. 1), in ants evolved through the gradual suppression of intermediate forms.

In her seminal book *Developmental Plasticity and Evolution* (West-Eberhard, 2003), Mary Jane West-Eberhard proposed that polyphenisms originate from an evolutionary process she called "genetic accommodation." Here, we propose a model for the origin of wing polyphenism in ants via genetic accommodation, which gave rise to two alternative phenotypes—winged queens and wingless workers—that are determined during development in response to specific environmental cues. Genetic accommodation starts with a novel phenotype that is recurrently induced by specific environmental conditions (Moczek et al., 2011; Suzuki & Nijhout, 2006; West-Eberhard, 2003). In ants, environmental induction of the novel wingless worker phenotype may have occurred in at least two alternative ways: (1) it may have been induced by a new or significant change in an existing environmental factor, such as poor larval nutrition, infections by parasites, or temperature fluctuations (Hunt, 1991; Wheeler, 1928a); or (2) it may have been induced by a mutation(s) in the genes that regulate the sensitivity of the hormones or wing GRN to the environmental cue (Billard, Vigne, & Braendle, 2020; Sieriebriennikov et al., 2018). If the novel, environmentally induced, phenotype is adaptive, then selection will act on the genes that control its frequency and form of expression to genetically accommodate it into the population as an alternative, conditionally expressed, phenotype (Moczek et al., 2011; Suzuki & Nijhout, 2006; West-Eberhard, 2003). In ants, this would have led to the accommodation and incorporation of the wingless worker phenotype into the colony giving rise to the origin of wing polyphenism with two alternative phenotypes expressed under specific environmental conditions; a winged queen phenotype adapted for dispersal and a wingless worker phenotype adapted for performing tasks underground.

Previous studies have shown that exposure to a new or to a significant change in an existing environment leads to the induction of a wide range of phenotypic variation (Gibson & Hogness, 1996; Rajakumar et al., 2012; Suzuki & Nijhout, 2006). For example, heat shocking larvae of the moth *Manduca sexta* gives rise to a wide range of color variants (Suzuki & Nijhout, 2006) and shocking fruit fly embryos with ether vapor gives rise to a wide range of variation in the growth of the haltere (hindwing) (Gibson & Hogness, 1996). This means that exposure to discretely different environments alone is not sufficient to produce discrete alternative phenotypes. Therefore, at a mechanistic level, the origin of a polyphenic trait via genetic accommodation also requires the evolution of a developmental threshold. Polyphenic traits have an underlying continuously distributed character (such as a hormone gradient) that has a threshold, which can be defined as the point of discontinuity that results in the expression of discrete traits (Falconer, 1960). For instance, individuals above a certain threshold develop into one morph, while individuals below the threshold develop into an alternative morph. A developmental threshold therefore transforms an underlying continuous character into one with discontinuous phenotypic expression (Falconer, 1960). Hormones mediate environmental cues and affect gene expression, which in turn affects phenotype.

How is a concentration gradient of a hormone, or any other substance, translated into discrete alternative phenotypes without intermediates? Here, we propose two possible mechanisms: the first mechanism is based on a model by Lewis, Slack, and Wolpert (1977). Their mechanism provides an explanation for how discrete phenotypes develop through a sharp transition in gene activity. In this model, production of a gene product is controlled by a continuous substance (e.g., hormone gradient) in a sigmoidal fashion through positive feedback. As a result, changes in hormone levels lead to sharp transitions in the production of the gene product. Depending on the parameters of the model, the shape of the sigmoidal curve can change (Nijhout, 1991)—if the sigmoidal curve between the substance and the gene product has a gentle slope it may result in a broad threshold that gives rise to discrete phenotypes, including intermediates (Fig. 9A–A'). If, however, the sigmoidal curve has a steep slope, then it may result in a narrow threshold that gives rise to discrete phenotypes with no intermediates (Fig. 9B–B').

The second mechanism is based on a toggle binary switch, which is composed of two components that mutually inhibit each other (Gardner, Cantor, & Collins, 2000; Lugagne et al., 2017). These components can be genes, endocrine hormones, and developmental or physiological factors.

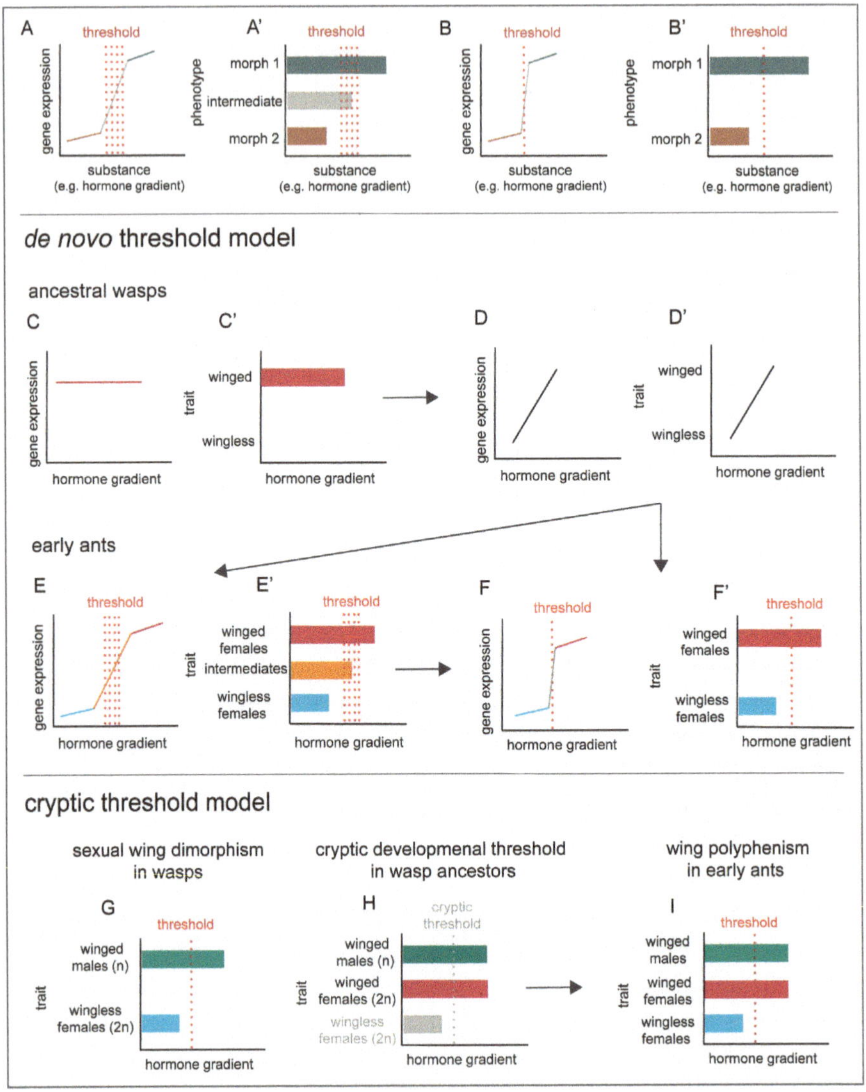

Fig. 9 A proposed conceptual model for the evolution of wing polyphenism via an environmentally regulated developmental threshold. (A) a sigmoidal relationship between a continuous substance, such as a hormone gradient, and gene expression. A broad threshold (illustrated by multiple lines) forms when the sigmoidal curve has a gentle slope. The broad threshold results in the development of discrete phenotypes with intermediates (A′). (B) a sigmoidal curve with a steep slope gives rise to a narrow threshold that results in the development of discrete phenotypes with no intermediates (B′). (C–F′) depict a scenario where a threshold evolved *de novo* in the early ant lineage. (C) in the ancestral wasps, gene expression levels were robust and did not respond to hormonal levels. (C′) the robustness of gene expression resulted in winged phenotype

For instance, a key gene in the wing GRN may be responsible for the activation of wing development, whereas another key gene may be responsible for the inhibition of wing development. These two genes mutually suppress each other resulting in a discrete switch. Two genes from the wing GRN—*dpp* and *brk*—are promising candidates for these mutually inhibiting genes: *dpp* is essential for wing development and it down-regulates *brk* expression; *brk* prevents wing disc growth, and although it does not suppress *dpp*, it suppresses the activity of its targets (Fig. 7A; Campbell & Tomlinson, 1999; Martin et al., 2004; Minami et al., 1999; Winter & Campbell, 2004). The induction of one of these genes over the other may be under the control of hormonal signals which mediate environmental cues (Shbailat et al., 2010).

Based on these two threshold mechanisms, we propose two of many possible models for the origin of wing polyphenism in ants, which we call the "*de novo* threshold model" and the "cryptic threshold model." In the *de novo* threshold model, the developmental threshold that facilitated the origin of wing polyphenism evolved *de novo* in the ancestral lineages leading to the ants. In contrast, in the cryptic threshold model, wing polyphenism in ants evolved from a cryptic developmental threshold that evolved originally to produce winged males and wingless females in their solitary wasp ancestors and was later co-opted to regulate winged queens and wingless workers.

regardless of hormonal levels. (D–D′) a mutation caused gene expression, and the resulting trait, to become sensitive to hormonal levels resulting in a linear reaction norm. (E) in the early ant lineage, a mutation or genetic variation (perhaps gain of a positive feedback mechanism) changed the relationship between gene expression and hormonal levels from linear to sigmoidal. This sigmoidal relationship might have created a broad threshold boundary, indicated by multiple threshold lines, that allowed the development of intermediate phenotypes (E′). Gene expression color in (E) corresponds with phenotype color in (E′). Subsequently, a more narrow threshold, indicated by one threshold line, formed resulting in two discontinuous phenotypes (F–F′). The transition from a broad to a narrow threshold may have not occurred; instead a sigmoidal relationship with a narrow threshold (F–F′) may have directly followed the linear reaction norm (D–D′). (G–I) depict a scenario where wing polyphenism evolved from a pre-existing threshold. (G) sexual wing dimorphism in wasps is controlled by a hormone gradient with a threshold. Genetic factors, such as ploidy levels, may control whether hormone levels fall above or below the threshold. (H) the developmental potential to generate a sexual wing dimorphism was retained but not realized in the wasp ancestors of ants. (I) during the early evolution of ants, the hormone levels of the females became environmentally-sensitive resulting in some individuals falling below or above the pre-existing threshold for wing development. For panels A–F′, gene expression color corresponds with phenotype/trait colors.

In the *de novo* threshold model, we propose that the ancestral wasps did not have a threshold for wing development, instead, this threshold specifically evolved in early ants. In the ancestral wasps, we propose that expression levels of genes in the wing GRN were environmentally insensitive, and therefore, did not vary based on hormonal levels, which mediate environmental conditions (Fig. 9C). This robust gene expression to environmental variation would have produced the constitutive expression of a winged phenotype regardless of hormone levels (Fig. 9C'). Next, we propose that a mutation rendered gene expression levels sensitive to the environmental/hormone gradient resulting in a linear relationship between gene expression and hormonal levels (Fig. 9D). If we assume that gene expression levels also had a positive linear relationship with the trait that they control, then different gene expression levels (which are based on hormone gradient) would generate different phenotypes creating a continuous reaction norm (Fig. 9D'). All the phenotypes in the reaction norm may have been produced in the ancestral wasp lineage, or perhaps the hormone concentration always reached a level that induced sufficient gene expression to produce fully winged individuals. Regardless, in the early ants, the hormone gradient may have evolved a threshold as a result of a mutation or genetic variation that caused one (or multiple) key wing genes to gain a positive feedback mechanism for the synthesis of their protein products. Based on Lewis et al.'s (1977), a sigmoidal relationship between the hormone gradient and a gene product is achieved through the gain of a positive feedback mechanism. Initially, this sigmoidal curve has a gentle slope, and therefore, in addition to producing discrete winged and wingless phenotypes, would also result in the production of individuals with a range of intermediate wing lengths (Fig. 9E'). Subsequently, a transition to a sigmoidal curve with a narrow slope evolves producing two discrete phenotypes with no intermediates (Fig. 9F–F'). The transition from a broad threshold boundary to a narrow one may have evolved very rapidly, reducing the likelihood that intermediate wing phenotypes were preserved in the fossil record.

This gradual transition from environmental insensitivity, to a linear reaction norm, to a sigmoidal curve with a broad threshold, and finally, to a sigmoidal curve with a narrow threshold is consistent with Wheeler's (1905, 1916) hypothesis that the origin of wing polyphenism (winged queens and wingless workers) evolved through the gradual suppression of intermediates. However, we cannot rule out that a *de novo* threshold originated in a single saltational step from a linear reaction norm (Fig. 9D–D') to a sigmoidal curve with a narrow threshold (Fig. 9F-F'). This can be achieved through Lewis et al. (1977) mechanism and also through the toggle binary switch mechanism.

The second model—the cryptic developmental threshold model—proposes that a cryptic threshold existed in the wasp lineages leading to ancestral lineages of ants. In the section titled "Hymenopteran clues for the origin of wing polyphenism in the Formicidae," we mapped on a phylogeny of the Hymenoptera the numerous wasp lineages that evolved a sexual wing dimorphism with winged males and wingless females. Here, we assume that this wing determination in wasps is controlled by a hormonal threshold, such that hormonal levels in male wasps pass the critical threshold resulting in wing development, whereas hormonal levels in female wasps fall below the threshold for wing development (Fig. 9G). Because sexual wing dimorphism is genetically determined, a likely genetic regulator of hormone levels may be the haplodiploidy sex-determination mechanism common to all Hymenoptera. In haplodiploidy, the females develop from fertilized eggs and are therefore diploid, whereas the males develop from unfertilized eggs and are haploid (Grimaldi & Engel, 2005). Consequently, the prevalence of sexual wing dimorphism among the Hymenoptera (Fig. 2) raises the possibility that the ancestral lineage of ants or their wasp ancestors possessed a cryptic developmental potential to generate a sexual wing dimorphism (Fig. 9H). In other words, this ancestral lineage possessed a threshold for the production of winged and wingless phenotypes, but the genetic factors controlling morph development always produced winged phenotypes. In the early ants, the genetically regulated cryptic threshold may have become environmentally sensitive. This environmental sensitivity now results in hormonal levels to fall below or above the threshold for wing development, therefore releasing the wingless alternative phenotype (Fig. 9I).

Both models illustrate how the expression of two discrete phenotypes—winged queen and wingless workers—originated through the creation of a new or modification of an existing developmental threshold. Both of these models involve changes to the concentration of a hormone. However, polyphenic traits can also arise in several different ways by modifying hormone concentration, threshold, the timing of hormone secretion, or the timing of the hormone-sensitive period (Nijhout, 1999; Nijhout, 2003). All that is required for these modifications to control polyphenic development is for them to become sensitive to environmental variables. We, therefore, acknowledge that many other possible models exist and present these two models as a starting point for the exploration of the evolutionary steps into wing polyphenism. Regardless of which model is correct, the novel wingless worker phenotype, once environmentally induced, was genetically accommodated in a wing polyphenism in ants approximately 160 mya. Testing these models will be challenging, but not impossible. The evolution of a polyphenic trait via genetic accommodation has been demonstrated in lab settings and inferred from natural populations and several authors have provided frameworks for testing how developmental plasticity can result in evolutionary adaptation (Jones & Robinson, 2018; Levis & Pfennig, 2016; Moczek et al., 2011; Rajakumar et al., 2012; Schlichting & Wund, 2014; Suzuki & Nijhout, 2006).

7. The origin of wing polyphenism enabled subsequent morphological diversification in ants

In this section, we highlight how the origin of wing polyphenism in ants laid the foundation for subsequent morphological diversification within the worker caste and within the reproductive caste in ants. We first explore how the loss of wings in the worker caste allowed for subsequent modifications in thoracic architecture leading to the increased strength of individual worker ants. We then discuss how the origin of wing polyphenism facilitated the evolution of the novel soldier subcaste, and finally, we discuss the evolution of wing polyphenism/dimorphism in the reproductive castes and its possible association with the origin of queen-worker wing polyphenism in ants.

A universal feature of the thorax of the worker caste is an enlarged first thoracic segment (T1) and its associated muscles (Keller et al., 2014; Peeters et al., 2020). This is in contrast to queen ants where T1 and the third thoracic (T3) segments are reduced and the second thoracic (T2) segment is extensively enlarged. The reduction of T2 and enlargement of T1 in worker ants was enabled by the loss of wings from T2 and T3 segments during the origin of wing polyphenism in ants. Indeed, this exact pattern is also observed in almost all wingless female wasps (Reid, 1941), and wingless queen ants (Keller et al., 2014). Keller et al. (2014) and Peeters et al. (2020) proposed that the enlargement of the T1 segment in workers provided added strength and mobility to the neck muscles that control head movements, thereby allowing ants to carry large objects and exploit various trophic resources. Therefore, the origin of wing polyphenism in ants enabled subsequent thoracic modifications in the worker caste of ants to evolve, and this modification has contributed to the ecological success of ants.

Another factor that contributed to the diversity and success of ants is the evolution of "worker polymorphism," which refers to colonies with a striking degree of variation in size and head-to-body allometry within the worker caste (Wilson, 1953). Worker polymorphism is largely generated and regulated by environment cues (Alvarado, Rajakumar, Abouheif, & Szyf, 2015; Wills, Powell, Rivera, & Suarez, 2018). Wilson (1953) organized the variation in worker polymorphism into four categories, one of which is called "complete dimorphism." In completely dimorphic species, such as in the ant genus *Pheidole*, the worker caste is divided into two morphologically distinct worker subcastes with no intermediates (Wilson, 2003). These two subcastes are called "minor workers" and "soldiers or majors" (Fig. 1). Soldiers have disproportionately large heads relative to their bodies as compared to minor workers (Rajakumar et al., 2018; Wilson, 2003). Although minor workers and soldiers are completely wingless as adults, the soldiers develop vestigial or rudimentary wing discs during larval development that degenerate by apoptosis before becoming adults (Sameshima et al., 2004; Shbailat et al., 2010). By contrast, the minor workers lack any visible wing rudiments during larval development (Abouheif & Wray, 2002; Sameshima et al., 2004; Wheeler & Nijhout, 1981a). A recent study by Rajakumar et al. (2018) showed that these rudimentary wing discs in *Pheidole* soldiers have been co-opted to function in generating their larger size and disproportionately large heads. Therefore, the origin of wing polyphenism released the constraint on the wing discs of having to produce functional wings in workers. This allowed wing discs to acquire novel functions that led to the evolution of further morphological diversification in the worker caste in ants, such as the big-headed soldier subcaste. Future studies should investigate the role of the rudimentary wing discs in less-derived Poneroid species that evolved worker polymorphism, such as *Myopopone castanea* and *Amblyopone australis* (Ito, 2010; Peeters & Molet, 2010). This will reveal whether the role of the rudimentary wing discs in generating worker polymorphism is conserved across varied taxa or evolved only later during ant evolution in the more derived clades. Finally, the absence of wing polyphenism in other social insects (bees and wasps), may indeed be the reason why they never evolved dramatic worker polymorphism like that observed in the ants.

Following the origin of wing polyphenism, additional wing polyphenisms or dimorphisms evolved within the reproductive queen or male castes. The evolution of ergatoid (wingless) and brachyapterous (short-winged) queens evolved independently in more than 50 ant genera across all subfamilies (for a list of genera see: Heinze & Tsuji, 1995; Peeters, 2012). The evolution of wing polyphenism in the male caste, on the other hand, is less common than wing polyphenism in the female queen caste and has been observed in only half a dozen genera, most notably in the genus

Cardiocondyla (Heinze, 2017; Heinze & Tsuji, 1995). Typically, ergatoids or brachyapterous queens, in addition to wingless males, occur together with winged queens and males, however, there are instances where some species have completely lost the production of winged queens and males (Heinze et al., 2005; Molet, Peeters, & Fisher, 2007; Peeters, 2012). Environmental conditions and social regulation play an instrumental role in wing morph determination in the reproductive caste (Cremer & Heinze, 2003; Ito, 1996; Schrempf & Heinze, 2006). However, a few studies have also found evidence for a strong genetic influence on the development of ergatoids (Buschinger & Schreiber, 2002; Fersch, Buschinger, & Heinze, 2000; Heinze & Buschinger, 1989). Fave et al. (2015) and Oettler et al. (2018) have explored the wing GRN and revealed interruption points in the wing GRN of wingless queens and males. Although these interruption points in the wingless reproductive caste occur further downstream relative to the interruption points found in wingless workers of the same species, they show that wing GRN interruption is a general mechanism that was co-opted to regulate polyphenic wing disc growth in reproductive queens and males. Collectively, these examples not only demonstrate the evolutionary role wing polyphenism played in the subsequent morphological diversification of ants, but they also pave the way for new avenues in ant research.

8. Conclusions

The goal of this review is to synthesis the ecological, developmental, and evolutionary factors that led to the origin of wing polyphenism in ants. Figure 10 briefly summarizes the multiple models and hypotheses proposed by this work. Briefly, we propose that there may have existed an ancestral developmental potential in Hymenoptera to generate sexual wing dimorphism (Fig. 10A). This potential—a preexisting cryptic threshold for wing development present in the ancestral wasps—may have facilitated the origin of wing polyphenism in ants (Fig. 10B). Alternatively, the wing polyphenism in ants may have originated by evolving a threshold *de novo* (Fig. 10C). The fossil record provides clear evidence that wing polyphenism and eusociality evolved together in early ants (Fig. 10D). However, it remains unclear whether wing polyphenism evolved prior to eusociality and contributed to its origin, or it originated after eusociality and contributed to its maintenance (Fig. 10E). Once wing polyphenism originated, interactions between the environment, hormones, wing GRN, and apoptosis were

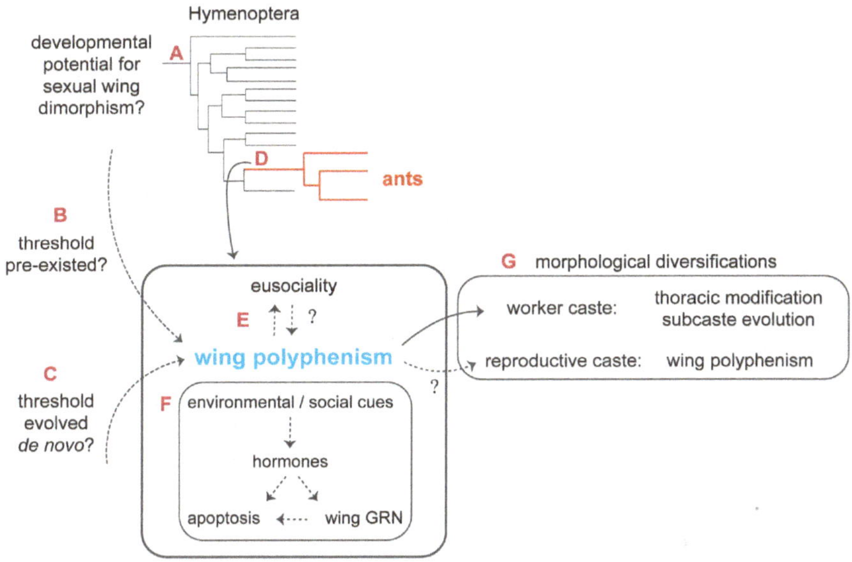

Fig. 10 A visual illustration of the main models and hypotheses proposed in this work. (A) Hymenoptera may possess a developmental potential for the generation of sexual wing dimorphism. (B) this developmental potential (in the form of a threshold) may have led to the generation of wing polyphenism in ants; alternatively, the threshold may have evolved *de novo* (C). (D) based on fossil evidence, wing polyphenism and eusociality originated early during the evolution of ants. (E) wing polyphenism may have originated prior to and drove the evolution of eusociality, or wing polyphenism followed eusociality but led to its reinforcement. (F) a proposed model for the interactions between the environment, social cues, hormones, GRN interruption, and apoptosis to generate a wingless worker caste. (G) wing polyphenism facilitated morphological diversifications in the worker and reproductive caste.

likely present and regulated the polyphenic switch between winged and wingless females in the early ants (Fig. 10F). Because wing polyphenism in ants evolved just once and is a universal feature of ants, these interactions evolved and contributed to the remarkable morphological diversity of ants, facilitating several morphological innovations: it enabled the evolution of thoracic modifications that increased the strength of ants and enabled the evolution of novel worker and reproductive castes in ants, including the evolution of soldiers, wingless queens and males (Fig. 10G).

Undoubtedly, the factors that facilitated the evolution of wing polyphenism are complex and intricately linked. Much remains to be elucidated about how wing polyphenism in ants originated – from ecological factors to hormonal regulation. In Table 1, we have summarized some of the main outstanding questions to be investigated by future work. Since polyphenism

Table 1 A list of outstanding questions for investigating the evolution of wing polyphenism in ants.

What are the ancestral characteristics and phylogenetic correlations between wing development and life history traits?	Section 2
What is the developmental mechanism for sexual wing dimorphism in wasps?	Section 2
Is there an ancestral developmental potential for a sexual wing dimorphism in Hymenoptera?	Sections 2 and 6
To help understand the relationship between the origins of eusociality and wing polyphenism, what is the relationship between wing development and reproductive capacity?	Section 4
What is the function of the metapleural gland in early branching ant lineage?	Section 4
What are the biomechanical constraints placed on the metapleural gland by the presence of wings?	Section 4
Is there an interruption in the wing GRN in taxa from the Poneroid clade?	Section 5
Is apoptosis a conserved mechanism in wing disc degeneration in Formicidae?	Section 5
What is the functional role of *dsx* and sphingolipid synthesis genes in ant wing development?	Section 5
What are the roles of ecdysone and IIS in regulating wing polyphenism in ants, and are these roles conserved among taxa?	Section 5
Is there a positive feedback mechanism between JH and IIS?	Section 5
Do JH levels regulate gene transcription in the wing GRN? Precisely, are low JH levels in worker destined larvae unable to activate the wing GRN?	Section 5
Does JH control *dsx* expression and ecdysone levels similar to other insects with polyphenic traits?	Section 5
Does ecdysone induce apoptosis in the rudimentary wing discs?	Section 5
Did ants evolve a discrete winged and wingless phenotypes through gradual transitions or through a single saltational step?	Section 6
Did wing polyphenism in ants evolve via genetic accommodation?	Section 6
Did the development al threshold for wing polyphenism in ants evolve *de novo* or through a cryptic development al threshold that	Section 6

Table 1 A list of outstanding questions for investigating the evolution of wing polyphenism in ants.—cont'd

was present in a wasp ancestor for the generation of sexual wing dimorphi sm?	
Are the re evidence for positive feedback or toggle mechanism in the wing GRN?	Section 6
Do the rudimentary wing discs facilitate polymorphic work er development in less-derived species?	Section 7
Is the evolution of wing polyphenism in queens and males attained through co-option of the mechanisms controlling queen-worker wing polyphenism?	Section 7

is associated with several major evolutionary transitions in individuality, understanding the origin and elaboration of wing polyphenism in ants can help us understand how other major evolutionary transitions in individuality arose and diversified.

Acknowledgments

We thank Scott Gilbert, H.F. Nijhout, Philip Ward, Mathieu Molet, Brendon Boudinot, and members of the Abouheif lab for discussion and comments. This work was supported by Natural Sciences and Engineering Research Council (NSERC) Discovery Grant to E.A.

References

Abouheif, E., Fave, M. J., Ibarraran-Viniegra, A. S., Lesoway, M. P., Rafiqi, A. M., & Rajakumar, R. (2014). Eco-evo-devo: The time has come. *Advances in Experimental Medicine and Biology*, *781*, 107–125. https://doi.org/10.1007/978-94-007-7347-9_6.

Abouheif, E., & Wray, G. A. (2002). Evolution of the gene network underlying wing polyphenism in ants. *Science*, *297*(5579), 249–252. https://doi.org/10.1126/science.1071468.

Adachi-Yamada, T., Gotoh, T., Sugimura, I., Tateno, M., Nishida, Y., Onuki, T., et al. (1999). De novo synthesis of sphingolipids is required for cell survival by down-regulating c-Jun N-terminal kinase in Drosophila imaginal discs. *Molecular and Cellular Biology*, *19*(10), 7276–7286. https://doi.org/10.1128/mcb.19.10.7276.

Agosti, D., Grimaldi, D., & Carpenter, J. M. (1998). Oldest known ant fossils discovered. *Nature*, *391*, 447.

Alvarado, S., Rajakumar, R., Abouheif, E., & Szyf, M. (2015). Epigenetic variation in the Egfr gene generates quantitative variation in a complex trait in ants. *Nature Communications*, *6*, 6513. https://doi.org/10.1038/ncomms7513.

Ayoade, O., Morooka, S., & Tojo, S. (1999). Enhancement of short wing formation and ovarian growth in the genetically defined macropterous strain of the brown planthopper, Nilaparvata lugens. *Journal of Insect Physiology*, *45*(1), 93–100. https://doi.org/10.1016/s0022-1910(98)00103-6.

Barden, P. (2016). Fossil ants (Hymenoptera: Formicidae): Ancient diversity and the rise of modern lineages. *Myrmecological News*, *24*, 1–30.

Barden, P., & Grimaldi, D. (2012). Rediscovery of the bizarre Cretaceous ant Haidomyrmex Dlussky (Hymenoptera: Formicidae), with two new species. *American Museum Novitates*, *3755*, 1–16.

Barden, P., & Grimaldi, D. (2014). A diverse ant fauna from the mid-cretaceous of Myanmar (Hymenoptera: Formicidae). *PLoS One*, *9*(4), e93627. https://doi.org/10.1371/journal.pone.0093627.

Barden, P., & Grimaldi, D. A. (2016). Adaptive radiation in socially advanced stem-group ants from the cretaceous. *Current Biology*, *26*(4), 515–521. https://doi.org/10.1016/j.cub.2015.12.060.

Behague, J., Fisher, B. L., Peronnet, R., Rajakumar, R., Abouheif, E., & Molet, M. (2018). Lack of interruption of the gene network underlying wing polyphenism in an early-branching ant genus. *Journal of Experimental Zoology. Part B, Molecular and Developmental Evolution*, *330*(2), 109–117. https://doi.org/10.1002/jez.b.22794.

Bertuso, A. G., Morooka, S., & Tojo, S. (2002). Sensitive periods for wing development and precocious metamorphosis after precocene treatment of the brown planthopper, Nilaparvata lugens. *Journal of Insect Physiology*, *48*(2), 221–229. https://doi.org/10.1016/s0022-1910(01)00167-6.

Billard, B., Vigne, P., & Braendle, C. (2020). A natural mutational event uncovers a life history trade-off via hormonal pleiotropy. *Current Biology*, *30*(21), 4142–4154. https://doi.org/10.1016/j.cub.2020.08.004.

Bolton, B. (2003). *Synopsis and classification of Formicidae*. The American Entomological Institude.

Boomsma, J. J., & Gawne, R. (2018). Superorganismality and caste differentiation as points of no return: How the major evolutionary transitions were lost in translation. *Biological Reviews of the Cambridge Philosophical Society*, *93*(1), 28–54. https://doi.org/10.1111/brv.12330.

Boomsma, J. J., Schmid-Hempel, P., & Hughes, W. O. H. (2005). Life histories and parasite pressure across the major groups of social insects. In M. D. E. Fellowes, G. J. Holloway, & J. Rolff (Eds.), *Insect evolutionary ecology* CABI Publishing.

Borowiec, M. L. (2019). Convergent evolution of the army ant syndrome and congruence in big-data phylogenetics. *Systematic Biology*, *68*(4), 642–656. https://doi.org/10.1093/sysbio/syy088.

Boudinot, B. E., Perrichot, V., & Chaul, J. C. M. (2020). *Camelosphecia* gen. nov., lost ant-wasp intermediates from the mid-Cretaceous (Hymenoptera, Formicoidea). *ZooKeys*, *1005*, 21–55. https://doi.org/10.3897/zookeys.1005.57629. https://zookeys.pensoft.net.

Brady, S. G., Schultz, T. R., Fisher, B. L., & Ward, P. S. (2006). Evaluating alternative hypotheses for the early evolution and diversification of ants. *Proceedings of the National Academy of Sciences of the United States of America*, *103*(48), 18172–18177. https://doi.org/10.1073/pnas.0605858103.

Branstetter, M. G., Danforth, B. N., Pitts, J. P., Faircloth, B. C., Ward, P. S., Buffington, M. L., et al. (2017). Phylogenomic insights into the evolution of stinging wasps and the origins of ants and bees. *Current Biology*, *27*(7), 1019–1025. https://doi.org/10.1016/j.cub.2017.03.027.

Branstetter, M. G., Longino, J. T., Ward, P. S., & Faircloth, B. C. (2017). Enriching the ant tree of life: Enhanced UCE bait set for genome-scale phylogenetics of ants and other Hymenoptera. *Methods in Ecology and Evolution*, *8*, 768–776.

Brisson, J. A., Ishikawa, A., & Miura, T. (2010). Wing development genes of the pea aphid and differential gene expression between winged and unwinged morphs. *Insect Molecular Biology*, *19*, 63–73.

Buschinger, A., & Schreiber, M. (2002). Queen polymorphism and queen-morph related facultative polygyny in the ant, Myrmecina graminicola (Hymenoptera, Formicidae). *Insectes Sociaux*, *49*, 344–353.

Buss, L. W. (2014). *The evolution of individuality*. Princeton University Press.
Campbell, G., & Tomlinson, A. (1999). Transducing the Dpp morphogen gradient in the wing of Drosophila: Regulation of Dpp targets by brinker. *Cell, 96*(4), 553–562. https://doi.org/10.1016/s0092-8674(00)80659-5.
Carroll, S. B. (2008). Evo-devo and an expanding evolutionary synthesis: A genetic theory of morphological evolution. *Cell, 134*(1), 25–36. https://doi.org/10.1016/j.cell.2008.06.030.
Cassill, D. L., & Tschinkel, W. R. (1995). Allocation of liquid food to larvae via trophallaxis in colonies of the fire ant, Solenopsis invicta. *Animal Behaviour, 50*(3), 801–813.
Chandra, V., Fetter-Pruneda, I., Oxley, P. R., Ritger, A. L., McKenzie, S. K., Libbrecht, R., et al. (2018). Social regulation of insulin signaling and the evolution of eusociality in ants. *Science, 361*(6400), 398–402. https://doi.org/10.1126/science.aar5723.
Chihara, C. J., Petri, W. H., Fristrom, J. W., & King, D. S. (1972). The assay of ecdysones and juvenile hormones on Drosophila imaginal disks in vitro. *Journal of Insect Physiology, 18*(6), 1115–1123. https://doi.org/10.1016/0022-1910(72)90146-1.
Cohen, S. M. (1993). Imaginal disc development. In M. Bate, & A. Martinez-Arias (Eds.), *The development of Drosophila melanogaster* (pp. 747–843). Cold Spring Harbor Laboratory Press.
Cohen, S. M. (1996). Controlling growth of the wing: Vestigial integrates signals from the compartment boundaries. *BioEssays, 18*(11), 855–858. https://doi.org/10.1002/bies.950181102.
Colombani, J., Bianchini, L., Layalle, S., Pondeville, E., Dauphin-Villemant, C., Antoniewski, C., et al. (2005). Antagonistic actions of ecdysone and insulins determine final size in Drosophila. *Science, 310*(5748), 667–670. https://doi.org/10.1126/science.1119432.
Consoli, F. L., & Vinson, S. B. (2004). Wing morph development and reproduction of the ectoparasitoid Melittobia digitata: Nutritional and hormonal effects. *Entomologia Experimentalis et Applicata, 112*, 47–55.
Cook, J. M., Compton, S. G., Allen Herre, E., & West, S. A. (1997). Alternative mating tactics and extreme male dimorphism in fig wasps. *Proceedings of the Royal Society of London, 264*, 747–754.
Couso, J. P., Knust, E., & Martinez Arias, A. (1995). Serrate and wingless cooperate to induce vestigial gene expression and wing formation in Drosophila. *Current Biology, 5*(12), 1437–1448. https://doi.org/10.1016/s0960-9822(95)00281-8.
Cremer, S., & Heinze, J. (2003). Stress grows wings: Environmental induction of winged dispersal males in Cardiocondyla ants. *Current Biology, 13*(3), 219–223. https://doi.org/10.1016/s0960-9822(03)00012-5.
Crespi, B. J. (1992). Eusociality in Australian gall thrips. *Nature, 359*, 724–726.
Darwin, C. (1859). *On the Origin of Species by Means of Natural Selection, or the Preservation of Favoured Races in the Struggle for Life*. John Murray.
Dayel, M. J., Alegado, R. A., Fairclough, S. R., Levin, T. C., Nichols, S. A., McDonald, K., et al. (2011). Cell differentiation and morphogenesis in the colony-forming choanoflagellate Salpingoeca rosetta. *Developmental Biology, 357*(1), 73–82. https://doi.org/10.1016/j.ydbio.2011.06.003.
de Azevedo, S. V., & Hartfelder, K. (2008). The insulin signaling pathway in honey bee (Apis mellifera) caste development - differential expression of insulin-like peptides and insulin receptors in queen and worker larvae. *Journal of Insect Physiology, 54*(6), 1064–1071. https://doi.org/10.1016/j.jinsphys.2008.04.009.
de Celis, J. F., & Barrio, R. (2000). Function of the spalt/spalt-related gene complex in positioning the veins in the Drosophila wing. *Mechanisms of Development, 91*(1-2), 31–41. https://doi.org/10.1016/s0925-4773(99)00261-0.
DeFalco, T. J., Verney, G., Jenkins, A. B., McCaffery, J. M., Russell, S., & Van Doren, M. (2003). Sex-specific apoptosis regulates sexual dimorphism in the Drosophila embryonic gonad. *Developmental Cell, 5*(2), 205–216. https://doi.org/10.1016/s1534-5807(03)00204-1.

Dewitz, H. (1878). Beiträge zur postembryonalen Gliedmassenbildung bei den Insecten. *Zeitschrift für Wissenschaftliche Zoologie*, *30*, 78–105.

Deyrup, M. (1988). Review of adaptations of velvet ants (Hymenoptera: Mutillidae). *The Great Lakes Entomologist*, *21*(1), 1–4.

Diaz-Benjumea, F. J., & Cohen, S. M. (1993). Interaction between dorsal and ventral cells in the imaginal disc directs wing development in Drosophila. *Cell*, *75*(4), 741–752. https://doi.org/10.1016/0092-8674(93)90494-b.

Diaz-Benjumea, F. J., & Cohen, S. M. (1995). Serrate signals through Notch to establish a Wingless-dependent organizer at the dorsal/ventral compartment boundary of the Drosophila wing. *Development*, *121*(12), 4215–4225. https://www.ncbi.nlm.nih.gov/pubmed/8575321.

Dlussky, G. M. (1983). A new family of Upper Cretaceous Hymenoptera: An "intermediate link" between the ants and the scolioids. *Paleontological Journal*, *3*, 63–76.

Dlussky, G. M. (1999). The first find of the Formicoidea (Hymenoptera) in the Lower Cretaceous of the Northern Hemisphere. *Paleontological Journal*, *33*(3), 274–277.

Edgar, B. A. (2006). How flies get their size: Genetics meets physiology. *Nature Reviews. Genetics*, 7(12), 907–916. https://doi.org/10.1038/nrg1989.

Emery, C. (1894). Die Entstehung und Ausbildung des Arbeiterstandes bei den Ameisen. *Biologisches Zentralblatt*, *14*, 53–59.

Emlen, D. J., & Allen, C. E. (2004). Genotype to phenotype: Physiological control of trait size and scaling in insects. *Integrative and Comparative Biology*, *43*(5), 617–634.

Engel, M. S., & Grimaldi, D. (2005). Primitive new ants in Cretaceous amber from Myanmar, New Jersey, and Canada (Hymenoptera: Formicidae). *American Museum Novitates*, *3485*.

Evans, H. E. (1969). Phoretic copulation in Hymenoptera. *Entomological News*, 113–124.

Falconer, D. S. (1960). *Introduction to quantitative genetics*. The Ronald Press Company.

Fave, M. J., Johnson, R. A., Cover, S., Handschuh, S., Metscher, B. D., Muller, G. B., et al. (2015). Past climate change on Sky Islands drives novelty in a core developmental gene network and its phenotype. *BMC Evolutionary Biology*, *15*, 183. https://doi.org/10.1186/s12862-015-0448-4.

Fernandez-Marin, H., Zimmerman, J. K., Rehner, S. A., & Wcislo, W. T. (2006). Active use of the metapleural glands by ants in controlling fungal infection. *Proceedings of the Biological Sciences*, *273*(1594), 1689–1695. https://doi.org/10.1098/rspb.2006.3492.

Fersch, R., Buschinger, A., & Heinze, J. (2000). Queen polymorphism in the Australian ant Monomorium sp.10. *Insectes Sociaux*, *47*, 280–284.

Fjerdingstad, E. J., & Crozier, R. H. (2006). The evolution of worker caste diversity in social insects. *The American Naturalist*, *167*(3), 390–400. https://doi.org/10.1086/499545.

Foster, W. A. (1990). Experimental evidence for effectiveand altruistic colony defence against natural predatorsby soldiersof the gall-forming aphid Pemphiguspyrothec e(Hemiptera: Pemphigidae). *Behavioral Ecology and Sociobiology*, *27*, 421–430.

Fristrom, D. (1969). Cellular degeneration in the production of some mutant phenotypes in Drosophila melanogaster. *Molecular & General Genetics*, *103*(4), 363–379. https://doi.org/10.1007/BF00383486.

Gardner, T. S., Cantor, C. R., & Collins, J. J. (2000). Construction of a genetic toggle switch in Escherichia coli. *Nature*, *403*(6767), 339–342. https://doi.org/10.1038/35002131.

Gibson, G., & Hogness, D. S. (1996). Effect of polymorphism in the Drosophila regulatory gene Ultrabithorax on homeotic stability. *Science*, *271*(5246), 200–203. https://doi.org/10.1126/science.271.5246.200.

Gilbert, S. F., & Epel, D. (2015). *Ecological developmental biology: The environmental regulation of development, health, and evolution*. Oxford University Press.

Gilbert, S. F., Opitz, J. M., & Raff, R. A. (1996). Resynthesizing evolutionary and developmental biology. *Developmental Biology*, *173*(2), 357–372. https://doi.org/10.1006/dbio.1996.0032.

Giraldez, A. J., & Cohen, S. M. (2003). Wingless and Notch signaling provide cell survival cues and control cell proliferation during wing development. *Development*, *130*(26), 6533–6543. https://doi.org/10.1242/dev.00904.

Gotoh, H., Miyakawa, H., Ishikawa, A., Ishikawa, Y., Sugime, Y., Emlen, D. J., et al. (2014). Developmental link between sex and nutrition; doublesex regulates sex-specific mandible growth via juvenile hormone signaling in stag beetles. *PLoS Genetics*, *10*(1), e1004098. https://doi.org/10.1371/journal.pgen.1004098.

Goulet, H., & Huber, J. T. (1993). *Hymenoptera of the world: An identification guide to families*. Canada Communication Group.

Grieder, N. C., Morata, G., Affolter, M., & Gehring, W. J. (2009). Spalt major controls the development of the notum and of wing hinge primordia of the Drosophila melanogaster wing imaginal disc. *Developmental Biology*, *329*(2), 315–326. https://doi.org/10.1016/j.ydbio.2009.03.006.

Grimaldi, D., & Agosti, D. (2000). A formicine in New Jersey cretaceous amber (Hymenoptera: Formicidae) and early evolution of the ants. *Proceedings of the National Academy of Sciences of the United States of America*, *97*(25), 13678–13683. https://doi.org/10.1073/pnas.240452097.

Grimaldi, D., Agosti, D., & Carpenter, J. M. (1997). New and rediscovered primitive ants (Hymenoptera: Formicidae) in Cretaceous amber from New Jersey, and their phylogenetic relationships. *American Museum Novitates*, *3208*, 1–43.

Grimaldi, D. A., & Engel, M. S. (2005). *Evolution of the Insects*. Cambridge University Press.

Halder, G., Polaczyk, P., Kraus, M. E., Hudson, A., Kim, J., Laughon, A., et al. (1998). The Vestigial and Scalloped proteins act together to directly regulate wing-specific gene expression in Drosophila. *Genes & Development*, *12*(24), 3900–3909. https://doi.org/10.1101/gad.12.24.3900.

Hamilton, W. D. (1979). Wingless and fighting males in fig wasps and other insects. In M. S. Blum, & N. A. Blum (Eds.), *Sexual selection and reproduction? Competition in insects* (pp. 167–220). Academic Press.

Hartfelder, K., & Emlen, D. J. (2012). Endocrine control of insect polyphenism. In L. I. Gilbert (Ed.), *Insect Endocrinology* (pp. 464–522). Academic Press.

Heinze, J. (2017). Life-history evolution in ants: The case of Cardiocondyla. *Proceedings of the Biological Sciences*, *284*(1850). https://doi.org/10.1098/rspb.2016.1406.

Heinze, J., & Buschinger, A. (1989). Queen polymorphism in Leptothorax Spec.A: Its genetic and ecological background (Hymenoptera: Formicidae). *Insectes Sociaux*, *36*(2), 139–155.

Heinze, J., Trindl, A., Seifert, B., & Yamauchi, K. (2005). Evolution of male morphology in the ant genus Cardiocondyla. *Molecular Phylogenetics and Evolution*, *37*(1), 278–288. https://doi.org/10.1016/j.ympev.2005.04.005.

Heinze, J., & Tsuji, K. (1995). Ant reproductive strategies. *Researches on Population Ecology*, *37*(2), 135–149.

Held, L. I. J. (2002). *Imaginal discs: The genetic and cellular logic of pattern formation*. Cambridge University Press.

Holldobler, B., & Wilson, E. O. (1990). *The ants*. Harvard University Press.

Hunt, J. (1991). Nourishment and the evolution of the social Vespidae. In K. G. Ross & R. W. Matthews (Eds.), *The social biology of wasps* (pp. 426-450). Cornell University Press.

Innocent, T. M., Abe, J., West, S. A., & Reece, S. E. (2010). Competition between relatives and the evolution of dispersal in a parasitoid wasp. *Journal of Evolutionary Biology*, *23*(7), 1374–1385. https://doi.org/10.1111/j.1420-9101.2010.02015.x.

Ishikawa, A., Gotoh, H., Abe, T., & Miura, T. (2013). Juvenile hormone titer and wing-morph differentiation in the vetch aphid Megoura crassicauda. *Journal of Insect Physiology*, *59*(4), 444–449. https://doi.org/10.1016/j.jinsphys.2013.02.004.

Ito, F. (1996). Colony characteristics of the Indonesian Myrmicine ant Myrmecina sp. (Hymenoptera, Formicidae, Myrmicinae): Polygynous reproduction by ergatoid queens. *Entomological Society of America*, *89*(4), 550–554.

Ito, F. (2010). Notes on the biology of the Oriental amblyoponine ant Myopopone castanea: Queen-worker dimorphism, worker polymorphism and larval hemolymph feeding by workers (Hymenoptera: Formicidae)ens_384 199. 204. *Entomological. Science*, *13*, 199–204.

Iwanaga, K., & Tojo, S. (1986). Effects of juvenile hormone and rearing density of wing dimorphism and oocyte development in the brown planthopper, Nilaparvata lugens. *Journal of Insect Physiology*, *32*(6), 585–590.

James, A. A., & Bryant, P. J. (1981). Mutations causing pattern deficiencies and duplications in the imaginal wing disk of Drosophila melanogaster. *Developmental Biology*, *85*(1), 39–54. https://doi.org/10.1016/0012-1606(81)90234-7.

Johnson, B. R., Borowiec, M. L., Chiu, J. C., Lee, E. K., Atallah, J., & Ward, P. S. (2013). Phylogenomics resolves evolutionary relationships among ants, bees, and wasps. *Current Biology*, *23*(20), 2058–2062. https://doi.org/10.1016/j.cub.2013.08.050.

Jones, B. M., & Robinson, G. E. (2018). Genetic accommodation and the role of ancestral plasticity in the evolution of insect eusociality. *The Journal of Experimental Biology*, *221*(Pt 23). https://doi.org/10.1242/jeb.153163.

Keller, R. A., Peeters, C., & Beldade, P. (2014). Evolution of thorax architecture in ant castes highlights trade-off between flight and ground behaviors. *eLife*, *3*, e01539. https://doi.org/10.7554/eLife.01539.

Kent, D. S., & Simpson, J. A. (1992). Eusociality in the beetle Austroplatypus incompertus (Coleoptera: Curculiomdae). *Naturwissenschaften*, *79*, 86–87.

Kijimoto, T., Andrews, J., & Moczek, A. P. (2010). Programed cell death shapes the expression of horns within and between species of horned beetles. *Evolution & Development*, *12*(5), 449–458. https://doi.org/10.1111/j.1525-142X.2010.00431.x.

Kijimoto, T., Moczek, A. P., & Andrews, J. (2012). Diversification of doublesex function underlies morph-, sex-, and species-specific development of beetle horns. *Proceedings of the National Academy of Sciences of the United States of America*, *109*(50), 20526–20531. https://doi.org/10.1073/pnas.1118589109.

Kim, J., Irvine, K. D., & Carroll, S. B. (1995). Cell recognition, signal induction, and symmetrical gene activation at the dorsal-ventral boundary of the developing Drosophila wing. *Cell*, *82*(5), 795–802. https://doi.org/10.1016/0092-8674(95)90476-x.

Kim, J., Magee, J., & Carroll, S. B. (1997). Intercompartmental signaling and the regulation of vestigial expression at the dorsoventral boundary of the developing Drosophila wing. *Cold Spring Harbor Symposia on Quantitative Biology*, *62*, 283–291. https://www.ncbi.nlm.nih.gov/pubmed/9598362.

Kimsey, L. S. (1991). Relationships among the tiphiid wasp subfamilies (Hymenoptera). *Systematic Entomology*, *16*, 427–438.

Klein, A., Schultner, E., Lowak, H., Schrader, L., Heinze, J., Holman, L., et al. (2016). Evolution of social insect polyphenism facilitated by the sex differentiation cascade. *PLoS Genetics*, *12*(3), e1005952. https://doi.org/10.1371/journal.pgen.1005952.

Korb, J. (2008). Termites, hemimetabolous diploid white ants? *Frontiers in Zoology*, *5*, 15. https://doi.org/10.1186/1742-9994-5-15.

Kraut, R. (2011). Roles of sphingolipids in Drosophila development and disease. *Journal of Neurochemistry*, *116*(5), 764–778. https://doi.org/10.1111/j.1471-4159.2010.07022.x.

LaPolla, J. S., Dlussky, G. M., & Perrichot, V. (2013). Ants and the fossil record. *Annual Review of Entomology*, *58*, 609–630. https://doi.org/10.1146/annurev-ento-120710-100600.

Levis, N. A., & Pfennig, D. W. (2016). Evaluating 'Plasticity-First' evolution in nature: key criteria and empirical approaches. *Trends in Ecology & Evolution*, *31*(7), 563–574. https://doi.org/10.1016/j.tree.2016.03.012.

Lewis, J., Slack, J. M., & Wolpert, L. (1977). Thresholds in development. *Journal of Theoretical Biology*, *65*(3), 579–590. https://doi.org/10.1016/0022-5193(77)90216-8.

Li, T. R., & White, K. P. (2003). Tissue-specific gene expression and ecdysone-regulated genomic networks in Drosophila. *Developmental Cell*, *5*(1), 59–72. https://doi.org/10.1016/s1534-5807(03)00192-8.

Libbrecht, R., Corona, M., Wende, F., Azevedo, D. O., Serrao, J. E., & Keller, L. (2013). Interplay between insulin signaling, juvenile hormone, and vitellogenin regulates maternal effects on polyphenism in ants. *Proceedings of the National Academy of Sciences of the United States of America*, *110*(27), 11050–11055. https://doi.org/10.1073/pnas.1221781110.

Lobbia, S., Niitsu, S., & Fujiwara, H. (2003). Female-specific wing degeneration caused by ecdysteroid in the Tussock Moth, Orgyia recens: Hormonal and developmental regulation of sexual dimorphism. *Journal of Insect Science*, *3*, 11. https://doi.org/10.1093/jis/3.1.11.

Loehlin, D. W., Oliveira, D. C., Edwards, R., Giebel, J. D., Clark, M. E., Cattani, M. V., et al. (2010). Non-coding changes cause sex-specific wing size differences between closely related species of Nasonia. *PLoS Genetics*, *6*(1), e1000821. https://doi.org/10.1371/journal.pgen.1000821.

Lu, H. L., & Pietrantonio, P. V. (2011). Insect insulin receptors: Insights from sequence and caste expression analyses of two cloned hymenopteran insulin receptor cDNAs from the fire ant. *Insect Molecular Biology*, *20*(5), 637–649. https://doi.org/10.1111/j.1365-2583.2011.01094.x.

Lucky, A., Trautwein, M. D., Guenard, B. S., Weiser, M. D., & Dunn, R. R. (2013). Tracing the rise of ants - out of the ground. *PLoS One*, *8*(12), e84012. https://doi.org/10.1371/journal.pone.0084012.

Lugagne, J. B., Sosa Carrillo, S., Kirch, M., Kohler, A., Batt, G., & Hersen, P. (2017). Balancing a genetic toggle switch by real-time feedback control and periodic forcing. *Nature Communications*, *8*(1), 1671. https://doi.org/10.1038/s41467-017-01498-0.

Maliet, O., Shelton, D. E., & Michod, R. E. (2015). A model for the origin of group reproduction during the evolutionary transition to multicellularity. *Biology Letters*, *11*(6), 20150157. https://doi.org/10.1098/rsbl.2015.0157.

Marquez-Zacarias, P., Pineau, R. M., Gomez, M., Veliz-Cuba, A., Murrugarra, D., Ratcliff, W. C., et al. (2020). Evolution of cellular differentiation: From hypotheses to models. *Trends in Ecology & Evolution*, *36*(1), 49–60. https://doi.org/10.1016/j.tree.2020.07.013.

Martin, F. A., Perez-Garijo, A., Moreno, E., & Morata, G. (2004). The brinker gradient controls wing growth in Drosophila. *Development*, *131*(20), 4921–4930. https://doi.org/10.1242/dev.01385.

Martin, P. M., & Shearn, A. (1980). Development of drosophila imaginal discs in vitro: Effects of ecdysone concentration and insulin. *The Journal of Experimental Zoology*, *211*, 291–301.

Masuko, K. (1990). Behavior and ecology of the enigmatic ant Leptanilla japonica Baroni Urbani (Hymenoptera: Formicidae: Leptanillinae). *Insectes Sociaux*, *37*(1), 31–57.

Matthews, R. W., & Gonzalez, J. M. (2008). The acasta conundrum: Polymorphism and taxonomic confusion within the parasitoid genus Melittobia (Hymenoptera: Eulophidae). *Zootaxa*, *1854*, 45–54.

Matthews, R. W., Gonzalez, J. M., Matthews, J. R., & Deyrup, L. D. (2009). Biology of the parasitoid Melittobia (Hymenoptera: Eulophidae). *Annual Review of Entomology*, *54*, 251–266. https://doi.org/10.1146/annurev.ento.54.110807.090440.

Maynard-Smith, J., & Szathmary, E. (1997). *The major transitions in evolution*. Oxford University Press.

Mayr, E. (1963). *Animal species and evolution*. Cambridge: Belknap Press of Harvard University Press.

McCulloch, G. A., Oliphant, A., Dearden, P. K., Veale, A. J., Ellen, C. W., & Waters, J. M. (2019). Comparative transcriptomic analysis of a wing-dimorphic stonefly reveals candidate wing loss genes. *EvoDevo*, *10*, 21. https://doi.org/10.1186/s13227-019-0135-4.

Michener, C. D. (1969). Comparative social behavior of bees. *Annual Review of Entomology*, *14*, 299–342.

Michener, C. D. (2000). *The Bees of the world*. JHU Press.

Mikhailov, K. V., Konstantinova, A. V., Nikitin, M. A., Troshin, P. V., Rusin, L. Y., Lyubetsky, V. A., et al. (2009). The origin of Metazoa: A transition from temporal to spatial cell differentiation. *BioEssays*, *31*(7), 758–768. https://doi.org/10.1002/bies.200800214.

Miller, J. S., & Reeve, H. K. (2019). Feedback loops in the major evolutionary transition to eusociality: The status and potential of theoretical approaches. *Current Opinion in Insect Science*, *34*, 85–90. https://doi.org/10.1016/j.cois.2019.04.008.

Minami, M., Kinoshita, N., Kamoshida, Y., Tanimoto, H., & Tabata, T. (1999). brinker is a target of Dpp in Drosophila that negatively regulates Dpp-dependent genes. *Nature*, *398*(6724), 242–246. https://doi.org/10.1038/18451.

Miner, A. L., Rosenberg, A. J., & Nijhout, H.F. (2000). Control of growth and differentiation of the wing imaginal disk of Precis coenia (Lepidoptera: Nymphalidae). *Journal of Insect Physiology*, *46*(3), 251–258. https://doi.org/10.1016/s0022-1910(99)00177-8.

Mirth, C. K., Tang, H. Y., Makohon-Moore, S. C., Salhadar, S., Gokhale, R. H., Warner, R. D., et al. (2014). Juvenile hormone regulates body size and perturbs insulin signaling in Drosophila. *Proceedings of the National Academy of Sciences of the United States of America*, *111*(19), 7018–7023. https://doi.org/10.1073/pnas.1313058111.

Mirth, C. K., Truman, J. W., & Riddiford, L. M. (2009). The ecdysone receptor controls the post-critical weight switch to nutrition-independent differentiation in Drosophila wing imaginal discs. *Development*, *136*(14), 2345–2353. https://doi.org/10.1242/dev.032672.

Moczek, A. P., Sultan, S., Foster, S., Ledon-Rettig, C., Dworkin, I., Nijhout, H. F., et al. (2011). The role of developmental plasticity in evolutionary innovation. *Proceedings of the Biological Sciences*, *278*(1719), 2705–2713. https://doi.org/10.1098/rspb.2011.0971.

Molet, M., Peeters, C., & Fisher, B. L. (2007). Permanent loss of wings in queens of the ant Odontomachus coquereli from Madagascar. *Insectes Sociaux*, *54*, 183–188.

Moran, N. A. (2007). Symbiosis as an adaptive process and source of phenotypic complexity. *Proceedings of the National Academy of Sciences of the United States of America*, *104*(Suppl 1), 8627–8633. https://doi.org/10.1073/pnas.0611659104.

Moreau, C. S., & Bell, C. D. (2013). Testing the museum versus cradle tropical biological diversity hypothesis: Phylogeny, diversification, and ancestral biogeographic range evolution of the ants. *Evolution*, *67*(8), 2240–2257. https://doi.org/10.1111/evo.12105.

Moreau, C. S., Bell, C. D., Vila, R., Archibald, S. B., & Pierce, N. E. (2006). Phylogeny of the ants: Diversification in the age of angiosperms. *Science*, *312*(5770), 101–104. https://doi.org/10.1126/science.1124891.

Mutti, N. S., Dolezal, A. G., Wolschin, F., Mutti, J. S., Gill, K. S., & Amdam, G. V. (2011). IRS and TOR nutrient-signaling pathways act via juvenile hormone to influence honey bee caste fate. *The Journal of Experimental Biology*, *214*(Pt 23), 3977–3984. https://doi.org/10.1242/jeb.061499.

Nahmad, M., Glass, L., & Abouheif, E. (2008). The dynamics of developmental system drift in the gene network underlying wing polyphenism in ants: A mathematical model. *Evolution & Development*, *10*(3), 360–374. https://doi.org/10.1111/j.1525-142X.2008.00244.x.

Niitsu, S., Lobbia, S., Izumi, S., & Fujiwara, H. (2008). Female-specific wing degeneration is triggered by ecdysteroid in cultures of wing discs from the bagworm moth, Eumeta variegata (Insecta: Lepidoptera, Psychidae). *Cell and Tissue Research*, *333*(1), 169–173. https://doi.org/10.1007/s00441-008-0615-7.

Niitsu, S., Toga, K., Tomizuka, S., Maekawa, K., Machida, R., & Kamito, T. (2014). Ecdysteroid-induced programmed cell death is essential for sex-specific wing degeneration of the wingless-female winter moth. *PLoS One*, *9*(2), e89435. https://doi.org/10.1371/journal.pone.0089435.

Nijhout, H. F. (1991). *The development and evolution of butterfly wing patterns*. Smithsonian Institution Press.

Nijhout, H. F. (1994). *Insect hormones*. Princeton University Press.

Nijhout, H. F. (1999). Control mechanism of polyphenic development in insects. *BioScience*, *49*(3), 181–192.

Nijhout, H. F. (2003). Development and evolution of adaptive polyphenisms. *Evolution & Development*, *5*(1), 9–18. https://doi.org/10.1046/j.1525-142x.2003.03003.x.

Nijhout, H. F., & McKenna, K. Z. (2018). The distinct roles of insulin signaling in polyphenic development. *Current Opinion in Insect Science*, *25*, 58–64. https://doi.org/10.1016/j.cois.2017.11.011.

Nijhout, H. F., Smith, W. A., Schachar, I., Subramanian, S., Tobler, A., & Grunert, L. W. (2007). The control of growth and differentiation of the wing imaginal disks of Manduca sexta. *Developmental Biology*, *302*(2), 569–576. https://doi.org/10.1016/j.ydbio.2006.10.023.

Nijhout, H. F., & Wheeler, D. E. (1982). Juvenile hormone and the physiological basis of insect polymorphisms. *The Quarterly Review of Biology*, *57*(2), 109–133.

Nowak, M. A., Tarnita, C. E., & Wilson, E. O. (2010). The evolution of eusociality. *Nature*, *466*(7310), 1057–1062. https://doi.org/10.1038/nature09205.

Oettler, J., Platschek, T., Schmidt, C., Rajakumar, R., Fave, M. J., Khila, A., et al. (2018). Interruption points in the wing gene regulatory network underlying wing polyphenism evolved independently in male and female morphs in Cardiocondyla ants. *Journal of Experimental Zoology. Part B, Molecular and Developmental Evolution*, *332*(1-2), 7–16. https://doi.org/10.1002/jez.b.22834.

Oettler, J., Suefuji, M., & Heinze, J. (2010). The evolution of alternative reproductive tactics in male Cardiocondyla ants. *Evolution*, *64*(11), 3310–3317.

Okada, Y., Miyazaki, S., Miyakawa, H., Ishikawa, A., Tsuji, K., & Miura, T. (2010). Ovarian development and insulin-signaling pathways during reproductive differentiation in the queenless ponerine ant Diacamma sp. *Journal of Insect Physiology*, *56*(3), 288–295. https://doi.org/10.1016/j.jinsphys.2009.10.013.

Orme, M. H., & Leevers, S. J. (2005). Flies on steroids: The interplay between ecdysone and insulin signaling. *Cell Metabolism*, *2*(5), 277–278. https://doi.org/10.1016/j.cmet.2005.10.005.

Passera, L., & Suzzoni, J. P. (1979). The role of the queen of Pheidole pallidula (Nyl.) (Hymenoptera, Formicidae) in the brood sexualization after JH [Juvenile Hormone] treatment. *Insectes Sociaux*.

Patel, A., Fondrk, M. K., Kaftanoglu, O., Emore, C., Hunt, G., Frederick, K., et al. (2007). The making of a queen: TOR pathway is a key player in diphenic caste development. *PLoS One*, *2*(6), e509. https://doi.org/10.1371/journal.pone.0000509.

Peeters, C. (1991). The occurrence of sexual reproduction among ant workers. *Biological Journal of the Linnean Society*, *44*, 141–152.

Peeters, C. (1997). Morphologically 'primitive' ants: Comparative review of social characters, and the importance of queen-worker dimorphism. In J. C. Choe, & B. J. Crespi (Eds.), *The evolution of social behavior in insects and Arachnids* Cambridge University Press.

Peeters, C. (2012). Convergent evolution of wingless reproductives across all subfamilies of ants, and sporadic loss of winged queens (Hymenoptera: Formicidae). *Myrmecological News*, *16*, 75–91.

Peeters, C., Keller, R. A., Khalife, A., Fisher, G., Katzke, J., Blanke, A., et al. (2020). The loss of flight in ant workers enabled an evolutionary redesign of the thorax for ground labour. *Frontiers in Zoology*, *17*(33), 1–13.

Peeters, C., & Molet, M. (2010). Evolution of advanced social traits in phylogenetically basal ants: Striking worker polymorphism and large queens in Amblyopone australis. *Insectes Sociaux*, *57*, 177–183.

Penick, C. A., & Liebig, J. (2012). Regulation of queen development through worker aggression in a predatory ant. *Behavioral Ecology*, 992–998.

Penick, C. A., Prager, S. S., & Liebig, J. (2012). Juvenile hormone induces queen development in late-stage larvae of the ant Harpegnathos saltator. *Journal of Insect Physiology*, *58*(12), 1643–1649. https://doi.org/10.1016/j.jinsphys.2012.10.004.

Pepperl, J., Reim, G., Luthi, U., Kaech, A., Hausmann, G., & Basler, K. (2013). Sphingolipid depletion impairs endocytic traffic and inhibits Wingless signaling. *Mechanisms of Development*, *130*(9-10), 493–505. https://doi.org/10.1016/j.mod.2013.04.001.

Perrichot, V. (2019). New Cretaceous records and the diversification of crown-group ants (Hymenoptera: Formicidae). In *8th International Congress on Fossil Insects, Arthropods & Amber*. Dominica: Saint Domingue.

Perrichot, V., Lacau, S., Neraudeau, D., & Nel, A. (2008). Fossil evidence for the early ant evolution. *Naturwissenschaften*, *95*(2), 85–90. https://doi.org/10.1007/s00114-007-0301-8.

Perrichot, V., Nel, A., Neraudeau, D., Lacau, S., & Guyot, T. (2008). New fossil ants in French Cretaceous amber (Hymenoptera: Formicidae). *Naturwissenschaften*, *95*(2), 91–97. https://doi.org/10.1007/s00114-007-0302-7.

Perrichot, V., Wang, B., & Engel, M. S. (2016). Extreme morphogenesis and ecological specialization among cretaceous basal ants. *Current Biology*, *26*(11), 1468–1472. https://doi.org/10.1016/j.cub.2016.03.075.

Peters, R. S., Krogmann, L., Mayer, C., Donath, A., Gunkel, S., Meusemann, K., et al. (2017). Evolutionary history of the hymenoptera. *Current Biology*, *27*(7), 1013–1018. https://doi.org/10.1016/j.cub.2017.01.027.

Quicke, D. J. (2015). *The braconid and ichneumonid parasitoid wasps*. John Wiley & Sons, Ltd.

Rabeling, C., Brown, J. M., & Verhaagh, M. (2008). Newly discovered sister lineage sheds light on early ant evolution. *Proceedings of the National Academy of Sciences of the United States of America*, *105*(39), 14913–14917. https://doi.org/10.1073/pnas.0806187105.

Rafiqi, A. M., Rajakumar, A., & Abouheif, E. (2020). Origin and elaboration of a major evolutionary transition in individuality. *Nature*, *585*(7824), 239–244. https://doi.org/10.1038/s41586-020-2653-6.

Rajakumar, R., Koch, S., Couture, M., Fave, M. J., Lillico-Ouachour, A., Chen, T., et al. (2018). Social regulation of a rudimentary organ generates complex worker-caste systems in ants. *Nature*, *562*(7728), 574–577. https://doi.org/10.1038/s41586-018-0613-1.

Rajakumar, R., San Mauro, D., Dijkstra, M. B., Huang, M. H., Wheeler, D. E., Hiou-Tim, F., et al. (2012). Ancestral developmental potential facilitates parallel evolution in ants. *Science*, *335*(6064), 79–82. https://doi.org/10.1126/science.1211451.

Reid, J. A. (1941). The thorax of the wingless and short-winged Hymenoptera. *The Royal Entomological Society*, *91*(8), 367–446.

Richard, D. S., & Gilbert, L. I. (1991). Reversible juvenile hormone inhibition of ecdysteroid and juvenile hormone synthesis by the ring gland of Drosophila melanogaster. *Experientia*, *47*(10), 1063–1066. https://doi.org/10.1007/BF01923343.

Ruiz-Trillo, I., & Nedelcu, A. M. (2015). *Evolutionary transitions to multicellular life: Principles and mechanisms*. Springer.

Salt, G. (1952). Trimorphism in the Ichneumonid Parasite Gelis corruptor. *Quarterly Journal ol Microscopical Science*, *93*(4), 453–474.

Sameshima, S. Y., Miura, T., & Matsumoto, T. (2004). Wing disc development during caste differentiation in the ant Pheidole megacephala (Hymenoptera: Formicidae). *Evolution & Development*, *6*(5), 336–341. https://doi.org/10.1111/j.1525-142X.2004.04041.x.

Sann, M., Niehuis, O., Peters, R. S., Mayer, C., Kozlov, A., Podsiadlowski, L., et al. (2018). Phylogenomic analysis of Apoidea sheds new light on the sister group of bees. *BMC Evolutionary Biology*, *18*(1), 71. https://doi.org/10.1186/s12862-018-1155-8.

Schlichting, C. D., & Wund, M. A. (2014). Phenotypic plasticity and epigenetic marking: An assessment of evidence for genetic accommodation. *Evolution*, *68*(3), 656–672. https://doi.org/10.1111/evo.12348.

Schmieder, R. G. (1933). The polymorphic forms of melittobia chalybii ashmead and the determining factors involved in their production. (Hymenoptera: Chalcidoidea, Eulophidæ). *Biological Bulletin*, *65*(2), 338–354.

Schrader, L., Simola, D. F., Heinze, J., & Oettler, J. (2015). Sphingolipids, transcription factors, and conserved toolkit genes: Developmental plasticity in the ant cardiocondyla obscurior. *Molecular Biology and Evolution*, *32*(6), 1474–1486. https://doi.org/10.1093/molbev/msv039.

Schrempf, A., & Heinze, J. (2006). Proximate mechanisms of male morph determination in the ant Cardiocondyla obscurior. *Evolution & Development*, *8*(3), 266–272. https://doi.org/10.1111/j.1525-142X.2006.00097.x.

Schwander, T., Humbert, J. Y., Brent, C. S., Cahan, S. H., Chapuis, L., Renai, E., et al. (2008). Maternal effect on female caste determination in a social insect. *Current Biology*, *18*(4), 265–269. https://doi.org/10.1016/j.cub.2008.01.024.

Schwander, T., & Leimar, O. (2011). Genes as leaders and followers in evolution. *Trends in Ecology & Evolution*, *26*(3), 143–151. https://doi.org/10.1016/j.tree.2010.12.010.

Sedlak, B. J., Manzo, R., & Stevens, M. (1984). Localized cell death in Drosophila imaginal wing disc epithelium caused by the mutation apterous-blot. *Developmental Biology*, *104*(2), 489–496. https://doi.org/10.1016/0012-1606(84)90106-4.

Shbailat, S. J., & Abouheif, E. (2013). The wing-patterning network in the wingless castes of Myrmicine and Formicine ant species is a mix of evolutionarily labile and non-labile genes. *Journal of Experimental Zoology. Part B, Molecular and Developmental Evolution*, *320*(2), 74–83. https://doi.org/10.1002/jez.b.22482.

Shbailat, S. J., Khila, A., & Abouheif, E. (2010). Correlations between spatiotemporal changes in gene expression and apoptosis underlie wing polyphenism in the ant Pheidole morrisi. *Evolution & Development*, *12*(6), 580–591. https://doi.org/10.1111/j.1525-142X.2010.00443.x.

Shingleton, A. W., Das, J., Vinicius, L., & Stern, D. L. (2005). The temporal requirements for insulin signaling during development in Drosophila. *PLoS Biology*, *3*(9), e289. https://doi.org/10.1371/journal.pbio.0030289.

Shubin, N., Tabin, C., & Carroll, S. (2009). Deep homology and the origins of evolutionary novelty. *Nature*, *457*(7231), 818–823. https://doi.org/10.1038/nature07891.

Sieriebriennikov, B., Prabh, N., Dardiry, M., Witte, H., Roseler, W., Kieninger, M. R., et al. (2018). A Developmental Switch Generating Phenotypic Plasticity Is Part of a Conserved Multi-gene Locus. *Cell Reports*, *23*(10), 2835–2843. e2834. https://doi.org/10.1016/j.celrep.2018.05.008.

Simmonds, A. J., Liu, X., Soanes, K. H., Krause, H. M., Irvine, K. D., & Bell, J. B. (1998). Molecular interactions between Vestigial and Scalloped promote wing formation in Drosophila. *Genes & Development*, *12*(24), 3815–3820. https://doi.org/10.1101/gad.12.24.3815.

Smith, D. R. (1993). Systematics, life history, and distribution of sawflies. In M. R. Wagner, & K. K. Raffa (Eds.), *Sawfly life history adaptations to woody plants* (pp. 3–32). Academic Press, Inc.

Stearns, S. C. (1989). The evolutionary significance of phenotypic plasticity. *BioScience*, *39*, 436–445.

Sultan, S. E. (2015). *Organisms and environment: Ecological development, niche construction, and adaptation*. Oxford University Press.

Suzuki, Y., & Nijhout, H. F. (2006). Evolution of a polyphenism by genetic accommodation. *Science, 311*(5761), 650–652. https://doi.org/10.1126/science.1118888.

Suzzoni, J. P., Passera, L., & Strambi, A. (1980). Ecdysteroid titre and caste determination in the ant,Pheidole pallidula (Nyl.) (Hymenoptera: Formicidae). *Experientia, 36*, 1228–1229.

Suzzoni, J. P., Passera, L., & Strambi, A. (1983). Ecdysteroid production during caste differentiation in larvae of the ant, Plagiolepis pygmaea. *Physiological Entomology, 8*, 93–96.

Szathmary, E., & Smith, J. M. (1995). The major evolutionary transitions. *Nature, 374*(6519), 227–232. https://doi.org/10.1038/374227a0.

Tatar, M., Kopelman, A., Epstein, D., Tu, M. P., Yin, C. M., & Garofalo, R. S. (2001). A mutant Drosophila insulin receptor homolog that extends life-span and impairs neuroendocrine function. *Science, 292*(5514), 107–110. https://doi.org/10.1126/science.1057987.

Taylor, B. A., Reuter, M., & Sumner, S. (2019). Patterns of reproductive differentiation and reproductive plasticity in the major evolutionary transition to superorganismality. *Current Opinion in Insect Science, 34*, 40–47. https://doi.org/10.1016/j.cois.2019.02.007.

Tobler, A., & Nijhout, H. F. (2010). A switch in the control of growth of the wing imaginal disks of Manduca sexta. *PLoS One, 5*(5), e10723. https://doi.org/10.1371/journal.pone.0010723.

Tomoyasu, Y., Arakane, Y., Kramer, K. J., & Denell, R. E. (2009). Repeated co-options of exoskeleton formation during wing-to-elytron evolution in beetles. *Current Biology, 19*(24), 2057–2065. https://doi.org/10.1016/j.cub.2009.11.014.

Toth, A. L., & Rehan, S. M. (2017). Molecular evolution of insect sociality: An eco-evo--devo perspective. *Annual Review of Entomology, 62*, 419–442. https://doi.org/10.1146/annurev-ento-031616-035601.

Truman, J. W., Hiruma, K., Allee, J. P., Macwhinnie, S. G., Champlin, D. T., & Riddiford, L. M. (2006). Juvenile hormone is required to couple imaginal disc formation with nutrition in insects. *Science, 312*(5778), 1385–1388. https://doi.org/10.1126/science.1123652.

Truman, J. W., & Riddiford, L. M. (2002). Endocrine insights into the evolution of metamorphosis in insects. *Annual Review of Entomology, 47*, 467–500. https://doi.org/10.1146/annurev.ento.47.091201.145230.

Tu, M. P., Yin, C. M., & Tatar, M. (2005). Mutations in insulin signaling pathway alter juvenile hormone synthesis in Drosophila melanogaster. *General and Comparative Endocrinology, 142*(3), 347–356. https://doi.org/10.1016/j.ygcen.2005.02.009.

Vellichirammal, N. N., Madayiputhiya, N., & Brisson, J. A. (2016). The genomewide transcriptional response underlying the pea aphid wing polyphenism. *Molecular Ecology, 25*(17), 4146–4160. https://doi.org/10.1111/mec.13749.

Wang, X., Wei, K., Yang, Z., Jennings, D. E., & Duan, J. J. (2016). Effects of biotic and abiotic factors on phenotypic partitioning of wing morphology and development in Sclerodermus pupariae (Hymenoptera: Bethylidae). *Scientific Reports, 6*, 26408. https://doi.org/10.1038/srep26408.

Ward, P. S. (2007). Phylogeny, classification, and species-level taxonomy of ants (Hymenoptera: Formicidae)*. *Zootaxa, 1668*, 549–563.

Wei, K., Gao, S. K., Tang, Y. L., Wang, X. Y., & Yang, Z. Q. (2016). Determination of the optimal parasitoid-to-host ratio for efficient mass-rearing of the parasitoid, Sclerodermus pupariae (Hymenoptera: Bethylidae). *Journal of Applied Entomology, 141*, 181–188.

West, S. A., Fisher, R. M., Gardner, A., & Kiers, E. T. (2015). Major evolutionary transitions in individuality. *Proceedings of the National Academy of Sciences of the United States of America, 112*(33), 10112–10119. https://doi.org/10.1073/pnas.1421402112.

West-Eberhard, M. J. (1978). Polygyny and the evolution of social behavior in wasps. *Journal of the Kansas Entomological Society, 51*(4), 832–856.

West-Eberhard, M. J. (2003). *Developmental plasticity and evolution*. Oxford University Press.
Wheeler, W. M. (1905). Worker ants with vestiges of wings. *Bulletin of the American Museum of Natural History*, *21*, 405–408.
Wheeler, W. M. (1916). The Phylogenetic development of subapterous and apterous castes in the Formicidae. *Zoology*, 109–117.
Wheeler, W. M. (1928a). Mermis parasitism and intercastes among ants. *Journal of Experimental Zoology*, *50*(2), 165–237.
Wheeler, W. M. (1928b). *The social insects: Their origin and evolution*. Kegan Paul, Trench: Trubner & Co. Ltd.
Wheeler, D. E. (1986). Developmental and physiological determinants of caste in social hymenoptera: Evolutionary implications. *The American Naturalist*, *128*(1), 13–34.
Wheeler, D. E., Buck, N., & Evans, J. D. (2006). Expression of insulin pathway genes during the period of caste determination in the honey bee, Apis mellifera. *Insect Molecular Biology*, *15*(5), 597–602. https://doi.org/10.1111/j.1365-2583.2006.00681.x.
Wheeler, D. E., Buck, N. A., & Evans, J. D. (2014). Expression of insulin/insulin-like signalling and TOR pathway genes in honey bee caste determination. *Insect Molecular Biology*, *23*(1), 113–121. https://doi.org/10.1111/imb.12065.
Wheeler, D. E., & Nijhout, H. F. (1981a). Imaginal wing discs in larvae of the soldier caste of Pheidole bicarinata vinelandica Forel (Hymenoptera : Formicidae). *International Journal of Insect Morphology and Embryology*, *10*(2), 131–139. https://doi.org/10.1016/S0020-7322(81)80018-9.
Wheeler, D. E., & Nijhout, H. F. (1981b). Soldier determination in ants: New role for juvenile hormone. *Science*, *213*(4505), 361–363. https://doi.org/10.1126/science.213.4505.361.
Wheeler, D. E., & Nijhout, H. F. (1984). Soldier determination in Pheidole bicarinata: Inhibition by adult soldiers. *Journal of Insect Physiology*, *30*(2), 127–135. https://doi.org/10.1016/0022-1910(84)90116-1.
Whitfield, J. B. (1998). Phylogeny and evolution of host-parasitoid interactions in hymenoptera. *Annual Review of Entomology*, *43*, 129–151. https://doi.org/10.1146/annurev.ento.43.1.129.
Whitfield, J. B. (2003). Phylogenetic insights into the evolution of parasitism in hymenoptera. *Advances in Parasitology*, *54*, 69–100. https://doi.org/10.1016/s0065-308x(03)54002-7.
Williams, J. A., Bell, J. B., & Carroll, S. B. (1991). Control of Drosophila wing and haltere development by the nuclear vestigial gene product. *Genes & Development*, *5*(12B), 2481–2495. https://doi.org/10.1101/gad.5.12b.2481.
Wills, B. D., Powell, S., Rivera, M. D., & Suarez, A. V. (2018). Correlates and consequences of worker polymorphism in ants. *Annual Review of Entomology*, *63*, 575–598. https://doi.org/10.1146/annurev-ento-020117-043357.
Wilson, E. O. (1953). The origin and evolution of polymorphism in ants. *The Quarterly Review of Biology*, 136–156.
Wilson, E. O. (1954). A new interpretation of the frequency curves associated with ant polymorphism. *Insectes Sociaux*, *1*, 75–80.
Wilson, E. O. (1971). *The insect societies*. Harvard University Press.
Wilson, E. O. (1987). The earliest known ants: An analysis of the Cretaceous species and an inference concerning their social organization. *Paleobiology*, *13*(1), 44–53.
Wilson, E. O. (2003). *Pheidole in the New World: A dominant, hyperdiverse ant genus*. Harvard University Press.
Wilson, E. O., Carpenter, F. M., & Brown, W. L., Jr. (1967a). The first mesozoic ants. *Science*, *157*(3792), 1038–1040. https://doi.org/10.1126/science.157.3792.1038.
Wilson, E. O., Carpenter, F. M., & Brown, W. L., Jr. (1967b). The first mesozoic ants, with the description of a new subfamily. *Psyche*, *74*(1), 1–17.

Wilson, E. O., & Holldobler, B. (2005a). Eusociality: Origin and consequences. *Proceedings of the National Academy of Sciences of the United States of America*, *102*(38), 13367–13371. https://doi.org/10.1073/pnas.0505858102.

Wilson, E. O., & Holldobler, B. (2005b). The rise of the ants: A phylogenetic and ecological explanation. *Proceedings of the National Academy of Sciences of the United States of America*, *102*(21), 7411–7414. https://doi.org/10.1073/pnas.0502264102.

Winter, S. E., & Campbell, G. (2004). Repression of Dpp targets in the Drosophila wing by Brinker. *Development*, *131*(24), 6071–6081. https://doi.org/10.1242/dev.01538.

Wolschin, F., Mutti, N. S., & Amdam, G. V. (2011). Insulin receptor substrate influences female caste development in honeybees. *Biology Letters*, 7(1), 112–115. https://doi.org/10.1098/rsbl.2010.0463.

Wu, Q., & Brown, M. R. (2006). Signaling and function of insulin-like peptides in insects. *Annual Review of Entomology*, *51*, 1–24. https://doi.org/10.1146/annurev.ento.51.110104.151011.

Yamauchi, K., Kimura, Y., Corbara, B., Kinomura, K., & Tsuji, K. (1996). Dimorphic ergatoid males and their reproductive behavior in the ponerine ant Hypoponera bondroiti. *Insectes Sociaux*, *43*, 119–130.

Yang, Z.-Q., Wang, X.-Y., Yao, Y.-X., Gould, J. R., & Cao, L.-M. (2012). A new species of sclerodermus (Hymenoptera: Bethylidae) parasitizing agrilus planipennis (Coleoptera: Buprestidae) From China, with a key to Chinese species in the genus. *Annals of the Entomological Society of America*, *105*(5), 619–627. https://doi.org/10.1603/an12017.

Yek, S. H., & Mueller, U. G. (2011). The metapleural gland of ants. *Biological Reviews of the Cambridge Philosophical Society*, *86*(4), 774–791. https://doi.org/10.1111/j.1469-185X.2010.00170.x.

Zecca, M., Basler, K., & Struhl, G. (1995). Sequential organizing activities of engrailed, hedgehog and decapentaplegic in the Drosophila wing. *Development*, *121*(8), 2265–2278. https://www.ncbi.nlm.nih.gov/pubmed/7671794.

Zera, A. J. (2004). The endocrine regulation of wing polymorphism in insects: State of the art, recent surprises, and future directions. *Integrative and Comparative Biology*, *43*, 607–616.

Zera, A. J. (2006). Evolutionary genetics of juvenile hormone and ecdysteroid regulation in Gryllus: A case study in the microevolution of endocrine regulation. *Comparative Biochemistry and Physiology. Part A, Molecular & Integrative Physiology*, *144*(3), 365–379. https://doi.org/10.1016/j.cbpa.2005.11.026.

Zhang, C. X., Brisson, J. A., & Xu, H. J. (2019). Molecular Mechanisms of Wing Polymorphism in Insects. *Annual Review of Entomology*, *64*, 297–314. https://doi.org/10.1146/annurev-ento-011118-112448.

CHAPTER TEN

Genetic assimilation and accommodation: Models and mechanisms

H. Frederik Nijhout*, Anna M. Kudla, and Caleb C. Hazelwood
Department of Biology, Duke University, Durham, NC, United States
*Corresponding author: e-mail address: hfn@duke.edu

Contents

Abstract

Genetic assimilation and genetic accommodation are mechanisms by which novel phenotypes are produced and become established in a population. Novel characters may be fixed and canalized so they are insensitive to environmental variation, or can be plastic and adaptively responsive to environmental variation. In this review we explore the various theories that have been proposed to explain the developmental origin and evolution of novel phenotypes and the mechanisms by which canalization and phenotypic plasticity evolve. These theories and models range from conceptual to mathematical and have taken different views of how genes and environment contribute to the development and evolution of the properties of phenotypes. We will argue that a deeper and more nuanced understanding of genetic accommodation requires a recognition that

Current Topics in Developmental Biology, Volume 141
ISSN 0070-2153
https://doi.org/10.1016/bs.ctdb.2020.11.006

phenotypes are not static entities but are dynamic system properties with no fixed deterministic relationship between genotype and phenotype. We suggest a mechanistic systems-view of development that allows one to incorporate both genes and environment in a common model, and that enables both quantitative analysis and visualization of the evolution of canalization and phenotypic plasticity.

1. Introduction

Genetic assimilation refers to the process of selection by which phenotypes that were originally expressed only after an environmental stimulus become genetically fixed (Waddington, 1942, 1959). Waddington originally called the environmentally-induced novel phenotypes phenocopies, because of their resemblance to certain mutant phenotypes. For instance, when Waddington exposed pupae of *Drosophila* to a heat shock, some of the adults showed a cross-veinless phenotype (Waddington, 1953). Selecting these phenocopied individuals for breeding and exposing the next generation to a similar heatshock increased the percentage of animals showing the cross-veinless phenotype. After continuing this selection for several generations, Waddington found that cross-veinless phenotype became fixed in the experimental population and appeared without the need for a heatshock. He provocatively called this the genetic assimilation of an acquired character (Waddington, 1953). When Waddington exposed *Drosophila* eggs to ether vapors, some of the offspring exhibited the bithorax phenotype (Waddington, 1956). Again, after generations of selection for this effect, the bithorax phenotype became established without further need of the ether stimulus (Waddington, 1956).

One can think of the environmental induction of phenocopies as a kind of phenotypic plasticity, although in most cases the environmental stimulus (ether, temperature shock) is not of the kind an organism would normally encounter in nature. Waddington noted that the cross-veinless and bithorax phenotypes he produced were unlikely to have an adaptive value in nature, so in order to examine the potentially evolutionary significance of genetic assimilation of more realistic phenotypes he exposed *Drosophila* eggs to near-lethal concentrations of sodium chloride, an environmental stimulus more likely to be encountered in nature, and selected for increased sizes of anal papillae, the organs used to control osmotic balance (Waddington, 1959). After selection of good responders Waddington found that enlarged papillae had become fixed in this experimental population, under normal sodium chloride concentrations. He interpreted this as the genetic assimilation of a novel trait that would be an adaptation for survival in a high osmotic environment (Waddington, 1959).

2. Genetic assimilation since Waddington

Following Waddington, numerous studies have aimed to reveal more examples of genetic assimilation and elucidate mechanisms. Bateman and Milkman used heat shock to genetically assimilate several different venation phenocopies in *Drosophila* (Bateman, 1959; Milkman, 1965). Schlichting and Wund (2014) review several putative instances of genetic assimilation, both in and out of the laboratory. For instance, the plastic capacities of the invasive herb *Prunella vulgaris* allow it to colonize the shady forest understory. However, prolonged exposure to this stable environment results in a loss of plasticity in traits such as water use efficiency. In the invasive population, the novel phenotype is canalized against variation in sunlight (Godoy, Saldaña, Fuentes, Valladares, & Gianoli, 2011). In the pale grass blue butterfly, *Zizeeria maha*, researchers were able to induce a polyphenic color pattern through artificial cold shock. A selective breeding experiment on the population demonstrated partial genetic assimilation within 10 generations, as animals began to express the environmentally induced color pattern in the absence of the cold-shock treatment (Otaki, Hiyama, Iwata, & Kudo, 2010). Finally, researchers demonstrated that Australian tiger snakes, *Notechis scutatus*, express varying degrees of plasticity in head size depending on how long ago the population colonized its island habitat. Tiger snakes are more plastic on islands where the founder events took place more recently. This suggests that, after sufficient exposure to stable selection pressures, genetic assimilation can occur, thereby enabling constitutive expression of the adaptive phenotype and eliminating the cost of plasticity (Aubret & Shine, 2009).

However, some studies are cited as examples of genetic assimilation when they only claim to reveal a loss of phenotypic plasticity. This is an important distinction to make. A loss of plasticity within a population can occur in at least two ways. One way is genetic assimilation. A second way, however, does not require genetic assimilation at all—only natural selection. Consider a population with a wide array of reaction norms. At one extreme, there are individuals whose reaction norms (plotted as a function of several environmental states) have a slope of 1. With each environmental state, the individual genotype expresses a different phenotype. Call this a specialist strategy. At the other extreme, there are individuals whose reaction norms have a slope of 0. With each environmental state, the individual genotype consistently expresses a single phenotype. Call this a generalist strategy. It is possible that, under certain conditions, the costs of a specialist strategy are high, and the fitness advantages conferred on plastic individuals is negligible. Under such conditions, selection would favor a generalist strategy, and we could expect to see an increase (and, ultimately, fixation) of non-plastic phenotypes. In other words, evolution has pushed the slope of the average reaction norm in this population closer and closer to zero. In this scenario, we observe a loss of plasticity, but we should not call it genetic assimilation. This is because the generalist phenotypes were never environmentally induced. They were always constitutively expressed, and the genetic, molecular, and physiological mechanisms that underlie a generalist strategy were simply more successful. For this scenario to be an instance of genetic assimilation, a phenotype would have to be environmentally induced, followed by the uncoupling of its expression from the environment. Future reviews should take care to distinguish between these two phenomena, as conflating them obscures an already elusive concept.

3. The role of phenotypic plasticity

Genetic assimilation results in the loss of phenotype plasticity. Phenotypic plasticity can be categorized in several ways. First there is the distinction between non-adaptive and adaptive plasticity. All traits are plastic to some degree, because biological processes are naturally sensitive to environmental variables like temperature and nutrients. Such natural variation can be neutral, deleterious or advantageous, depending on the phenotypic change and the environment in which it is induced. Advantageous plasticity can be refined by positive selection, and we recognize the product as adaptive plasticity. Disadvantageous plasticity can be mitigated by negative selection and neutral plasticity may be subject to drift.

The second distinction is between continuous plasticity, described by reaction norms, and discontinuous plasticity, or polyphenism, in which two (and occasionally three) distinct and stable alternative phenotypes develop. Reaction norms, of course, need not be adaptive, and many probably are not. If a reaction norm is continuous, that is, if there are a large number of intermediate phenotypes produced by a gradient of a specific environmental stimulus such as temperature or nutrition, it is difficult to see how each phenotype is a specific adaptation to that specific value of the environmental variable. This would require a large number of selective events at each value of the environmental variable. The actual shape and continuity of a reaction norm is seldom known. Typically, investigators test two (sometimes three) environmental values and then connect the resulting phenotypes with straight lines that sometimes serve as visual aids, but imply a linear continuity across environments.

The kind of plasticity Waddington encountered that led to assimilation of cross-veinless and bithorax is peculiar. It produced aberrant phenotypes that were not obviously on a continuum with the natural phenotype. Stress-induced phenotypic aberrations, using a variety of chemicals as well as temperature, have been described in many animals and plants (Badyaev, 2005; Hoffmann & Hercus, 2000). Stress during various developmental stages in *Drosophila* were studied by Goldschmidt, who first noted that these abnormalities resembled known mutations, and therefore named them phenocopies (Goldschmidt, 1935). In many cases, the kind of phenocopy that is induced depends on the exact time during development when the stressor is applied (Mitchell & Petersen, 1982), and different genetic strains can differ in the kinds of phenocopies they tend to produce (Goldschmidt & Piternick, 1957a, 1957b; Nair & Dearden, 2016).

Phenocopies are induced by brief near-lethal stressors such as cold-shock (a brief time at near-freezing temperature), heat shock (typically an hour at 38–42 °C), near lethal exposure to ether, high salt concentrations, and injections of borate or heavy metals (Goldschmidt & Piternick, 1957a; Sang & McDonald, 1954), heat-shock being the most commonly used. These are environmental shocks that typically kill a fraction of the organisms rather than environmental shifts that are within a physiologically manageable range. Environmental shocks that induce phenocopies stimulate the production of heat-shock proteins (HSPs). HSPs are molecular chaperones that protect certain proteins from degradation and denaturation (Marx, 1983; Morimoto, Kline, Bimston, & Cotto, 1997), and stop RNA and protein synthesis and development. This response is believed to have evolved as a mechanism to prevent or minimize cellular damage. Mitchell and Lipps (1978) showed that HSPs are degraded soon after heat-shock, and as RNA and protein synthesis gradually resume not all proteins reappear at the same time. As development resumes after a heat-shock, some essential proteins do not become active in time (Chomyn, Moller, & Mitchell, 1979; Mitchell & Lipps, 1978), and that leads to the developmental aberrations we recognize as phenocopies, that resemble mutations in the genes that code for the missing proteins.

Genetic assimilation does not require an extreme aberrant phenocopy response, but should be able to operate on any environmentally-induced trait. Genetic assimilation requires phenotypic plasticity, and can, in general, be thought of as the selection process by which a plastic trait becomes canalized, or becomes robust to a particular kind of environmental variation. Which kinds of plastic traits can become assimilated depends on the cause of the plasticity. Phenotypic plasticity extends the range of phenotypes that are physiologically and developmentally possible, given a particular genetic makeup, in a particular environment, and there is no reason to believe that it would not be possible to canalize any environmentally-induced phenotype through selective assimilation.

Genetic assimilation results in the fixation, or canalization, of a previously plastic trait. The trait thus becomes insensitive to variation in the environmental variable that originally induced it. Whether genetic assimilation to one environmental variable also gives robustness to others, or to genetic variation is an open question. It is conventional to make a distinction between genetic canalization/robustness and environmental canalization/robustness, although it is easy to think of examples where they are probably the same thing. For instance, if an enzyme requires a vitamin as a cofactor, as many do (Scott-Boyer et al., 2016), then genetic variation in the activity of the enzyme and environmental variation in the availability of the vitamin would have indistinguishable effects on the phenotype, and canalization that produces insensitivity of the system to variation in the activity of that enzyme would convey both genetic and environmental robustness/canalization.

However, Waddington's interpretation of his experimental results in *Drosophila*, the role of genetic assimilation in adaptive evolution, and its underlying mechanisms have all been called into question. For instance, Theodosius Dobzhansky and Ernst Mayr—two prominent figures of the modern evolutionary synthesis—doubted that genetic assimilation occurs at all. Instead, they argued, the phenotypic shifts that Waddington observed could be explained by selection on novel genetic mutations (Gilbert, 1991). Many, however, accept that genetic assimilation is a real phenomenon, but doubt that it is relevant to the evolution of natural populations. Skeptics cite the fact that most cases of genetic assimilation have been shown in laboratory populations under artificial selection (Ehrenreich & Pfennig, 2015). As the loss of plasticity can occur within a handful of generations, as Waddington's experiments demonstrate, it has been argued that the transition from environmental dependence to constitutive expression can "occur so rapidly as to pass below the radar screen of evolutionary biologists, unless they were explicitly looking for it" (Pigliucci, Murren, & Schlichting, 2006). In the absence of sufficient evidence, some have argued that genetic assimilation ought not to be taken seriously as an evolutionary mechanism (de Jong, 2005; Orr, 1999). Others, however, remain optimistic. Recent reviews (e.g., Schlichting and Wund (2014)) demonstrate that the results of some field studies are consistent with the loss of plasticity that one expects to observe in a population undergoing genetic assimilation. Excellent examples and plausible scenarios for genetic assimilation and the role of phenotypic plasticity in the evolution of novel traits come from the work of Moczek, Pfennig and their associated (Casasa & Moczek, 2018b; Casasa & Moczek, 2019; Ehrenreich & Pfennig, 2015; Moczek, 2007, 2008; Moczek et al., 2011; Pfennig et al., 2010; Pfennig & Ehrenreich, 2014). Those who are optimistic about the potential of genetic assimilation to incite evolutionary change argue that the available evidence is sufficient to reject claims to the contrary. Genetic assimilation is thought to be a significant mode of phenotypic evolution (Pigliucci et al., 2006; Schlichting & Pigliucci, 1998; West-Eberhard, 2003, 2005).

4. Plasticity, the Baldwin effect and genetic accommodation

Genetic assimilation results in the fixation, or canalization, of a previously plastic trait. The trait thus becomes insensitive to variation in the environmental variable that originally induced it. It is also possible to for selection to improve and stabilize plasticity, if the plastic phenotypes are suited to alternative environments. This idea was first proposed by Baldwin (Baldwin, 1896, 1902), and assumes that selection can improve condition-sensitivity, so that in a variable environment there is increased survival of individual that can respond plastically. West-Eberhard (2003) notes that this does not imply fixation of a plastic trait, but rather, a genetically improved ability to respond appropriately to an environment.

West-Eberhard (West-Eberhard, 2003, 2005) developed a broad and inclusive view of phenotypic evolution through the concept of genetic accommodation. This view encompasses both genetic assimilation and the Baldwin effect. The mechanisms involved in each of these concepts, in addition to mutation and selection, are considered in Table 1. The principles behind genetic accommodation are relatively straightforward (although they have been widely misunderstood). The idea centers around the origin of phenotypes. It assumes that a new phenotype can arise through either a genetic change (mutation) or an environmental change. The new alternative phenotype can then persist if it is advantageous and if a recurring environment, plus selection, increases its frequency of expression. The new trait can then be further modified and improved by selection. Among the subsequent possibilities are fixation of the trait, so that it becomes the only alternative that is expressed. This encompasses genetic assimilation. Another possibility is that the trait's sensitivity to the environment is increased,

Table 1 Comparison of assumptions and content of different mechanisms for the origin and evolution of novel traits.

	Mutation-selection	Genetic assimilation	Baldwin effect	Genetic accommodation
Initiated by mutation	X	–	–	X
Initiated by environmental shift	–	X	X	X
Modifies standing genetic variation	–	X	X	X
Improves form/function of trait	–	–	–	X
Increases genetic regulation of trait	–	X	X	X
Improves integration of a trait	–	–	–	X
Removes harmful pleiotropic effects	–	–	–	X
Eliminates disadvantageous traits	X	–	–	X
Alters frequency of expression	X	X	X	X
Refines conditions of expression	–	–	–	X
Leads to fixation of novel trait	X	X	–	X
Leads to improved plasticity of trait	–	–	X	X

After (Nijhout, 2015).

and that there is a refinement of the specific environmental signals that induce the trait. This describes the Baldwin effect. Genetic accommodation accounts for both of these processes of selection, as well as the processes by which initially deleterious traits (arising by mutation or environment) are eliminated or modified and rendered neutral or even advantageous. In other words, genetic accommodation refers to the increase and decrease of plasticity, canalization, genetic influence, and environmental influence (Pfennig & Ehrenreich, 2014). It describes the continuum of possible evolutionary outcomes for a plastic phenotype. One such outcome is genetic assimilation: the decrease of plasticity and environmental influence, and the increase of canalization. It is not clear, however, that genetic assimilation necessarily results in an increase of genetic influence. Highly plastic genotypes may be very sensitive to genetic influence if alternative genetic variants are not as sensitive to environmental variables. Similarly, if a phenotype is canalized against genetic variation as well as environmental variation, then evolution of the system might not increase the influence of the genetic component of the phenotype. Instead, it is possible that a physiological mechanism responsible for the trait is buffered against environmental and genetic variation, resulting in decreased influence of both variables.

5. Models and mechanisms for the evolution of assimilation, canalization and accommodation

To understand the mechanism by which plasticity, robustness, canalization, assimilation and accommodation work, and how they can evolve, we need a mechanistic systems-level understanding of the phenotype. This is an exceptionally difficult problem and several different conceptual and quantitative approaches have been taken. We survey some of the proposed models of assimilation and accommodation below. However, many of the models proposed thus far rely on scant empirical evidence or on overly simplistic pictures of development.

5.1 Single gene mutation pathways to genetic assimilation and accommodation

The first mechanism we consider for genetic assimilation occurs when a single gene mutation fixes one of the forms of what is normally a plastic polyphenism. There are at least two cases of this mechanism in the literature in which a single gene mutation fixes one of the forms of what is normally a plastic polyphenism. In *Caenorhabditis elegans*, the matricidal form of larval development, in which hatchling larvae consume the mother from within, is induced by nutrient-deficient environments. Vigne et al. (2020) describe a mutation in the gene kcnl-1 discovered in populations with near-constitutive matricidal hatching. This gene encodes a small-conductance calcium-activated potassium channel subunit, and a single base substitution leads to reduced vulval muscle excitability, and thus reduced expulsion of hatchlings, irrespective of environment. Thus larvae cannot hatch and consume the mother. A single gene mutation appears to have fixed one of the extremes of a plastic trait.

A similar single gene effect occurs in the *rosa-linea* color polyphenism of the Buckeye butterfly, *Junonia* (formerly *Precis*) *coenia*. The red-brown *rosa* autumn form is induced by short daylengths and low temperatures during late larval life, whereas long-days and warm temperatures induce the pale tan *linea* form (Rountree & Nijhout, 1995b; Smith, 1991). The underlying developmental cause is a shift in the timing of the secretion of ecdysone during the early pupal stage. There is an ecdysone-sensitive period for color morph induction between 24 and 48 h after pupation; high levels of ecdysone during this time induce development of the *linea* form, whereas the *rosa* form develops if ecdysone is low or absent during this critical period (Rountree & Nijhout, 1995b). Under long-day conditions ecdysone secretion begins about 20 h after pupation, but under short-days ecdysone section is delayed and begins after 48 h, thus missing the developmental sensitive period. The ecdysone-induced switch during the critical period is associated with a massive change in the pattern of gene expression (Daniels, Murad, Mortazavi, & Reed, 2014). Some populations of *Junonia* have a low percentage of individuals that develop the *rosa* form even under warm long-day conditions (Rountree & Nijhout, 1995a). Selecting on those individuals produced a genetic strain that was 100% *rosa* under all environmental conditions, and genetic analysis showed the trait was inherited as a single gene recessive. Under long-day warm conditions, the constitutive *rosa* strain had an ecdysone secretion profile like that of the *linea* form (Rountree & Nijhout, 1995a), so the effect on morphology was not due to a shift in ecdysone secretion, like that of the plastically induced *rosa* form, but must lie somewhere downstream of the ecdysone signal.

In these two cases the mutation that fixes one of the alternative forms that already existed at low frequencies in natural populations, so it can be thought of as a stepping stone to the evolution from phenotypic plasticity to genetic fixation of one of the alternative phenotypes. A somewhat different single gene step to fixation of a polyphenism may exists in the pea aphid, *Acyrthosiphon pisum*. This species has an environmentally induced winged-wingless polyphenism in females, and a genetic winged-wingless polymorphism in males that is controlled by a single quantitative trait locus (Braendle, Davis, Brisson, & Stern, 2006; Ogawa & Miura, 2013). If the conditions were to arise in which this same gene also affected wing development in females, then this would result in the genetic fixation of what was originally a plastic phenotype. A single-gene mutation in the Hsp90 heat-shock protein can reveal an enormous amount of cryptic genetic variation on which selection can act to genetically fix a novel phenotype (Rutherford & Lindquist, 1998; Sangster et al., 2008). These are not exactly the scenarios that Waddington and others have had in mind, which assume that genetic assimilation occurs by gradualistic evolution involving many genes, but these single gene extremes show that simple genetic changes can be sufficient, and may suggest it might be worth investigating how common genetic assimilation by single mutations might be.

Genetic assimilation results in a fixed phenotype that is insensitive to environment. By contrast, the evolution of adaptive phenotypic plasticity, for instance a polyphenisms from a previously canalized phenotype, requires the acquisition of environmental sensitivity. Genetic accommodation suggest that the initiation of a novel phenotype can start with either a mutation or an environmental shift. A single-gene stepping stone to the genetic accommodation of an environmentally-sensitive plastic phenotype is provided by the artificial evolution of a color polyphenism in *Manduca sexta.*

Temperature shocks of larvae of the black-larva genetic strain of *Manduca sexta* caused variation in coloration after the molt to the last larval instar that ranged from black to a grayish green. Selection for the greenest response to high-temperature shock (42 °C) while maintaining the black phenotype at normal temperatures (25 °C), resulted in the evolution of a strain that was black at temperatures below 25 °C, but with a sharp transition to green at temperatures above 30 °C (Suzuki & Nijhout, 2006). Green is the normal larval color of *Manduca*, and the black strain is due to a single gene mutation that reduced the level of juvenile hormone (JH) at the time of the molt (Safranek & Riddiford, 1975). The green phenotype of black larvae can be rescued with an injection or a topical application of JH (Truman, Riddiford, & Safranek, 1973).

Evolution of the black/green temperature-dependent polyphenism was due to the fact that JH levels became temperature-sensitive (Suzuki & Nijhout, 2006, 2008). The interesting result was also that after selection, the reaction norm of color to variation in a temperature gradient (20 to 40 °C) was sharply sigmoidal, as noted above. The threshold temperature (between 25 and 30 °C) was well below the inducing temperature shock (42 °C) used during the selection. A model of how genetic accommodation of this polyphenism worked is shown in Fig. 1. The essential component was an enabling mutation (the black larva mutation) that moved the phenotype into a region where temperature shock could have a phenotypic effect, followed by selection that stabilized high and low JH levels, depending on temperature. Backcrosses of the polyphenic strain with the parental strain showed that the polyphenism was regulated by one sex-linked gene of large effect and many smaller effect modifier genes (Suzuki & Nijhout, 2008). Thus, genetic accommodation appeared to occur through an initial change in one gene with large phenotypic effect, followed by changes in genes of lesser effect, just as is assumed to occur in conventional cases of adaptive evolution.

5.2 An epigenetics model

Another proposed mechanism of genetic assimilation is epigenetic modification (i.e., cytosine or adenine methylation, histone modification, micro RNAs, etc.) and inheritance (i.e., the persistence of modifications across generations). Richards et al. (2017) review several studies where this principle is demonstrated and, in doing so, argue that DNA methylation can mediate phenotypic plasticity and maintain the expression of a novel phenotype in

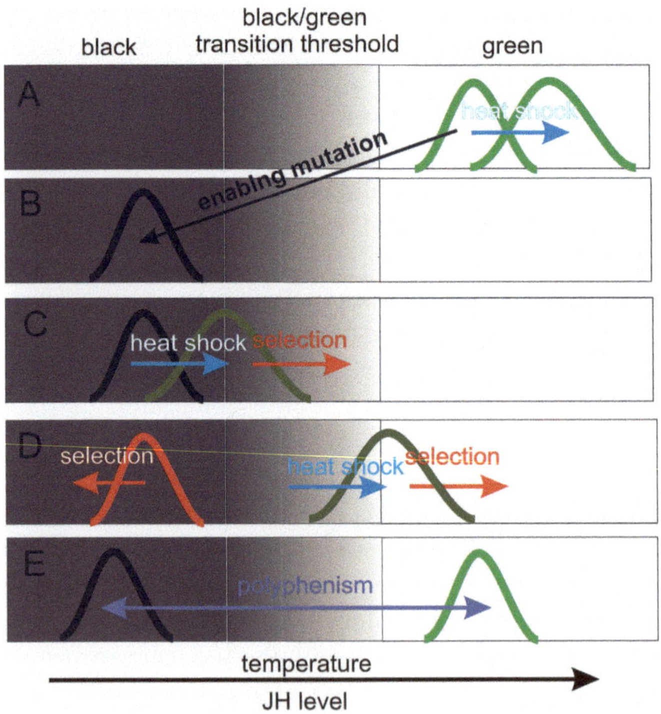

Fig. 1 Mechanism for the genetic accommodation of a plastic threshold trait in *Manduca sexta*. (A) Wild type green phenotype in response to heats hock remains green. (B) Enabling single-gene mutation that lowers juvenile hormone levels and turns larva black. (C) Heat shock of black mutants move some individuals into the black/green transition region. (D) Selection on genetic background to produce ever greener individuals after heat shock and black individuals without heat shock. (E) Polyphenic response with black developing at low temperatures and green at higher temperatures. After (Suzuki & Nijhout, 2006).

a variable environment (Richards et al., 2017). The epigenetics hypothesis is intuitively consistent with phenotypic plasticity and genetic assimilation. After all, epigenetic modifications are environmentally induced, and they can differ between individuals with identical genotypes, which we would detect as plasticity. However, if these modifications are retained across generations, they can also become uncoupled from the original environmental stimulus, resulting in genetic assimilation. Houri-Zeevi and Rechavi (2017) acknowledge that, in many cases, genetic assimilation that is induced by epigenetic modifications will predate selection. Donohue acknowledges that the epigenetics hypothesis provides a promising context for studying the "ecological and genetic conditions under which stably inherited epigenetic changes are most likely to be adaptive" (Donohue, 2014). A challenge to this hypothesis is the variation that exists in the extent to which epigenetic modifications can persist across generations. In some mammals the epigenetic slate is wiped clean during meiosis or in the zygote, but in *Drosophila melanogaster*, the nematode *Caenorhabditis elegans,* and some plants, DNA methylation patterns can be transmitted to offspring, and it is thought transgenerational inheritance of methylation patterns is common in plants (Cedar & Bergman, 2012; Feng, Jacobsen, & Reik, 2010; Herman & Sultan, 2011, 2016; Nilsson, Sadler-Riggleman, & Skinner, 2018; Richards, 2006). Thus, the epigenetics hypothesis may be a candidate mechanism of genetic assimilation in some organisms but not in others. Further research into epigenetic modification and inheritance will elucidate the extent to which it is responsible for genetic assimilation across taxonomic kingdoms.

5.3 Quantitative genetics approaches

A quantitative genetic interpretation of genetic assimilation was offered by (Bateman, 1959) and (Falconer & Mackay, 1996). They treated genetic assimilation as a polygenic threshold trait, regulated by many genes each with a small effect, and assumed that a genotype can be defined by its liability or propensity to produce a particular phenotype (see Fig. 2). Normally, the propensity to produce cross-veinless, for instance, lies below the threshold required to produce this phenotype, but an environmental shock can lower that threshold, so that a small proportion of a population now falls above the threshold and expresses the phenotype. Selection on that sub-population will shift the liability of the population in the direction of the threshold, and with repetition, eventually more or most of the population will be above the natural liability threshold and will now exhibit the phenotype (Fig. 2).

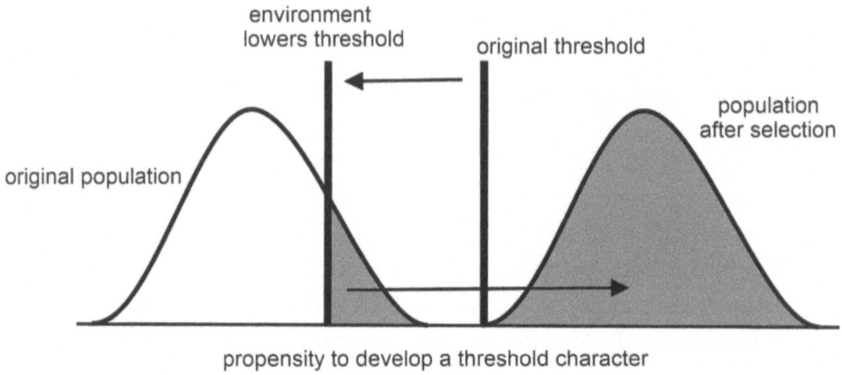

Fig. 2 Model for the evolution of genetic assimilation of a threshold trait. The model assumes a threshold trait, and that the propensity to develop the trait is always below threshold. If an environmental stimulus lowers the threshold then the most sensitive individuals in a population will express the trait (gray region under left-most distribution). Selection on these individuals over generations, with continual environmentally-induced lowering of the threshold, will move the population to the right, and eventually beyond the original threshold, so that when the environmental stimulus stops and the threshold returns to its original value, the entire population will express the novel trait value. (After Bateman (1959) and Falconer and Mackay (1996)).

A mechanism by which genetic assimilation occurred by permanently shifting a threshold of induction was described by Sikkink, Reynolds, Ituarte, Cresko, and Phillips (2014).

Instead of inducing a specific phenotype, environmental stress can also simply increase phenotypic variability around a pre-existing mean, if the stressor unbalances or destroys a homeostatic mechanism that stabilizes the phenotype (Badyaev, 2005; Gibson & Dworkin, 2004; Nijhout, Best, & Reed, 2019; Nijhout & Reed, 2014; Nijhout, Sadre-Marandi, Best, & Reed, 2017). Homeostatic mechanisms that stabilize the phenotype against genetic variation allow the accumulation of mutations in genes that affect the mechanism that produces the phenotype. These mutations can accumulate because they have little or no effect on the phenotype that is realized (for examples see (Nijhout et al., 2017; Nijhout et al., 2019)). This is referred to as cryptic variation (Gibson & Dworkin, 2004; Gibson & Reed, 2008). A mutation or environmental stressor that reduces the homeostatic effect would "release" this genetic variation, which then becomes manifest as increased phenotypic variation (Gibson, 2009; Gibson & Dworkin, 2004; Nijhout et al., 2019). An example of such a releasing mutation comes from the work of Rutherford and Lindquist (1998), who discovered a role for heat-shock proteins as a mechanism that confers robustness to genetic variation. They found that in *Drosophila*, mutations that partially inactivate Hsp90 result in the production of a host of developmental abnormalities in many body parts. The type of abnormality depended on both the genetic background and the temperature at which flies were reared. Selection on specific types of phenotypic defects increased their frequency in the populations, and Rutherford and Lindquist found that after several generations of selection the phenotype became independent of the Hsp90 mutation. The strong response to selection and the variability of the selected phenotypes indicated that the selected phenotypes were polygenic, and that there must exist a large amount of previously cryptic genetic variation that was masked by the normal function of Hsp90. Heat-shock proteins, therefore, allow the accumulation of genetic variation much like a capacitors store electrical energy, and function as a robustness mechanism that stabilize phenotypes against genetic variation. In this case, aberrant phenotypes were induced, not by an environmental stimulus, like a heat-shock, but by a mutation in a robustness mechanism, and fixation of aberrant phenotypes came about by selection on previously cryptic genetic variation. Standard selection on polygenic phenotypic variation that is released when a robustness mechanism is disrupted can thus lead to changes in the mean phenotype and canalization around that mean.

Lande (2009) took the approach of studying the evolution of canalization through the quantitative genetic analysis of linear reaction norms. A reaction norm describes the sensitivity of a phenotype to environment. Linear reaction norms have a slope, that describes the sensitivity to environment, and an elevation (often called an intercept) which, in terms of quantitative genetics, is the additive genetic effect or breeding value. Lande assumed that genetic assimilation and canalization result from stabilizing selection and defined them as the reduction of genetic and phenotypic variance and the loss of correlation between slope and elevation of the reaction norm, in the average environment of the population. There is still a reaction norm, however, so the phenotype can still respond to an environmental shift. This raises an interesting question about whether the loss of variance in one environment is both a necessary and sufficient condition to infer genetic assimilation and canalization has occurred, or whether the canalized phenotype must also be insensitive to either genetic variation, or environmental variation, or both.

Scheiner, Barfield, and Holt (2017) developed an individual-based population genetics model in which individuals phenotypes are due to the additive effects of four kinds of factors: non-plastic loci, plastic loci, the environment and a random factor. In this model, environment-sensitive plastic loci, non-plastic loci, and environment could be given different weights or values in a survival or fitness function. With this model they studied the probability of survival of populations with different relative amounts of plastic loci after either step-like changes or continuous changes in the value of environment. They also assigned a variable "cost" function to plasticity and studied the conditions under which it is possible to eliminate plastic loci by selection, thus producing genetic assimilation. Their results show that genetic assimilation will not occur if there is no cost to plasticity. If plasticity is costly, then genetic fixation can occur, but would take hundreds of generations, not nearly as rapid as Waddington's experiments suggest, supporting earlier simulation studies by Lande (2009) that showed elimination of plasticity to be very slow. If the selective environment does not stay constant for many hundreds of generations, genetic assimilation is unlikely to occur. Genetic assimilation and loss of plasticity would require a high cost to plasticity and a stable selective environment. The conclusion is that genetic assimilation requires strong selection *against* plasticity, and that if plasticity is not costly, then plasticity will be favored in evolution.

Obviously, Waddington was able to achieve genetic assimilation in 15–20 generations, which implies that, in his experiments, selection against plasticity was exceptionally strong, or that there was something about the induced *Drosophila* phenotype that was not captured by the simple assumptions that reaction norms are linear and that there exist genetic loci that somehow encode or promote plasticity. It is worth noting that quantitative and statistical models of evolution have mostly dealt with stabilizing selection rather than genetic assimilation. Selection for stable phenotypes tends to select for homozygosity (Roff, 1997), so the stable phenotypes are not canalized because they are not necessarily robust to new mutations, nor to environmental variation.

An alternative view to the assumption that specific genes or alleles somehow encode plasticity is the idea that plasticity is a property of gene regulatory networks (GRNs) (Pfennig & Ehrenreich, 2014; Schneider & Meyer, 2017). This implies that there are GRNs for plastic traits and GRNs for fixed non-plastic traits, and that there are transcription factors that are influenced by the environment (Schneider & Meyer, 2017). There are conceptual models of how this might operate, using simple motifs, but no quantitative models that explicitly deal with how plasticity and selection could operate in a GRN. Although GRNs in themselves may not be inherently either plastic or non-plastic, the regulation of their activity can include environmental factors, which can lead to plastic outcomes. The GRN that patterns wing development across holometabolous insects has been largely conserved for 325 million years (Abouheif & Wray, 2002). Many species of holometabolous insects have environmentally induced wing polyphenisms that involve alterations of this GRN. In female ants, the winged-wingless polyphenism is one of the characteristic differences between queens and workers and is induced by differences in nutrition and pheromones (Wheeler, 1991). Although this polyphenism has a single evolutionary origin, the GRN for wing patterning is interrupted in different locations in different species, to produce wingless workers (Abouheif & Wray, 2002). Male ants are typically winged, but *Cardiocondyla obscurior* evolved a male wing polyphenism that produces wingless males that fight to gain mates, and winged males that disperse (Oettler et al., 2019). This novel polyphenism is due to an interruption of the wing patterning GRN that is much farther downstream in the GRN than the interruption that causes the female polyphenism in the same species (Oettler et al., 2019). The fact that the wing patterning GRN has been interrupted in many different places during the evolution of a plastic polyphenism, suggest there is little developmental constraint in how this GRN can be manipulated. If gene inactivation in the wing patterning GRN via an environmental signal produces a winged-wingless polyphenism, then a mutation in any of the genes of the GRN could fix the wingless state, as has happened in the aphid, *Acyrtosiphon* (Braendle et al., 2006; Ogawa & Miura, 2013), and produce genetic assimilation of a previously environmentally induced trait.

5.4 Systems models

Neither the "gene for…" and "GRN for…" models for plasticity and genetic assimilation capture the complexity and non-linearity of the processes that lead to phenotypes. Genes code for proteins, not traits, and GRNs are essential, but they are but a small component of the complex of developmental processes that lead to phenotypes. Phenotypes do not map readily to genotypes or gene regulatory networks. Although phenotypic variation can be statistically related to genetic variation, there is no simple deterministic relationship between genes and traits.

The vast majority of molecular and developmental processes that produce traits are non-linear, so it is unlikely that the relationships between cause and effect, or input and output, or genotype and phenotype, or environment and phenotype is linear. Moreover, phenotypes are constructed through the interaction of many diverse and complex processes that operate at different levels of organization (molecular, biochemical, cellular, hormonal, physiological, organismal), most of which are contingent. Coming to grips with the non-linearity and complexity of biological processes is both conceptually and theoretically challenging. It is always simpler and more convenient to that assume only genes are involved and that they act linearly and additively in a purely deterministic fashion to produce a phenotype.

This, unfortunately, is never the case. Phenotypes are system properties and, excepting phenotypes of dead structures like horns and insect exoskeletons, are dynamically maintained. Even apparent phenotypic stability requires continuous maintenance as cells die and are replaced and molecules break down and are recycled. This is seldom evident in a macroscopic view of the phenotype, but goes on, unnoticed, under the hood. The importance of dynamic maintenance of phenotypes is most readily seen by what happens after an organism dies. It is therefore unlikely that a static pattern of gene expression can produce and maintain a phenotype. Phenotypes are not steady-state or equilibrium processes but operate far from steady state.

To better understand the system properties of phenotypes, and how genes, environment, and developmental processes play a role in the production of plasticity and genetic assimilation, we need concrete examples of such systems in which we can explicitly see and study the underlying mechanisms that generate the phenotype and test methods of analysis. We will look at two such systems: a morphogen diffusion-threshold mechanism for pattern formation in development, and the insulin signaling network in the control of growth. These are chosen because there are good mathematical models available for these systems that produce the relevant phenotypes.

6. Developmental models

Diffusion-threshold pattern formation is a mechanism that establishes the location and dimensions of features during development. For instance, where along the limb a joint will form, where segment boundaries in insects will be, or how big an eyespot on a butterfly wing will be. Diffusion gradient threshold mechanisms operate throughout development (Affolter & Basler, 2007; Barkai & Shilo, 2009; Chaudhary et al., 2019; Eldar, Shilo, & Barkai, 2004; Grimm, Coppey, & Wieschaus, 2010; Monteiro, French, Smit, Brakefield, & Metz, 2001; O'Connor, Umulis, Othmer, & Blair, 2006; Schwank & Basler, 2010; Strigini & Cohen, 1999; Wartlick, 2010). A model of the mechanism is shown in Fig. 3. Source cells produce a morphogen that moves from cell to cell by diffusion. At some time in development all cells that have a morphogen above a threshold adopt one fate, the

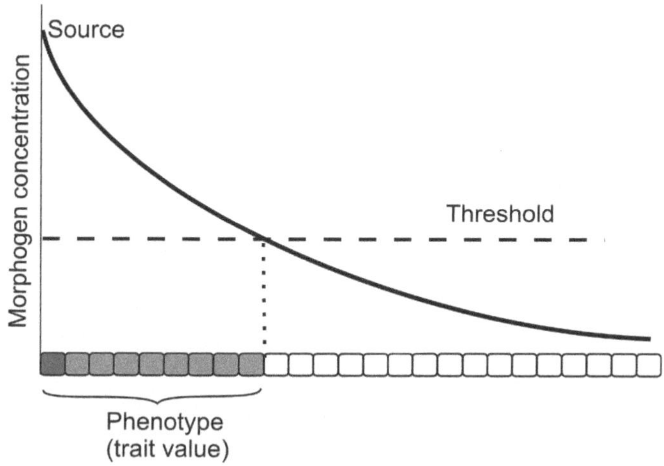

Fig. 3 Morphogen diffusion gradient-threshold mechanism for generating phenotypes during development. Morphogen is produced by a source cell and diffuses away from cell to cell. Cells with morphogen concentrations above a threshold will activate a particular pattern of gene expression that leads to a "phenotype."

remainder another (in addition, a new morphogen could be synthesized by cells that experience a specific concentration of the first morphogen at a defined time in development). The phenotype is the size of the field of cells that are above the threshold (Fig. 3). The phenotype is thus determined by several variables: the source value, the diffusion coefficient, the duration of diffusion (i.e., the time at which the gradient is "read"), and the threshold. Each of these is a complex variable that has both genetic and environmental components (or, rather, whose value is determined by genetic and environmental factors). For instance, the source value includes the rate of transcription and translation and modification of the morphogen; diffusion includes, among others, the properties of gap junctions, hydration state of the morphogen; the threshold includes the properties of receptor of the morphogen and the signaling pathway by which gene transcription is activated.

This system should be sensitive to environmental variation, especially temperature, since the underlying biochemical and molecular processes will vary with temperature. How would canalization evolve in this simple system? The equation for the phenotype value of a diffusion-gradient-threshold trait is

$$trait = k_1 * \left(k_2 * \sqrt{D * t} + \sqrt{D} * \sqrt{k_3 * t - k_4 * t * \log\left(\frac{threshold}{source}\right)} \right),$$

where D is the diffusion coefficient, t is the time at which the gradient is read, source is the value of morphogen at the constant-level source, threshold is the concentration of the morphogen that establishes the boundary between two fields, and k_1–k_5 are constants. Fig. 4 shows how the value of the trait varies with variation in each of the underlying generating factors. The relationships are nonlinear in each case.

To obtain each of the curves in Fig. 4 it was necessary to pick specific values for the other three variables in the system that are not being plotted. Different choices of values for these factors result in different curves for how the target variable affects the trait value. To illustrate this, we plot the graphs for the value of the trait when two factors, time and diffusion coefficient (Fig. 5A), and source and threshold (Fig. 5B), are varied. The resulting graphs are the phenotypic landscapes that show how the phenotype depends on variation in two underlying factors.

The values of the four system variables in Fig. 3 (source, threshold, time, and the diffusion coefficient) are determined by both genetic and environmental factors. If variation in the values of source and threshold in Fig. 5B are due to genetic variation, then the phenotypic surface would be a

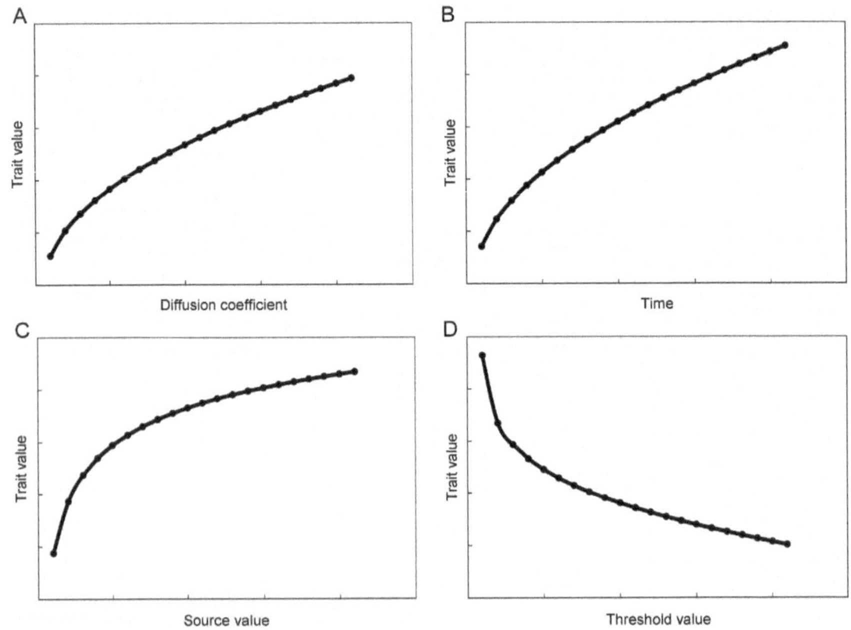

Fig. 4 Dependence of phenotype on the four underlying parameters in the morphogen diffusion-threshold mechanism shown in Fig. 3.

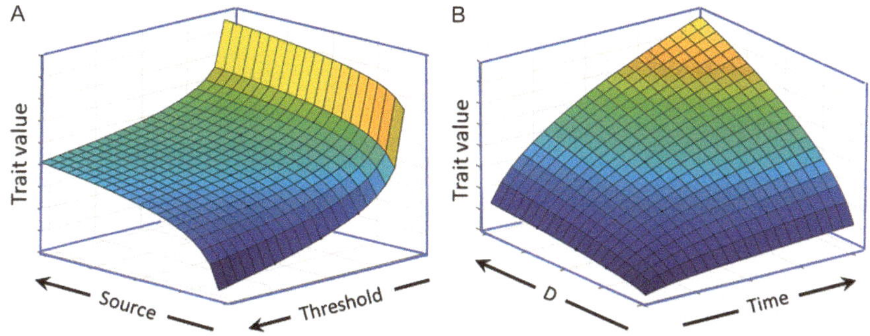

Fig. 5 Phenotypic landscapes for a morphogen diffusion-threshold mechanism. Each panel shows the simultaneous dependence of the phenotype on the values of two of the underlying variables.

genotype-phenotype map. If the variation, instead, is due to environment, such as temperature, then the phenotypic surface would be a reaction norm. This is an important point about systems phenotypic landscapes. They emerge from the underlying mechanism, and can be used to investigate hypotheses about how either genes, or environment, or both, will affect the phenotype. Even for a very simple system, the phenotypic landscapes are non-linear in every dimension. This is true also for more complex systems (see, for instance, Fig. 8).

It is possible to visualize how selection would operate on the phenotypic landscape. If the factors are independent and determined by genetic variation, then, following Lande and Arnold (1983), directional selection for an increased phenotype would take the direction of the steepest slope on the landscape. The gradient of the surface is shown by arrows in Fig. 6A, and several trajectories, starting from different initial "genotypes," are shown by red lines.

6.1 Evolution of robustness

It is unlikely that selection would favor ever increasing phenotypic values, but that an intermediate phenotypic value has highest fitness. Thus, selection would not be for an ever-increasing phenotype, but toward a target phenotypic value. A specific phenotype would be depicted by a specific value on the z axis, and therefore a contour on the landscape. Phenotypic values both above and below the target phenotype would have lowered fitness. A surface in which one of the contours is optimal is shown in Fig. 6B. The contour is now a ridge, and the z axis now represents deviations from optimality.

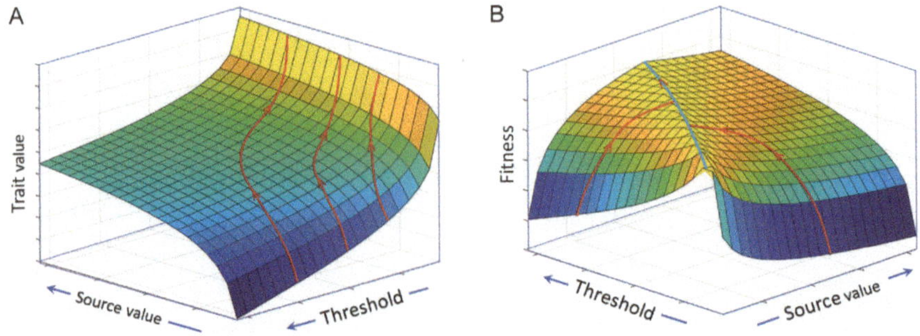

Fig. 6 (A) Evolution on a systems phenotypic landscape. Red traces show theoretical paths of change in the values of Source and Threshold over time with directional selection for larger trait value. Traces follow the local steepest slope of the phenotypic landscape. (B) Evolution on a systems fitness landscape. Blue line is one of the contours in A. This is an isophenotype line. If that phenotype has the highest fitness, then the deviations from this contour (up or down) have less fitness. Red traces show theoretical changes in Source and Threshold for selection and evolution to the optimal phenotype staring from two different initial conditions. At the optimal phenotype, selection will proceed along the blue trace to values of Source and Threshold that define the flattest (orthogonal to the z axis) region of the landscape, where genetic variation will have the least effect in the phenotype.

This implies that there are many combinations of x and y values that will give the same optimal phenotype. This is the systems view of a fitness landscape. It is different from traditional fitness landscape where the x and y axes represent allele frequencies. Rather, the axes represent "activities" of the underlying factors which, as noted above, can represent genetic variation as well as environmental variation. In a system-fitness landscape, there are no discrete adaptive peaks, but rather long equi-phenotypic ridges. Selection will take a phenotype toward the ridge via the path with the steepest slope. But that is not the end of it. Phenotypes at a location on the ridge where the slopes fall off steeply would not be robust to genetic and environmental variation (i.e., variation along the x and y axes) as those will reduce fitness. So, selection will gradually carry the mean phenotype along the ridge to a location where the fitness landscape is flattest, meaning that genetic and environmental variation will have little effect on the phenotype. Evolution along the ridge is thus an example of the evolution of canalization.

Plasticity in this simple network would exist if the Q_{10} (the fold-change in the activity of a process with 10 degrees Celsius change in temperature) values of the underlying factors were not perfectly matched. For instance, if variation in temperature affects source more than any of the other factors (i.e., if source has a higher Q_{10}), then the resulting reaction norm would look like the graph shown in Fig. 4C (or like a section of the graph in Fig. 5B parallel to the source axis).

The phenotypic equation for a diffusion-gradient-threshold trait is written in terms of a source/threshold quotient. Independent of their absolute values, if the source/threshold quotient remains the same, it will not alter the phenotype. So, if both source and threshold have the same Q_{10}, then variation in temperature would not affect the quotient, nor the value of the trait. This would be a natural way to canalize the trait against environmental temperature variation. Evolution of canalization would be by convergence of the Q_{10} for source and threshold. The diffusion coefficient appears as the square root of D^*t, so trait constancy would require this value to also be constant with temperature. Increasing temperature would increase D (diffusion is faster at higher temperatures), so if higher temperatures reduced timing (the time at which the gradient is read) proportionally, the product would remain constant. A doubling in D would have to be matched by a halving of time. If the processes that determine timing run faster at higher temperatures, then matching their Q_{10} to that of D would keep the product constant with varying temperature. Evolution of canalization here could also be accomplished by convergence of the Q_{10}s. Most biological processes have a Q_{10} of about 2 (Schmidt-Nielsen, 1997), so it is possible that robustness to temperature variation is inherent in this phenotypic mechanism, and perhaps that the similarities of the Q_{10}s are evolved adaptations that protect against temperature variation.

7. Signaling pathways

We've used a simple example of how systems phenotypic landscapes can be used to visualize GP maps, phenotypic plasticity, and the evolution of robustness. We now explore how they can also be used to understand the evolution of genetic accommodation and polyphenism (a bi-stable phenotypic plasticity), using a model for insulin signaling (Nijhout & Callier, 2013). Insulin signaling is intimately involved in the control of growth in animals ranging from mammals to insects and nematodes, where it acts as a growth stimulating hormone (Colombani, Bianchini, Layalle, & Leopold, 2006; Essers et al., 2016; Hoopes, Rimbault, Liebers, Ostrander, & Sutter, 2012; McKenna, Tao, & Nijhout, 2019; Nijhout & Grunert, 2002; Nijhout, Laub, & Grunert, 2018; Oldham & Hafen, 2003; Oldham et al., 2002; Puig & Tjian, 2006; Ruaud, Katic, & Bessereau, 2011; So, Miyahara, & Ohshima, 2011). Insulin acts through a complex but conserved pathway to stimulate glucose and amino acid uptake and protein synthesis. A diagrammatic representation of the insulin signaling pathway is shown in Fig. 7.

We assume the phenotype of interest is protein synthesis capacity. A phenotypic landscape for protein synthesis as a function of AKT and FOXO activity is shown in Fig. 8A.

What do the axes mean? They are the underlying factors that give rise to the phenotype. They are themselves system properties. For instance, the axes shown here represent the activities of proteins and enzymes such as kinases. Enzymes are gene products, so in one sense these could represent genes or gene activity. But enzyme activity is also determined by the degree of its expression (i.e., its concentration), the presence of cofactors, and the activities of allosteric activators and inhibitors. Variation in any of these will produce quantitative variation in activity. Evolution of the phenotype is due to

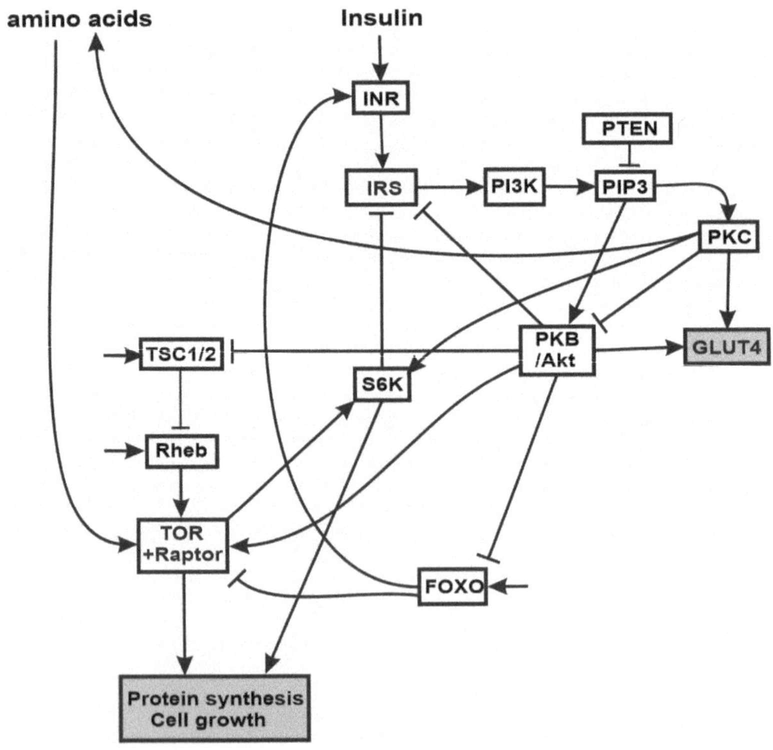

Fig. 7 Model of insulin signaling in the control of growth. Inputs are insulin and amino acids. Outputs of the model are protein synthesis and activation of the GLUT4 glucose transporter. (After Nijhout and Callier (2013)). The level of protein synthesis is assumed to be the mechanism by which variation in insulin signaling regulates growth.

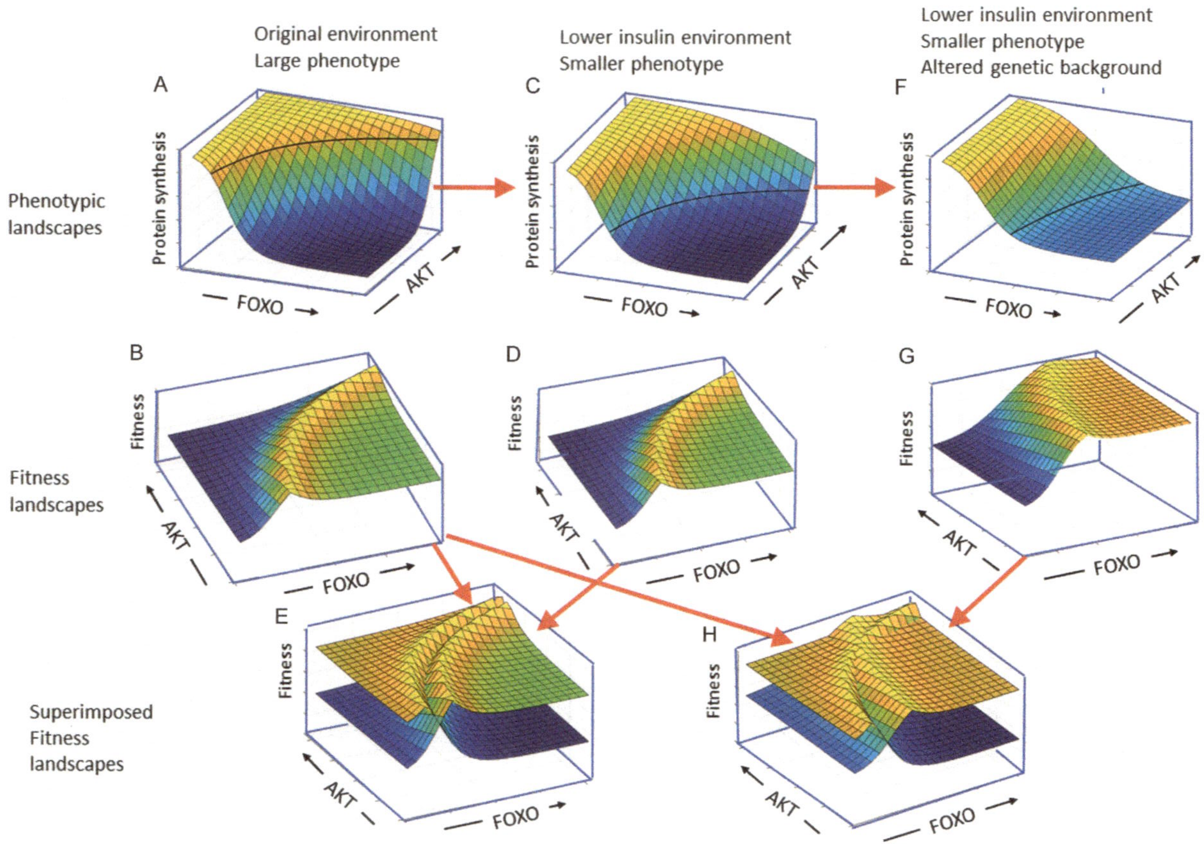

Fig. 8 See figure legend on next page.

quantitative evolutionary changes in the underlying factors. Evolution of protein activity could occur by evolution of the gene, but also by evolution of any of the factors that are required for and contribute to its activity. These factors themselves are often gene products whose activities, like those of enzymes, also depend on their expression level and the many things they interact with.

We assume that an intermediate level of protein synthesis, given a particular strength of insulin signaling, is the optimal phenotype. As before, this produces a system-fitness landscape with a ridge of optimal fitness (Fig. 8B). The long ridge shows that there are many combinations of AKT and FOXO activity that will produce the same optimal phenotype. Depending on the starting point, a population will move up the steepest slope to a location on the ridge. Depending on their history, populations with equally fit genotype-phenotype combinations can be located at different points on the fitness ridge.

Now suppose there is an environmental shift, perhaps an environment with lower nutrient availability that reduces the level of insulin signaling (e.g., Snell-Rood and Moczek (2012), Casasa and Moczek (2018a)). A different (lower) phenotypic value will be optimal in that environment, and its system-fitness landscape shown in Fig. 8C and D. Its ridge is parallel to that of the original optimal ridge (Fig. 8E). If the shifted environment persists, the prediction is that genotypes will evolve toward this second ridge.

But if the environment is not stable but alternates between the two alternatives, as might happen with seasonal changes, then there will be no combination of values of the two underlying factors that is optimal in both

Fig. 8 Model for genetic accommodation of a plastic phenotype based on variation in insulin signaling. (A) The phenotypic landscape for variation in AKT and FOXO activity in the network shown in Fig. 7. Black line is a contour at a trait value of 0.75. (B) System fitness landscape for trait value of 0.75. (C) Phenotypic landscape in a different environment with lower insulin signaling with contour at 0.45 indicating the lower optimal phenotype. (D) System fitness landscape for low insulin and trait value at 0.45. (E) Superimposition of fitness landscapes B and D, showing optimal fitness in the two different environmental conditions. The fitness ridges run roughly parallel to each other, and no combination of AKT and FOXO is optimal in both environments. (F) Phenotypic landscape for low insulin with changed value for two interactions in the network. Black line shows contour at 0.45. (G) Fitness system fitness landscape for trait value 0.45 under these conditions. (H) Superimposition of fitness landscapes B and G, showing there is a unique value (at the crossing point) for AKT and FOXO that produce optimal phenotypes in the two different environments. All values are scaled from 0 (no activity) to 1 (maximal activity).

environments. However, the shape of the phenotypic landscape is determined by the values of all the other factors in the system. Increases in the strength of activation of S6K by PKC and of inhibition of AKT by PKC change the shape of the phenotypic landscape and of the fitness ridge of the new phenotype in the new environment (Fig. 8F and G). We now see that the fitness ridges of the old phenotype in the old environment and the new phenotype in the new environment cross at a specific genotype for AKT and FOXO (Fig. 8H). This means that there exists a genotype in which the two alternative phenotypes can each have optimal fitness in their respective environment. The expectation is that selection would move genotypes to this location. Of course, small changes in other parameters of the system could have similar qualitative effects in the shapes of the fitness ridges, so there may be many paths to bi-stability of phenotypes in a stable genotype in this system. This would represent evolution of adaptive phenotypic plasticity—specifically, the evolution of a polyphenism—as would occur during genetic accommodation.

8. Epilogue

There is now abundant evidence that phenotypic plasticity, genetic assimilation and genetic accommodation are mayor forces in the origin and the evolution of phenotypes.

Phenotypic plasticity can generate a large range of complex phenotypes, and genetic assimilation experiments have shown that these can be canalized and rapidly fixed in a population. Canalization and robustness mechanisms that protect the phenotype from genetic variation, enable the accumulation of large amounts of cryptic genetic variation that, in turn, provide the substrate for rapid adaptive evolution when these mechanisms are perturbed.

Although a novel trait can arise by a new mutation, the conditions for this to be successful (the mutation should have a big phenotypic effect, be dominant and improve fitness) are so stringent that this is unlikely to happen, particularly for complex morphological traits. Phenotypic plasticity and robustness are therefore likely to be a general precondition for the origin and evolution of complex phenotypes.

Phenotypic plasticity and robustness are pervasive and eliminate a clear and unambiguous relationship between genotype and phenotype. This has led to the idea of phenotypes-first evolution of new traits, proposed by West-Eberhard (2003). The idea is that an environmental stimulus or shift can induce a range of phenotypes. If some of those phenotypes are well suited to their environment, and if the environmental stimulus is repeated over generations, then there will be selection to stabilize that phenotype and make its development independent of the environmental stimulus that originally induced it. The evolution of a new trait does not depend on a fortuitous mutation, but emerges by standard gradualistic selection from the rich and mostly cryptic genetic background present in a population. This helps explain why the evolution of new traits can be extremely rapid, far more rapid than could be accounted for by the successive accumulation of rare beneficial mutations that initially induce and then refine and stabilize a phenotype. Such mutations are still required, but they have been accumulating without phenotypic effect over eons. A detailed knowledge of the many molecular, genetic, developmental and physiological processes that produce phenotypes, and not just the mutations that alter them, is essential for developing a deep understanding of how phenotypes develop and evolve.

Acknowledgments

We thank members of the Nijhout lab for many useful discussions, and Mary Jane West-Eberhard for advice in constructing Table 1. This work was supported by grant IOS-1557341 from the National Science Foundation.

References

Abouheif, E., & Wray, G. (2002). Evolution of the gene network underlying wing polyphenism in ants. *Science*, *297*, 249–252.

Affolter, M., & Basler, K. (2007). The decapentaplegic morphogen gradient: From pattern formation to growth regulation. *Nature Reviews. Genetics*, *8*, 663–674.

Aubret, F., & Shine, R. (2009). Genetic assimilation and the postcolonization erosion of phenotypic plasticity in island tiger snakes. *Current Biology*, *19*, 1932–1936.

Badyaev, A. V. (2005). Stress-induced variation in evolution: From behavioural plasticity to genetic assimilation. *Proceedings of the Royal Society B: Biological Sciences*, *272*(1566), 877–886.

Baldwin, J. M. (1896). A new factor in evolution. *American Naturalist*, *30*, 536–553.

Baldwin, J. M. (1902). *Development and evolution*. New York: Macmillan Co.

Barkai, N., & Shilo, B.-Z. (2009). Robust generation and decoding of morphogen gradients. *Cold Spring Harbor Perspectives in Biology*, *1*(5), a001990.

Bateman, K. G. (1959). The genetic assimilation of four venation phenocopies. *Journal of Genetics*, *56*, 443–474.

Braendle, C., Davis, G. K., Brisson, J. A., & Stern, D. L. (2006). Wing dimorphism in aphids. *Heredity*, *97*(3), 192–199.

Casasa, S., & Moczek, A. P. (2018a). Insulin signalling's role in mediating tissue-specific nutritional plasticity and robustness in the horn-polyphenic beetle *Onthophagus taurus*. *Proceedings of the Royal Society B: Biological Sciences*, *285*(1893), 20181631.

Casasa, S., & Moczek, A. P. (2018b). The role of ancestral phenotypic plasticity in evolutionary diversification: Population density effects in horned beetles. *Animal Behaviour*, *137*, 53–61.

Casasa, S., & Moczek, A. P. (2019). Evolution of, and via, developmental plasticity: Insights through the study of scaling relationships. *Integrative and Comparative Biology, 59*(5), 1346–1355.

Cedar, H., & Bergman, Y. (2012). Programming of DNA methylation patterns. *Annual Review of Biochemistry, 81*(1), 97–117.

Chaudhary, V., Hingole, S., Frei, J., Port, F., Strutt, D., & Boutros, M. (2019). Robust Wnt signaling is maintained by a Wg protein gradient and Fz2 receptor activity in the developing *Drosophila* wing. *Development, 146*(15), dev174789.

Chomyn, A., Moller, G., & Mitchell, H. K. (1979). Patterns of protein synthesis following heat shock in pupae of *Drosophila melanogaster. Developmental Genetics, 1*(1), 77–95.

Colombani, J., Bianchini, L., Layalle, S., & Leopold, P. (2006). Steroids, insulin and growth: The flies dope the research. *M S-Medecine Sciences, 22*(3), 241–243.

Daniels, E. V., Murad, R., Mortazavi, A., & Reed, R. D. (2014). Extensive transcriptional response associated with seasonal plasticity of butterfly wing patterns. *Molecular Ecology, 23*(24), 6123–6134.

de Jong, G. (2005). Evolution of phenotypic plasticity: Patterns of plasticity and the emergence of ecotypes. *New Phytologist, 166*, 101–118.

Donohue, K. (2014). The epigenetics of adaptation: Focusing on epigenetic stability as an evolving trait. *Evolution, 68*, 617–619.

Ehrenreich, I. M., & Pfennig, D. W. (2015). Genetic assimilation: A review of its potential proximate causes and evolutionary consequences. *Annals of Botany, 117*(5), 769–779.

Eldar, A., Shilo, B.-Z., & Barkai, N. (2004). Elucidating mechanisms underlying robustness of morphogen gradients. *Current Opinion in Genetics & Development, 14*(4), 435–439.

Essers, P., Tain, L. S., Nespital, T., Goncalves, J., Froehlich, J., & Partridge, L. (2016). Reduced insulin/insulin-like growth factor signaling decreases translation in *Drosophila* and mice. *Scientific Reports, 6*(1), 30290.

Falconer, D. S., & Mackay, T. F. C. (1996). *Introduction to quantitative genetics* (4th ed.). Harlow: Addison Wesley Longman.

Feng, S., Jacobsen, S. E., & Reik, W. (2010). Epigenetic reprogramming in plant and animal development. *Science, 330*(6004), 622–627.

Gibson, G. (2009). Decanalization and the origin of complex disease. *Nature Reviews. Genetics, 10*, 134–140.

Gibson, G., & Dworkin, I. (2004). Uncovering cryptic genetic variation. *Nature Reviews. Genetics, 5*, 681–690.

Gibson, G., & Reed, L. K. (2008). Cryptic genetic variation. *Current Biology, 18*, R989–R990.

Gilbert, S. F. (1991). *A conceptual history of modern embryology*. New York: Pleanum.

Godoy, O., Saldaña, A., Fuentes, N., Valladares, F., & Gianoli, E. (2011). Forests are not immune to plant invasions: Phenotypic plasticity and local adaptation allow *Prunella vulgaris* to colonize a temperate evergreen rainforest. *Biological Invasions, 13*(7), 1615–1625.

Goldschmidt, R. B. (1935). Gen und Ausseneigenschaft. (Untersuchungen an Drosophila). *Z. indukt. Abstamm.- u. Vererb.-Lehre, 69*, 38–131.

Goldschmidt, R. B., & Piternick, L. K. (1957a). The genetic background of chemically induced phenocopies in Drosophila. *Journal of Experimental Zoology, 135*(1), 127–202.

Goldschmidt, R. B., & Piternick, L. K. (1957b). The genetic background of chemically induced phenocopies in Drosophila II. *Journal of Experimental Zoology, 136*(2), 201–228.

Grimm, O., Coppey, M., & Wieschaus, E. (2010). Modelling the Bicoid gradient. *Development (Cambridge, England), 137*(14), 2253–2264.

Herman, J., & Sultan, S. (2011). Adaptive transgenerational plasticity in plants: Case studies, mechanisms, and implications for natural populations. *Frontiers in Plant Science, 2*, 102.

Herman, J., & Sultan, S. (2016). DNA methylation mediates genetic variation for adaptive transgenerational plasticity. *Proceedings of the Royal Society B: Biological Sciences*, *283*(1838), 20160988.

Hoffmann, A. A., & Hercus, M. J. (2000). Environmental stress as an evolutionary force. *Bioscience*, *50*(3), 217–226.

Hoopes, B. C., Rimbault, M., Liebers, D., Ostrander, E. A., & Sutter, N. B. (2012). The insulin-like growth factor 1 receptor (IGF1R) contributes to reduced size in dogs. *Mammalian Genome: Official Journal of the International Mammalian Genome Society*, *23*(11 – 12), 780–790.

Houri-Zeevi, L., & Rechavi, O. (2017). A. Matter of time: Small RNAs regulate the duration of epigenetic inheritance. *Trends in Genetics*, *33*, 46–57.

Lande, R. (2009). Adaptation to an extraordinary environment by evolution of phenotypic plasticity and genetic assimilation. *Journal of Evolutionary Biology*, *22*(7), 1435–1446.

Lande, R., & Arnold, S. J. (1983). The measurement of selection on correlated characters. *Evolution*, *37*(6), 1210–1226.

Marx, J. L. (1983). Surviving heat shock and other stresses. *Science*, *221*(4607), 251.

McKenna, K. Z., Tao, D., & Nijhout, H. F. (2019). Exploring the role of insulin signaling in relative growth: A case study on wing-body scaling in Lepidoptera. *Integrative and Comparative Biology*, *59*, 1324–1337.

Milkman, R. D. (1965). The genetic basis of natural variation. VI. Selection of a crossveinless strain of drosophila by phenocopying at high temperature. *Genetics*, *51*, 87–96.

Mitchell, H. K., & Lipps, L. S. (1978). Heat shock and phenocopy induction in *Drosophila*. *Cell*, *15*(3), 907–918.

Mitchell, H. K., & Petersen, N. S. (1982). Developmental abnormalities in *Drosophila* induced by heat shock. *Developmental Genetics*, *3*(2), 91–102.

Moczek, A. P. (2007). Developmental capacitance, genetic accommodation, and adaptive evolution. *Evolution & Development*, *9*(3), 299–305.

Moczek, A. P. (2008). On the origins of novelty in development and evolution. *BioEssays*, *30*(5), 432–447.

Moczek, A. P., Sultan, S., Foster, S., Ledón-Rettig, C., Dworkin, I., Nijhout, H. F., et al. (2011). The role of developmental plasticity in evolutionary innovation. *Proceedings of the Royal Society B: Biological Sciences*, *278*(1719), 2705–2713.

Monteiro, A., French, V., Smit, G., Brakefield, P., & Metz, J. (2001). Butterfly eyespot patterns: Evidence for specification by a morphogen diffusion gradient. *Acta Biotheoretica*, *49*, 77–88.

Morimoto, R. I., Kline, M. P., Bimston, D. N., & Cotto, J. J. (1997). The heat-shock response: Regulation and function of heat-shock proteins and molecular chaperones. *Essays in Biochemistry*, *32*, 17–29.

Nair, A., & Dearden, P. (2016). A systems view of Waddington's genetic assimilation. *International Journal of Bioinformatics and Biological Systems*, *1*(1), 10–17.

Nijhout, H. F. (2015). A developmental-physiological perspective on the development and evolution of phenotypic plasticity. In A. Love (Ed.), *Conceptual change in biology* (pp. 147–173). Dordrecht: Springer.

Nijhout, H. F., Best, J. A., & Reed, M. C. (2019). Systems biology of robustness and homeostatic mechanisms. *Wiley Interdisciplinary Reviews: Systems Biology and Medicine*, *11*(3), e1440.

Nijhout, H. F., & Callier, V. (2013). A new mathematical approach for qualitative modeling of the insulin-TOR-MAPK network. *Frontiers in Physiology*, *4*, 245.

Nijhout, H. F., & Grunert, L. W. (2002). Bombyxin is a growth factor for wing imaginal disks in Lepidoptera. *Proceedings of the National Academy of Sciences*, *99*(24), 15446–15450.

Nijhout, H. F., Laub, E., & Grunert, L. W. (2018). Hormonal control of growth in the wing imaginal disks of *Junonia coenia*: The relative contributions of insulin and ecdysone. *Development*, *145*(6), dev160101.

Nijhout, H. F., & Reed, M. C. (2014). Homeostasis and dynamic stability of the phenotype link robustness and plasticity. *Integrative and Comparative Biology*, *54*(2), 264–275.

Nijhout, H. F., Sadre-Marandi, F., Best, J., & Reed, M. C. (2017). Systems biology of phenotypic robustness and plasticity. *Integrative and Comparative Biology*, *57*(2), 171–184.

Nilsson, E. E., Sadler-Riggleman, I., & Skinner, M. K. (2018). Environmentally induced epigenetic transgenerational inheritance of disease. *Environmental Epigenetics*, *4*(2), 1–13.

O'Connor, M. B., Umulis, D., Othmer, H. G., & Blair, S. S. (2006). Shaping BMP morphogen gradients in the *Drosophila* embryo and pupal wing. *Development*, *133*(2), 183–193.

Oettler, J., Platschek, T., Schmidt, C., Rajakumar, R., Favé, M.-J., Khila, A., et al. (2019). Interruption points in the wing gene regulatory network underlying wing polyphenism evolved independently in male and female morphs in Cardiocondyla ants. *Journal of Experimental Zoology Part B: Molecular and Developmental Evolution*, *332*(1–2), 7–16.

Ogawa, K., & Miura, T. (2013). Two developmental switch points for the wing polymorphisms in the pea aphid Acyrthosiphon pisum. *EvoDevo*, *4*(1), 30.

Oldham, S., & Hafen, E. (2003). Insulin/IGF and target of rapamycin signaling: A TOR de force in growth control. *Trends in Cell Biology*, *13*(2), 79–85.

Oldham, S., Stocker, H., Laffargue, M., Wittwer, F., Wymann, M., & Hafen, E. (2002). The *Drosophila* insulin/IGF receptor controls growth and size by modulating PtdInsP(3) levels. *Development*, *129*(17), 4103–4109.

Orr, H. A. (1999). An evolutionary dead end? *Science*, *285*, 343–344.

Otaki, J. M., Hiyama, A., Iwata, M., & Kudo, T. (2010). Phenotypic plasticity in the range-margin population of the lycaenid butterfly *Zizeeria maha*. *BMC Evolutionary Biology*, *10*, 252.

Pfennig, D. W., & Ehrenreich, I. M. (2014). Towards a gene regulatory network perspective on phenotypic plasticity, genetic accommodation and genetic assimilation. *Molecular Ecology*, *23*(18), 4438–4440.

Pfennig, D. W., Wund, M. A., Snell-Rood, E. C., Cruickshank, T., Schlichting, C. D., & Moczek, A. P. (2010). Phenotypic plasticity's impacts on diversification and speciation. *Trends in Ecology & Evolution*, *25*(8), 459–467.

Pigliucci, M., Murren, C. J., & Schlichting, C. D. (2006). Phenotypic plasticity and evolution by genetic assimilation. *Journal of Experimental Biology*, *209*(12), 2362–2367.

Puig, O., & Tjian, R. (2006). Nutrient availability and growth: Regulation of insulin signaling by dFOXO/FOXO1. *Cell Cycle*, *5*, 503–505.

Richards, E. (2006). Inherited epigenetic variation—Revisiting soft inheritance. *Nature Reviews. Genetics*, *7*, 395–401.

Richards, C. L., Alonso, C., Becker, C., Bossdorf, O., Bucher, E., Colomé-Tatché, M., et al. (2017). Ecological plant epigenetics: Evidence from model and non-model species, and the way forward. *Ecology Letters*, *20*(12), 1576–1590.

Roff, D. A. (1997). *Evolutionary quantitative genetics*. New York: Chapman and Hall.

Rountree, D. B., & Nijhout, H. F. (1995a). Genetic control of a seasonal morph in *Precis coenia* (Lepidoptera: Nymphalidae). *Journal of Insect Physiology*, *41*, 1141–1145.

Rountree, D. B., & Nijhout, H. F. (1995b). Hormonal control of a seasonal polyphenism in *Precis coenia* (Lepidoptera: Nymphalidae). *Journal of Insect Physiology*, *41*, 987–992.

Ruaud, A.-F., Katic, I., & Bessereau, J.-L. (2011). Insulin/insulin-like growth factor signaling controls non-dauer developmental speed in the nematode *Caenorhabditis elegans*. *Genetics*, *187*(1), 337.

Rutherford, S. L., & Lindquist, S. (1998). Hsp90 as a capacitor for morphological evolution. *Nature*, *396*, 336.

Safranek, L., & Riddiford, L. M. (1975). The biology of the black larval mutant of the tobacco hornworm, Manduca sexta. *Journal of Insect Physiology*, *21*(12), 1931–1938.

Sang, J., & McDonald, J. (1954). Production of phenocopies in *Drosophila* using salts, particularly sodium metaborate. *Journal of Genetics*, *52*, 392–412.

Sangster, T. A., Salathia, N., Undurraga, S., Milo, R., Schellenberg, K., Lindquist, S., et al. (2008). HSP90 affects the expression of genetic variation and developmental stability in quantitative traits. *Proceedings of the National Academy of Sciences of the United States of America*, *105*(8), 2963–2968.

Scheiner, S. M., Barfield, M., & Holt, R. D. (2017). The genetics of phenotypic plasticity. XV. Genetic assimilation, the Baldwin effect, and evolutionary rescue. *Ecology and Evolution*, 7(21), 8788–8803.

Schlichting, C. D., & Pigliucci, M. (1998). *Phenotypic evolution: A reaction norm perspective*. Sunderland, MA: Sinauer Inc.

Schlichting, C. D., & Wund, M. A. (2014). Phenotypic plasticity and epigenetic marking: An assessment of evidence for genetic accommodation. *Evolution*, *68*(3), 656–672.

Schmidt-Nielsen, K. (1997). *Animal physiology* (5th ed.). Cambridge: Cambridge University Press.

Schneider, R. F., & Meyer, A. (2017). How plasticity, genetic assimilation and cryptic genetic variation may contribute to adaptive radiations. *Molecular Ecology*, *26*(1), 330–350.

Schwank, G., & Basler, K. (2010). Regulation of organ growth by morphogen gradients. *Cold Spring Harbor Perspectives in Biology*, *2*(1), a001669.

Scott-Boyer, M. P., Lacroix, S., Scotti, M., Morine, M. J., Kaput, J., & Priami, C. (2016). A network analysis of cofactor-protein interactions for analyzing associations between human nutrition and diseases. *Scientific Reports*, *6*(1), 19633.

Sikkink, K. L., Reynolds, R. M., Ituarte, C. M., Cresko, W. A., & Phillips, P. C. (2014). Rapid evolution of phenotypic plasticity and shifting thresholds of genetic assimilation in the nematode *Caenorhabditis remanei*. *G3: Genes|Genomes|Genetics*, *4*(6), 1103–1112.

Smith, K. (1991). The effects of temperature and daylength on the rosa polyphenism in the buckeye butterfly, *Precis coenia* (Lepidoptera: Nymphalidae). *Journal of Research on the Lepidoptera*, *30*, 237–244.

Snell-Rood, E. C., & Moczek, A. P. (2012). Insulin signaling as a mechanism underlying developmental plasticity: The role of FOXO in a nutritional polyphenism. *PLoS One*, 7(4), e34857.

So, S., Miyahara, K., & Ohshima, Y. (2011). Control of body size in *C. elegans* dependent on food and insulin/IGF-1 signal. *Genes to Cells*, *16*(6), 639–651.

Strigini, M., & Cohen, S. (1999). Formation of morphogen gradients in the *Drosophila* wing. *Seminars in Cell and Developmental Biology*, *10*, 335–344.

Suzuki, Y., & Nijhout, H. F. (2006). Evolution of a polyphenism by genetic accommodation. *Science*, *311*(5761), 650–652.

Suzuki, Y., & Nijhout, H. F. (2008). Genetic basis of adaptive evolution of a polyphenism by genetic accommodation. *Journal of Evolutionary Biology*, *21*(1), 57–66.

Truman, J. W., Riddiford, L. M., & Safranek, L. (1973). Hormonal control of cuticle coloration in the tobacco hornworm, *Manduca sexta*: Basis of an ultrasensitive bioassay for juvenile hormone. *Journal of Insect Physiology*, *19*(1), 195–203.

Vigne, P., Gimond, C., Ferrari, C., Vielle, A., Hallin, J., Pino-Querido, A., et al. (2020). A single nucleotide change underlies the genetic assimilation of a plastic trait. *bioRxiv*. 2020.2006.2029.176990.

Waddington, C. H. (1942). Canalization of development and the inheritance of acquired characters. *Nature*, *150*(3811), 563–565.

Waddington, C. H. (1953). Genetic assimilation of an acquired character. *Evolution*, 7(2), 118–126.

Waddington, C. H. (1956). Genetic assimilation of the bithorax phenotype. *Evolution*, *10*(1), 1–13.

Waddington, C. H. (1959). Canalization of development and genetic assimilation of acquired characters. *Nature*, *183*(4676), 1654–1655.

Wartlick, O. (2010). *Control of proliferation by the Dpp gradient in the wing imaginal disc of Drosophila melanogaster.* Université de Genève. Thèse.
West-Eberhard, M. J. (2003). *Developmental plasticity and evolution.* New York: Oxford University Press.
West-Eberhard, M. J. (2005). Phenotypic accommodation: Adaptive innovation due to developmental plasticity. *Journal of Experimental Zoology Part B: Molecular and Developmental Evolution, 304B*(6), 610–618.
Wheeler, D. E. (1991). The developmental basis of worker caste polymorphism in ants. *The American Naturalist, 138*(5), 1218–1238.

CHAPTER ELEVEN

Animal development in the microbial world: The power of experimental model systems

Margaret McFall-Ngai[a,]* and Thomas C.G. Bosch[b]

[a]Pacific Biosciences Research Center, Kewalo Marine Laboratory, University of Hawai'i at Mānoa, Honolulu, HI, United States

[b]Zoological Institute, Christian-Albrechts-University Kiel, Kiel, Germany

*Corresponding author: e-mail address: mcfallng@hawaii.edu

Contents

Abstract

The development of powerful model systems has been a critical strategy for understanding the mechanisms underlying the progression of an animal through its ontogeny. Here we provide two examples that allow deep and mechanistic insight into the development of specific animal systems. Species of the cnidarian genus *Hydra* have provided excellent models for studying host-microbe interactions and how metaorganisms function in vivo. Studies of the Hawaiian bobtail squid *Euprymna scolopes* and its

Current Topics in Developmental Biology, Volume 141
ISSN 0070-2153
https://doi.org/10.1016/bs.ctdb.2020.10.002

luminous bacterial partner *Vibrio fischeri* have been used for over 30 years to understand the impact of a broad array of levels, from ecology to genomics, on the development and persistence of symbiosis. These examples provide an integrated perspective of how developmental processes work and evolve within the context of a microbial world, a new view that opens vast horizons for developmental biology research. The *Hydra* and the squid systems also lend an example of how profound insights can be discovered by taking advantage of the "experiments" that evolution had done in shaping conserved developmental processes.

1. Models systems as a strategy for understanding conserved features of complex processes like development

"For a large number of problems there will be some animal of choice on which it can be most conveniently studied ... I am afraid that most of them are unknown ... Zoologists need to find them and lay their hands on them" (Krogh, 1929). Even before this insight by Krogh and in the years to follow, much has been learned about development through the study of animal models, a small set of diverse metazoans that have particular advantages for laboratory research. Research with model systems (Fig. 1) has shown a remarkable degree of similarity in the developmental mechanisms of all animals (see A. McKenna et al., this volume). These findings demonstrate that, although the embryology of simpler animals might appear superficially very different from that of humans, knowledge gained from those models can often be applied directly to understanding human developmental mechanisms.

Fortuitously, many of the important model organisms for developmental biology have also been adopted for the study of host-microbe interactions, including *Hydra*, the fruit fly, worm, zebrafish and mouse (Fig. 1) (reviewed in Bosch, Guillemin, & McFall-Ngai, 2019). In addition to these systems, other experimental model associations have come from a decades-long history in the field of symbiosis (e.g., the parasitic nematode-*Xenorhabdus* association, the squid-vibrio association, the leech) or have developed anew (e.g., bean bug, honeybee, starlet anemone) (Fig. 1). These models are being used to explore the development of the many animals that exist and function only in cooperation with an abundance of microorganisms. While some animals may be just an accidental place for microbial settlement, others appear to have coevolved with microbes over the course of the earth's history. Information on microbial colonization of animal tissue and documentation

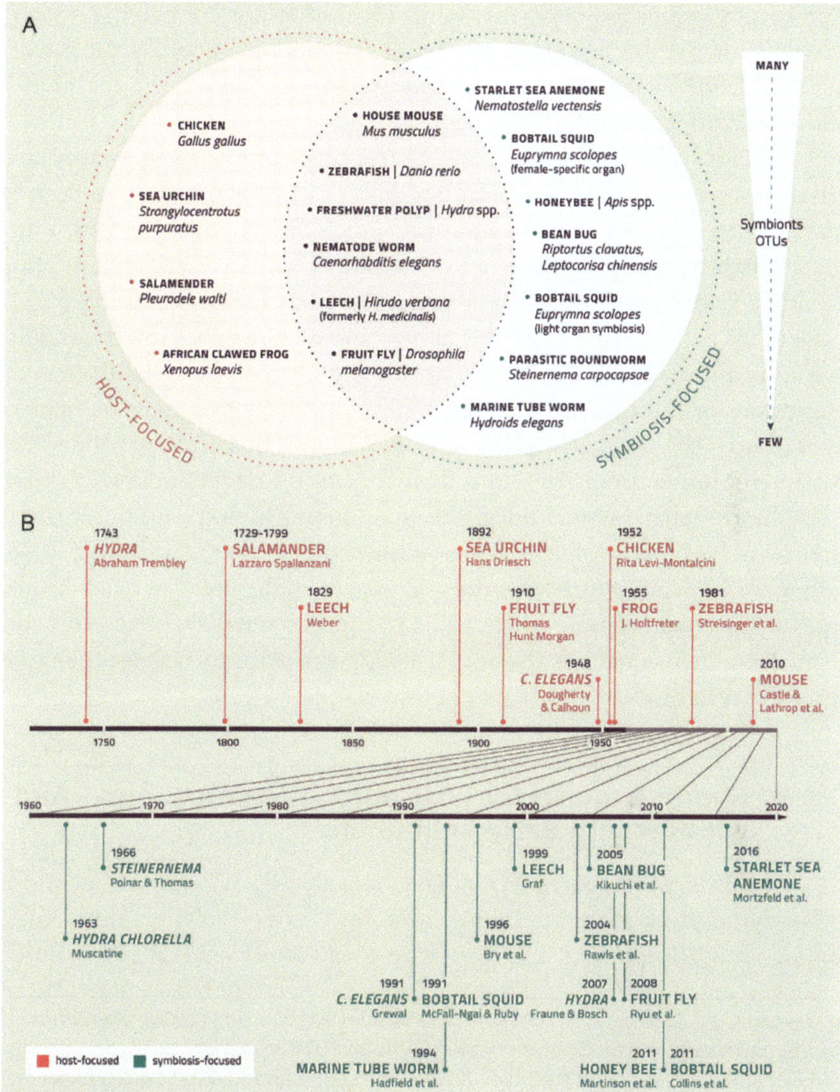

Fig. 1 Models of animal development. (A) Some traditional and new experimental model systems for the study of host-microbe interactions. Traditional systems (left circle) not widely used or widely used to study how symbiosis influences development; new models (right circle) with a major focus on the impact of symbiosis on development. (B) The timeline shows the date of publication of the corresponding model animal along with the discoverer (red lines: animal models for host-focused research; green lines: animal models for symbiosis focused research). *Traditional models of general development/model of microbe-induced development*: *Mus musculus*, mouse, Davisson & Linder (2004)/Bry, Falk, Midtvedt, et al. (1996); *Danio rerio*, zebrafish, Streisinger, Walker,

(Continued)

of physically associated taxa, including those that are coevolving with or codependent on the animal host, is still fragmentary because the field is young and only a few animal/plant/fungal species have been analyzed in depth. Microbes certainly are abundant and important to host fitness in corals, aphids, cows and in humans. Some other animal groups appear to harbor few or no resident microbes (table 1 in Hammer, Sanders, & Fierer, 2019). And in other animal taxa (including nematomorphs, entoprocts, gnathostomulids, gastrotrichs, brachiopods, phoronids, onychophora and others) nothing thus far is known at all about the microbiome.

Here we provide two examples of how interactions with microbes influence animal development. Our intention here is not to be exhaustive in the descriptions of the findings for each system, but rather to give the reader a flavor of how the study of models is providing insight into basic processes of host-microbiome associations and their impact on development. It should be emphasized that, as in traditional developmental biology, no single model system will be capable of addressing all questions in the field (Ruby, 2008). Rather the most powerful outcome will arise from the use of as many models of developmental processes that can be found in nature to reveal both conserved elements as well as those that create diversity in developmental of symbiotic systems.

2. *Hydra*: A model for host-microbe interactions—And a window into early animal life

The freshwater polyp *Hydra* (Fig. 2A and B) is an excellent model for studying host-microbe interactions and how metaorganisms function in vivo (Bosch, 2013, 2014; Klimovich & Bosch, 2018; Schröder

Fig. 1—Cont'd Dower, et al. (1981)/Rawls, Samuel, and Gordon (2004); *Caenorhabditis elegans*, nematode worm, Dougherty and Calhoun (1948)/Grewal (1991), Schulenburg and Félix (2017); *Hydra* spp., Trembley (1744)/Muscatine and Lenhoff (1963) (algal symbiosis), Fraune and Bosch (2007) (bacterial symbioses); *Hirudo medicinalis*, leech, Weber (1829)/Graf (1999); *Drosophila melanogaster*, fruit fly, Bellen, Tong, and Tsuda (2010)/Ryu, Kim, Lee, et al. (2008). *Symbiosis focused models*: *Nematostella vectensis*, starlet sea anemone, Mortzfeld, Urbanski, Reitzel, et al. (2016); *Euprymna scolopes*, bobtail squid (consortial symbiosis in females), Collins and Nyholm (2011); *Apis* spp., honey bee, Martinson, Danforth, Minckley, et al. (2011); *Steindernema* (formerly *Neoaplectana*) *carpocapsae*, nematode worm, Poinar & Thomas, 1966; *Riptortus clavatus* and *Leptocorisa chinensis*, bean bug, Kikuchi, Meng, and Fukatsu (2005); *E. scolopes*, bobtail squid (light-organ symbiosis), McFall-Ngai & Ruby (1991); *Hydroides elegans*, marine tube worm, Hadfield, Unabia, Smith, and Michael (1994).

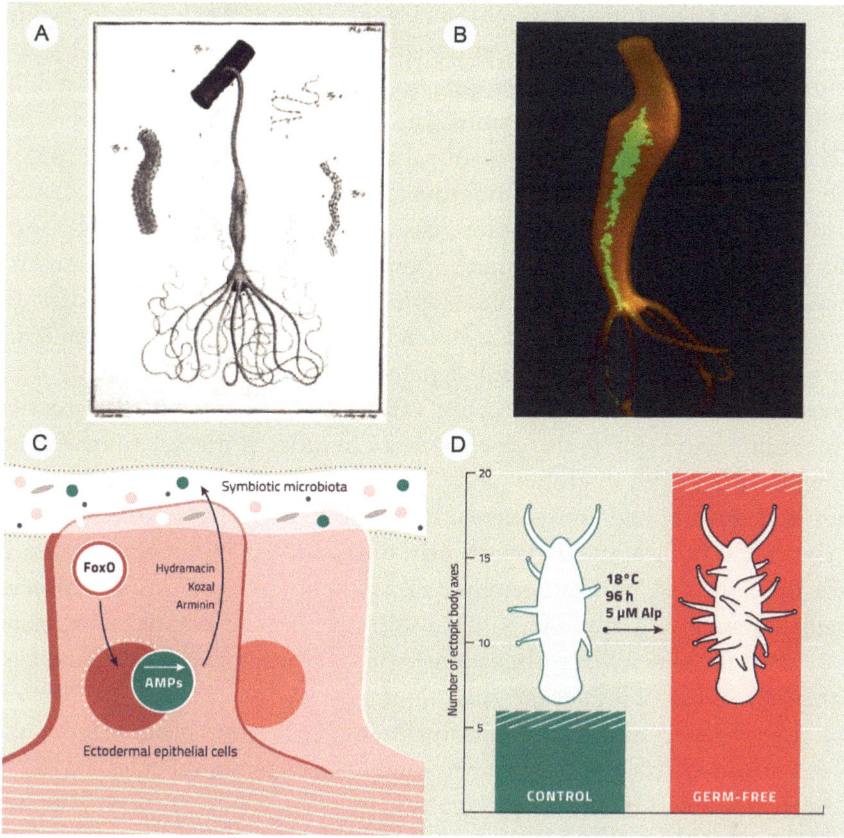

Fig. 2 The freshwater polyp *Hydra*. (A) A *Hydra* as depicted in Trembley (1744) book. (B) Transgenic *H. vulgaris* (AEP) polyp with a mosaic expression of EGFP in some of its endodermal epithelial stem cells. (C) Schematic representation showing the effect of transcription factor FoxO on expression of antimicrobial peptides (AMPs) and maintenance of the symbiotic microbiota. (D) WNT signaling-mediated developmental processes are dependent on microbial colonization.

& Bosch, 2016). In *Hydra*'s simple tube-like body structure, the single layer of ectodermal epithelial cells covered by a multilayered glycocalyx represents a physical barrier toward the environment, whereas a single layer of endodermal epithelial cells separates the body from the content of the gastric cavity. Both, ecto- and endodermal epithelial cells produce a rich repertoire of antimicrobial peptides that regulate the microbiome. In addition, ectodermal epithelial cells synthesize an extracellular cuticle ("glycocalyx") with a complex layered structure consisting of several proteins and glycosaminoglycans. The outer layer of the glycocalyx has mucus-like properties and provides the habitat for the symbiotic bacterial community (Fig. 1C). This extracellular pattern—an inner, firmly adherent, layer with stratified organization devoid of bacteria, beneath an outer loose layer colonized by symbionts—is similar to the mammalian colon. This pattern of tissue layers may be a conserved principle in animal anatomy and points to the epithelium as a crucial regulator of microbial homeostasis (Schröder & Bosch, 2016). Conversely, the microbiota has a profound impact on both form and functioning of the epithelium. Observations on long-term germ-free animals indicate, that *Hydra*'s developmental pathways are dependent on the microbiota (J. Je & T.C.G. Bosch, unpubl.). Since this dependence includes transcription factor and signaling pathways that are highly conserved in the animal kingdom, this reliance on microbes points to two important principles: (1) that development in most, if not all, animals relies on interactions with microbes, and (2) that some conserved developmental pathways initially had components that were induced by microbes.

In *Hydra* both epithelial layers play different and independent roles in development and morphogenesis. Classic experiments using chimeric strains (Wanek & Campbell, 1982) indicated that the ectoderm regulates body morphology, while the endoderm controls body size (Wanek & Campbell, 1982). This decades-old observation is interesting because body size in *Hydra* is determined not only by internal regulatory processes, but also by environmental factors (Mortzfeld, Taubenheim, Klimovich, et al., 2019). Abiotic influences include temperature, which directly affects Wnt and TGF-beta signaling. The observation that germ-free polyps are considerably larger than conventionalized animals (He & Bosch, unpubl.) gives an interesting indication of the importance of the "microbiome" in size determination (Beard & Blaser, 2002).

The epithelial surface is densely colonized by a stable multi-species bacterial community (Fraune & Bosch, 2007). The *Hydra* host shapes the specific microbiome by means of the innate immune system and a rich repertoire of antimicrobial peptides (Franzenburg, Walter, Künzel, et al., 2013). The presence and structure of *Hydra*'s microbiota is critical for the tissue homeostasis and health of the polyps. Remarkably, each *Hydra* species supports long-term associations with a different set of bacteria, suggesting that the host imposes specific selection pressure onto its microbiome (Franzenburg et al., 2013; Fraune & Bosch, 2007). The findings reveal that epithelia and components of the innate immune system play an active role in selecting the inhabitant microbiota via a complex genetic network. This work has contributed to a paradigm shift in evolutionary immunology:

components of the innate immune system with its host-specific antimicrobial peptides appear to have evolved in early metazoans because of the need to control the resident beneficial microbes rather than to fight invasive pathogens (Bosch, 2014).

The *Hydra* model system also has provided insights of general significance with the discovery of the role of interactions between symbiotic bacteria in colonization resistance (Fraune, Anton-Erxleben, Augustin, et al., 2015). In previous studies we discovered that these early evolutionary divergent animals not only select their bacterial communities, but also select for viral communities in a species-specific manner (Bosch, Grasis, & Lachnit, 2015). Viruses are an important part of the *Hydra* holobiont by their influence on the species-specific microbiome (Li, Lachnit, Fraune, et al., 2017).

Although the epithelial cells of *Hydra* had been considered as prime regulators of the microbiome, recent data uncovered a previously underappreciated role of the nervous system with its rich repertoire of neuropeptides that participate in controlling resident beneficial microbes (Augustin, Schroeder, Murillo Rincon, et al., 2017). Recent findings also show that microbes affect the animal's behavior by directly interfering with neuronal receptors (Murillo-Rincon, Klimovich, Permoller, et al., 2017). These observations provide new insights into the original role of the nervous system, and suggest that it emerged to orchestrate multiple functions including host-microbiome interactions (Klimovich & Bosch, 2018). Below we illustrate that: (i) the presence and structure of this microbiota is critical for patterning processes and tissue homeostasis; (ii) the transcription factor FoxO provides a direct link between stem cell proliferation, tissue homeostasis, and the microbiota; and (iii) *Hydra*'s microbiome contributes to tumor formation.

2.1 FoxO is a key regulator of epithelial homeostasis and host-microbiome crosstalk

Besides its well-known conserved function as major tissue regulator and aging antagonist, the transcription factor FoxO modulates the innate immune system in various model organisms including mouse, *Drosophila melanogaster, Caenorhabditis elegans* and *Hydra* spp. (reviewed in Mortzfeld & Bosch, 2017). In response to environmental or bacterial signals, FoxO can be shuttled between the transcriptionally inactive state in the cytoplasm and the active form in the nucleus. Binding of FoxO to conserved regulatory genomic regions can modulate the expression of genes involved with stem cell and immune functions (AMPs) and thus has a crucial impact on the aging process.

In *Hydra*, FoxO is regulated by symbiotic microbes, involving it in a multi-species set of interactions that regulates the aging and maintenance of the metaorganism. In *Hydra*, the microbiome is selectively assembled by a species-specific combination of AMPs that are predominantly expressed in epithelial cells (Franzenburg et al., 2013). FoxO-deficient *Hydra* polyps show, in addition to defects in stem cell maintenance (Boehm et al., 2012), a severe change of the immune status, and drastically altered expression of AMPs. Since FoxO-deficient polyps are impaired in selection for bacteria resembling the native microbiome, and more susceptible to colonization by foreign bacteria (Mortzfeld, Taubenheim, Fraune, et al., 2018), FoxO signaling not only plays a key role in the communication between host and microbiota, but also incorporates the evolutionarily conserved longevity factor FoxO into the metaorganism/holobiont concept (Fig. 3C).

2.2 WNT/β-catenin signaling is influenced by microbial colonization

Hydra has provided important insights into general principles of biological pattern formation. Regeneration and transplantation experiments performed as early as 1909 revealed that the tip of the *Hydra* head has organizer function (Browne, 1909), and the formation of the head organizer involves the canonical Wnt pathway (Broun, Gee, Reinhardt, & Bode, 2005; Gee, Hartig, Law, et al., 2010; Hobmayer, Rentzsch, Kuhn, et al., 2000). Treatment with alsterpaullone (ALP), a GSK3 inhibitor, results in increased activity of the head organizer system in the body column (Gee et al., 2010). Intriguingly, germ-free *Hydra* polyps are significantly more sensitive to ALP than control polyps (Taubenheim, Willoweit-Ohl, Knop, et al., 2020) (Fig. 1D). Germ free animals have four times more ectopic tentacles after ALP treatment compared to control animals. It therefore seems that the presence of the microbiota promotes the activity of the proliferation zone and inhibits terminal differentiation. Screening for genes affected by the bacterial colonizers resulted in the identification of four *Hydra* genes that were upregulated by bacterial colonizers and showed spatially and temporally regulated expression patterns (Taubenheim et al., 2020). Gene expression analyses led to the identification of a small, secreted protein, named Eco1a, being upregulated in the response to both bacterial colonizers and low temperatures. Loss-of-function studies showed that this novel microbe-dependent effector gene is indeed involved in the regulation of pattern formation and has an antagonistic function to Wnt signaling in *Hydra* (Taubenheim et al., 2020).

2.3 *Hydra*'s microbiome contributes to tumor formation

Development requires precise spatial and temporal activation of developmental signaling pathways so that cells do find their way into organized and "well behaved" structures. Cancer is primarily caused by the disruption of normal cellular functions through genetic, epigenetic, and paracrine changes. Once a tumor is initiated, however, complex interactions between cancer cells and their surrounding environment—known as the tumor microenvironment—can contribute to disease progression.

The recognition that certain bacterial and viral pathogens have long been considered as potential causes of cancer begs the question: does the healthy microbiome play a role in the development of cancer? Several years ago, Domazet-Lošo and colleagues used *Hydra* as phylogenetically ancient multicellular organism to demonstrate that in principle all multicellular animals can form tumors (Domazet-Lošo, Klimovich, Anokhin, et al., 2014). Cancer is now considered a legacy of transition to multicellularity early in the evolution of life. Using the experimental accessibility of the *Hydra* holobiont, recent research has demonstrated that after an environmental disruption of the normal bacterial community, the tissue of a polyp may be colonized by bacteria from the environment. The contact with already present microbes then leads to the production of bacterial factors that have a damaging effect on the *Hydra* cell structures, disturb tissue homeostasis and developmental pathways, and ultimately trigger tumor formation (Rathje, Mortzfeld, Hoeppner, et al., 2020).

Hydra develop cancer if a specific type of foreign bacteria from the phylum *Spirochaetes* becomes increasingly prevalent in the microbiome, and thereby disrupts the balance of the bacterial colonization in their tissue. Interestingly, these bacteria exert their harmful influence only in the presence of certain other bacteria from the genus *Pseudomonas*, which are part of the normal composition of the microbiome. The invading bacteria successfully colonize the host tissue only if the *Hydra* tissue has already been weakened by changing factors in the environment. These factors include a change in temperature, and the resulting change in microbial colonization. When they encounter each other, the bacteria modify their movement patterns and seek direct contact with one another. As a result, they also begin to express different genetic information and activate factors that have a pathogenic effect for the host organism. Due to these changes, the microbiota composition becomes drastically altered, which is followed by structural changes in the *Hydra* epithelial cells and ultimately leading to tumor formation. How these interactions occur at the molecular level, and which specific biochemical mechanisms are involved in this form of cancer development, are the subject of currently ongoing investigations.

In light of these observations, the pathogenesis of cancer can be seen as a comprehensive interaction of genetic and environmental factors, including specific microbial influences. Since it was only the common presence of certain bacteria interacting with each other within a disturbed microbiome that enabled the formation of tumors, it is probably not a single malicious intruder, but the malfunction of the microbiome as a protective barrier for the body as a whole what can promote the development of cancer. These findings offer a promising perspective because the protective function of the microbiome could possibly be used in future: The microbial colonization of the body normally balances out and protects the organism against harmful influences, potentially even against carcinogenic influences. Future research will show whether this ability of the microbiome to maintain a healthy barrier, which protects the body from colonization by harmful microorganisms, may also be used for the prevention of cancer.

3. The binary light organ squid-vibrio model sheds light on basic mechanisms of symbiosis development

The association of the Hawaiian bobtail squid *Euprymna scolopes* and the luminous vibrio bacterium *Vibrio fischeri* has been studied for over 30 years (for companion reviews on host and symbiont see: Nyholm & McFall-Ngai, 2020 and Stabb, Visick, & Ruby, 2020 *Nature Reviews Microbiology*). At the time of this writing, 18 labs with a principal focus on the squid-vibrio system have opened in 14 states or US territories. This natural model symbiosis has been used to identify and characterize the conserved features of the chronic colonization of animal epithelia by gram-negative bacteria, perhaps the most common type of host-symbiont interaction across the animal kingdom. [The female of this species also has a consortial symbiosis, which has been studied over the last few years (Kerwin, Gromek, Suria, et al., 2019), that will not be discussed in detail here.]

Being a one-host/one microbial species association (Fig. 1), the squid-vibrio partnership is similar to the rhizobia-legume symbioses in plants; these relationships have provided unparalleled insights into conservation of symbiosis characters, particularly those shared in the development of alliances between animal and plant hosts and their microbes (Hirsch & McFall-Ngai, 2000). These binary associations offer high resolution for determination of the precise dialog between partners. They do not address mechanisms of symbiosis with consortia of microbial species, but rather provide a window into the responses of small clones of bacteria associated with host cells, such as in the mammalian gut. Further, they provide low "noise" for studies focused upon determining the impact of the enormous symbiont strain variation that has been observed in a variety of symbiotic systems, including the squid-vibrio and human associations (Bongrand, Koch, Moriano-Gutierrez, et al., 2016; Koehler, Gaedeke, Thompson, et al., 2019; Yan, Nguyen, Franzosa, et al., 2020). A most critical advantage of binary systems is the opportunity to genetically manipulate all naturally occurring partners of the symbiosis and interpret the influence of these manipulations on symbiosis onset, development and persistence in the host. Such powerful methods have been available in the rhizobia-legume symbioses for over 25 years (for review see Roy, Liu, Nandety, et al., 2020), but, until recently, no animal models existed where we knew the detailed genetics of both partners. In the last few years, we've come to understand the genetic systems in both partners of the model system between the bean bug *Riptortus pedestris* and its bacterial partner *Burkholderia* sp. (Takeshita & Kikuchi, 2017), and researchers at the Marine Biological Laboratory (University of Chicago, Woods Hole, MA, USA) are analyzing the genetic systems of squids (Crawford, Diaz Quiroz, Koenig, et al., 2020), including the bobtail squid. These breakthroughs bring the power of genetic tools in all partners of the symbiosis to the animal world.

Several features specific to the squid-vibrio system make it an ideal candidate for the study of the developmental biology of symbiosis from ecology and evolution to physiology, biochemistry and molecular biology. *V. fischeri* is the only species that can colonize the light organ of *E. scolopes*. As such, the animals can be raised "aposymbiotically," i.e., exposed to the typical one million environmental bacteria per milliliter of seawater without *V. fischeri* present, and the organ will remain uncolonized. In addition, as noted in other cephalopods (Lutz, Ramírez-Puebla, Abbo, et al., 2019), male *E. scolopes* does not harbor bacteria in other regions of the body; strong antimicrobials keep the tissue microbe-free (Chen, Krasity, Peyer, et al., 2017). Further, several characters make this system ideal for study of development, including: (i) the bobtail-squid females mate and lay serial clutches of ~200 eggs/clutch in captivity throughout the year; as such, a facility with ~12 females and 8 male adults will yield ~50,000 juveniles/year for experiments with a large "n"; (ii) the embryonic period is ~20 days, and the juveniles hatch a dusk in large cohorts, which allows synchronization of symbiont inoculation; (iii) juveniles average ~2 mm in mantle length, and the juvenile light organ is ~400 μm across, so the whole system is an ideal subject for confocal microscopy; (iv) first full colonization of the light-organ crypts, where the symbionts reside, takes place over ~12 h, so events can be visualized in real time; (v) successful symbiosis onset can be monitored non-invasively by monitoring symbiont light output from the animal with a photometer; and, (vi) because the "currency" of the symbiosis is light production for host camouflaging, the animal can be raised aposymbiotically through its life in the laboratory, i.e., the symbiosis is ecologically obligate, but, unlike symbioses that are nutritional or involved in tissue defense, the host is physiologically unharmed by the aposymbiotic state; as such, the impact of interactions with microbes during development can be characterized over the trajectory of host ontogeny, from symbiosis onset through maturation.

3.1 Embryonically-developed biophysical and biochemical features work in concert to achieve specificity of the association

V. fischeri cells, at a few hundred per mL of seawater, only represent a small fraction of the bacterial cells in the environment (Lee & Ruby, 1995). Evolution has selected biophysical and biochemical determinants in host and symbiont that promote efficient recruitment and selection of the specific symbiont cells. Following embryogenesis the juvenile host is poised to interact with the symbiont immediately upon hatching (Fig. 3A and B, upper and lower left). The tissues that give rise to the light organ form in the hindgut-ink sac complex. Invaginations on the surface form the six independent crypts, and swellings on the surface form the juvenile-specific ciliated fields that potentiate light-organ colonization (Montgomery & McFall-Ngai, 1993). Also during embryogenesis, the biochemistry of the organ is established that will promote engagement of *V. fischeri* at the exclusion of other bacteria in the environment; acidic mucus stores are laid down in the cells of the ciliated appendages, that contain a complex cocktail of antimicrobials (Davidson, Koropatnick, Kossmehl, et al., 2004; Heath-Heckman, Gillette, Augustin, et al., 2014; Kremer, Schwartzman, Augustin, et al., 2014; Troll, Bent, Pacquette, et al., 2010).

Shortly after hatching, the long cilia on the light-organ surface begin to metachronally beat (Fig. 3B, center; Nawroth, Guo, Koch, et al., 2017) and, within the first 0.5 h, the acidic mucus is shed in a non-specific response of the tissues to free cell-wall materials, or peptidoglycan (PGN), of environmental

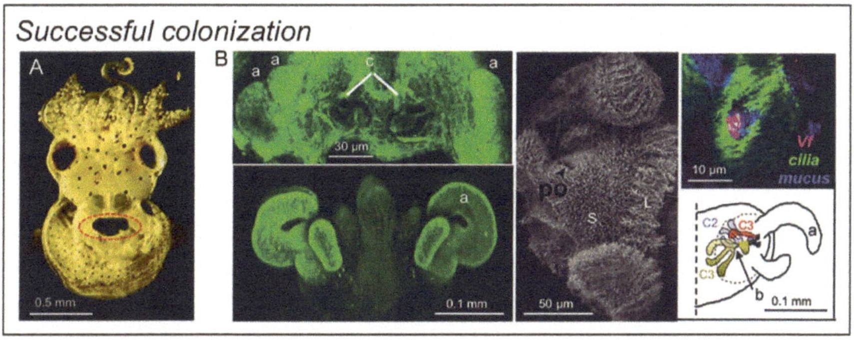

Fig. 3 Early development of the association. (A) Newly hatched *E. scolopes*; red circle, location of the nascent light organ in the mantle cavity. (B) *Left, upper*: a confocal image of the light organ about 2/3 of the way through embryogenesis; white lines indicate invaginated regions where the epithelium-lined crypts are forming. *Left, lower*: confocal image of the complex ciliated fields of the hatchling; a fully developed appendages on the light organ surface. *Middle*, a scanning-electron-microscope image of the ciliated surface, the metachronal waves can be seen along the long cilia (L) of the appendages and medial region of the ciliated field; surrounding the pores are short cilia (s) that randomly beat to mix biomolecules above the pores (po). *Right, upper*: a confocal image showing an aggregate of *V. fischeri* entering a pore on the organ surface. *Right, lower*: the three independent regions colonized on each side of the organ; after the bacteria enter the pores, *V. fischeri* cells travel through three biochemically and biophysically distinct regions over about 100 μm; at the medial portion of the biogeography of each of the three regions is a bottleneck (b) that allows, on average, a single *V. fischeri* cell to enter and grow out in one of the three crypts (C1, C2, C3). a, (left two images and right bottom image), superficial light organ appendages.

bacteria (Fig. 3B, upper right; Kremer et al., 2014; Nyholm, Deplancke, Gaskins, et al., 2002; Nyholm, Stabb, Ruby, et al., 2000). *V. fischeri* cells aggregate above the pores for minutes to hours, depending on the strain (Koehler et al., 2019; Nyholm et al., 2000); the vast majority of the research has been done with *V. fischeri* ES114, which is a strain that aggregates for about 3 h before entering host tissues. The subsequent discussion will be restricted to consideration of colonization by strain ES114.

Because many, if not most, bacteria in the seawater are around the size of *V. fischeri*, the action of the ciliated fields cannot be the only selection determinant. Detailed study of the system showed that, although in the absence of *V. fischeri* other gram-negative bacteria will aggregate on the hatchling organ surface, gram-positive bacteria do not aggregate (Koehler et al., 2019; Nyholm et al., 2000). Many of the antimicrobials shed with the mucus at hatching, including a peptidoglycan-recognition protein with amidase activity (EsPGRP2; Troll et al., 2010) and lysozyme (R. Augustin & M. McFall-Ngai, unpublished data), are specific for gram-positives, which is likely to contribute to their absence in the aggregates. When *V. fischeri* is present, it becomes the competitive dominant over other gram-negative bacteria for that space through the aggregation process. Studies of the symbiont-cell behavior and host transcriptome at this time, showed that ~3 h, with a natural aggregate of ~5 *V. fischeri* cells, these cells attach to the cilia near the pores (Altura, Heath-Heckman, Gillette, et al., 2013); the host responds to the presence of this small number of *V. fischeri* cells, associating with only one or two host cells, with a detectable change in host gene expression across the ~10,000-cell hatchling light organ, a testament to the host sensitivity to its specific symbiont. Studies of the system provide evidence that this change has two effects: (i) an increase in the antimicrobials, which creates a "cocktail" that favors *V. fischeri*; and (ii) the breakdown of polymeric chitin, also in the mucus, into chitobiose (Kremer, Philipp, Carpentier, et al., 2013).

Genetic studies with *V. fischeri* have demonstrated that, to colonize normally, they must be motile (Aschtgen, Brennan, Nikolakakis, et al., 2019) and express genes that confer acid (Schwartzman, Lynch, Flores Ramos, et al., 2019) and nitric oxide tolerance (Dunn, Karr, Wang, et al., 2010; Wang, Dunn, Wilneff, et al., 2010), conditions more acute in the light-organ ducts (Davidson et al., 2004). Also while aggregating they express genes for chemotaxis to chitobiose, which is also more abundant in the ducts (Kremer et al., 2013). Thus, during the aggregation process the symbiont cells become primed to resist the stressful biochemical environment of the host tissues and to chemotax into the organ. After some residence in the aggregate, *V. fischeri* cells express a gene that mediates detachment from the cilia (Christensen, Marsden, Hodge-Hanson, et al., 2020), and then they move into host tissues (Fig. 3, upper right). Only *V. fischeri* can be found living in host tissues beyond the pores. Once inside one of the pores, an aggregate of *V. fischeri* cells begins a ~100-μm journey (Fig. 3, lower right; Essock-Burns, Bongrand, Goldman, et al., 2020) through a duct that opens into larger antechamber. The entrance to a narrow bottleneck is on the medial side of antechamber. This bottleneck has a critical function as a gatekeeper. Although many cells can make it to the bottleneck, on average a single *V. fischeri* cell enters a crypt and grows out to populate the space. This gatekeeper also serves to restrict *V. fischeri* to the regions of the crypts.

As mentioned, the above findings were the result of experiments with *V. fischeri* ES114. Studies of strain variation have shown efficiency of aggregation and colonization are highly strain dependent (Bongrand et al., 2016;

Koehler et al., 2019). These new data offer an exciting horizon for the study of variation in host development in response to colonization by different symbiont strains.

3.2 Symbiont molecules and light production induce early morphogenesis of the host organ

At about 12h following first exposure to *V. fischeri* cells, the populations of symbionts have grown out to fill the crypt spaces. At this point, they begin to luminesce in response to the buildup of their homoserine-lactone molecules associated with quorum sensing (Lupp & Ruby, 2005). Concomitantly, the symbionts induce host morphogenesis (Fig. 4) (for review see McFall-Ngai, 2014). The most dramatic reflections of this phenomenon are the induction of hemocyte trafficking into and the widespread apoptosis of the superficial ciliated fields on the organ, which occurs several cell layers away from the colonizing symbiont. These events are triggered by microbe-associated molecular patterns (MAMPs) (Fig. 4A). Specifically, the morphogens are the lipid A of the lipopolysaccharide molecule (LPS) of the *V. fischeri* outer membrane and the peptidoglycan (PGN) fragments of the symbiont's cell wall; both LPS and PGN are bacterial molecules typically associated with perturbing homeostasis of animal cells. Shortly after the irreversible bacterial signaling of morphogenesis, the juvenile squid begins to "tame" the symbiont by detoxifying the MAMPs (Rader, Kremer, Apicella, et al., 2012; Troll et al., 2010) and also by attenuating the production of antimicrobials (Davidson et al., 2004). Although the phenomenon of bacterial MAMPs-induced morphogenesis was first discovered in the squid-vibrio system, these molecules also are now known to mediate microbiome-induced development and maturation in mammals (Bouskra, Brézillon, Bérard, et al., 2008; Rakoff-Nahoum, Paglino, Eslami-Varzaneh, et al., 2004).

In addition to driving morphogenesis of the surface epithelium, *V. fischeri* colonization of the organ induces development and maturation of the internal tissues (Fig. 4B; for review see McFall-Ngai, 2014). With actin rearrangements, the duct diameter decreases, as does the bottleneck (Essock-Burns et al., 2020), although the symbiont's driver of this phenotype remains unknown. In addition, the symbionts induce cell swelling of the crypt epithelial cells and an increase in microvillar density, which is mediated by lipid A (Heath-Heckman, Foster, Apicella, et al., 2016). The bottleneck closes down and only opens each day with the dawn venting of the symbionts into the surrounding seawater (see discussion below section on daily rhythms of the symbiosis); then, it closes down again within an hour

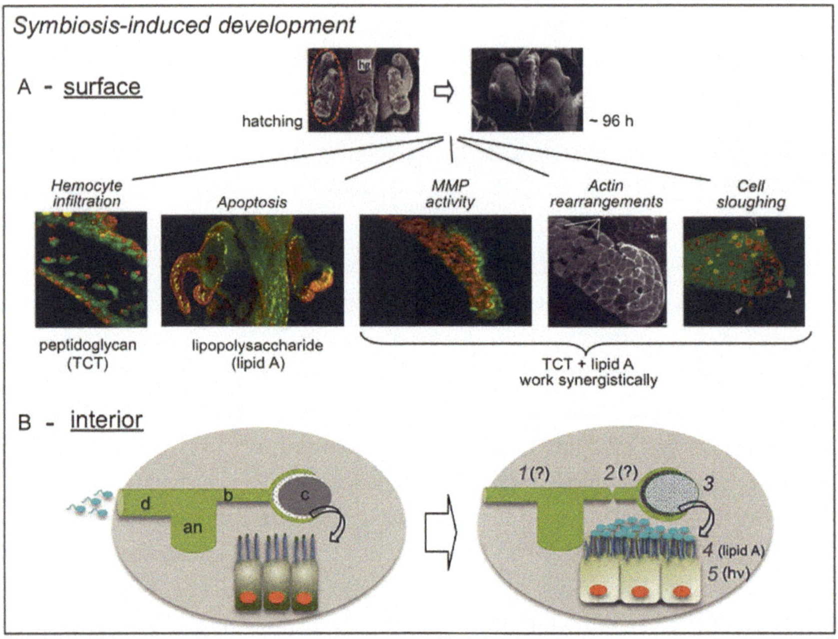

Fig. 4 Symbiont induced early development in the squid-vibrio symbiosis. (A) Colonization-induced loss of the superficial ciliated field that fosters colonization. Upper images: scanning electron micrographs of this process. Lower images: confocal images of the cellular events associated with morphogenesis and the symbiont MAMPs that are largely responsible. (B) Symbiont-induced changes in the light organ interior. Left, diagram showing the microbiogeography of the organ and the path of the symbionts to the crypts (c); arrow indicates an abstraction of the cells the line the crypt to show that they are columnar at hatching; an, antechamber, b, bottleneck, d, duct. Right, the five regions that undergo developmental change: 1, 2—duct and bottleneck respectively; symbiont inducers for these changes have not been identified; 3—the crypts become filled with a dense population of *V. fischeri* cells; 4—crypt epithelial cells change in microvillar density, due to exposure to symbiont lipid A; and 5—crypt epithelial cells swell, in response to symbiont luminescence, respectively.

after venting, restricting the regrowth to the crypt region. Taken together, the organ transforms from an anatomy and morphology that encourages symbiont recruitment to one that discourages subsequent colonization and tightly restricts *V. fischeri* cells to a specific region of the tissues.

It remains unknown how the presence of symbiont MAMPs, or other active symbiont molecules in the crypts, is communicated to the light organ surface to induce developmental events. The MAMPs are presented both as freely diffusing molecules and as constituents of outer membrane vesicles (OMVs) that are shed by *V. fischeri* (Aschtgen, Lynch, Koch, et al., 2016;

Koropatnick, Goodson, Heath-Heckman, & McFall-Ngai, 2014). These OMVs also carry other materials that are imported into the host cells, including symbiont non-coding RNAs, which change host gene expression, as well as proteins (Lynch, Schwartzman, & Bennett, 2019); and, some of the imported materials target the nucleolus and euchromatin (Cohen, Aschtgen, Lynch, et al., 2020). A current research focus explores which symbiont molecules in the OMVs are critical for normal developmental responses in the host.

One of the greatest drivers of symbiont-induced host development is not a biomolecule, but rather light production by the bacteria (for review see McFall-Ngai, Heath-Heckman, Gillette, et al., 2012), the principal "currency" of the association. In a *V. fischeri* mutant defective in light production (*delta-lux*), all of the above mentioned developmental events are delayed. The most conspicuous defect is the lack of cell swelling in *delta-lux* colonized animals. Comparing the 24-h transcriptomes between apo- and symbiotic organs, either colonized by wild-type or *delta-lux* strains (Moriano-Gutierrez, Koch, Bussan, et al., 2019), revealed that symbiont light production is the *principal* driver of colonization-induced change in light-organ gene expression. Interestingly, in this same study, it was discovered that the gene expression of remote tissues, specifically eyes and gills, was also influenced by colonization of the light organ; however, whereas gills had no difference in this phenotype by colonization with the *delta-lux* strain, all symbiont-induced gene expression in the eye was abrogated when the light organ was colonized with the *delta-lux* strain. These data, along with several other studies, have shown that the eyes and light organ are highly convergent in form and function, from their biochemistry and the control of their gene expression to their anatomical features.

3.3 Symbiosis persistence—The symbiosis is on a profound daily rhythm that undergoes maturation in the weeks following establishment of the partnership

Once a symbiosis is established, how is it maintained throughout the life of the host? In many symbioses, particularly those in which the microbial cells are extracellular, the symbiont cells usually continue to grow, so control of symbiont number is critical. Years ago, the squid-vibrio system reported the first daily rhythm on a bacterial symbiosis (Boettcher, Ruby, & McFall-Ngai, 1996). It has now been well established that the gut symbioses of mammals, including those of humans, are on a profound circadian rhythm, with strong evidence that the bacteria actually drive those rhythms (for review see Zheng, Ratiner, & Elinav, 2020). The squid host is a night-active predator, burying in the sand during the day and emerging at night into the water column to forage for prey. Symbiont light production peaks at night (Boettcher et al., 1996), when the host uses the bacterial luminescence for counterillumination, a camouflaging behavior common in marine environments. This day-night behavior is controlled by the variation in oxygen provision to the symbiont by the host; as in all bioluminescence reactions, oxygen is required for bacterial luminescence. While the juveniles do not exhibit a strong daily rhythm of foraging/burying during the night/day as adults, this cycling of luminescence begins with the first day of colonization (Boettcher et al., 1996). Analysis of the underlying drivers of this behavior showed that the expression of the light-organ-specific clock gene, *escry1* (*E. scolopes* cryptochrome 1), is high when luminescence is high, i.e., at night. In contrast, another cry gene, *escry2*, shows highest expression during the day in the eye when environmental light is bright. While the *escry2* gene of the eye cycles independent of colonization state of the light organ, the *escry1* gene cycles only when the symbionts are present, and this cycling requires light production by the symbionts, i.e., colonization with mutants defective in light production does not induce cycling of *escry1*. Finally, experimental manipulation of the system showed that light alone, however, will not induce cycling of *escry1* in the light organ; the luminescence works in synergy with MAMPs. The other behavior that begins with symbiosis onset is the daily venting of ~90% of the symbiont population into the surrounding seawater. At this time, the microvilli are effaced, but the cells remain more polarized in the juvenile organ, whereas in the adults there is marked disruption of the crypt epithelial polarization.

Does the mature symbiosis differ in its rhythms and, if so, when does that change begin in host ontogeny? A decade ago, a study of the symbiosis showed a profound change in the transcriptomes over the day-night cycle of both the adult host central-core epithelium, which supports the symbiont population, and the symbionts cells in the crypts (Wier, Nyholm, Mandel, et al., 2010). In the hours before dawn, all recognizable host cytoskeleton transcripts are upregulated; and after dawn they are all down-regulated. Transmission electron microscopy revealed that, before dawn, the crypt cells are disrupted, with microvillar effacement, shortening or absence of the zonula occludens and other junctional complexes between cells and the apical surfaces are blebbing into the crypt spaces. TEM showed that in the hours after dawn, the cells completely repolarize, regaining their microvillar surfaces and cell junctions. Analyses of the symbiont transcriptomes revealed that they respond to the presentation of host membranes by expressing genes associated with the use of membranes, by anaerobic respiration, as a nutrient source. Once this food source is exhausted, and the bacteria have regrown to fill the crypt spaces, they go into type of stationery phase called a "metabolically active growth stasis." The transcriptional profile of the adult organ tissues provided evidence that the host then begins to feed the *V. fischeri* cells another nutrient, chitin, which they metabolize by aerobic fermentation.

Recent studies of the system have demonstrated that early symbiosis does not have such profound cycling; only after 3–4 weeks following the onset of the symbiosis do the daily cycles take on this complexity. This period is concomitant with the onset of discrete host behaviors, i.e., the typical burying in the sand during the day and foraging at night. Further analysis of the juveniles with symbiont mutants defective in chitin utilization showed that juveniles do not require this nutrient source; these mutants are lost from the symbiosis at 3–4 weeks (Schwartzman, Koch, Heath-Heckman, et al., 2014). In adult animals, however, chitin-rich host hemocytes traffic to the crypt spaces at night releasing their chitin stores for the nocturnal aerobic fermentation by the symbionts. Thus, only the adult animal metabolism exhibits pronounced day-night cycles of anaerobic respiration and aerobic fermentation, respectively.

The low level of hemocytes in the light-organ crypts during the day and high levels at night are controlled by a highly conserved cytokine (Koch et al., 2020), Macrophage Migration Inhibitory Factor (MIF), whose gene expression and protein production in the light-organ epithelium oscillate, with high levels during the day and low levels at night. In addition, when MIF is low in the organ, the symbionts are producing fermentation products and releasing bacterial surface molecules that are chemoattractive to the hemocytes. Also important here is that the blood pigment of the host, hemocyanin, is brought to the crypts loaded with the oxygen. The fermentation of chitin drives the pH of the crypt space down, which causes the efficient offloading of oxygen from the hemocyanin by the classic Bohr effect, which enhances luminescence of the symbionts (Kremer et al., 2014). As such, the host controls the metabolism of the symbionts so that luminescence is fostered during the night, when the host is active in the water column. In response to the dawn light cue, as mentioned above, the host animals vent much of their symbiont population, and the cycle begins again.

Although daily rhythms on the mammalian gut symbioses are now well documented, possible changes over ontogeny or maturation of the rhythms have not yet been studied. Intriguing, however, are the high levels of MIF that have been reported in the mammalian intestine, where it functions in concert with the microbiota to enhance barrier function (Manes, Shulzhenko, Nuccio, et al., 2017; Vujicic, Saksida, Despotovic, et al., 2018), but whether it oscillates on a daily rhythm or controls the well-known circadian rhythms of blood cell migration to the gut is not yet known. In the squid-vibrio system, this cytokine controls nutrient provision to the symbionts, but its high levels during the day may also enhance barrier function as in mammals. Thus, certain controls on daily rhythms of symbioses may also be conserved across the animal kingdom.

In summary, the high resolution of this binary symbiosis has provided a window into the conserved mechanisms underlying the establishment and maintenance of symbiotic associations along the apical surfaces of animal epithelia. The research on squid-vibrio model has shed light on how embryogenesis prepares the juvenile animal of a horizontally transmitted symbiosis to recruit specific bacteria from the environment each generation into a permanent association. In addition, it has described how the elements of the partners, previously associated with microbial pathogenesis (e.g., microbe-associated molecular patterns, pattern recognition receptors, anti-microbial peptides), are actually drivers of development. Further, the system has led the way in defining circadian rhythms as the mechanism that is developed in the symbiosis to maintain the association in homeostasis. Using this symbiosis, the community of biologists now studying this system will continue to provide insights into how microbes shape animal biology.

4. Conclusion

Biology is a complex science at every level. As such, it demands mechanisms by which to make sense of and find patterns in its diversity. Since the advent of next-generation sequencing, biologists have become increasingly aware that animals are typically complex assemblages of organisms, i.e., the host and its associated microbial partners. To understand their development, we will need to bring on all available models. In an insightful essay on the nature of models [https://sites.google.com/a/ncsu.edu/emily-griffiths/whatisamodel.pdf], Emily Griffiths posits that four types of models can be constructive: conceptual, such as a diagram; in silico, i.e., mathematical models; in vitro, e.g., cell culture or tissue/organ extracts; and, in vivo, such as the animal model systems for developmental biology that we have discussed here. Although our focus here has been on in vivo models, we believe that two approaches will be essential for making progress in our understanding of development of the holobiont: (i) an exploitation and integration of all four of the above mentioned model types; and (ii) continued exploration into the diversity of animal symbioses and an expansion of the array of new model systems for this subdiscipline. This frontier is indeed vast and rich for the community of developmental biologists.

Acknowledgments

We thank Scott Gilbert for the opportunity to contribute to this volume and for critically reading the manuscript. M.M.-N. is currently funded by National Institutes of Health (NIH) R37 AI50661 (to M.M.-N. and E.G. Ruby); NIH, Center for Biomedical Research Excellence P20 GM125508 (to M.M.-N. and E.G. Ruby); NIH R01 OD11024 and GM135254 (to E.G. Ruby and M.M.-N.); and a grant from the Gordon and Betty Moore Foundation (to A. Briegel, U. Leiden, M.M.-N., E.G. Ruby and A. Septer, U. North Carolina Chapel Hill). This work was supported in part by grants from the Deutsche Forschungsgemeinschaft and the CRC 1182 ("Origin and Function of Metaorganisms") (to T.C.G.B.). T.C.G.B. appreciates support from the Canadian Institute for Advanced Research.

References

Altura, M. A., Heath-Heckman, E. A., Gillette, A., et al. (2013). The first engagement of partners in the *Euprymna scolopes-Vibrio fischeri* symbiosis is a two-step process initiated by a few environmental symbiont cells. *Environmental Microbiology*, *15*, 2937–2950. https://doi.org/10.1111/1462-2920.12179.

Aschtgen, M. S., Brennan, C. A., Nikolakakis, K., et al. (2019). Insights into flagellar function and mechanism from the squid-vibrio symbiosis. *NPJ Biofilms and Microbiomes*, *5*, 32. https://doi.org/10.1038/s41522-019-0106-5.

Aschtgen, M. S., Lynch, J. B., Koch, E., et al. (2016). Rotation of *Vibrio fischeri* flagella produces outer membrane vesicles that induce host development. *Journal of Bacteriology*, *198*, 2156–2165. https://doi.org/10.1128/JB.00101-16.

Augustin, R., Schroeder, K., Murillo Rincon, A. P., et al. (2017). A secreted antibacterial neuropeptide shapes the microbiome of *Hydra*. *Nature Communications*, *8*, 698.

Beard, A. S., & Blaser, M. J. (2002). The ecology of height—The effect of microbial transmission on human height. *Perspectives in Biology and Medicine*, *45*, 475–498. https://doi.org/10.1353/pbm.2002.0064.

Bellen, H. J., Tong, C., & Tsuda, H. (2010). 100 years of Drosophila research and its impact on vertebrate neuroscience: A history lesson for the future. *Nature Reviews Neuroscience*, *11*, 514–522. https://doi.org/10.1038/nrn2839.

Boehm, A. M., Hemmrich, G., Khalturin, K., Puchert, M., Anton-Erxleben, F., Wittlieb, J., et al. (2012). FoxO is a critical regulator of stem cell maintenance and immortality in Hydra. *Proceedings of the National Academy of Sciences of the United States of America*, *109*(48), 19697–19702.

Boettcher, K. J., Ruby, E. G., & McFall-Ngai, M. J. (1996). Bioluminescence in the symbiotic squid *Euprymna scolopes* is controlled by a daily biological rhythm. *Journal of Comparative Physiology B*, *179*, 65–73.

Bongrand, C., Koch, E. J., Moriano-Gutierrez, S., et al. (2016). A genomic comparison of 13 symbiotic *Vibrio fischeri* isolates from the perspective of their host source and colonization behaviour. *The ISME Journal*, *10*, 2907–2917. https://doi.org/10.1038/ismej.2016.69.

Bosch, T. C. G. (2013). Cnidarian-microbe interactions and the origin of innate immunity in metazoans. *Annual Review of Microbiology, 6*, 499–518. https://doi.org/10.1146/annurev-micro-092412-155626.

Bosch, T. C. G. (2014). Rethinking the role of immunity: Lessons from *Hydra. Trends in Immunology, 35*, 495–502. https://doi.org/10.1016/j.it.2014.07.008.

Bosch, T. C. G., Grasis, J. A., & Lachnit, T. (2015). Microbial ecology in *Hydra*: Why viruses matter. *Journal of Microbiology, 53*, 193–200. https://doi.org/10.1007/s12275-015-4695-2.

Bosch, T. C. G., Guillemin, K., & McFall-Ngai, M. (2019). Evolutionary "experiments" in symbiosis: The study of model animals provides insights into the mechanisms underlying the diversity of host–microbe interactions. *BioEssays: News and Reviews in Molecular, Cellular and Developmental Biology, 41*, e1800256. https://doi.org/10.1002/bies.201800256.

Bouskra, D., Brézillon, C., Bérard, M., et al. (2008). Lymphoid tissue genesis induced by commensals through NOD1 regulates intestinal homeostasis. *Nature, 456*, 507–510. https://doi.org/10.1038/nature07450.

Broun, M., Gee, L., Reinhardt, B., & Bode, H. R. (2005). Formation of the head organizer in *Hydra* involves the canonical Wnt pathway. *Development, 132*, 2907–2916. https://doi.org/10.1242/dev.01848.

Browne, E. N. (1909). The production of new hydranths in *Hydra* by the insertion of small grafts. *Journal of Experimental Zoology, 7*, 1–24. https://doi.org/10.1002/jez.1400070102.

Bry, L., Falk, P. G., Midtvedt, T., et al. (1996). A model of host-microbial interactions in an open mammalian ecosystem. *Science, 273*, 1380–1383. https://doi.org/10.1126/science.273.5280.1380.

Chen, F., Krasity, B. C., Peyer, S. M., et al. (2017). Bactericidal permeability-increasing proteins shape host-microbe interactions. *MBio, 8*, e00040–17. https://doi.org/10.1128/mBio.00040-17.

Christensen, D. G., Marsden, A. E., Hodge-Hanson, K., et al. (2020). LapG mediates biofilm dispersal in *Vibrio fischeri* by controlling maintenance of the VCBS-containing adhesin LapV (published online ahead of print, 2020 Jul 12). *Molecular Microbiology*. https://doi.org/10.1111/mmi.14573.

Cohen, S. K., Aschtgen, M. S., Lynch, J. B., et al. (2020). Tracking the cargo of extracellular symbionts into host tissues with correlated electron microscopy and nanoscale secondary ion mass spectrometry imaging. *Cellular Microbiology, 22*, e13177. https://doi.org/10.1111/cmi.13177.

Collins, A. J., & Nyholm, S. V. (2011). Draft genome of *Phaeobacter gallaeciensis* ANG1, a dominant member of the accessory nidamental gland of *Euprymna scolopes. Journal of Bacteriology, 193*, 3397–3398. https://doi.org/10.1128/JB.05139-11.

Crawford, K., Diaz Quiroz, J. F., Koenig, K. M., et al. (2020). Highly efficient knockout of a squid pigmentation gene (published online ahead of print, 2020 Jul 26). *Current Biology, 30*(17), 3484–3490.e4. https://doi.org/10.1016/j.cub.2020.06.099. S0960-9822(20)30985-4.

Davidson, S. K., Koropatnick, T. A., Kossmehl, R., et al. (2004). NO means 'yes' in the squid-vibrio symbiosis: Nitric oxide (NO) during the initial stages of a beneficial association. *Cellular Microbiology, 6*, 1139–1151. https://doi.org/10.1111/j.1462-5822.2004.00429.x.

Davisson, M. T., & Linder, C. C. (2004). Chapter 2—Historical foundations. In H. Henrich (Ed.), *The laboratory mouse* (1st ed., pp. 15–24). Amsterdam: Elsevier.

Domazet-Lošo, T., Klimovich, A., Anokhin, B., et al. (2014). Naturally occurring tumours in the basal metazoan *Hydra. Nature Communications, 5*, 4222. https://doi.org/10.1038/ncomms5222.

Dougherty, E. C., & Calhoun, H. G. (1948). Possible significance of free-living nematodes in genetic research. *Nature, 161*, 29. https://doi.org/10.1038/161029a0.

Dunn, A. K., Karr, E. A., Wang, Y., et al. (2010). The alternative oxidase (AOX) gene in *Vibrio fischeri* is controlled by NsrR and upregulated in response to nitric oxide. *Molecular Microbiology*, *77*, 44–55. https://doi.org/10.1111/j.1365-2958.2010.07194.x.

Essock-Burns, T., Bongrand, C., Goldman, W. E., et al. (2020). Interactions of symbiotic partners drive the development of a complex biogeography in the squid-vibrio symbiosis. *MBio*, *11*, e00853–20. https://doi.org/10.1128/mBio.00853-20.

Franzenburg, S., Walter, J., Künzel, S., et al. (2013). Distinct antimicrobial tissue activity shapes host species-specific bacterial associations. *Proceedings of the National Academy of Sciences of the United States of America*, *110*, E3730–E3738. https://doi.org/10.1073/pnas.1304960110.

Fraune, S., Anton-Erxleben, F., Augustin, R., et al. (2015). Bacteria-bacteria interactions within the microbiota of the ancestral metazoan *Hydra* contribute to fungal resistance. *The ISME Journal*, *9*, 1543–1556. https://doi.org/10.1038/ismej.2014.239.

Fraune, S., & Bosch, T. C. G. (2007). Long-term maintenance of species-specific bacterial microbiota in the basal metazoan *Hydra*. *Proceedings of the National Academy of Sciences of the United States of America*, *104*, 13146–13151. https://doi.org/10.1073/pnas.0703375104.

Gee, L., Hartig, J., Law, L., et al. (2010). *beta*-catenin plays a central role in setting up the head organizer in *Hydra*. *Developmental Biology*, *340*, 116–124. https://doi.org/10.1016/j.ydbio.2009.12.036.

Graf, J. (1999). Symbiosis of *Aeromonas veronii* biovar sobria and *Hirudo medicinalis*, the medicinal leech: A novel model for digestive tract associations. *Infection and Immunity*, *67*, 1–7. https://doi.org/10.1128/IAI.67.1.1-7.1999.

Grewal, P. S. (1991). Relative contribution of nematodes (*Caenorhabditis elegans*) and bacteria towards the disruption of flushing patterns and losses in yield and quality of mushrooms (*Agaricus bisporus*). *Annals of Applied Biology*, *119*, 483–499.

Hadfield, M. G., Unabia, C. C., Smith, C. M., & Michael, T. M. (1994). Settlement preferences of the ubiquitous fouler Hydroides elegans. In M. F. Thompson, R. Nagabhushanam, R. Sarojini, & M. Fingerman (Eds.), *Recent Developments in Biofouling Control*. New Delhi: Oxford and IBH Publishing Co.

Hammer, T. J., Sanders, J. G., & Fierer, N. (2019). Not all animals need a microbiome. *FEMS Microbiology Letters*, *366*, fnz117. https://doi.org/10.1093/femsle/fnz117.

Heath-Heckman, E. A., Foster, J., Apicella, M. A., et al. (2016). Environmental cues and symbiont microbe-associated molecular patterns function in concert to drive the daily remodelling of the crypt-cell brush border of the *Euprymna scolopes* light organ. *Cellular Microbiology*, *18*, 1642–1652. https://doi.org/10.1111/cmi.12602.

Heath-Heckman, E. A., Gillette, A. A., Augustin, R., et al. (2014). Shaping the microenvironment: Evidence for the influence of a host galaxin on symbiont acquisition and maintenance in the squid-vibrio symbiosis. *Environmental Microbiology*, *16*, 3669–3682. https://doi.org/10.1111/1462-2920.12496.

Hirsch, A., & McFall-Ngai, M. (2000). Fundamental concepts in symbiotic interactions: Light and dark, day and night, squid and legume. *Journal of Plant Growth Regulation*, *19*, 113–130. https://doi.org/10.1007/s003440000025.

Hobmayer, B., Rentzsch, F., Kuhn, K., et al. (2000). WNT signalling molecules act in axis formation in the diploblastic metazoan *Hydra*. *Nature*, *407*, 186–189. https://doi.org/10.1038/35025063.

Kerwin, A. H., Gromek, S. M., Suria, A. M., et al. (2019). Shielding the next generation: Symbiotic bacteria from a reproductive organ protect bobtail squid eggs from fungal fouling. *MBio*, *10*, e02376–19. https://doi.org/10.1128/mBio.02376-19.

Kikuchi, Y., Meng, X. Y., & Fukatsu, T. (2005). Gut symbiotic bacteria of the genus *Burkholderia* in the broad-headed bugs *Riptortus clavatus* and *Leptocorisa chinensis* (Heteroptera: Alydidae). *Applied and Environmental Microbiology*, *71*, 4035–4043. https://doi.org/10.1128/AEM.71.7.4035-4043.2005.

Klimovich, A. V., & Bosch, T. C. G. (2018). Rethinking the role of the nervous system: Lessons from the *Hydra* holobiont. *BioEssays*, *40*, e1800060. https://doi.org/10.1002/bies.201800060.

Koch, E. J., Bongrand, C., Bennett, B. D., et al. (2020). The cytokine MIF controls daily rhythms of symbiont nutrition in the squid-vibrio association. *Proceedings of the National Academy of Sciences of the United States of America* (in press).

Koehler, S., Gaedeke, R., Thompson, C., et al. (2019). The model squid-vibrio symbiosis provides a window into the impact of strain- and species-level differences during the initial stages of symbiont engagement. *Environmental Microbiology*, *21*(9), 3269–3283. https://doi.org/10.1111/1462-2920.14392.

Koropatnick, T., Goodson, M. S., Heath-Heckman, E. A., & McFall-Ngai, M. (2014). Identifying the cellular mechanisms of symbiont-induced epithelial morphogenesis in the squid-Vibrio association. *The Biological Bulletin*, *226*(1), 56–68. https://doi.org/10.1086/bblv226n1p56.

Kremer, N., Philipp, E. E., Carpentier, M. C., et al. (2013). Initial symbiont contact orchestrates host-organ-wide transcriptional changes that prime tissue colonization. *Cell Host & Microbe*, *14*, 183–194. https://doi.org/10.1016/j.chom.2013.07.006.

Kremer, N., Schwartzman, J., Augustin, R., et al. (2014). The dual nature of haemocyanin in the establishment and persistence of the squid-vibrio symbiosis. *Proceedings of the Royal Society B: Biological Sciences*, *281*, 20140504. https://doi.org/10.1098/rspb.2014.0504.

Krogh, A. (1929). The progress of physiology. *American Journal of Physiology*, *90*, 243–251. https://doi.org/10.1126/science.70.1809.200.

Lee, K., & Ruby, E. G. (1995). Symbiotic role of the viable but nonculturable state of *Vibrio fischeri* in Hawaiian coastal seawater. *Applied and Environmental Microbiology*, *61*, 278–283. https://doi.org/10.1128/AEM.61.1.278-283.1995.

Li, X.-Y., Lachnit, T., Fraune, S., et al. (2017). Temperate phages as self-replicating weapons in bacterial competition. *Journal of the Royal Society Interface*, *14*, 20170563. https://doi.org/10.1098/rsif.2017.0563.

Lupp, C., & Ruby, E. G. (2005). *Vibrio fischeri* uses two quorum-sensing systems for the regulation of early and late colonization factors. *Journal of Bacteriology*, *187*, 3620–3629. https://doi.org/10.1128/JB.187.11.3620-3629.2005.

Lutz, H. L., Ramírez-Puebla, S. T., Abbo, L., et al. (2019). A simple microbiome in the European common cuttlefish, *Sepia officinalis*. *mSystems*, *4*, e00177–19. https://doi.org/10.1128/mSystems.00177-19.

Lynch, J. B., Schwartzman, J. A., & Bennett, B. D. (2019). Ambient pH alters the protein content of outer membrane vesicles, driving host development in a beneficial symbiosis. *Journal of Bacteriology*, *201*, e00319–19. https://doi.org/10.1128/JB.00319-19.

Manes, N. P., Shulzhenko, N., Nuccio, A. G., et al. (2017). Multi-omics comparative analysis reveals multiple layers of host signaling pathway regulation by the gut microbiota. *mSystems*, *2*, e00107–e00117. https://doi.org/10.1128/mSystems.00107-17.

Martinson, V. G., Danforth, B. N., Minckley, R. L., et al. (2011). A simple and distinctive microbiota associated with honey bees and bumble bees. *Molecular Ecology*, *20*, 619–628. https://doi.org/10.1111/j.1365-294X.2010.04959.x.

McFall-Ngai, M. J. (2014). The importance of microbes in animal development: Lessons from the squid-vibrio symbiosis. *Annual Review of Microbiology*, *68*, 177–194. https://doi.org/10.1146/annurev-micro-091313-103654.

McFall-Ngai, M., Heath-Heckman, E. A., Gillette, A. A., et al. (2012). The secret languages of coevolved symbioses: Insights from the *Euprymna scolopes-Vibrio fischeri* symbiosis. *Seminars in Immunology*, *24*, 3–8. https://doi.org/10.1016/j.smim.2011.11.006.

McFall-Ngai, M. J., & Ruby, E. G. (1991). Symbiont recognition and subsequent morphogenesis as early events in an animal-bacterial mutualism. *Science*, *254*, 1491–1494. https://doi.org/10.1126/science.1962208.

Montgomery, M. K., & McFall-Ngai, M. (1993). Embryonic development of the light organ of the pepiolid squid *Euprymna scolopes* Berry. *The Biological Bulletin, 184*, 296–308. https://doi.org/10.2307/1542448.

Moriano-Gutierrez, S., Koch, E. J., Bussan, H., et al. (2019). Critical symbiont signals drive both local and systemic changes in diel and developmental host gene expression. *Proceedings of the National Academy of Sciences of the United States of America, 116*, 7990–7999. https://doi.org/10.1073/pnas.1819897116.

Mortzfeld, B. M., & Bosch, T. C. (2017). Eco-aging: Stem cells and microbes are controlled by aging antagonist FoxO. *Current Opinion in Microbiology, 38*, 181–187. https://doi.org/10.1016/j.mib.2017.06.009.

Mortzfeld, B. M., Taubenheim, J., Fraune, S., et al. (2018). Stem cell transcription factor FoxO controls microbiome resilience in *Hydra*. *Frontiers in Microbiology, 9*, 629. https://doi.org/10.3389/fmicb.2018.00629.

Mortzfeld, B. M., Taubenheim, J., Klimovich, A. V., et al. (2019). Temperature and insulin signaling regulate body size in *Hydra* by the Wnt and TGF-beta pathways. *Nature Communications, 10*, 3257. https://doi.org/10.1038/s41467-019-11136-6.

Mortzfeld, B. M., Urbanski, S., Reitzel, A. M., et al. (2016). Response of bacterial colonization in *Nematostella vectensis* to development, environment and biogeography. *Environmental Microbiology, 18*, 1764–1781. https://doi.org/10.1111/1462-2920.12926.

Murillo-Rincon, A. P., Klimovich, A., Permoller, E., et al. (2017). Spontaneous body contractions are modulated by the microbiome of *Hydra*. *Scientific Reports*, 7, 15937. https://doi.org/10.1038/s41598-017-16191-x.

Muscatine, L., & Lenhoff, H. M. (1963). Symbiosis: On the role of algae symbiotic with *Hydra*. *Science, 142*, 956–958. https://doi.org/10.1126/science.142.3594.956.

Nawroth, J. C., Guo, H., Koch, E., et al. (2017). Motile cilia create fluid-mechanical microhabitats for the active recruitment of the host microbiome. *Proceedings of the National Academy of Sciences of the United States of America, 114*, 9510–9516. https://doi.org/10.1073/pnas.1706926114.

Nyholm, S. V., Deplancke, B., Gaskins, H. R., et al. (2002). Roles of *Vibrio fischeri* and nonsymbiotic bacteria in the dynamics of mucus secretion during symbiont colonization of the *Euprymna scolopes* light organ. *Applied and Environmental Microbiology, 68*, 5113–5122. https://doi.org/10.1128/aem.68.10.5113-5122.2002.

Nyholm, S. V., & McFall-Ngai, M. J. (2020). Beyond the winnowing: The impact of a squid's life long journey with symbionts. *Nature Reviews Microbiology, 2*, 632–642. in press.

Nyholm, S. V., Stabb, E. V., Ruby, E. G., et al. (2000). Establishment of an animal-bacterial association: Recruiting symbiotic vibrios from the environment. *Proceedings of the National Academy of Sciences of the United States of America, 97*, 10231–10235. https://doi.org/10.1073/pnas.97.18.10231.

Poinar, G. O., Jr., & Thomas, G. M. (1966). Significance of *Achromobacter nematophilus* Poinar and Thomas (Achromobacteraceae: Eubacteriales) in the development of the nematode, DD-136 (*Neoaplectana* sp. Steinernematidae). *Parasitology, 56*, 385–390. https://doi.org/10.1017/s0031182000070980.

Rader, B. A., Kremer, N., Apicella, M. A., et al. (2012). Modulation of symbiont lipid A signalling by host alkaline phosphatases in the squid-vibrio symbiosis. *MBio, 3*, 300093–12. https://doi.org/10.1128/mBio.00093-12.

Rakoff-Nahoum, S., Paglino, J., Eslami-Varzaneh, F., et al. (2004). Recognition of commensal microflora by toll-like receptors is required for intestinal homeostasis. *Cell, 118*, 229–241. https://doi.org/10.1016/j.cell.2004.07.002.

Rathje, K., Mortzfeld, B., Hoeppner, M. P., et al. (2020). Dynamic interactions within the host-associated microbiota cause tumor formation in the basal metazoan *Hydra*. *PLoS Pathogens, 163*, e1008375. https://doi.org/10.1371/journal.ppat.1008375.

Rawls, J. F., Samuel, B. S., & Gordon, J. I. (2004). Gnotobiotic zebrafish reveal evolutionarily conserved responses to the gut microbiota. *Proceedings of the National Academy of Sciences of the United States of America, 101*, 4596–4601. https://doi.org/10.1073/pnas.0400706101.

Roy, S., Liu, W., Nandety, R. S., et al. (2020). Celebrating 20 years of genetic discoveries in legume nodulation and symbiotic nitrogen fixation. *The Plant Cell, 32*, 15–41. https://doi.org/10.1105/tpc.19.00279.

Ruby, E. G. (2008). Symbiotic conversations are revealed under genetic interrogation. *Nature Reviews Microbiology, 6*, 752–762. https://doi.org/10.1038/nrmicro1958.

Ryu, J. H., Kim, S. H., Lee, H. Y., et al. (2008). Innate immune homeostasis by the homeobox gene caudal and commensal-gut mutualism in *Drosophila. Science, 319*, 777–782. https://doi.org/10.1126/science.1149357.

Schröder, K., & Bosch, T. C. G. (2016). The origin of mucosal immunity: Lessons from the holobiont *Hydra. MBio*, 7, e01184–16. https://doi.org/10.1128/mBio.01184-16.

Schulenburg, H., & Félix, M. A. (2017). The natural biotic environment of *Caenorhabditis elegans. Genetics, 206*, 55–86. https://doi.org/10.1534/genetics.116.195511.

Schwartzman, J. A., Koch, E., Heath-Heckman, E. A., et al. (2014). The chemistry of negotiation: Rhythmic, glycan-driven acidification in a symbiotic conversation. *Proceedings of the National Academy of Sciences of the United States of America, 112*, 566–571. https://doi.org/10.1073/pnas.1418580112.

Schwartzman, J. A., Lynch, J. B., Flores Ramos, S., et al. (2019). Acidic pH promotes lipopolysaccharide modification and alters colonization in a bacteria-animal mutualism. *Molecular Microbiology, 112*, 1326–1338. https://doi.org/10.1111/mmi.14365.

Stabb, E. V., Visick, K., & Ruby, E. G. (2020). *Nature Reviews Microbiology*. in press.

Streisinger, G., Walker, C., Dower, N., et al. (1981). Production of clones of homozygous diploid zebra fish (*Brachydanio rerio*). *Nature, 291*, 293–296. https://doi.org/10.1038/291293a0.

Takeshita, K., & Kikuchi, Y. (2017). Riptortus pedestris and Burkholderia symbiont: an ideal model system for insect-microbe symbiotic associations. *Research in Microbiology, 168*(3), 175–187. https://doi.org/10.1016/j.resmic.2016.11.005.

Taubenheim, J., Willoweit-Ohl, D., Knop, M., et al. (2020). Bacteria- and temperature-regulated peptides modulate β-catenin signaling in *Hydra. Proceedings of the National Academy of Sciences of the United States of America, 117*, 202010945. https://doi.org/10.1073/pnas.2010945117.

Trembley, A. (1744). *Mémoires, pour servir à l'histoire d'un genre de polypes d'eau douce, à bras en forme de cornes*. Leiden: Verbeek.

Troll, J. V., Bent, E. H., Pacquette, N., et al. (2010). Taming the symbiont for coexistence: A host PGRP neutralizes a bacterial symbiont toxin. *Environmental Microbiology, 12*, 2190–2203. https://doi.org/10.1111/j.1462-2920.2009.02121.x.

Vujicic, M., Saksida, T., Despotovic, S., et al. (2018). The role of macrophage migration inhibitory factor in the function of intestinal barrier. *Scientific Reports, 8*, 6337. https://doi.org/10.1038/s41598-018-24706-3.

Wanek, N., & Campbell, R. D. (1982). Roles of ectodermal and endodermal epithelial cells in hydra morphogenesis. *Journal of Experimental Zoology, 22*, 37–47. https://doi.org/10.1002/jez.1402210107.

Wang, Y., Dunn, A. K., Wilneff, J., et al. (2010). *Vibrio fischeri* flavohaemoglobin protects against nitric oxide during initiation of the squid-vibrio symbiosis. *Molecular Microbiology, 78*, 903–915. https://doi.org/10.1111/j.1365-2958.2010.07376.x.

Weber, E. H. (1829). Ueber die Entwicklung des medicinischen Blutegels. *Meckel's Archiv für die Physiologie*, 366–418.

Wier, A. M., Nyholm, S. V., Mandel, M. J., et al. (2010). Transcriptional patterns in both host and bacterium underlie a daily rhythm of anatomical and metabolic change in a beneficial symbiosis. *Proceedings of the National Academy of Sciences of the United States of America*, *107*, 2259–2264. https://doi.org/10.1073/pnas.0909712107.

Yan, Y., Nguyen, L. H., Franzosa, E. A., et al. (2020). Strain-level epidemiology of microbial communities and the human microbiome. *Genome Medicine*, *12*, 71. https://doi.org/10.1186/s13073-020-00765-y.

Zheng, D., Ratiner, K., & Elinav, E. (2020). Circadian influences of diet on the microbiome and immunity. *Trends in Immunology*, *41*, 512–530. https://doi.org/10.1016/j.it.2020.04.005.

CHAPTER TWELVE

Animal development in the microbial world: Re-thinking the conceptual framework

Thomas C.G. Bosch[a,*] and Margaret McFall-Ngai[b]
[a]Zoological Institute, Christian-Albrechts-University Kiel, Kiel, Germany
[b]Pacific Biosciences Research Center, Kewalo Marine Laboratory, University of Hawai'i at Mānoa, Honolulu, HI, United States
*Correspondence author: e-mail address: tbosch@zoologie.uni-kiel.de

Contents

Abstract

Animals have evolved within the framework of the microbes and are constantly exposed to diverse microbiota. This dominance of the microbial world is forcing all fields of biology to question some of their most basic premises, with developmental biology being no exception. While animals under laboratory conditions can develop and live

Current Topics in Developmental Biology, Volume 141
ISSN 0070-2153
https://doi.org/10.1016/bs.ctdb.2020.11.007

without microbes, they are far from normal, and would not survive under natural conditions, where their fitness would be strongly compromised. Since much of the undescribed biodiversity on Earth is microbial, any consideration of animal development in the absence of the recognition of microbes will be incomplete. Here, we show that animal development may never have been autonomous, rather it requires transient or persistent interactions with the microbial world. We propose that to formulate a comprehensive understanding of embryogenesis and post-embryonic development, we must recognize that symbiotic microbes provide important developmental signals and contribute in significant ways to phenotype production. This offers limitless opportunities for the field of developmental biology to expand.

1. Introduction

1.1 What do we mean by "development"?

Classical animal developmental biology has a tradition of (or has focused upon) studying the process by which animals grow and develop through ontogeny. Within this construct, the main processes involved include regional specification, morphogenesis, cell differentiation, growth, and the overall control of timing, much of which occurs in the embryo, the starting point for all animal species (Gilbert, 2001). As an animal develops, different combinations of genes are switched on giving the species different traits. Various factors can change the combination of genes that are switched on. Critical stages of embryonic development are defined by key epigenetic modifications including global demethylation of the early embryo and remodeling of 5′-methylcytosine throughout embryo development (Morgan et al., 2005). Such changes allow the species to adapt and survive depending on the circumstance.

Developmental biologists have long been fascinated by this relationship between development and evolution and the fact that homologous genes often play similar roles during the development of very different organisms. In the last three decades we have learnt that the underlying regulatory machinery is ancient because many of the genes and physiological pathways that govern life are similar among all living organisms from bacteria to humans (Cheatle Jarvela & Pick, 2016; Domazet-Loso & Tautz, 2010). In that context evolutionary developmental biologists also realized that early diverging animals such as sponges and corals are much more complex and "vertebrate" like than previously thought. It is clear that the origin and evolution of developmental novelties lies in developmental processes.

The fundamental concern of evolutionary developmental biology (evo-devo) is to understand how development is tweaked over time. How do developmental pathways evolve? What are the evolutionarily conserved or universal codes of development? And what processes of development create biological diversity?

These concepts beg the question whether nuclear DNA is the only source of developmental instructions. As pointed out by John Tyler Bonner (1974), each life cycle is governed by some immediate instructions from the genes, by many gene-initiated instructions that were given in previous life cycles, and by signals originating from outside the organism. All of these instructions taken together govern the development of a new embryo. Further, through development, an animal's genotype is expressed as a phenotype, exposing genes to the action of natural selection. Thus, the evolution of animal development is sensitive to environmental factors.

Evolutionary developmental biology (evo-devo) contends that evolution occurs through changes in developmental processes. Ecological developmental biology (eco-devo) has shown that microbes are critically involved in developmental processes (Bosch, Guillemin, & McFall-Ngai, 2019; Gilbert, Bosch, & Ledon-Rettig, 2015; McFall-Ngai, 2002). If both of these statements are true, then changes in the developmental functions of symbiotic microbes can cause changes in evolution. This yields the new science of ecological evolutionary biology, eco-evo-devo (Gilbert, 2001; Gilbert & Epel, 2015; Skúlason et al., 2019). The life cycles of hosts and their microbial symbionts are intimately integrated (Gilbert 2019).

Anticipating this change of perspective from the "evolution of phenotypes" toward an "ecology of ontogenies", John Bonner (1965, 1974) proposed that the unit of evolutionary selection was the life cycle, and he recognized that signals for developmental change could come from the environment, not only from the genome. Moreover, Leigh van Valen remarked in the early 1970s that "A plausible argument could be made that evolution is the control of development by ecology" (Van Valen, 1973, p 488). Studying these environmental factors and ecological forces have largely been restricted to analyses of developmental plasticity due to abiotic forces (e.g., environmental temperature, pressure, salinity, oxygen concentration) or the biotic forces of predation, crowding, and competition.

Very recent work, however, has revealed a new element in the "equation": the microbial world. What, then, is the role of microbes in animal development? This chapter seeks to address that question by expanding the eco-devo perspective. To do so, it explores how a number of different animal groups use microbial cues to orchestrate developmental processes and how animal hosts and microbes interact at the molecular level.

1.2 What makes us rethink the classic field of development?

Development in animals, plants, and fungi is not just the product of one genome. On the contrary, the nuclear genome of an animal comprises a relatively small fraction of the organism's total genetic repertoire, and the majority of genes is provided by microbes. These microbes can be obtained vertically (from the host's parents) or horizontally (from the environment, other host species, or other members of the same species). The phenotype is a consequence of host-microbe interactions, and these multi-organismic associations fundamentally alter the host's evolution and ecology (Roughgarden et al., 2017). Taking an integrated view, we therefore see the host–microbiome system as a holobiont (Rosenberg & Zilber-Rosenberg, 2013), or a compound meta-organism consisting of the macroscopic host along with its all of its symbiotic microbes (Morris, 2018; Theis et al., 2016). Indeed, the physiology and immunity of most animals has been shown to be an intimate entanglement of host and microbial interactions. Functional nervous systems, digestive systems, and immune systems are often the product of both the host and the microbes (Gilbert et al., 2012; McFall-Ngai et al., 2013; Pradeau, 2012; Tauber, 2017). Important in the context of this review is the recognition that microbial populations and communities are integrated into animal developmental programs often over the entire trajectory of an animal's ontogeny.

This new perspective radically changes our conception of the "Hierarchy of Life" (Fig. 1), our concept of what an individual is, and our concept of life cycles. The multispecies developmental program of most animals results, not in a monogenomic "individual", but in a holobiont with tissues and organ systems that are created by the host genome and an organ system that comprises the interacting communities of microbes coevolved with the host. These microbial communities occur in discrete regions of organ systems across the landscape of the host "Bauplan" (body plan). Each animal is both an individual and biomes of multiple interacting ecosystems (Lee & Mazmanian, 2010) We are teams. As such, the microbiome organ system enriches the individual animal with levels of the microbial hierarchy, from organism to populations and communities, rendering the metaorganism (or holobiont) a chimera of nested hierarchical levels.

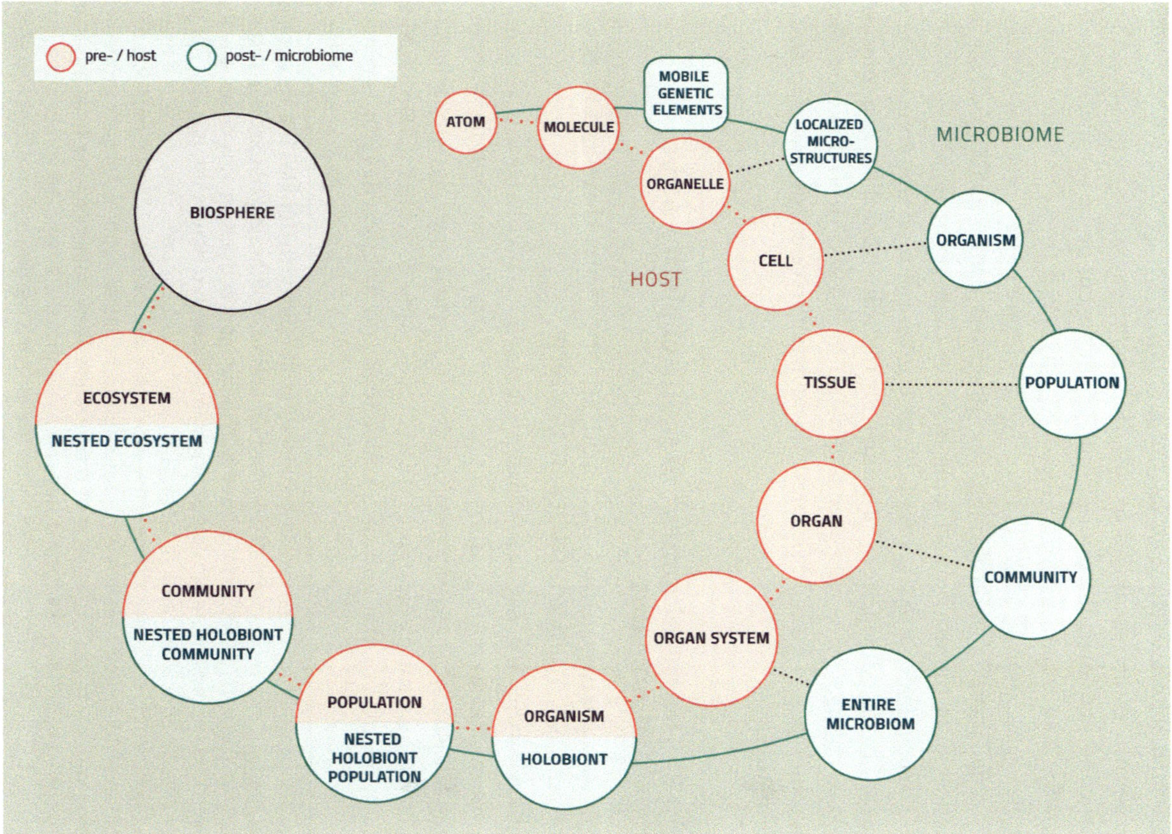

Fig. 1 See figure legend on next page.

Within this conceptual framework, the development of vertically transmitted symbionts occurs in tandem with host embryonic development. In horizontally acquired symbioses, metaorganism development is more sequential. In these latter cases, the embryo develops features that allow the animal to recognize and recruit the specific coevolved microbes; then, at birth or hatching, the microbiome organ system develops and integrates with the host organ systems. However, this is not to say that horizontally acquired microbes do not affect early development. In mice, for instance, the population of microbes in the pregnant dam's gut can affect the phenotype of the mice developing in utero (Kimura et al., 2020).

To keep pace with the paradigm of animals as holobiont consortia constructed by the interactions of numerous microbes and their genomes, we need to construct a developmental biology and an evo-devo that recognizes microbial agency from fertilization, through embryogenesis, and throughout maturation.

2. The development of multicellularity across all domains

2.1 Autonomous and non-autonomous development of multicellularity in bacteria

Multicellularity is a developmental phenomenon, involving the regulation of mitosis and (often) the differentiation of parts within a whole. Complex multicellularity has arisen by various developmental strategies multiple times in both the prokaryotes and eukaryotes (Lyons & Kolter, 2015; Rokas, 2008). While, to the authors' knowledge, no animals develop normally

Fig. 1 Incorporating coevolved symbioses into the hierarchy of life. The recognition that many, if not most, animals are metaorganisms, comprising host and associated microbial partners, drives a conceptual shift in our view of the hierarchy of life, one that reflects the nested nature of biological systems. Illustrated here is the hierarchy of the fully mature individual and its associated ecological context. Prior to the dawn of next-generation sequencing (pre-), the increasing complexity of the hierarchical organization of biological systems was presumed to be linear, with "emergent properties" arising with each successive level of organization. New data, resulting from advances in sequencing technology (post-), have revealed that the biosphere comprises a more complex hierarchy, one that incorporates the inherent nesting of the macro- and microbial worlds. A holobiont is created through dynamic interactions between the host and microbial elements, through the trajectory of development, from fertilization through ontogeny.

under natural conditions without interactions with the microbial world, many examples of autonomous development of multicellularity have been characterized in the prokaryotes. Mechanisms used to generate asymmetry in a predivisional/parental cell are well known in microbes. For example, cell division in the single-celled aquatic bacterium *Caulobacter crescentus* results in two daughter cells which have different fates—one becomes a "stalked" cell, resembling the parental cell; the other grows a flagellum, becoming a "swarmer" cell that, when mature, swims away. *C. crescentus* thus accomplishes a fundamental function that underlies all development: it generates daughter cells that differ from the parent cell (Lasker et al., 2020; Zhou et al., 2019). Interestingly, *C. crescentus* development is obligate, i.e., it is on a perpetual "clock" that drives the differentiation independently from environmental conditions.

Other bacterial phylotypes undergo development in a response to environmental conditions, most often resulting in the production of spores. The most classic case of a system of bacterial multicellularity is illustrated by the development of *Myxococcus xanthus*, a social predatory bacterium that hunts for prey cooperatively (Bretl & Kirby, 2016). Under conditions of starvation, large populations of *M. xanthus* cells coordinate to migrate into a mass that develops a stalk. That stalk elongates and eventually differentiates a fruiting body with spores that disperse and germinate when prey is again plentiful.

Bacillus subtilis also has another sort of development, specifically within a biofilm (Vlamakis et al., 2013). Microbial biofilms comprise either populations of a single microbial species or are mixed species assemblages. Typically of the integrity the biofilm is held together by biomolecules exported by the biofilm members. Recent approaches using phylotranscriptomics revealed (Futo et al., 2020) that *B. subtilis* biofilm formation shares many properties of embryonic development with discrete reproducible stages, with the oldest genes being expressed first and the newest last. The study concludes that biofilm formation is a true multicellular developmental process, indeed one in which its "ontogeny" appears to be recapitulating "phylogeny."

Distinct spatial distribution patterns of cells are also observed in experimentally established biofilm communities of other species, and particular processes of their evolution, metabolic capabilities, and resistance toward antimicrobials have been revealed (Haagensen et al., 2015). In other words, some microbes seem to have developed all of the genetic architecture that is required for spatial-temporal controlled embryonic development. The underlying evolutionary and developmental processes of these communities remain elusive, but this topic represents a rich arena for future exploration of the origins of developmental mechanisms.

Although the examples of prokaryotic multicellularity described thus far occur through autonomous development, some species require more than one genotype. A striking example can be found in the two-species phototrophic consortium "Chlorochromatium aggregatum" (Overmann, 2010). It consists of one central non-photosynthetic motile cell surrounded by ~15 associated cells of another species that are non-motile photosynthetic cells. The phototactic behavior of the consortium is the result of coordination between these cell types, with the photosynthetic cells sensing the light and the motile cells driving the group toward the light. The cell division of the developing consortium occurs in a highly coordinated manner to promote this "division of labor" between the two species. The development of other highly structured multispecies biofilms among prokaryotic species have been reported, although most widely studied are those associated with an animal host. For example, the development and maintenance of "corncob"-like and "hedgehog"-like nine-taxon structures of the microbial plaques of the mammalian oral cavity are the result of highly orchestrated developmental programs (Mark Welch et al., 2016).

2.2 Development of multicellularity in eukaryotes may never have been autonomous, rather requiring transient or persistent interactions with the microbial world

Eukaryotic multicellularity has arisen more than 25 times (Grosberg & Strathmann, 2007; Wegener Parfrey et al., 2011), and it has been accomplished through two major strategies: aggregation of individual cells as in *Dictyostelium,* and division of a single cell (zygote or spore) as in animals, plants and fungi (Bonner, 1998). As Leo Buss (1987) emphasized in "The Evolution of Individuality", the very existence of integrated multicellular organisms is an outcome of evolutionary processes, not a starting condition. One major clade of unicellular eukaryotes of particular interest here are the Opisthokonta, the group that in evolution gave rise to the animal and fungal clades. Within the ophistokonts, the closest living relatives of animals have unambiguously been shown to be the choanoflagellates (King et al., 2008; Ruiz-Trillo et al., 2008; Shalchian-Tabrizi et al., 2008; Torruella et al., 2015). Choanoflagellates are a globally distributed group of marine and freshwater protozoans with a highly distinctive morphology characterized by a whip-like flagellum and a collar of shorter hairs, resembling the food-filtering "collar" cells that line the channels of sponges.

Research on marine choanoflagellates over the last decade has demonstrated that their life history and in particular their development and reproduction is closely linked with the presence of developmental signals originating in symbiotic bacteria. In *Salpingoeca rosetta*, the development of multicellular colonies is dependent on a novel sulfonolipid signaling molecule, RIF-1 (Rosette-Inducing Factor-1) from the bacterium *Algoriphagus* sp. (Alegado et al., 2012). This is of interest not only because choanoflagellates are the closest living relatives of animals but also because Bacteroidetes bacteria, of which *Algoriphagus* is a member, predominate the microbiomes of animal intestines.

In addition to RIF-1, *Algoriphagus* also produces lipid activators, synergistic enhancers, and inhibitors that regulate rosette development (Woznica et al., 2017). Interestingly, in the life cycle of *S. rosetta* bacteria are not only needed as developmental cues to trigger colony formation but also for mating. *Vibrio fischeri* bacteria induce the full sexual cycle in *S. rosetta* by producing a protein, a chondroitin lyase named EroS (Extracellular Regulator of Sex). This protein induces fusion and meiosis in the choanoflagellates.

Similarly, microbial communities associated with the surfaces of plants are essential for their growth and morphogenesis (Wichard, 2015). The morphogenetic factor thallusin, for example, is an amino acid derivative produced by the bacterium *Zobellia uliginosa*, that is sufficient to induce thallus development in the marine macroalgae *M. oxyspermum* and partially promote thallus development in Ulva species. Studies of *Ulva* have also shown that morphogenesis may require multiple bacterial molecules. On their own, bacteria belonging to *Cytophaga* and *Roseobacter* genera induce incomplete *Ulva mutabilis* development, promoting either cell division or thallus differentiation, respectively. Only the combined activities of these bacteria fully restore normal morphogenesis, showing that synergistic interactions are required at the molecular level (Matsuo et al., 2005; Spoerner et al., 2012).

Taken together, bacteria appear as master regulators of the life history of some choanoflagellate species and induce morphogenesis in marine algae indicating that certain bacteria groups have a high level of influence on the development of marine eukaryotes. Apparently, from the very beginnings, our animal ancestors were able to interpret developmental cues from diverse microbes.

3. Evolution of development in animals

3.1 Evolution of transmission mode—Horizontal vs. vertical transmission

The initial stages of animal development rely on the production of male and female gametes that, upon fertilization, produce the single-cell zygote. In light of this reduction to a single cell, how do coevolved symbiotic systems ensure persistence of the host-microbe association between generations? While a spectrum of transmission modes occurs cross the animal kingdom, most holobionts maintain their integrity between generations by one of two mechanisms—horizontal (environmental) or vertical (transovarian) transmission, which present very different developmental challenges for maintenance of the holobiont (Bright & Bulgheresi, 2010; McFall-Ngai, 2002). In a horizontally transmitted symbiosis, the juvenile host recruits its specific symbionts from the surrounding habitat with each generation. As such, the microbial partners are not present to interact directly with host cells during embryogenesis (although, for indirect agency, see (Kimura et al., 2020).

However, evolution has selected for determinants developed during embryogenesis that allow the host to engage and recognize the coevolved, i.e., specific, symbionts, on hatching or birth. In these associations, the bacterial partners occur most often as extracellular consortia colonizing polarized epithelia, such as in the vertebrate alimentary canal (Sharpton, 2018; Song et al., 2020), the surface of *Hydra* (Schröder & Bosch, 2016) and the accessory nidamental gland of various squid species (Kerwin et al., 2019). However, some monospecific and/or intracellular alliances that exhibit environmental transmission do occur [e.g., the squid–vibrio mutualism (McFall-Ngai & Ruby, 1991) and the associations between vent tube worms and their sulfur-oxidizing bacterial symbionts (Nussbaumer, Fisher, & Bright, 2006). Typically following initial establishment of a horizontally transmitted symbiosis, characters are developed post-embryonically that promote symbiotic persistence. For example, in systems as divergent as the squid-vibrio association and the ruminant microbial consortia, which have conspicuous anatomical features associated with symbiont recruitment, the juvenile specific morphology that facilitates horizontal transmission undergoes dramatic morphogenesis following symbiosis onset (see Sections 3.3.4 and 6. More subtle changes can also be observed, for example in the maturation of the gut-associated lymphoid tissue mammals, which requires exposure to Gram-negative symbionts (Bouskra et al., 2008).

Resident bacteria also play important roles in the maturation of the zebrafish intestine, including promoting intestinal epithelial cell proliferation and recruiting innate immune cells in the gut (Bates et al., 2006; Cheesman et al., 2011; Stephens et al., 2016). These complex maturation and persistence processes mediate the mechanisms to maintain specificity, control of community diversity, prevention of symbiont loss or overgrowth, and finally a seeding of the environment with symbionts for the next generation.

In vertically transmitted symbioses, bacterial partners are carried in or on the eggs by the female parent (Douglas, 2014). These types of associations are widespread, particular among the insects and certain mollusk clades (Cary, 1994) e.g., at deep-sea vent and seep environments), and restricted to the invertebrates. Rather than being superficial, the bacterial partners in such symbioses are most often intracellular constituents of organs located deep in the body cavity. While these alliances are difficult to study experimentally, they are not only very common but they are also often ecologically and economically very important associations and have become the subjects of a rich literature. Because the symbionts are provided in or on the egg, the microbial partners are incorporated into the host's embryonic program. As such, evolutionary selection pressure has fostered mechanisms for targeting of the symbionts to specific tissues/organs during embryogenesis and restricting the growth of the symbionts to these areas. Similar to horizontally transmitted symbioses, features are developed during maturation that promote a stable and controlled symbiosis.

3.2 The generation of two or more germ layers and the origin of gastrulation as major milestones in animal evolution

Perhaps the most common types of interactions that eumetazoans have with microbes occur at the apical surfaces of the epithelia, either along the body surface or along the gut. Below we consider the evolutionary and developmental origin of these tissues and our current knowledge concerning the role of microbes as potential drivers of these processes.

3.2.1 The endoderm as the first germ layer to evolve

According to the germ layer hypothesis (Remak, 1850; von Baer, 1828), all tissues and organs in our body derive from one of three germ layers that are established during early embryogenesis. Ancestral metazoans such as the Cnidaria have only two cell layers, an ectoderm and an endoderm. Following Thomas Henry Huxley (1871), the two layers of adult Cnidaria are homologous to two of the germ layers in vertebrate embryos—ectoderm and endoderm, i.e., they are conserved across the eumetazoa. Recent studies of the temporal and spatial components of developmental gene expression in several species (Hashimshony et al., 2015; Yanai, 2018) show that the mesoderm of the Bilateria was the last germ layer to evolve because the gene expression program of the mesoderm was found to be induced after those of the ectoderm and endoderm. Interestingly, the endoderm expression programs (and the endodermally active genes, themselves) preceded the ectodermal genes in nematodes (e.g., *Caenorhabditis elegans*), frogs (e.g., *Xenopus tropicalis*), cnidarians (e.g., *Nematostella vectensis*), and sponge (e.g., *Amphimedon queenslandica*) embryogenesis. Therefore, it may well be that the endoderm program dates back to the origin of multicellularity, while the ectoderm originated as a secondary germ layer.

3.2.2 Epithelia in "pre-bilaterians" are in close contact with colonizing microbes

Work in metazoan animal models such as the freshwater polyp *Hydra* has shown that much of the behavior of the epithelial layers is impacted by the interactions with colonizing microbes. Both, ecto- and endodermal epithelial cells produce a rich repertoire of antimicrobial peptides which regulate the microbiome (Franzenburg et al., 2013). Conversely, as we will describe later (Chapter "Animal development in the microbial world: The power of experimental model systems" by McFall-Ngai and Bosch, this volume), the microbiota has a profound impact on both form and functioning of the epithelium. Observations on long-term germ free pre-bilaterian animals such as *Hydra* indicate that developmental pathways are influenced by the microbiota (He and Bosch, unpubl.). Since this includes transcription factors and signaling pathways that are highly conserved in the animal kingdom, it may well be that development in all animals is relying on an interaction with microbes in a specific manner.

3.2.3 Thoughts about microbial influences on the origin of gastrulation and the evolution of the gut

An important step in animal embryology is the laying down of a second and a third layer of cells so that the body has an outside of cells (ectoderm), an inside (endoderm), and a middle one (mesoderm). This process begins with gastrulation, which can occur either by invagination, by the ingression of cells from the outer blastula layer to fill the blastula cavity. In both cases, gastrulation leads to the formation of the archenteron, which will give rise to the endoderm and the mesoderm. Whatever drove the evolutionary pressure for gastrulation was certainly some abiotic or biotic environmental feature. In his gastraea theory, Ernst Haeckel (1874) imagined a gastrula that was like a planula larva on a biofilm, where an intrusion is created that produces a restricted, nutritive, and relatively stable environment. Such an environment is the ideal situation for the first interactions with the microbial world, particularly with the recognition that benthic biofilms are typically bacterial. Recent research has indicated that one the strongest and most widely shared features across the animal kingdom is the carriage of microbes the gut (Brody, 2020; Moeller & Sanders, 2020). It is likely that these interactions had a role in driving the form and function of the gut over evolutionary time and are also likely to do so within the life history of an individual animal.

The ruminants offer a striking example of the role of microbes in the evolutionary selection for a complex gut (Chiu & Gilbert, 2020). Herbivory comes late into evolutionary history. Vertebrates, for instance (and most herbivorous insects) have no gene in their genome encoding enzymes capable of digesting cellulose, hemicellulose, or pectins. This capacity is given these animals by their bacterial symbionts. Herbivores evolved as complex associations between animal and symbiotic microbes. Perhaps the most successful of these associations can be seen in ruminants, whose taxon is defined by this portion of their stomach (Hofmann, 1989). The rumen is a complex, dynamic ecosystem composed of mainly anaerobic bacteria, protozoa, anaerobic fungi, methanogenic archaea and phages (Jami et al., 2013); this fermenter produces volatile fatty acids (VFAs) that provide ~70% of the cow's daily energy requirement from their fiber-rich diets (Yeoman & White, 2014). However, the calf is born functionally monogastric, without the fully developed rumen, and the rumen remains non-functional and underdeveloped until weaning. Once it gets fed grain and grass, anatomical alterations allow the fiber-rich diet to enter the immature rumen (Jami et al., 2013). The bacteria in the rumen, which had migrated there during the first week of the calf's life, can now breakdown the grain into short-chained fatty acids, including butyrate. The butyrate is the developmental signal for the growth and differentiation of the rumen (Baldwin & Connor, 2017; Sander et al., 1959). The regulation of rumen development is thought to be achieved through changes in DNA methylation and microRNA production (Li et al., 2019; Malmuthuge, Liang, & Guan, 2019). By 3 months, the rumen comprises about 85% of the calf's stomach volume. Thus, through developmental symbiosis, the gut bacteria construct their own niche, the rumen, which will then perform the nutritive symbiosis that will enable the cow to survive. Here, developmental symbiosis, sympoiesis, has enabled herbivory–an entire new evolutionary trajectory.

The evolution of herbivory in both invertebrates and vertebrates depended on the interactions between gut microbes and the gut epithelia (Gilbert, 2020). Bacterial symbionts not only affect the development and evolution of the vertebrate gut system, but they may also be the reason that some animal hosts do not develop a gut system at all. More than 100 species of gutless worms have been found in marine sediments around the world, and they have formed symbiotic relationships with bacteria that provide them with nutrition and have enabled them to colonize nutrient-poor environments. The symbionts are so efficient at feeding their hosts and recycling their waste compounds that the worms have completely reduced their digestive and excretory systems. In one species, the well-studied vent tubeworm *Riftia pachyptila* (Felbeck, 1981; Nussbaumer et al., 2006) the juvenile animal starts with a gut. However, early in development the planktonic form of the bacterial symbiont bores into the body wall of the host and takes up residence and proliferates in the regions of the body that will become the trophosome. The associations between gutless marine worms and their bacterial symbiont communities are ideal model systems for the study of beneficial animal-microbe symbioses (Hinzke et al., 2019; Jäckle et al., 2019).

4. The influence of microbes on development through ontogeny

4.1 The role of bacteria in developmental programs of animals with complex life-history strategies

As mentioned above, the formation of biofilms is a common behavior among the prokaryotes, either in clones of an individual genotype or as mixed species assemblages. Animals have taken advantage of this microbial behavior under a variety of circumstances. Perhaps the most widespread and ecologically critical biofilms are those created on the benthic substrates of marine environments. Such biofilms are drivers of the settlement and metamorphosis of marine invertebrate larvae. Planktonic larvae of a wide variety of taxa in at least 10 phyla, including sponges, and sessile cnidarians, annelids, mollusks and chordates species, require cues from specific bacterial species in biofilms to settle and metamorphose in a particular spot on the substrate (Hadfield, 2011).

It has become increasingly clear that this phenomenon shapes the benthic ecology of the world's oceans. However, while invertebrate biologists had long known that larvae do not settle randomly, because of the historical gulf between the fields of macro- and microbiology, it wasn't until the 1970s that the first characterization of bacterial biofilms as inducers of larval settlement was published (for review see Hadfield & Paul, 2001).

The developmental trajectory of the serpulid polychaete *Hydroides elegans*, has been under intense study for more than 25 years and offers a striking example of how bacterial biofilms influence animal development (Freckelton, Nedved, & Hadfield, 2017). Early studies confirmed that the presence of bacteria within biofilms was essential for settlement and metamorphosis of *H. elegans*, and they demonstrated that bacteria known to be inductive often represent a very small fraction of the bacteria within a given biofilm. Further, within a bacteria species, there is strain specificity to the response. For example, the bacterial species *Pseudoalteromonas luteoviolacea* is a potent inducer of *H. elegans* settlement and metamorphosis, but not all strains of the species are inductive. Further, this triggering of development in *H. elegans*, as with other invertebrate larvae, can occur by exposure to several different bacteria in the biofilm. The bacteria *P. luteoviolacea* combines hundreds of phage-tail-like elements to make complex structures effectively induce the settlement and metamorphosis of *H. elegans*. In another inductive bacterium, *Cellulophaga lytica*, the outer membrane lipopolysaccharide of this species appears to be the active inducer of *H. elegans* development (Freckelton et al., 2020). Clearly an entire research horizon has been presented by these new and exciting findings in the developmental biology of marine invertebrates.

The above-described phenomenon of bacterial biofilm-induced development represents a situation where animal-bacterial interactions are essential, but they are not symbiotic systems where the organisms are living together (Freckelton & Nedved, 2020). Some symbiotic systems do, however, require bacterial biofilm formation during development. For example, in the onset of symbiosis in the bobtail squid, candidate symbiont microbes aggregate in biofilms above the sites of colonization. For this behavior, the cells of the symbiont, *Vibrio fischeri*, must export an exopolysaccharide that promotes cell-cell interaction during aggregation; mutants in this behavior do not colonize the host animal (Shibata et al., 2012). In the more complex mammalian systems, biofilms of the oral cavity and gut are recognized, but the determination of their positive and negative benefits in these situations is far from resolved. Further, the extent to which biofilm formation is an important element of development in terrestrial animals, i.e., along the mucosal surfaces, has not been widely studied as yet.

Microbial influences on the developmental programs of other animals with complex life history strategies are also being explored. Many marine invertebrates have multiple larval stages, but the microbiomes of only a few of these species have been examined. A recent laboratory study of the developmental stages of the sea urchin *Strongylocentrotus purpuratus* revealed that bacteria occur with the zygote, embryo, and early larval stages, but the largest concentration of bacteria is recruited after feeding begins and localizes to the gut lumen (Schuh et al., 2020). Future work will need to both confirm this in the field and to determine which subset of these microbes are stable symbionts vs. transient microbial "tourists", e.g., passing through with the food (Carrier & Reitzel, 2019; Carrier, Dupont, & Reitzel, 2019). Nonetheless, bacterial exposure in the early development of *S. purpuratus* is critical for the induction of normal developmental changes in morphology and in the maturation of the immune system. Specifically, the normal growth of the animal and its ability to resist pathogens are compromised in animals without the exposure to the dense populations of microbes typical to their seawater environment (Schuh et al., 2020). While the study of the impact of symbiosis on complex life history strategies is in its infancy in marine invertebrates, this topic has been extensive studied in insects for many decades (for review, see Hammer & Moran, 2019). As mentioned above, many insects have vertical transmission of symbionts. Whether the insect is hemi- or holometabolous can strongly impact the symbiosis itself in a variety of ways and can be impacted by the symbiosis, and new molecular tools are providing opportunities to define the mechanisms underlying these impacts. For example, a recent study of a weevil *Sitophilus oryzae* provides a striking example (Maire et al., 2020). These insects are holometabolous and the harbor symbiont *Sodalis pierantonius* in organs called bacteriomes. In this system, the bacteriome of the larvae, which contains the symbiont-containing bacteriocyte cells, dissociates. The bacteriocytes change gene expression and migrate through the tissues to the site where the symbionts will reside in the adult animals. Simultaneously, the symbionts change gene expression to allow them to invade host stems cells, move into the nuclei of these cells and induce bacteriocyte development to form the adult bacteriome. The host and symbionts have complex cellular and molecular dialogs that ensures the provision of the symbionts to the next life stage and the development of the productive association in a precise location.

4.2 Host-microbe interactions drive conserved developmental programs

4.2.1 Symbionts educate the immune system

Over the past few years, the field of immunology has been revolutionized by the growing understanding of the fundamental role of the microbiota in the induction, education, and function of the mammalian immune system (Belkaid & Hand, 2014). Studies in mice have covered the critical role of colonization by maternal microbes in early life for normal immune development, including innate immune cell production, antibody production, immune regulatory functions, epithelial barrier integrity, and even tryptophan and nucleotide metabolism in offspring. As the early-life microbiota is seeded by passage through the birth canal, the education of the immunity in offspring is critically dependent on the maternal microbiota. In addition, the maternal microbiome can also have an impact on disease development later in life. It seems that an altered microbiota can increase intestinal permeability and thus promote the development of a proinflammatory milieu that stimulates beta-cell autoimmunity in the predisposed host. For example, differences in human milk oligosaccharide-utilizing bacteria provide a route to differences in immune education that may predispose the body to immune and metabolic related conditions such as diabetes (Garidou et al., 2015); an altered gut microbiota during the weaning period (due to antibiotics, high fat diet etc.) may hinder adequate development of host immune responses, which leads to increased susceptibility to immune-related diseases such as allergy in later life. Interestingly, microbial exposure after weaning does not alter the development of immune responses, highlighting the critical window of weaning for preventing immunopathologies in adult life.

Obviously we are witnessing a paradigm shift in immunology: symbiotic microbes are essential for the development of the immune system; and conversely, immune systems control the symbiotic microbiota in addition to defending against invasive pathogens. Immunity is not just a host function. It is a holobiont function (Chiu et al., 2016; Gilbert & Tauber, 2016; Pradeau, 2012; Tauber, 2017). This is true for the adaptive immune system where there is increasing support for the view (McFall-Ngai, 2007) that a memory-based immune system may have evolved because of the need to recognize and manage complex communities of beneficial microbes. And this seems also true for the innate immune system with its host-specific antimicrobial peptides and rich repertoire of pattern recognition receptors, which control the resident beneficial microbes rather than invasive pathogens (Bosch, 2014).

4.2.2 Symbionts and neurogenesis

Similar to the immune system, the development and evolution of the nervous system must be considered in the context of host-microbe interactions. Studies of germ-free animals or animals treated with broad-spectrum antibiotics show that specific microbiota can impact central nervous system (CNS) physiology and neurochemistry (Sharon et al., 2016). GF mice that are devoid of associated microflora exhibit neurological deficiencies in learning, memory, recognition, and emotional behaviors (Foster, Rinaman, & Cryan, 2017; Gareau, 2014). Proliferation of neurons in the dorsal hippocampus is greater in GF mice than in conventional mice. However, post-weaning exposure of GF mice to microbial clones did not influence neurogenesis, suggesting that neuronal growth is stimulated by microbiota at an early stage (Ogbonnaya et al., 2015). The connection between microbiota and hippocampal neuronal generation is further strengthened by the findings that deficient neurogenesis can be counteracted by a probiotic combination of specific bacterial strains (Ait-Belgnaouiet al., 2014; Möhle et al., 2016).

While it was known for some time that NF-κB signaling participates in microbiota-neuron axis, recent studies indicate that microbiota disturbance leads to increased NF-κB activation and TNF-α expression with induced memory impairment in animal models, and the restoration of microbiota composition alleviates neuroinflammation in hippocampus and ameliorates relevant symptoms (Jang et al., 2018; Ma et al., 2019).

Similar observations on the role of microbiota in training the nervous system were made in the enteric nervous system (ENS). Germ-free mice show an abnormal proportion of calbindin and nitrergic myenteric neurons, suggesting that the microbiota influences the maturation of intestinal neural networks (De Palma et al., 2014; McVey Neufeld et al., 2015; McVey Neufeld et al., 2016). The functional activity of the ENS can be regulated by the microbiota either directly through Toll-like receptors (Brun et al., 2013) or by inducing the secretion of BMP2 by macrophages (Muller et al., 2014). Recent studies show that the immature enteric nervous system of germ-free mice can be normalized upon colonization with a normal microbiota (De Vadder et al., 2018); and that the gut microbiota can induce maturation of the adult ENS through release of 5-HT and activation of 5-HT4 receptors (De Vadder et al., 2018). From an evolutionary point of view, the discovery of an interaction of microbes with the mammalian CNS and also ENS comes as no surprise.

Functional transcriptomic data in *Hydra* indicate very clearly that the nervous system evolved to sense and communicate via "antimicrobial peptides" (AMPs) with the microbes (Klimovich & Bosch, 2018). Indeed, prior to the evolution of the mesoderm, neurons seem to have evolved as immune cells (Klimovich et al., 2020) to control the symbiotic microbiota. And conversely, it looks like since the beginning of animal evolution that microbes directly affect the functional activity of the nervous system in *Hydra* by controlling spontaneous body contractions. The hypothesis that gut bacteria are integral contributors to development and function of the nervous system (Sharon et al., 2016) and that a "gut microbiota-brain axis" exists is getting support from both experimental and clinical sources (Cryan et al., 2019; Desbonnet et al., 2014; Heijtz et al., 2011; Hsiao, 2013; Sampson & Mazmanian, 2015; Sharon et al., 2016).

4.2.3 Gut microbiota affects juvenile development, growth, and maturation

The first evidence that the microbiota directly contributed to developmental processes came from observations revealing that bacteria provide developmental signals to the intestinal epithelia. Bacteria-induced expression of mammalian genes was first demonstrated by Umesaki (1984); see also Bry et al. (1996) who noticed that a particular fucosyl transferase enzyme characteristic of mouse intestinal villi was induced by bacteria. Other studies have shown that the intestines of germ-free mice can initiate, but not complete, their differentiation. Germ-free mice have smaller intestines, with a paucity of enteroendocrine and goblet cells (Bates et al., 2006). Microbial colonization is accompanied by profound changes in the intestinal transcriptional programs (Rakoff-Nahoum et al., 2015). Moreover, microbiota also play a key role in constructing the microvascular network (Stappenbeck, Hooper, & Gordon, 2002). Here, it appears that bacteria such as *Bacteroides thetaiotamicron* are inducing the intestinal Paneth cells to activate the genes encoding Angiogenin-4. When Angiogenin-4 is secreted by the cells, it instructs the mesoderm surrounding the gut to form capillaries (Hooper et al., 2001; Rhee et al., 2004).

Mammals aren't the only animals whose development of gut and immune system depend on microbial symbionts. By tagging bacteria with a fluorescent molecule to observe how and when they take up residence in the developing fish, University of Oregon scientists have discovered that bacteria colonize the immature gut soon after the larvae hatches, and before the animal is fully mature (Bates et al., 2006). In the absence of microbiota, the zebrafish gut is arrested in specific aspects of differentiation, and the animals have trouble absorbing nutrients. All these defects can be reversed by the introduction of bacteria later in development. In zebrafish, microbial symbionts use the beta-catenin signaling pathway to initiate cell division in the intestinal stem cells (Rawls, Samuel, & Gordon, 2004).

Extensive evidence has also indicated that symbiotic microbes are critical to providing numerous functions in developing insects. As illustrated clearly by comparing germ-free and conventionalized *Drosophila* larvae cultured on protein-poor food, gut microbiota can contribute to the host juvenile development and maturation by positively impact insulin/IGF-1 and steroid hormone signaling (Shin et al., 2011; Storelli et al., 2011). While conventional larvae grow more slowly than on protein-rich food because of a decrease in TOR signaling and the consequent disarray in the insulin and steroid hormonal signaling (Boulan, Milan, & Leopold, 2015), germ-free larvae are much more delayed. *Wolbachia* bacteria accomplish numerous functions–from immunity to sex determination—in several species of insects (see Gilbert & Epel, 2015), and in the nematode *Brugia malayi*, *Wolbachia* bacteria are responsible for the correct anterior-posterior polarity of the second mitotic division (Landmann, 2014).

Microbes, therefore, must be considered as an integral part of development. In comparing the zebrafish gut microbe interactions to those of mice, some responses appear conserved across millions of years of evolution (Rawls et al., 2006), indicating that animals may possess a conserved program of interactions with the microbes with which they have coevolved.

5. Conclusions, opportunities and future perspectives: A paradigm shift and a new way of exploring development

Developmental biology has advanced remarkably and substantially enough for scientists to be confident for the first time that some aspects of development in some organisms are understood at the molecular level. Protein components are identified, their functions in developmental processes are known, and the time and place in the embryo of expression of the genes encoding them are known. Developmental biology however needs a 21st century upgrade that incorporates the impacts of the microbial world. Within this arena, many expansive vistas of research are presented that promise to drive developmental biology to the next levels of enquiry. This sampling of the open questions, some expansive and some more focused, includes:

- How do the interactions with the microbial world influence the developmental biology of an animal, from its molecular form and function to its ecological position within the biosphere?
- How does animal development ensure the deployment of cellular and molecular mechanisms for the integration of microbial influences across the holobiont?
- How does embryogenesis and postembryonic development promote microbiome-mediated communication across the body?
- How is symbiosis reflected in the host and symbiont genes that are expressed throughout the trajectory of development; does some portion of these genes designated as 'hypothetical' encode the genes that drive formation of the holobiont? What portion is borrowed, or the result of evolutionary tinkering, and what portion is unique to that type of symbiotic organ, i.e., among the orphan genes of that organism?
- How will developmental biology of animals with their microbiomes be affected by climate change? How do environmental perturbations (e.g., changes in temperature, environmental pH) affect the normal development of the holobiont?
- Will disruption of symbiosis drive extinctions of the host and/or microbial partners?
- What role do animal-microbe interactions in development (e.g., metamorphosis by particular biofilm communities) play to drive invasions? (Haeckel, 1874)
- How has the development of a symbiosis been modified during radiation into new habitats or niches (e.g., in the water-to-land transition)?
- To what extent does phenotype variation across species depends on species-specific microbiota? Since it is known that at least some animal species acquire and maintain their species-specific microbiota, do in turn microbiota contribute to phenotypes that are unique to each species?

Some of these questions are already under study in laboratories around the globe, but a more focused effort is required. The evidence is now irrefutable—the developmental biology of animals is in the Anthropocene, and has been over evolutionary time, shaped by interactions with the microbial world. As such, to be effective and full contributors to the field of biology, animal developmental biologists must switch from an isolated view of a discipline to one that includes this dimension.

In conclusion, this chapter sought to explore the current state of developmental biology in the microbiome era. The stunning range of opportunities now available for an integrative evolutionary developmental biology so that we have an excellent chance to move toward a future comprehensive understanding of animal development.

Acknowledgments

We thank Scott Gilbert for both the invitation to write this chapter and for his careful and critical reading of the manuscript. We also are thankful to Tyler Carrier for his comments on the text. This work was supported in part by grants from the Deutsche Forschungsgemeinschaft and the CRC 1182 ("Origin and Function of Metaorganisms") (to T.C.G.B.). T.C.G.B. appreciates support from the Canadian Institute for Advanced Research. McFall-Ngai is currently funded by National Institutes of Health (NIH) R37 AI50661 (to M.M.-N. and E. G. Ruby); NIH, Center for Biomedical Research Excellence P20 GM125508 (to M.M.-N. and E.G. Ruby PIs); NIH R01 OD11024 and GM135254 (to E. G. Ruby and M.M.-N.); and a grant from the Gordon and Betty Moore Foundation (to A. Briegel, U Leiden, M.M.-N., E.G. Ruby and A. Septer, U North Carolina Chapel Hill).

References

Ait-Belgnaoui, A., Colom, A., Braniste, V., et al. (2014). Probiotic gut effect prevents the chronic psychological stress-induced brain activity abnormality in mice. *Neurogastroenterology and Motility*, *26*, 510–520. https://doi.org/10.1111/nmo.12295.

Alegado, R. A., Brown, L. W., Cao, S., et al. (2012). A bacterial sulfonolipid triggers multicellular development in the closest living relatives of animals. *eLife*, *1*, e00013. https://doi.org/10.7554/eLife.00013.

Baldwin, R. L., & Connor, E. (2017). Rumen function and development. *Veterinary Clinics of North America Food Anima*, *33*(3).

Bates, J. M., Mittge, E., Kuhlman, J., Baden, K. N., et al. (2006). Distinct signals from the microbiota promote different aspects of zebrafish gut differentiation. *Developmental Biology*, *297*, 374–386. https://doi.org/10.1016/j.ydbio.2006.05.006.

Belkaid, Y., & Hand, T. W. (2014). Role of the microbiota in immunity and inflammation. *Cell*, *157*, 121–141. https://doi.org/10.1016/j.cell.2014.03.011.

Bonner, J. T. (1965). *Size and cycle. An Essay on the structure of Biology*. Princeton UP.

Bonner, J. T. (1974). *On development. The biology of form*. Commonwealth Fund Publications. ISBN 9780674634121.

Bonner, J. T. (1998). The origins of multicellularity. *Integrative Biology*, *1*, 27–36. https://doi.org/10.1002/(SICI)1520-6602(1998)1:1<27::AID-INBI4>3.0.CO;2-6.

Bosch, T., Guillemin, K., & McFall-Ngai, M. (2019). Evolutionary "experiments" in symbiosis: The study of model animals provides insights into the mechanisms underlying diversity of host-microbe interactions. *BioEssays, e1800256*.

Boulan, L., Milan, M., & Leopold, P. (2015). The systemic control of growth. *Cold Spring Harbor Perspectives in Biology*, 7, a019117. https://doi.org/10.1101/cshperspect.a019117.

Bouskra, D, et al. (2008). Lymphoid tissue genesis induced by commensals through NOD1 regulates intestinal homeostasis. *Nature*, *456*(7221).

Bretl, D. J., & Kirby, J. R. (2016). Molecular mechanisms of signaling in *Myxococcus xanthus* development. *Journal of Molecular Biology*, *428*, 3805–3830. https://doi.org/10.1016/j.jmb.2016.07.008.

Bright, M., & Bulgheresi, S. (2010). A complex journey: Transmission of microbial symbionts. *Nature Reviews Microbiology*, *8*, 218–230. https://doi.org/10.1038/nrmicro2262.
Brody, H. (2020). The gut microbiome. *Nature*, *577*, S5. https://doi.org/10.1038/d41586-020-00194-2.
Brun, P., Giron, M. C., Qesari, M., et al. (2013). Toll-like receptor 2 regulates intestinal inflammation by controlling integrity of the enteric nervous system. *Gastroenterology*, *145*, 1323–1333. https://doi.org/10.1053/j.gastro.2013.08.047.
Bry, L., Falk, P. G., Midtvedt, T., et al. (1996). A model of host-microbial interactions in an open mammalian ecosystem. *Science*, *273*, 1380–1383. https://doi.org/10.1126/science.273.5280.1380.
Buss, L. W. (1987). *The evolution of individuality*. Princeton University Press.
Carrier, TJ, & Reitzel, AM. (2019). Bacterial community dynamics during embryonic and larval development of three confamilial echinoids. *Marine Ecology Progress Series*, *611*.
Carrier, T. J., Dupont, S., & Reitzel, A. M. (2019). Geographic location and food availability offer differing levels of influence on the bacterial communities associated with larval sea urchins. *Federation of European Microbiological Societies Microbiology Ecology*, *95*. https://doi.org/10.1093/femsec/fiz103, fiz103.
Cary, S. (1994). Vertical transmission of a chemoautotrophic symbiont in the protobranch bivalve. *Solemya reidi*. *Molecular Marine Biology and Biotechnology*, *3*, 121–130.
Cheatle Jarvela, A. M., & Pick, L. (2016). Evo-Devo: Discovery of diverse mechanisms regulating development. *Current Topics in Developmental Biology*, *117*, 253–274. https://doi.org/10.1016/bs.ctdb.2015.11.014.
Cheesman, S. E., Neal, J. T., Mittge, E., et al. (2011). Epithelial cell proliferation in the developing zebrafish intestine is regulated by the Wnt pathway and microbial signaling via Myd88. *Proceedings of the National Academy of Sciences of the United States of America*, *108*(Suppl. 1), 4570–4577. https://doi.org/10.1073/pnas.1000072107.
Chiu, L., et al. (2016). Microorganisms as scaffolds of host individuality: An eco-immunity account of the holobiont. *Biology & Philosophy*, *31*(6).
Chiu, L, & Gilbert, SF. (2020). Niche construction and the transition to herbivory: Phenotype switching and the organization of new nutritional modes. *Phenotypic Switching*, 459–482.
Cryan, J. F., et al. (2019). The microbiota-gut-brain axi. *Physiological Reviews*, *99*(4).
De Palma, G., Collins, S. M., Bercik, P., et al. (2014). The microbiota-gut-brain axis in gastrointestinal disorders: Stressed bugs, stressed brain or both. *Journal of Physiology*, *592*, 2989–2997. https://doi.org/10.1113/jphysiol.2014.273995.
De Vadder, F., Grasset, E., Mannerås Holm, L., et al. (2018). Gut microbiota regulates maturation of the adult enteric nervous system via enteric serotonin networks. *Proceedings of the National Academy of Sciences of the United States of America*, *115*, 6458–6463. https://doi.org/10.1073/pnas.1720017115.
Desbonnet, L., et al. (2014). Microbiota is essential for social development in the mouse. *Molecular Psychiatry*, *19*(2).
Domazet-Loso, T., & Tautz, D. (2010). Phylostratigraphic tracking of cancer genes suggests a link to the emergence of multicellularity in metazoa. *BMC biology*, *8*, 66. https://doi.org/10.1186/1741-7007-8-66.
Douglas, A. E. (2014). *The symbiotic habit*. Princeton University Press.
Felbeck, H. (1981). Chemoautotrophic potential of the hydrothermal vent tube worm, *Riftia pachyptila* Jones (Vestimentifera). *Science*, *213*, 336–338. https://doi.org/10.1126/science.213.4505.336.
Foster, J. A., Rinaman, L., & Cryan, J. F. (2017). Stress & the gut-brain axis: Regulation by the microbiome. *Neurobiology of Stress*, 7, 124–136. https://doi.org/10.1016/j.ynstr.2017.03.001.
Franzenburg, S, et al. (2013). Distinct antimicrobial peptide expression determines host species-specific bacterial associations. *PNAS*, *110*(39).

Freckelton, M. L., & Nedved, B. T. (2020). *Bosch & Hadfield*. CRC Press.
Freckelton, M. L., Nedved, B. T., Cai, Y.-S., et al. (2020). Bacterial lipopolysaccharide induces settlement and metamorphosis in a marine larva. *bioRxiv*. https://doi.org/10.1101/851519.
Freckelton, M. L., Nedved, B. T., & Hadfield, M. G. (2017). Induction of invertebrate larval settlement: Different bacteria, different mechanisms? *Scientific Reports*, 7, 42557. https://doi.org/10.1038/srep42557.
Futo, M., Opašić, L., Koska, S., et al. (2020). Embryo-like features in developing Bacillus subtilis biofilms. *bioRxiv*, 2020.03.09.983718. https://doi.org/10.1101/2020.03.09.983718.
Gareau, M. G. (2014). Microbiota-gut-brain axis and cognitive function. *Advances in Experimental Medicine and Biology*, *817*, 357–371. https://doi.org/10.1007/978-1-4939-0897-4_16.
Garidou, L., Pomié, C., Klopp, P., et al. (2015). The gut microbiota regulates intestinal CD4 T cells expressing RORγt and controls metabolic disease. *Cell Metabolism*, *22*, 100–112. https://doi.org/10.1016/j.cmet.2015.06.001.
Gilbert, S. F., et al. (2012). A symbiotic view of life: We have never been individuals. *The Quarterly Review of Biology*, *87*(4).
Gilbert, S.F. (2019). Evolutionary transitions revisited: Holobiont evo-devo. *J Exp Zool B Mol Dev Evol*. *332*(8):307–314. doi: https://doi.org/10.1002/jez.b.22903.
Gilbert, SF. (2020). Developmental symbiosis facilitates the multiple origins of herbivory. *Evolution and Development*, *22*(1–2).
Gilbert, S. F., & Tauber, A. I. (2016). Rethinking individuality: The dialectics of the holobiont. *Biology And Philosophy*, *31*(6).
Gilbert, S. F. (2001). Ecological developmental biology: Developmental biology meets the real world. *Developmental Biology*, *233*, 1–12. https://doi.org/10.1006/dbio.2001.0210.
Gilbert, S. F., Bosch, T. C. G., & Ledon-Rettig, C. (2015). Eco-Evo-Devo: Developmental symbiosis and developmental plasticity as evolutionary agents. *Nature Reviews Genetics*, *16*, 611–622. https://doi.org/10.1038/nrg3982.
Gilbert, S. F., & Epel, D. (2015). *Ecological developmental biology: The environmental regulation of development, health, and evolution*. Sunderland: Sinauer Associates.
Grosberg, R. K., & Strathmann, R. R. (2007). The evolution of multicellularity: A minor major transition? *Annual Review of Ecology, Evolution, and Systematics*, *38*, 621–654. https://doi.org/10.1016/S0169-5347(97)01313-X.
Haagensen, J. A., Hansen, S. K., Christensen, B. B., et al. (2015). Development of spatial distribution patterns by biofilm cells. *Applied and Environmental Microbiology*, *81*, 6120–6128. https://doi.org/10.1128/AEM.01614-15.
Hadfield, M. G. (2011). Biofilms and marine invertebrate larvae: What bacteria produce that larvae use to choose settlement sites. *Annual Review of Marine Science*, *3*, 453–470. https://doi.org/10.1146/annurev-marine-120709-142753.
Hadfield, M. G., & Paul, V. J. (2001). Natural chemical cues for settlement and metamorphosis of marine-invertebrate chapter 13. In J. B. McClintock, & B. J. Baker (Eds.), *Larvae marine chemical ecology* (pp. 432–450). CRC Press.
Haeckel, E. (1874). Die Gastraea-Theorie, die phylogenetische Classification des Thierreichs und die Homologie der Keimblätter. *Zeitschrift für Naturwissenschaften*. 81–55.
Hammer, T. J., & Moran, N. A. (2019). Links between metamorphosis and symbiosis in holometabolous insects. *Philosophical Transactions of the Royal Society of London B, Biological Science*, *374*, 20190068. https://doi.org/10.1098/rstb.2019.0068.
Hashimshony, T., Feder, M., Levin, M., et al. (2015). Spatiotemporal transcriptomics reveals the evolutionary history of the endoderm germ layer. *Nature*, *519*, 219–222. https://doi.org/10.1038/nature13996.
Heijtz, RD, et al. (2011). Normal gut microbiota modulates brain development and behavior. *PNAS*, *108*.

Hinzke, T., Kleiner, M., Breusing, C., et al. (2019). Host-microbe interactions in the chemosynthetic *Riftia pachyptila* symbiosis. *mBio*, *10*, e02243-19. https://doi.org/10.1128/mBio.02243-19.

Hofmann, R. R. (1989). Evolutionary steps of ecophysiological adaptation and diversification of ruminants: a comparative view of their digestive system. *Oecologia*, *78*, 443M57. https://doi.org/10.1007/BF00378733.

Hooper, LV, et al. (2001). Molecular analysis of commensal host-microbial relationships in the intestine. *Science*, *291*(5505).

Hsiao, E. Y. (2013). Microbiota modulate behavioral and physiological abnormalities associated with neurodevelopmental disorders. *Cell*, *155*(7).

Huxley, T. H. (1871). *A manual of the anatomy of Vertebrated animals*. London: Churchill.

Jäckle, O., Seah, B. K. B., Tietjen, M., et al. (2019). Chemosynthetic symbiont with a drastically reduced genome serves as primary energy storage in the marine flatworm *Paracatenula*. *Proceedings of the National Academy of Sciences of the United States of America*, *116*, 8505–8514. https://doi.org/10.1073/pnas.1818995116.

Jami, E., Israel, A., Kotser, A., et al. (2013). Exploring the bovine rumen bacterial community from birth to adulthood. *International Society for Microbial Ecology Journal*, *7*, 1069–1079. https://doi.org/10.1038/ismej.2013.2.

Jang, H. M., Lee, H. J., Jang, S. E., et al. (2018). Evidence for interplay among antibacterial-induced gut microbiota disturbance, neuro-inflammation, and anxiety in mice. *Mucosal Immunology*, *11*, 1386–1397. https://doi.org/10.1038/s41385-018-0042-3.

Kerwin, A. H., et al. (2019). Shielding the next generation: Symbiotic bacteria from a reproductive organ protect bobtail squid eggs from fungal fouling. *mBio*, *10*(e02376-19).

Kimura, I, et al. (2020). Maternal gut microbiota in pregnancy influences offspring metabolic phenotype in mice. *Science*, *367*(6481).

King, N., Westbrook, M. J., Young, S. L., et al. (2008). The genome of the choanoflagellate *Monosiga brevicollis* and the origin of metazoans. *Nature*, *451*, 783–788. https://doi.org/10.1038/nature06617.

Klimovich, AV, et al. (2020). Prototypical pacemaker neurons interact with the resident microbiota. *PNAS*, *117*(30).

Klimovich, A. V., & Bosch, T. C. G. (2018). Rethinking the role of the nervous system: Lessons from the hydra holobiont. *BioEssays*, *1800060*.

Landmann, F. (2014). Co-evolution between an endosymbiont and its nematode host: Wolbachia asymmetric posterior localization and AP polarity establishment. *PLOS Neglected Tropical Diseases*, *8*(8).

Lasker, K., von Diezmann, L., Zhou, X., et al. (2020). Selective sequestration of signalling proteins in a membraneless organelle reinforces the spatial regulation of asymmetry in *Caulobacter crescentus*. *Nature Microbiology*, *5*, 418–429. https://doi.org/10.1038/s41564-019-0647-7.

Lee, Y. K., & Mazmanian, S. K. (2010). Has the microbiota played a critical role in the evolution of the adaptive immune system? *Science*, *330*(6012).

Li, F, et al. (2019). Host genetics influence the rumen microbiota and heritable rumen microbial features associate with feed efficiency in cattle. *Microbiome*, 7(92).

Lyons, N. A., & Kolter, R. (2015). On the evolution of bacterial multicellularity. *Current Opinion in Microbiology*, *24*, 21–28. https://doi.org/10.1016/j.mib.2014.12.007.

Ma, Q., Xing, C., Long, W., et al. (2019). Impact of microbiota on central nervous system and neurological diseases: The gut-brain axis. *Journal of Neuroinflammation*, *16*, 53. https://doi.org/10.1186/s12974-019-1434-3.

Maire, J., Parisot, N., Galvao Ferrarini, M., et al. (2020). Spatial and morphological reorganization of endosymbiosis during metamorphosis accommodates adult metabolic requirements in a weevil. *Proceedings of the National Academy of Sciences of the United States of America*, *117*, 19347–19358. https://doi.org/10.1073/pnas.2007151117.

Malmuthuge, N., Liang, G., & Guan, L. L. (2019). Regulation of rumen development in neonatal ruminants through microbial metagenomes and host transcriptomes. *Genome Biology*, *20*, 172. https://doi.org/10.1186/s13059-019-1786-0.

Mark Welch, J. L., Rossetti, B. J., Rieken, C. W., et al. (2016). Biogeography of a human oral microbiome at the micron scale. *Proceedings of the National Academy of Sciences of the United States of America*, *113*, E791–E800. https://doi.org/10.1073/pnas.1522149113.

Matsuo, Y., Imagawa, H., Nishizawa, M., et al. (2005). Isolation of an algal morphogenesis inducer from a marine bacterium. *Science*, *307*, 1598. https://doi.org/10.1126/science.1105486.

McFall-Ngai, M. J. (2002). Unseen forces: The influence of bacteria on animal development. *Developmental Biology*, *242*, 1–14. https://doi.org/10.1006/dbio.2001.0522.

McFall-Ngai, M. (2007). Adaptive immunity: Care for the community. *Nature*, *44*, 153. https://doi.org/10.1038/445153a.

McFall-Ngai, M., Hadfield, M., Bosch, T. C. G., et al. (2013). Animals in a bacterial world, a new imperative for the life sciences. *Proceedings of the National Academy of Sciences of the United States of America*, *110*, 3229–3236. https://doi.org/10.1073/pnas.1218525110.

McFall-Ngai, MJ, & Ruby, EG. (1991). Symbiont recognition and subsequent morphogenesis as early events in an animal-bacterial mutualism. *Science*, *254*(5037).

McVey Neufeld, K. A., Luczynski, P., Dinan, T. G., et al. (2016). Reframing the teenage wasteland: Adolescent microbiota-gut-brain axis. *Canadian Journal of Psychiatry*, *61*, 214–221. https://doi.org/10.1177/0706743716635536.

McVey Neufeld, K. A., Perez-Burgos, A., Mao, Y. K., et al. (2015). The gut microbiome restores intrinsic and extrinsic nerve function in germ-free mice accompanied by changes in calbindin. *Journal of Neurogastroenterology & Motility*, *27*, 627–636. https://doi.org/10.1111/nmo.12534.

Moeller, A. H., & Sanders, J. G. (2020). Roles of the gut microbiota in the adaptive evolution of mammalian species. *Philosophical Transactions Royal Society London B Biological Science*, *375*, 20190597. https://doi.org/10.1098/rstb.2019.0597.

Möhle, L., Mattei, D., Heimesaat, M. M., et al. (2016). Ly6C(hi) monocytes provide a link between antibiotic-induced changes in gut microbiota and adult hippocampal neurogenesis. *Cell Reports*, *15*, 1945–1956. https://doi.org/10.1016/j.celrep.2016.04.074.

Morgan, H. D., Santos, F., Green, K., et al. (2005). Epigenetic reprogramming in mammals. *Human Molecular Genetics*, *14*, R47–R58. https://doi.org/10.1093/hmg/ddi114. Spec No 1.

Morris, J. J. (2018). What is the hologenome concept of evolution? *F1000Research*, *7*, 1664. https://doi.org/10.12688/f1000research.14385.1.

Muller, P. A., Koscsó, B., Rajani, G. M., et al. (2014). Crosstalk between muscularis macrophages and enteric neurons regulates gastrointestinal motility (published correction appears in *Cell*, 158:1210. Dosage error in article text). *Cell*, *158*, 300–313. https://doi.org/10.1016/j.cell.2014.04.050.

Nussbaumer, A. D., Fisher, C. R., & Bright, M. (2006). Horizontal endosymbiont transmission in hydrothermal vent tubeworms. *Nature*, *441*, 345–348. https://doi.org/10.1038/nature04793.

Ogbonnaya, E. S., Clarke, G., Shanahan, F., et al. (2015). Adult hippocampal neurogenesis is regulated by the microbiome. *Biological Psychiatry*, *78*, e7–e9. https://doi.org/10.1016/j.biopsych.2014.12.023.

Overmann, J. (2010). The phototrophic consortium "*Chlorochromatium aggregatum*"—A model for bacterial heterologous multicellularity. *Advances in Experimental Medicine & Biology*, *675*, 15–29. https://doi.org/10.1007/978-1-4419-1528-3_2.

Pradeau, T. (2012). *The limits of the selfImmunology and biological identity*. Oxford University Press.

Rakoff-Nahoum, S., Kong, Y., Kleinstein, S. H., et al. (2015). Analysis of gene-environment interactions in postnatal development of the mammalian intestine. *Proceedings of the National Academy of Sciences of the United States of America*, *112*, 1929–1936. https://doi.org/10.1073/pnas.1424886112.

Rawls, J. F., Mahowald, M. A., Ley, R. E., et al. (2006). Reciprocal gut microbiota transplants from zebrafish and mice to germ-free recipients reveal host habitat selection. *Cell, 127*, 423–433. https://doi.org/10.1016/j.cell.2006.08.043.

Rawls, J. F., Samuel, B. S., & Gordon, J. I. (2004). Gnotobiotic zebrafish reveal evolutionarily conserved responses to the gut microbiota. *Proceedings of the National Academy of Sciences of the United States of America, 101*, 4596–4601. https://doi.org/10.1073/pnas.0400706101.

Remak, R. (1850). *Untersuchungen über die Entwickelung der Wirbelthiere*. Berlin: Reimer G.

Rhee, K-J, et al. (2004). Roles of commensal bacteria in development of gut-associated lymphoid tissues and preimmune antibody repertoire. *Journal of Immunology, 172*.

Rokas, A. (2008). The origins of multicellularity and the early history of the genetic toolkit for animal development. *Annual Review of Genetics, 42*, 235–251. https://doi.org/10.1146/annurev.genet.42.110807.091513.

Rosenberg, E., & Zilber-Rosenberg, I. (2013). *The hologenome concept: Human, animal and plant microbiota*. Berlin: Springer.

Roughgarden, J., et al. (2017). Holobionts as units of selection and a model of their population dynamics and evolution. *Biological Theory, 13*(5).

Ruiz-Trillo, I., Roger, A. J., Burger, G., et al. (2008). A phylogenomic investigation into the origin of metazoa (2008). *Molecular Biology and Evolution, 25*, 664–672. https://doi.org/10.1093/molbev/msn006.

Sampson, T. R., & Mazmanian, S. K. (2015). Control of brain development, function, and behavior by the microbiome. *Cell Host & Microbe, 17*(5).

Sander, E. G., et al. (1959). The stimulatory effect of sodium butyrate and sodium propionate on the development of rumen mucosa in the young calf. *The Journal of Dairy Science, 42*, 1600.

Schröder, K., & Bosch, T. (2016). The origin of mucosal immunity: Lessons from the holobiont hydra. *mBio*, 7(6).

Schuh, N. W., Carrier, T. J., Schrankel, C. S., et al. (2020). Bacterial exposure mediates developmental plasticity and resistance to lethal *Vibrio lentus* infection in purple sea urchin (*Strongylocentrotus purpuratus*) larvae. *Frontiers in Immunology, 10*, 3014. https://doi.org/10.3389/fimmu.2019.03014.

Shalchian-Tabrizi, K., Minge, M. A., Espelund, M., et al. (2008). Multigene phylogeny of Choanozoa and the origin of animals. *PLoS One, 3*, e2098. https://doi.org/10.1371/journal.pone.0002098.

Sharon, G., Sampson, T. R., Geschwind, D. H., et al. (2016). The central nervous system and the gut microbiome. *Cell, 167*, 915–932. https://doi.org/10.1016/j.cell.2016.10.027.

Sharpton, T. J. (2018). Role of the gut microbiome in vertebrate evolution. *mSystems, 3*. https://doi.org/10.1128/msystems.00174-17, e00174-17.

Shibata, S., Yip, E. S., Quirke, K. P., et al. (2012). Roles of the structural symbiosis polysaccharide (*syp*) genes in host colonization, biofilm formation, and polysaccharide biosynthesis in *Vibrio fischeri*. *Journal of Bacteriology, 194*, 6736–6747. https://doi.org/10.1128/JB.00707-12.

Shin, S. C., Kim, S.-H., You, H., et al. (2011). Drosophila microbiome modulates host developmental and metabolic homeostasis via insulin signaling. *Science, 334*, 670–674. https://doi.org/10.1126/science.1212782.

Skúlason, S., et al. (2019). A way forward with eco evo devo: An extendedtheory of resource polymorphism withpostglacial fishes as model systems. *Biological Reviews, 94*, 1786.

Song, S. J., Sanders, J. G., Delsuc, F., et al. (2020). Comparative analyses of vertebrate gut microbiomes reveal convergence between birds and bats. *mBio, 11*, e02901–e02919. https://doi.org/10.1128/mBio.02901-19.

Spoerner, M., Wichard, T., Bachhuber, T., et al. (2012). Growth and thallus morphogenesis of *Ulva mutabilis* (Chlorophyta) depends on a combination of two bacterial species excreting regulatory factors. *Journal of Phycology, 48*, 1433–1447. https://doi.org/10.1111/j.1529-8817.2012.01231.x.

Stappenbeck, T. S., Hooper, L. V., & Gordon, J. I. (2002). Developmental regulation of intestinal angiogenesis by indigenous microbes via Paneth cells. *Proceedings of the National Academy of Sciences of the United States of America, 99*, 15451–15455. https://doi.org/10.1073/pnas.202604299.

Stephens, W. Z., Burns, A. R., Stagaman, K., et al. (2016). The composition of the zebrafish intestinal microbial community varies across development. *International Society for Microbial Ecology Journal, 10*, 644–654. https://doi.org/10.1038/ismej.2015.140.

Storelli, G., Defaye, A., Erkosar, B., et al. (2011). *Lactobacillus plantarum* promotes *Drosophila* systemic growth by modulating hormonal signals through TOR-dependent nutrient sensing. *Cell Metabolism, 14*, 403–414. https://doi.org/10.1016/j.cmet.2011.07.012.

Tauber, A. I. (2017). *Immunity: The evolution of an idea.* Oxford University Press.

Theis, K. R., et al. (2016). Getting the hol-ogenome concept right: An eco-evolutionary framework for hosts and their microbiomes. *mSystem, 1*(2), e00028–16.

Torruella, G., de Mendoza, A., Grau-Bové, X., et al. (2015). Phylogenomics reveals convergent evolution of lifestyles in close relatives of animals and fungi. *Current Biology, 25*, 2404–2410. https://doi.org/10.1016/j.cub.2015.07.053.

Umesaki, Y. (1984). Immunohistochemical and biochemical demonstration of the change in glycolipid composition of the intestinal epithelial cell surface in mice in relation to epithelial cell differentiation and bacterial association. *Journal of Histochemistry and Cytochemistry, 32*, 299–304. https://doi.org/10.1177/32.3.6693758.

Van Valen, L. (1973). Festschrift. *Science, 488*, 180.

Vlamakis, H., Chai, Y., Beauregard, P., et al. (2013). Sticking together: Building a biofilm the *Bacillus subtilis* way. *Nature Reviews Microbiology, 11*, 157–168. https://doi.org/10.1038/nrmicro2960.

von Baer, K. E. (1828). *Über Entwicklungsgeschichte der Thiere. Beobachtung und Reflexion.* Königsberg: Gebrüder, Bornträger.

Wegener Parfrey, L, et al. (2011). Estimating the timing of early eukaryotic diversification with multigene molecular clocks. *PNAS, 108*(33).

Wichard, T. (2015). Exploring bacteria-induced growth and morphogenesis in the green macroalga order Ulvales (Chlorophyta). *Frontiers in Plant Science, 6*, 86. https://doi.org/10.3389/fpls.2015.00086.

Woznica, A., Gerdt, J. P., Hulett, R. E., et al. (2017). Mating in the closest living relatives of animals is induced by a bacterial chondroitinase. *Cell, 170*, 1175–1183.e11. https://doi.org/10.1016/j.cell.2017.08.005.

Yanai, I. (2018). Development and evolution through the lens of global gene regulation. *Trends in Genetics, 34*, 11–20. https://doi.org/10.1016/j.tig.2017.09.011.

Yeoman, C. J., & White, B. A. (2014). Gastrointestinal tract microbiota and probiotics in production animals. *Annual Review of Animal Biosciences, 2*, 469–486. https://doi.org/10.1146/annurev-animal-022513-114149.

Zhou, X., Wang, J., Herrmann, J., et al. (2019). Asymmetric division yields progeny cells with distinct modes of regulating cell cycle-dependent chromosome methylation. *Proceedings of the National Academy of Sciences of the United States of America, 116*, 15661–15670. https://doi.org/10.1073/pnas.1906119116.

Further reading

Anastas, J., & Moon, R. (2013). WNT signalling pathways as therapeutic targets in cancer. *Nature Reviews Cancer, 13*, 11–26. https://doi.org/10.1038/nrc3419.

Blanton, L. V., Charbonneau, M. R., Salih, T., et al. (2016). Gut bacteria that prevent growth impairments transmitted by microbiota from malnourished children. *Science, 351*, 1533. https://doi.org/10.1126/science.aad3311.

Boehm, A. M., Khalturin, K., Anton-Erxleben, F., et al. (2012). FoxO is a critical regulator of stem cell maintenance in immortal Hydra. *Proceedings of the National Academy of Sciences of the United States of America, 109*, 19697–19702. https://doi.org/10.1073/pnas.1209714109.

Dorsky, R. I., Moon, R. T., & Raible, D. W. (1998). Control of neural crest cell fate by the Wnt signalling pathway. *Nature, 396*, 370–373. https://doi.org/10.1038/24620.

Douglas, A. E. (2015). Multiorganismal insects: Diversity and function of resident microorganisms. *Annual Review of Entomology, 60*, 17–34. https://doi.org/10.1146/annurev-ento-010814-020822.

Efstratiadis, A. (1998). Genetics of mouse growth. *International Journal of Developmental Biology, 42*, 955–976.

Flint, H. J., Scott, K. P., Duncan, S. H., et al. (2012). Microbial degradation of complex carbohydrates in the gut. *Gut Microbes, 3*, 289–306. https://doi.org/10.4161/gmic.19897.

Knoll, A. H. (2011). The multiple origins of complex multicellularity. *Annual Reviews of Earth and Planetary Sciences, 39*, 217–239. https://doi.org/10.1146/annurev.earth.031208.100209.

Loebel, D. A., Watson, C. M., De Young, R. A., et al. (2003). Lineage choice and differentiation in mouse embryos and embryonic stem cells. *Developmental Biology, 264*, 1–14. https://doi.org/10.1016/s0012-1606(03)00390-7.

Pereira, F. C., & Berry, D. (2017). Microbial nutrient niches in the gut. *Environmental Microbiology, 19*, 1366–1378. https://doi.org/10.1111/1462-2920.13659.

Ridley, E. V., Wong, A. C., Westmiller, S., et al. (2012). Impact of the resident microbiota on the nutritional phenotype of *Drosophila melanogaster*. *PLoS One*, 7, e36765. https://doi.org/10.1371/journal.pone.0036765.

Schwarzer, M., Makki, K., Storelli, G., et al. (2016). *Lactobacillus plantarum* strain maintains growth of infant mice during chronic undernutrition. *Science, 351*, 854–857. https://doi.org/10.1126/science.aad8588.

Schwarzer, M., Strigini, M., & Leulier, F. (2018). Gut microbiota and host juvenile growth. *Calcified Tissue International, 102*, 387–405. https://doi.org/10.1007/s00223-017-0368-y.

Shaani, Y., Zehavi, T., Eyal, S., et al. (2018). Microbiome niche modification drives diurnal rumen community assembly, overpowering individual variability and diet effects. *International Society for Microbial Ecology Journal, 12*, 2446–2457. https://doi.org/10.1038/s41396-018-0203-0.

Shank, E. A., & Kolter, R. (2011). Extracellular signaling and multicellularity in *Bacillus subtilis*. *Current Opinion in Microbiology, 14*, 741–747. https://doi.org/10.1016/j.mib.2011.09.016.

Skúli, S., Parsons, K. J., Svanbäck, R., et al. (2019). A way forward with eco evo devo: An extended theory of resource polymorphism with postglacial fishes as model systems. *Biological Reviews, 94*, 1786–1808. https://doi.org/10.1111/brv.12534.

Smith, M. I., Yatsunenko, T., Manary, M. J., et al. (2013). Gut microbiomes of Malawian twin pairs discordant for kwashiorkor. *Science, 339*, 548–554. https://doi.org/10.1126/science.1229000.

Takeshita, K., & Kikuchi, Y. (2017). *Riptortus pedestris* and *Burkholderia* symbiont: An ideal model system for insect-microbe symbiotic associations. *Research in Microbiology, 168*, 175–187. https://doi.org/10.1016/j.resmic.2016.11.005.

Vijayan, N., Lema, K. A., Nedved, B. T., et al. (2019). Microbiomes of the polychaete *Hydroides elegans* (Polychaeta: Serpulidae) across its life-history stages. *Marine Biology, 166*, 19. https://doi.org/10.1007/s00227-019-3465-9.

Wong, A. C., Dobson, A. J., & Dougla, A. E. (2014). Gut microbiota dictates the metabolic response of *Drosophila* to diet. *Journal of Experimental Biology, 217*, 1894–1901. https://doi.org/10.1242/jeb.101725.

Zhang, Z., Xu, D., Wang, L., et al. (2016). Convergent evolution of rumen microbiomes in high-altitude mammals. *Current Biology, 26*, 1873–1879. https://doi.org/10.1016/j.cub.2016.05.012.

Kolibri 360 GmbH
Ferdinand-Jühlke-Straße 7
99095 Erfurt, Deutschland
produktsicherheit@kolibri360.de